FUNDAMENTALS OF
INSECT PHYSIOLOGY

FUNDAMENTALS OF INSECT PHYSIOLOGY

Edited by

MURRAY S. BLUM
University of Georgia
Athens, Georgia

A Wiley-Interscience Publication

JOHN WILEY & SONS

New York   ·   Chichester   ·   Brisbane   ·   Toronto   ·   Singapore

Library of Congress Cataloging in Publication Data:

Main entry under title:

Fundamentals of insect physiology.

  "A Wiley-Interscience publication."
  Includes index.
  1. Insects--Physiology. I. Blum, Murray Sheldon,
1929–

QL495.F86   1985   595.7'01   84-26936
ISBN 0-471-05468-2

Printed in the United States of America

10 9 8 7 6 5 4 3 2 1

CONTRIBUTORS

S. J. BERRY, Department of Biology, Wesleyan University, Middletown, Connecticut

M. S. BLUM, Department of Entomology, University of Georgia, Athens, Georgia

D. G. COCHRAN, Department of Entomology, Virginia Polytechnic and State University, Blacksburg, Virginia

J. L. EATON, Department of Entomology, Virginia Polytechnic and State University, Blacksburg, Virginia

J. L. FRAZIER, Biochemicals Department, E. I. DuPont de Nemours and Co., Wilmington, Delaware

S. FRIEDMAN, Department of Entomology, University of Illinois, Urbana, Illinois

H. R. HEPBURN, Department of Physiology, University of the Witwatersrand, Johannesburg, Republic of South Africa

J. E. MCFARLANE, Department of Entomology, Macdonald Campus of McGill University, Quebec, Canada

J. L. NATION, Department of Entomology and Nematology, University of Florida, Gainesville, Florida

H. OBERLANDER, Insect Attractants, Behavior, and Basic Biology Research Laboratory, Agricultural Research, Science and Education Administration, USDA, Gainesville, Florida

D. L. SHANKLAND, Department of Entomology and Nematology, University of Florida, Gainesville, Florida

T. SMYTH, JR., Department of Entomology, The Pennsylvania State University, University Park, Pennsylvania

W. R. TSCHINKEL, Department of Biological Science, Florida State University, Tallahassee, Florida

J. P. WOODRING, Department of Zoology and Physiology, Louisiana State University, Baton Rouge, Louisiana

To the late
Howard Everest Hinton,
whose intellect,
imagination, and curiosity
combined to render
insects more wondrous
than ever

PREFACE

Insect physiology has come of age. This discipline began to gain real momentum in the early 1950s when it became evident that insects possessed a host of novel physiological and biochemical systems that made them eminently suitable as experimental animals. Since that time the research pace has been dizzying and exciting, and journals devoted exclusively to insect physiology have appeared to punctuate this investigative explosion. However, as fruitful as this topic has been, it is safe to say that the best is yet to come!

Insects are now closely identified with major subjects that were physiologically embryonic only a few decades ago. For example, the ecdysteroidal hormones have been structurally characterized only during that last 25 years, a development that has matured the dynamic topic of insect endocrinology. This analytic tour de force has led to investigations on the hormonal regulation of gene puffing with their concomitant implications as powerful tools of molecular biology. Recent understanding of how these hormones regulate cuticular-tanning reactions, metamorphosis, and reproduction has made them especially valuable research tools for investigating a variety of facets of insect development. But the ecdysteroids are only part of the hormonal scenario elucidated in the last few decades.

Novel animal hormones, the juvenile hormones, working in delicate balance with the ecdysteroids, have emerged as the *deus ex machina* of insect metamorphosis. In addition, their structural elucidation (which continues!) has led to the development of new synthetic insect growth regulators (IGRs), which promise to revolutionize insect control strategies. Beyond that, illuminating the roles of these terpenoids in the reproduction of female insects has proven to be especially significant in comprehending the subtle factors that regulate both mating and fecundity in insects.

It would be no exaggeration to state that most of the traditional topics identified with insect physiology have yielded a plethora of new findings that have drastically altered our perceptions of the metabolic capabilities of those arthropods. Subjects as disparate as cuticle, nitrogen excretion, hemolymph, reproduction (and now sexual selection!), and biochemistry have been demonstrated to be considerably more complex (and elegant) than heretofore suspected, and the informational avalanche continues unabated. Complementing the traditional topics are new ones, such as exocrinology, which promise to expand our physiological horizons even more. Indeed, the natural products identified with exocrinology—pheromones and allomones—have insinuated themselves into a wide range of physiological topics that include reproduction, biochemistry, morphology, and behavior. More surprises undoubtedly await us.

A challenge of any textbook is to present a contemporary view of the subject under consideration, a view with lucid explications and meaningful syntheses. The authors of *Fundamentals of Insect Physiology* have accepted this challenge and endeavored to treat their subjects with contemporary brushes whose strokes cover this major physiological discipline. What emerges are exciting state-of-the-art discussion treatments that address the major issues while at the same time emphasizing the multitude of questions that beg to be answered by tomorrow's research. We believe that for both students and nonstudents, this book will provide an exciting journey into the physiological world of these most wondrous of invertebrates.

MURRAY S. BLUM

Athens, Georgia
March 1985

ACKNOWLEDGMENTS

I wish to thank all the authors for their accommodating attitudes in dealing with a multitude of editorial problems. I am grateful to Drs. D. W. Whitman and L. J. Orsak for comments on terminology. Special thanks are due to Ms. Elaine Huff for her outstandingly accurate typing of most of the manuscript.

The outstanding quality of this book would not have been possible without the editorial input of my wife Ann. She graciously made a major commitment to this textbook by thoroughly editing all of the chapters, thus ensuring that each was developed with uniform excellence. Her expertise as a professional editor made it possible to deal with a multitude of problems that confounded and frustrated me. I am deeply grateful to her for the sacrifices that she made for insect physiology, and I love her for it.

M.S.B.

CONTENTS

FUNDAMENTALS OF
INSECT PHYSIOLOGY

INSECT PHYSIOLOGY: PROBLEMS IN TERMINOLOGY

M. S. BLUM
Department of Entomology
University of Georgia
Athens, Georgia

Physiology, like any scientific discipline, relies heavily on the exactness and uniform use of its terminology. Unfortunately, this is not the case for several terms commonly used in insect physiology, and entomology in general. Therefore, it seems highly desirable to present these terms in this introductory section and to briefly focus on their cardinal importance as indicators of key events in the physiology of Insecta.

Some biologists describe the development of different groups of insects as either simple metamorphosis or complete metamorphosis. These terms have been used to refer to insects with no apparent metamorphic changes (Ametabola), incomplete changes (Hemimetabola), gradual changes (Paurometabola), and complete changes (Holometabola). While this classification has proven to be of some phenotypic value, it is developmentally so inaccurate—and so full of exceptions—as to render its retention unjustified. For example, since species in so-called "ametabolous" orders (e.g., Thysanura, Collembola) undergo extensive developmental changes with each apolysis, especially the phased maturation of reproductive organs, these species can hardly be considered as without metamorphosis. For the remainder of the insect orders, this metamorphic classification is equally misleading and contradictory.

Hemimetabolous, paurometabolous, and holometabolous species supposedly differ from each other in terms of the degrees of their metamorphoses. However, about all that can be said with any developmental accuracy is that

1

species referred to as members of the Hemimetabola (e.g., earwigs, grass-hoppers, cockroaches) are not larviform as immatures, lack a pupal stage, and develop their wings externally. On the other hand, beetles, flies, and bees, which constitute so-called "holometabolous" species, possess larval and pupal stages and develop their wings internally until they are external-ized at the pupal molt. These differences notwithstanding, the fact of the matter is that the *metamorphoses of both Hemimetabola and Holometabola are characterized by drastic developmental changes*. Therefore, the adjec-tives *incomplete* and *complete* are terminologically meaningless for describ-ing the developmental changes of Hemimetabola and Holometabola, respec-tively.

On the other hand, one major characteristic clearly distinguishes the metamorphosis of hemimetabolous from that of holometabolous species—external versus internal wing development. By utilizing this basic metamor-phic character for contrasting these two insect groups developmentally, they can simply and accurately be referred to as Exopterygota (former Hemime-tabola) and Endopterygota (former Holometabola). In this book the epithets *exopterygote* and *endopterygote* thus replace the terms sometimes utilized to describe species with "simple" and "complete" metamorphoses, respec-tively. Utilizing this terminology also logically accommodates the so-called "Ametabola," which are then referred to as apterygotes (Apterygota).

Consistent with this metamorphic classification is the utilization of the term *larva* to refer to all immature insects. Retention of the term *nymph* can hardly be justified, especially since a large number of exceptions to its use occur in several orders that typically possess "nymphal" stages. For exam-ple, metamorphosis in thrips, whiteflies, and male scale insects is character-ized by apparent larval and pupal stages, notwithstanding the fact that these species are considered to be members of hemimetabolous orders. Nor is the term *nymph* considered especially appropriate for the immatures of Ephem-eroptera, Odonata, and Plecoptera, and they are often referred to as *naiads*. This terminological inconsistency is easily avoided by referring to all imma-ture insects as *larvae*. (It is also worth noting that the French noun for pupa is *la nymphe*).

Molt is another term that requires developmental exactitude. This word is generally used to refer to two quite different phenomena in the entomological literature, and there is little likelihood that this situation will change. Molt is used to describe (a) the separation of the old from the newly formed cuticle at the beginning of a new instar and (b) the shedding of the old cuticle by the new instar. These two events are quite distinct and usually of significant temporal separation. In this text, the term *apolysis* is utilized whenever specific reference is made to the molting fluid-catalyzed separation of the new and old cuticles. By contrast, the event describing the shedding of the old cuticle is referred to as *ecdysis*.

Developmental events subsequent to apolysis, but *before* ecdysis, de-serve some terminological comment. Often a larva or an adult enclosed in the cuticle of the previous instar is incorrectly perceived as representing the former instar. In the case of adults enclosed in the pupal cuticle, this has led to serious developmental misinterpretations, especially regarding when final adult structures are formed and how long the adult lives. In addition, serious errors in describing "remarkable" reproductive behaviors have been recorded. For example, some authors have referred to the paedogenetic reproduction of pupae of certain species, when in fact it is actually the adult, enclosed in the pupal cuticle, that is being described. In this text, a stage enclosed in the cuticle of the previous instar is referred to as a *pharate stage*. For example, an adult, enclosed in the pupal cuticle is described as a pharate adult.

Finally, the term *instar* deserves clarification. Instar, like molt, has two unrelated uses in the entomological literature. However, in contrast to the term *molt,* instar can be employed unambiguously to describe both uses. Instar refers to both the time interval between ecdyses (so does the word *stadium*) and the individual between ecdyses. Instar is used in both senses in this volume. On the other hand, it should be borne in mind that whereas the interval between successive ecdyses can be accurately described by the word *instar,* no term is available to refer to the period between successive apolyses.

1

CIRCULATORY SYSTEMS

J. P. WOODRING
Department of Zoology and Physiology
Louisiana State University
Baton Rouge, Louisiana

CONTENTS

SUMMARY

Insects have a simple tubular heart and a single blood vessel. The circulatory system is an open system with generalized pathways of flow, low pressure, and a large blood volume. The main functions of this system are transport and protection. Insect blood is not involved in oxygen transport, utilizing instead a system of tracheae for this function. Insects have a great tolerance for drastic changes in blood volume, and they effectively maintain osmotic balance in the face of rapid changes of blood volume.

The insect system may at first seem inefficient or inferior when compared to that of vertebrates, but the insect circulatory system works superbly well for the size and organization of the insect body, and a scaled-down version of the vertebrate system would likely not work at all. For example, when the organ for processing and storing metabolites (the fat body) is spread throughout the body cavity, an open circulatory system is very efficient.

Insect blood is a very complex and dynamic fluid. The composition of the blood is influenced by the age, sex, stage of development, and feeding conditions. The composition of the blood fluctuates with the molting cycle in all insects, and usually also with the reproductive cycle. There is a daily cycle of blood sugar titer in insects, and for some blood constituents (hormones or enzymes) there may be but one cycle or peak per instar or generation. The cycles of blood hormone titers during larval and adult development in insects control growth and development.

1.1. INTRODUCTION

1.1.1. Embryology of the Circulatory System

Though lacking in some specifics, the general pattern of the embryology of the circulatory system is known (Mori, 1979). The following outline is probably typical for most insects. After the blastoderm is completed, a small plaque of cells in the ventral surface multiplies and elongates to form a germ band that infolds to form a longitudinal mesodermal tube running the length of the embryo. The mesodermal tube thickens and spreads laterally, at which time it differentiates into a series of somatic pouches corresponding to the body segments. The pouches become filled with a clear coelomic fluid.

After the nerve cord differentiates from the epidermal invagination along the ventral midline, a distinct epineural space forms and separates the remaining yolk from the nerve cord. As the lateral coelomic pouches begin to grow dorsally on either side of the embryo, the pouches loosen up and open medially into the epineural space to form the *hemocoel,* which contains embryonic blood. As the embryo continues to develop, cells from the dorsal–lateral edge of coelomic pouches migrate in successive waves to form connective tissue, muscles, fat body, and hemocytes. The dorsal edge of the lateral mesoderm continues to grow dorsally, and eventually both sides meet at the dorsal midline to form the dorsal vessel, dorsal diaphragm, alary muscles, and pericardial cells. In most insects the heart starts to beat as soon as it is formed, which may be long before egg hatch. The first hemocytes appearing in the embryonic blood are generally of one type, but in *Gerris* three kinds of hemocytes are distinguishable just prior to hatch.

1.1.2. Gross Morphology of the Circulatory System

The tubular heart is wider than the aorta, contains lateral ostia, is usually supported by the dorsal diaphragm, and in most insects is restricted to the abdomen (Fig. 1-1). The diameter of the heart is about $\frac{1}{15}$th the diameter of

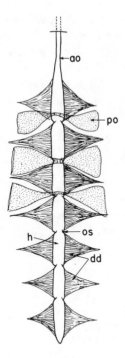

FIGURE 1-1. Dorsal vessel of *Acheta* (ventral view) stained with nile blue sulfate to make the membranous dorsal diaphragm (dd) and phagocytic organs (po) more conspicuous. See legend to Fig. 1-2 for abbreviations.

the abdomen (Fig. 1-2), and the heart wall is very thin (Fig. 1-3). With a few exceptions, the posterior end of the heart is closed. What appear to be paired lateral vessels arise from the heart in some Orthoptera and run through the dorsal diaphragm to the pleural regions. In all other insects the median aorta is the only vessel arising from the heart. The aorta is very thin, lacks ostia or diaphragm connections, and passes under the brain (on top of the esophagus) to empty into a frontal sinus. In all exopterygotes and in all larval endopterygotes the aorta is a simple straight tube, but in adult Lepidoptera and Coleoptera the aorta forms a dorsoventral loop (Fig. 1-4), and in some adult Hymenoptera the aorta is tightly coiled as it passes through the petiole.

Anywhere from 2 to 12 pairs of ostia penetrate the heart wall at roughly equal intervals, generally between insertions of the alary muscles. The typical number of ostia is 6 to 8 pairs, with reduction in the more advanced orders. The incurrent ostia are in the form of vertical slits within lateral depressions in the heart wall. The lips of the slits project into the heart lumen (Fig. 1-5), often projecting anteriorly. This creates an automatic one-way valve that closes during systole, preventing back flow, and the blood normally flows anteriorly through the heart. The ostia open during diastole, permitting blood from the pericardial cavity to enter the heart. Excurrent ostia have been described, especially in those cockroaches with lateral vessels, but how they function is not understood. Viewed from the dorsal aspect, the inflection of the incurrent ostia gives the heart the appearance of

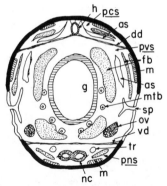

FIGURE 1-2. Diagrammatic cross section through abdomen to show size and position of heart relative to other organs and to show division of hemocoel into pericardial (psc), perivisceral (pvs), and perineural (pns) spaces. Abbreviations for Figures 1-1 through 1-6: ao, aorta; as, air sac; br, brain; cc, *corpus cardiacum*; cn, cardiac nerve cord; cu, cuticle; dd, dorsal diaphragm; ep, epidermis; fb, fat body; g, gut; h, heart; he, head; l, leg; m, muscle; mt, mesothorax; mtb, Malpighian tubule; nc, ventral nerve cord; os, ostium; ov, ovary; pcs, pericardial space; pns, perineural space; po, phagocytic organ; pvs, perivisceral space; sp, spiracle; tr, trachea; vd, ventral diaphragm; wh, wing heart.

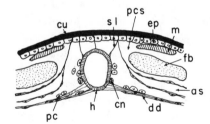

FIGURE 1-3. Enlarged cross section of heart to show attachment of dorsal diaphragm (which contains the alary muscles), the pericardial cells (pc), and the cardiac nerve cord (cn). See legend to Fig. 1-2 for abbreviations.

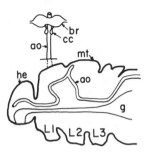

FIGURE 1-4. Diagrammatic sagittal section of a moth to show loop in aorta (ao). See legend to Fig. 1-2 for abbreviations.

FIGURE 1-5. Detail of a pair of ostia to show typical inflection in heart wall. See legend to Fig. 1-2 for abbreviations.

9

being chambered. However, with rare exceptions, there are no internal heart valves and therefore no functional heart chambers.

In addition to the dorsal diaphragm, the heart is supported by connective tissue strands that connect to the dorsal body wall (Fig. 1-3). The entire dorsal vessel may have thin connectives to the brain, somatic muscles, and fat body throughout its length. A dorsal diaphragm occurs in all insects and is typically restricted to the abdomen. In grasshoppers the dorsal diaphragm forms an almost continuous sheet, but in most insects the lateral edges of the diaphragm curve medially between each connection to the lateral body wall (Fig. 1-1).

The very thin alary muscles run from a narrow origin on the lateral body wall and expand fanlike to a broad insertion on the heart. The alary muscles are attached to the dorsal diaphragm, and the shape of the diaphragm often corresponds to the shape of the alary muscles (Fig. 1-1). The number of alary muscle pairs usually correspond to the number of pairs of ostia. Though some alary muscle fibers may pass under the heart without interruption from one side to the other and some fibers may parallel the heart for short distances, most fibers connect directly on the ventral or ventral–lateral walls of the heart.

A ventral diaphragm, mostly restricted to the abdomen, occurs in Odonata and some Orthoptera, Neuroptera, Lepidoptera, and Diptera (except Cyclorrhapha). Much like the dorsal diaphragm, the ventral diaphragm usually forms a delicate web of fine alary muscles mixed with a thin sheet of connective tissue. The width varies from a wide continuous sheet, or a more typical membrane curving medially between thin lateral connections in each segment, to a reduced form where a thin band of muscles crosses over each abdominal ganglion. In the typical insect, the abdominal hemocoel is divided by the dorsal and ventral diaphragms into a pericardial, perivisceral, and perineural space (Fig. 1-2).

There is no restructuring of the dorsal vessel and diaphragm during postembryonic development in exopterygotes and no change through the pupal stage in many Coleoptera and Nematocera. Partial or complete restructuring of the dorsal vessel and diaphragm occurs in the pupae of Lepidoptera and higher Diptera. For example, in *Calliphora* the posterior end of the heart and three pairs of alary muscles undergo complete histolysis and are reconstructed during the pupal stage.

Almost all insects have numerous large pericardial cells, often in strands paralleling the alary muscles and/or the heart (Fig. 1-3). The sessile pericardial cells phagocytize large inert particles from the blood (Section 1.3.3). In addition to pericardial cells, some insects have clusters or sheets of cells associated with the heart or dorsal diaphragm that may function in hemopoiesis (Section 1.3.2) and in some cases phagocytosis as well. A Ringer's buffer containing nile blue sulfate is useful for observing the living heart, dorsal diaphragm, alary muscles, pericardial cells, and phagocytic organs.

In many insects accessory pulsating organs in the form of thin-walled sacs

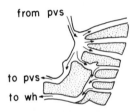

from pvs

to pvs

to wh

FIGURE 1-6. Diagrammatic wing base to show blood flow pattern as produced by the sucking action of the wing hearts (wh). See legend to Fig. 1-2 for abbreviations.

or membranous sheets aid in circulation of blood through legs, wings, antennae, and, in some cases, the cerci. Such an accessory pump never attaches to the dorsal vessel, and its beat is independent of that of the heart. Wing hearts are found at the base of the wings in most endopterygote insects, functioning to draw blood from the posterior wing veins (Fig. 1-6).

1.1.3. Histology of the Heart and Hemocoel

There are no large blood pools in the hemocoel because of the tight spacing of organs and the distribution of the dispersed tissues such as fat body, tracheae, and Malpighian tubules between organs. Nevertheless, the blood volume is large because the blood extends between all organs and into every fold and wrinkle, reaching the tips of all appendages and between individual fibers in some flight muscles. The blood does not come into direct contact with the plasma membrane of any cells in the body except for the hemocytes. Nervous tissue is wrapped in glial cell membranes, resulting in a thin (2-nm) perineural space which contains a fluid differing from that of the blood especially in regards to ion composition. All other body tissues secrete an acellular basement membrane, which seems to offer no barrier to the flow of metabolites or ions into or out of the cells. Therefore, except for nervous tissue, all tissues functionally are bathed in blood.

The heart has a thin outer layer of connective tissue, a one-cell-thick myocardium, and a very thin basement membrane facing the heart lumen. The connective tissue cells secrete various kinds of fibers into a membranous matrix, which forms an adventitia ensheathing the tracheoles and neural elements associated with the heart. The fibers of the adventitia connect the heart to the dorsal diaphragm (Fig. 1-4) and also hold the pericardial cells in place. The dorsal diaphragm is a connective tissue sheath only one cell thick.

Typically both circular and longitudinal fibers compose the myocardium, but in Hemiptera and some Diptera only circular fibers are reported. Intercalated disks have been identified in the heart muscle of some insect species, but appear to be lacking in others. Both cardiac and alary muscles are clearly striated. A tubular invagination of the plasma membrane (a T-system) is poorly developed in most insect cardiac muscle, and it is not always aligned with the Z bands. The sarcoplasmic reticulum is poorly developed compared to vertebrate skeletal muscle.

1.2. CARDIOPHYSIOLOGY AND CIRCULATION

1.2.1. Characteristics of the Heartbeat

The basic nature of the heartbeat is peristalsis, in which a wave of systole (contraction) starts at the posterior end of the heart and flows anteriorly. Systole never results in the complete occlusion of the heart lumen. Under certain conditions the peristaltic waves may travel very slowly (1 mm/min), and another systole may start at the posterior end before the previous systolic wave is halfway through the heart. Typically, the peristaltic waves move much faster, over 50 mm/sec in some species. In some species the heart seems to beat as one whole unit. Such a total beat need not be the most rapid because diastasis (the interval between beats) can be of long duration.

The heartbeat in insects is very complex (Miller, 1974), and great variations are apparent even within a single species. A typical recording (Fig. 1-7) shows the ascending line associated with systole, the descending diastole, and the resting period of diastasis. It also shows the presystolic notch that occurs in some insects. In *Schistocerca* the presystolic notch is caused by contraction of alary muscles. Diastasis and the notch are usually lacking in rapidly beating hearts. Although cardiac output has never been directly measured in insects, it can clearly be altered by changing either beat frequency or amplitude.

The average heartbeat rates for 104 species were tabulated by Jones (1977). From this tabulation some typical values at 22–25°C can illustrate the general range in the major orders: Odonata: *Anax* larvae, 60 beats/min (at 28°C); Orthoptera: *Acheta* larvae, 110 (25°); *Locusta* larvae, 74 (25°); *Periplaneta* larvae, 80 (26°); Hemiptera: *Rhodnius* larvae, 30 (26°); Coleoptera: *Leptinotarsa* larvae, 63 (22°); *Tenebrio* larvae, 20 (26°); Lepidoptera: *Bombyx* larvae, 40 (22°); *Galleria* larvae, 45 (23°); *Manduca* adults, 170 (26°); Diptera: *Aedes* adults, 315 (25°); *Calliphora* adults, 350 (25°); Hymenoptera: *Apis* adult workers, 90 (25°). In general, smaller species will have a faster heartbeat than larger species. However, this is not always true because more active species have a faster heartbeat than less active species and adults generally have a faster heartbeat than the larval stages.

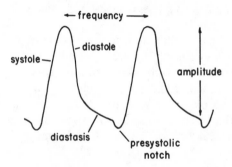

FIGURE 1-7. Idealized recording of the heartbeat.

Heartbeat reversals are characteristic of pupae and rarely occur in larvae. Among adults reversal is more common in the advanced orders of insects. Periods of reversed beating always alternate with periods of normal forward beating, and the duration of reversal is quite variable depending on the species. The backward beat rate is always slower than the forward beat rate. In adult *Calliphora* the fast forward beat lasts for about 14 sec, the heart stops, the slower reversed beat lasts for about 28 sec, stops, and the forward beat starts. Heart ligation indicates at least two pacemakers in *Calliphora*. Ligation of the heart of *Oncopeltus* results in peristaltic waves starting at both ends of the heart and meeting at the ligature. Ligation of the heart of adult *Acheta* produces different effects. The portion of the heart anterior to the ligature consistently beats forward, but the portion posterior to the ligature periodically reverses. The pacemaker mechanism must be quite different in different species, and the mechanism may change during development in any one species.

In most insects the aorta pulsates passively, but does not actively beat. However, under special circumstances in some species the aorta beats faster or slower than the heart. For example, this occurs in preflight warmup in moths or during pupal ecdysis in muscoid flies. During ecdysis in *Musca* the aorta may beat 300 beats/min and the heart only 180 beats/min.

The characteristics of the heartbeat appear to be regulated by general factors, including the alary muscles. Diastole occurs in most insects even when the dorsal alary muscles are cut, indicating a degree of resiliency in the myocardium. This raises the question of the function of alary muscles, which are well innervated in almost all species examined. In *Periplaneta* the alary muscles appear to be neither contractile nor excitable, but in *Chironomus* the alary muscles are always partially contracted. In adult *Hydrophilus* the right and left alary muscles alternate in their contractions, causing the heart to move from side to side. In many species of insects the alary muscles may not be required for diastole nor involved with the basic rhythm of heartbeats, but it is possible that contraction of the alary muscle controls the amplitude of the heartbeat. The amplitude would be increased when alary muscle contraction was in synchrony with diastole and decreased when asynchronous. This function of alary muscles has not been experimentally proven.

When present, the ventral alary muscles are usually innervated, and the contraction of these muscles produces undulations of the ventral diaphragm to aid in circulation.

1.2.2. Bioelectrogenesis of the Myocardium

According to Miller (1974), the resting potential of the myocardium in insects averages about 40 mV. In *Periplaneta* and several other species the potential is potassium dependent, but in several species of silkmoths the resting potential is not potassium dependent. In others it is only dependent if the external K^+ concentration is >20 mM, and it has been suggested that in these cases the myocardial resting potential is based on a complex of ion permeabilities.

Action potentials, resulting from ion influx, may reverse the polarity of the membrane by as much as 20 mV in some cases. The current-generating ion is not necessarily sodium because in denervated hearts of *Periplaneta* and *Hyalophora* an action potential is generated in the absence of sodium in the perfusion fluid. Although there is evidence that calcium ions may be required for depolarizing electrogenesis in the hearts of some insects, that of *Periplaneta* supports action potentials when perfused with zero calcium Ringer's. High magnesium stops the heart of *Periplaneta,* but in *Hyalophora* the heart stops if Mg^{2+} ions are omitted from the perfusion Ringer's.

The effect of inorganic ion composition on the semi-isolated heart (neural connections intact) is unpredictable from one species to another, but there is always some effect. One example suffices: Excess Na^+ or Ca^{2+} (up to 150% over normal) in *Locusta* increases the heartbeat rate and decreases the amplitude, an additional 10 mM K^+ decreases both heartbeat rate and amplitude, and any additional Mg^{2+} ions stop the heart.

1.2.3. Innervation of Heart and Neural Control

The two sources of heart innervation are the cardiac neurons and segmental nerves. All Odonata, Orthoptera, and Hemiptera examined have 30–80 very large cardiac neurons lying appressed to the lateral walls of the heart. Some of these neurons contain neurosecretory material, and both motor and neurosecretory cardiac neurons form synapses on the myocardium. The cell bodies of the segmental nerves are in the ventral ganglia; some are simple motor neurons and others contain neurosecretory material. Segmental nerves form synapses on the myocardium, but there is no direct evidence that they synapse on cardiac neurons. None of the segmental nerves appear to be sensory. In the Orthoptera the segmental nerves are mixed with the cardiac neurons to form a pair of lateral cardiac nerve cords that run from the retrocerebral complex to the posterior end of the heart. Though not proven, nerves from the *corpus cardiacum* may form the beginning of the cardiac nerve cords. In the higher orders (e.g., larval *Musca* and larval *Bombyx*) cardiac neurons and nerve cords are lacking, and the heart region in a segment is innervated by only the ganglion in that segment. In adult *Hyalophora* and in mosquito larvae there is no innervation of the heart at all, and segmental nerves innervate only the alary muscles.

When the heartbeat and the spontaneous discharge of the cardiac neurons are recorded simultaneously (Fig. 1-8) in *Periplaneta,* there may be no heartbeat with considerable neural activity (Fig. 1-8A), no relationship of the heartbeat to neural activity (Fig. 1-8B), or perfect synchrony of heartbeat and neural activity (Fig. 1-8C). However, where there was synchrony, the burst of neural activity occurs during diastole and not, as one would expect,

semi-isolated heart preparations and have no effect when injected into the whole insect.

Neurotransmitters that act on skeletal muscle, such as glutamate, aspartate, and GABA, have no effect when perfused over a semi-isolated heart (Miller, 1979). Acetylcholine stimulates the lateral nerve cord in *Periplaneta* but has no effect on the myocardium. Acetylcholine has no effect on larval heartbeat in *Manduca* or *Galleria* but does cause cardio-acceleration in the adults. The response is probably mediated via segmental nerves.

Serotonin, 5-hydroxytryptamine (5-HT), acts directly on the myocardium, and in most insects it is a powerful cardioaccelerator in trace amounts (15^{-9} M) when applied to a semi-isolated heart. *Calliphora* is exceptional because even a high concentration (10^{-3} M) has only a weak effect on the semi-isolated heart. An effective cardioaccelerator on semi-isolated locust hearts, 5-HT has no effect when injected into intact locusts. This would seem to indicate that 5-HT probably is not a normal neurotransmitter for insect hearts.

Adrenergic compounds, such as octopamine, synephrine, dopamine, and tryamine, are all powerful cardioaccelerators in low concentrations (10^{-9} M) on semi-isolated hearts. The pentapeptide proctolin, isolated from the cockroach hindgut, is also a powerful cardioaccelerator on the semi-isolated heart. Like 5-HT, most of the adrenergic drugs also increase rhythmic contractions of the gut and the Malpighian tubules. However, it is not clear if any of these adrenergic drugs are involved in cardioregulation in the intact insect.

1.2.5. Other Factors Affecting Heartbeat

1.2.5.1. Activity. Activity has little effect on the heartbeat. Sudden movements in some species cause a transient increase of heartbeat, but in other species the startle response is a slowing or stopping of the heartbeat. Long periods of stress in the form of crowding, physical abuse, or chasing usually have little effect on the heartbeat. Almost no form of increased activity, be it running, swimming, or flying, increases the heartbeat. For example, the heartbeat of a locust after 10 min of flight is the same as the unflown control.

1.2.5.2. Sex. Sexual differences are inconsistent. The heartbeats of male and female *Acheta* are identical, but in *Blattella* the heartbeat of the female is faster than that of the male, and in *Periplaneta* the male heartbeat exceeds that of the female. Sexual activity may temporarily alter the heartbeat. For example, in male silkmoths the male heartbeat increases during copulation and decreases and becomes irregular after ejaculation. The heartbeat of the female silkmoth reverses during oviposition.

1.2.5.3. Development. The heartbeat generally becomes slower during each succeeding larval instar. This seems to follow the tendency of smaller insects to have a faster heartbeat than larger ones. In several species the heartbeat slows just prior to ecdysis, then speeds up after ecdysis. The pupal heartbeat is generally slower than that of the larvae, and reversal is common in pupae. The adult heartbeat may be the same as the larval heartbeat (*Locusta*), slower than the larval heartbeat (*Blattella*), or faster than the larval heartbeat (*Anopheles*).

1.2.5.4. Feeding.   Feeding or fasting affects the heartbeat in some species but not in others. In *Anopheles* the heartbeat decreases if food is removed; in adult mosquitoes the heartbeat slows after a blood meal; in *Mamestra* the heartbeat increases after two days starvation, but in *Periplaneta* the heartbeat is not affected by starvation. In some species the heartbeat is more irregular during feeding, a phenomenon believed to be associated with violent movements of the gut. In most insects (*Acheta,* for example) feeding does not affect the heartbeat.

1.2.5.5. Rhythm.   A daily rhythm of heartbeat is reported in *Locusta,* where the heartbeat at night is about 20 beats/min slower than during the day. Since, when locusts are kept in continuous dark, the slower beat persists, the rhythm is not circadian. The effects of lights-on is thought to act directly on the heart, because the response persists when the nerve cord is severed. Some other experiments with the lights-on response indicate the requirement of intact *corpora allata,* but the mechanism is not clear.

1.2.5.6. Temperature.   Temperature has a direct and pronounced effect on heartbeat frequency. In general, the heartbeat in most insects approximately doubles for every 10°C increase of temperature. For example, the heartbeat of larval *Periplaneta* is 13 beats/min at 5°C and 149 beats/min at 40°C.

Typical of rate functions in ectothermic animals, the heartbeat Q_{10} decreases at higher temperature ranges. For example, in *Melanoplus* the Q_{10} is 1.8 at 5–10°C but 1.5 at 40–45°C.

There does not appear to be any temperature compensation (acclimation) of the heartbeat in insects. Groups of *Acheta* acclimated for 10 days at 25°, 30°, and 35°C all have the same heartbeat frequency when equilibrated and tested at 30°C.

1.2.6. Blood Circulation

Blood from the abdomen is pumped by the heart and carried via the aorta into the head. According to Jones (1977) blood enters the antennae and mandibles ventrally and exits dorsally, but in the maxillae and labium the blood enters dorsally and exits ventrally (assuming the head is flexed dorsally). Dorsal, lateral, and ventral currents flow posteriorly through the thorax. Wing circulation is similar in all insects, and most species have accessory pumps at the posterior margin of the base of the wings (Fig. 1-6).

Dorsolateral thoracic currents enter the anterior wing margins, flow to the wing tip via anterior wing veins, and return via posterior veins aided by sucking action of the wing pumps. Circulation in the legs often follows a twisting route, where distal flow is anterior in the coxa and posterior in the femur. In some insects leg movement is required, where a distal flow occurs with leg flexure and a proximal flow occurs with leg extension. Only in Hemiptera–Homoptera are pulsating membranes found in the legs to aid leg circulation. Abdominal blood currents flow posteriorly and dorsally, often aided by the ventral diaphragm. In *Bombyx* the ventral diaphragm is capable of coordinated undulation to push the blood posteriorly.

The time of circulation varies greatly among insects, being more rapid in smaller or more active species. After injecting a fluorescent dye into the sixth abdominal segment of *Periplaneta,* it takes only 30 sec to detect the dye in the head and 8 min to detect the dye in the mesothoracic tarsus. Even in very large species, such as *Schistocerca,* dye circulation is very rapid. Fluorescein injected into the abdomen appears in the thorax in 45 sec, in the femoral–tibial joint in 100 sec, and in all parts of the body in probably 5 min. Complete circulation does not mean complete mixing of all of the blood. Complete circulation occurs in minutes, but complete mixing may require hours. In estimating blood volume, complete mixing of injected [^{14}C]inulin requires at least 1 hr in most insects, and when the cpm of the blood no longer increases, complete mixing is assumed. For *Oncopeltus,* 2 hr is recommended for complete inulin mixing, and for *Hyalophora* 3 hr is recommended.

When at rest, a minimal circulation of blood is maintained by the beating of the heart. Indeed, to provide a minimum circulation when no skeletal muscles are contracting may be a major function of the insect heart. For certain functions (see Section 1.2.3) the circulatory rate is increased by increased amplitude or beat frequency of the heart. For the most part, however, increased circulation during increased activity is caused by the movements of leg and wing muscles and by the increased contraction frequency of the abdominal wall muscles used for tracheal ventilation, rather than by increasing the heartbeat. The increased blood circulation increases the supply of metabolites but not of oxygen to tissues. Complete circulation is not required as long as an increase in the back-and-forth movement of blood between muscles and fat body is achieved. Such limited circulation is facilitated by the dispersed nature of the fat body in insects. The blood is in very close proximity to individual muscle cells, especially flight muscles; the transverse tubules penetrate these muscles and blood is pumped into them by working muscles.

1.2.7. Blood Pressure

With the open circulatory system characteristic of insects, the heart has little effect on blood pressure and is in itself little affected by changing blood pressure. This is in contrast to a closed system, such as that of vertebrates. Blood pressure in insects is produced by contractions of body wall muscles, particularly those of the abdominal wall muscles. The only published report of intracardiac pressure is for *Locusta,* in which the diastolic pressure ranges from 0 to 2.4 mm Hg and systolic pressure is 6.3–6.9 mm Hg. The hemocoel pressure, the functional blood pressure, in the resting locust is −2.6 mm Hg. When a capillary is inserted into the hemocoel, the blood–air meniscus pulsates in response to respiratory movements and not to the beat of the heart. The blood pressure in resting *Bombyx* larvae is about 3 mm Hg. This increases to 10–15 mm Hg in walking larvae and is maximal at 50 mm Hg when the adult unfurls its wings. The average blood pressure in larval *Aeshna* is 3.3 mm Hg, but reaches 5.5 mm Hg during adult emergence and drops to zero after gut deflation. The blood pressure in pupal *Tenebrio* fluctuates from −1.8 to 4.4 mm Hg in direct response to respiratory contractions of abdominal muscles (Sláma et al., 1979).

Breathing and ecdysis are important functions that depend on blood pressure. Air is expelled from the spiracles by compression of air sacs and/or tracheae when the blood pressure increases. When the blood pressure decreases, the resiliency of the body wall and tracheae increases the air space volume and air is sucked into the tracheae. The pacemaker for respiration is generally located in the first abdominal ganglion. The expansive force to split the old cuticle results from swelling of the pharate form by swallowing air or water (Reynolds, 1980). Rhythmic peristaltic contraction of the body, starting at the posterior end, increases the pressure (via the blood) on special cuticular sutures and causes them to split open. In cyclorrhaphan flies the ptilinum is used to split the puparium, and as the ptilinum fills with blood, it pulsates in synchrony with rhythmic contractions of the abdomen.

In many cases the speculated uses of localized blood pressure are incorrect, especially in regards to expulsion of secretions (saliva, venom, etc.) or gametes. For example, erection of the aedeagus, which is often cited as being due to blood pressure, often occurs when the tip of the abdomen is cut off. Both the extension and retraction of the proboscis can be caused by direct muscle attachments rather than blood pressure (e.g., *Calliphora*) (Van der Starre and Ringrok, 1980). On the other hand, localized blood pressure is used for extension of the proboscis in Lepidoptera and for unfurling the wings of *Locusta* and *Manduca*. In some insects, severing the nerve cord before eclosion or ligation between the thorax and abdomen during eclosion prevents unfurling of the wings. This indicates a positive role for whole-body blood pressure mediated by nervous control of abdominal wall muscles. However, a wing severed from the body at eclosion expands as quickly and completely as an unsevered wing if the cut end of the wing is sealed (Glaser and Vincent, 1979). In *Locusta* the wing is pulled through a small hole in the larval wing pad, which flattens the basal third of the wing. The remaining tightly pleated wing is expanded, even in the amputated wing, by blood pressure *generated in the wing itself.* This localized blood pressure results from the rapid stiffening of the already expanded base of the wing. The extension of the proboscis in Lepidoptera depends at least partially on localized increased blood pressure, but it is uncertain if continuous blood pressure is required for complete extension. Proboscis recoil is based on the unloading of a resilin rod compressed during extension.

In *Manduca, Rhodnius,* and *Oncopeltus* a critical weight must be achieved early in a larval instar in order to initiate the next apolysis (Nijhout, 1979). Larvae of subcritical weight can be induced to molt by saline injections in the absence of any further feeding or growth. This suggests that molting hormone synthesis is triggered by stimulation of abdominal stretch receptors and that the blood acts as the transducer of growth to the stretch receptors.

1.3. CYTOPHYSIOLOGY AND DEFENSIVE FUNCTIONS

1.3.1. Hemocytes

Though insect hemocytes have been extensively studied for many years, there remains a great deal of confusion about the kinds of hemocytes, their numbers, and their functions (Rowley and Ratcliffe, 1981). Part of the problem is attributable to plasticity of hemocytes, which both develop and differentiate during the life of an insect. In addition, homologous hemocytes, even with similar functions, have dissimilar appearances in different species, especially species in different orders. The inconsistent methods used in studying hemocytes add to the confusion. Jones (1977) emphasizes how different kinds of fixation, staining, and microscopy can alter the interpretations of a hematogram.

Five recognized kinds of hemocytes occur in the blood of almost all insects. These are *prohemocytes,* which are specialized for reproduction; *plasmocytes,* specialized for phagocytosis; *granulocytes and spherulocytes,* specialized for secretion or storage; and *coagulocytes,* specialized for clotting. Unlike vertebrate cells which can be classified according to shape and function, insect hemocyte functions change as they mature. Because hemocyte transformation takes place in the circulating blood, intermediate types occur that are difficult to categorize. The prohemocyte–plasmocyte intermediate form is very common in the blood of many species. Figure 1-9 is a composite sketch of the five basic hemocyte types as they would appear after Wright's staining. Excellent photomicrographs abound in the literature (Crossley, 1975; Arnold, 1974; Price and Ratcliffe, 1974; Rowley and Ratcliffe, 1981), and recent electron microscopic (EM) studies have helped clarify our understanding of hemocyte functions.

1.3.1.1. Prohemocytes (PRs).   These cells, the smallest hemocytes (6–13 μm), are usually round. The nucleus occupies 70–80% of the cell. They have

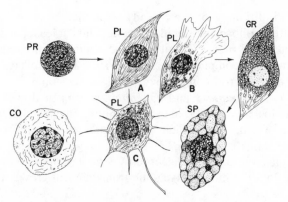

FIGURE 1-9. Idealized drawing of the five basic types of hemocytes occurring in insect blood. PR, prohemocyte; PL, plasmocyte; GR, granulocyte; SP, spherulocyte; CO, coagulocyte. Note the different forms of the plasmocytes (A, B, and C). The arrows indicate postulated direction of differentiation.

ribosomes and mitochondria, but sparse endoplasmic reticulum (ER) and little Golgi. The cytoplasm stains so dark with Wright's stain that the nucleus is often obscured. The number of mitosing PRs/100 hemocytes is the mitotic index (MI). Prohemocytes are nonphagocytic and nonmobile; they occur in all insects; and in most species they occur in all stages of development. Prohemocytes differentiate into plasmocytes.

1.3.1.2. Plasmocytes (PLs). These are pleomorphic amoeboid cells, and though the spindle shape is characteristic (Fig. 1-9A), they can assume any shape. When creeping, they often have a ruffled leading edge (Fig. 1-9B), and on glass slides they extend many fine pseudopodia (Fig. 1-9C). In electron micrographs the plasma membrane appears irregular with micropapillae and pinocytotic invaginations. Golgi and ER are well developed and many lysosomes are formed. With Wright's stain the nucleus is red-purple and the cytoplasm pale to dark blue. Plasmocytes occur in all insects examined and are generally the predominant hemocyte. While their primary function is phagocytosis, they are also involved in encapsulation (Section 1.3.3). There is some evidence that some PLs may differentiate into granulocytes, and in some species it is difficult to distinguish granulocytes from PLs. In some species, very large PLs are sometimes called podocytes.

1.3.1.3. Granulocytes (GRs). Characteristically oval or spindle shaped, GRs have a variable number of small to moderate-sized granules in the cytoplasm depending on the degree of differentiation. With Wright's stain the granules stain bright red, orange, or pink in a light blue cytoplasm. The cytoplasm is rich in Golgi, ER, lysosomes, and free ribosomes. Electron micrographs reveal at least three different kinds of granules, some of which contain lamellar structures or arrays of microtubules. Most of the granules are unstructured, and histochemical tests indicate that they contain glycoproteins or neutral mucopolysaccharides. Though not proven, it is assumed that the granule contents are released into the plasma. The exact function of GRs is uncertain but is usually given as storage or secretion. The GRs of some species are said to be phagocytic; for other species no phagocytic activity is indicated. Granulocytes appear to be involved in cellular defensive functions in some species.

1.3.1.4. Spherulocytes (SPs).   These cells are round or ovoid hemocytes, each with a small nucleus. The cytoplasm often is so packed with large vacuoles (1.5–5 μm) that the nucleus is obscured and the plasma membrane appears lumpy. With Wright's stain the spherules stain a dark red-purple. Some spherules are structured and others not. With EM the cytoplasm has limited Golgi, numerous microtubules, and elongate mitochondria. The function of SPs is uncertain, though some authors consider them to be storage cells. They are not motile and are nonphagocytic. Spherulocytes occur in the blood of all endopterygote orders and in cockroaches but seem to be lacking in the blood of other Orthoptera and in Hemiptera. Some authors suggest that SPs are a further differentiated stage of GRs.

1.3.1.5. Coagulocytes (COs).   These are round or oval cells, often with a hyaline cytoplasm and a large nucleus, and are extremely fragile. With Wright's stain the nucleus appears red-purple and often has a cartwheel appearance. The cytoplasm stains a pale blue or not at all and sometimes appears wrinkled or frothy. A few colorless granules may or may not be present. In EM very few organelles are seen, mainly ribosomes and small mitochondria. Coagulocytes are probably found in the blood of all insects, especially if the conclusion that what are called *oenocytoids* in some of the advanced orders are modified COs is accepted. Coagulocytes rupture immediately on contact with foreign particles or surfaces. It is generally believed that COs are not phagocytic, but it has recently been reported that COs in *Locusta* take up inert particles, become necrotic, and die. It has also been suggested that COs are required for blood clotting in many species.

Since the total hemocyte count (THC) is a concentration measurement, it is greatly affected by the blood volume and the ratio of circulating to sessile hemocytes. The THC can be corrected for blood volume changes by using an absolute hemocyte count (AHC), which is THC times blood volume. One to 2 min in 60°C water will kill an insect and fix the hemocytes with the shape with which they are circulating. The heat will also restore all adhering sessile hemocytes to the circulation. The THC of heat-fixed insects is always at least double the number of nonfixed specimens and is more consistent than the THC of blood collected from the latter. Jones (1977) lists the THC of about 100 species. A selection of representative species in the major orders in which the THC ($\times 10^3/\mu$l) is based on heat-fixed specimens follows: Odonata:

Aeshna larvae—10; Orthoptera: *Periplaneta*—70–120; *Locusta*—8–12, *Acheta*—45; Hemiptera: *Rhodnius*—0.3–5.0, *Oncopeltus*—25–40; Coleoptera: *Tenebrio* larvae—20–44, *Leptinotarsa* larvae and adults—5.0; Lepidoptera: *Bombyx* larvae—17—and adults—1.6, *Manduca* larvae—8.2, *Galleria* larvae—35, *Hyalophora* adults—0.4–5.0; Diptera: *Aedes* larvae—0, *Sarcophaga* larvae—20; Hymenoptera: *Apis* larvae—18, *Vespa* larvae—1.0.

Even considering circulating versus sessile hemocytes and blood volume changes, the THC does vary with age, stage, sex, activity, and physiological state. In general, larval stages have a higher THC than adults, especially in Lepidoptera. In most species the THC steadily increases during larval life, and in *Periplaneta, Bombyx,* and *Sarcophaga* there is a cycle of increased THC before ecdysis and a decrease after ecdysis. Sexual differences in THC are minor and are only indicated in a few species. Male *Periplaneta*, for example, seem to have a higher and less variable THC than females. The effect of starvation on THC usually reflects blood volume changes induced by starvation. The THC decreases in starved *Acheta* as the blood volume increases; likewise the THC in starved *Leptinotarsa* increases as the blood volume decreases. However, the blood volume does not change as the THC decreases in starved *Galleria* larvae. Most forms of trauma such as wounding, bleeding, or abuse increase the THC beyond that attributable to putting more sessile hemocytes into circulation because the mitotic index rapidly increases. On the other hand, toxic compounds or infectious agents lower the THC because the cellular response to these agents removes certain kinds of hemocytes from circulation (Section 1.3.3).

The differential hemocyte count (DHC) and mitotic index (MI) are not affected by changing blood volume. Some typical DHCs are given in Table 1-1. Ultimately, any change in DHC during the life of an insect must result from hemocyte differentiation in the blood or release of hemocytes from hemopoietic organs, but a variety of factors can influence these two sources of hemocytes. In endopterygotes the percent GRs usually increases just prior to pupation, and Arnold (1974) believes that the typical increase in THC during larval development is due mostly to an increase in PLs. In *Locusta* an injection of iron saccharate dramatically alters both THC and

TABLE 1-1. Typical Differential Hemocyte Counts (DHC) and Total Hemocyte Counts (THC) in Representative Species[a]

Order	Genus	%PRs	%PLs	%GRs	%SPs	%COs	THC $\times 10^3/\mu l$
Orthoptera	*Locusta*	1	28	18	—	53	8–12
	Acheta larvae	5	40	5	—	50	45
Lepidoptera	*Galleria* larvae	0.2	83	11	1.5	5	35
Hymenoptera	*Apis* larvae	12	60	4	3	21	1

[a] PL, plasmocyte; PR, prohemocyte; GR, granulocyte; SP, spherulocyte; CO, coagulocyte.

TABLE 1-2. Variations in the Hemogram during the Last Larval Stadium of *Locusta migratoria* after a Single Injection of 250-μg Iron Saccharate

	n	THC^a ± SE	%COs	%PLs	%GRs
Control 12 hr after ecdysis					
At moment of injection	18	8220 ± 944	53.3	28.1	18.5
1 day after injection	10	4528 ± 965	24.0	36.8	39.0
3 days after injection	10	1075 ± 495	1.5	12.1	86.2
5 days after injection	10	1680 ± 609	21.0	34.0	44.3
8 days after injection	10	12800 ± 1365	64.6	19.3	15.9
Control noninjected,					
8-day-old larvae	15	11337 ± 3188	58.0	20.8	19.3

^{a} THC, total hemocyte count; CO, coagulocyte; PL, plasmocyte; GR, granulocyte.
Source: Adapted from M. Brehelin and J. A. Hoffmann, *C. R. Acad. Sci. Paris(D)* **272,** 1409 (1971).

DHC (Table 1-2) and it is altered by starvation in *Bombyx* and *Tenebrio*. In *Calliphora* the percent PLs sharply increase when the larvae cease feeding prior to pupation, but if feeding larvae are injected with an ecdysteroid a premature increase in PLs results. There is not convincing evidence of a daily rhythm of DHC, THC, or MI in any insect.

The idea that hemocytes may contribute to membrane formation is based on circumstantial evidence. Some hemocyte types are rich in vacuoles containing fibrous, sometimes collagen-type, material, and during inactive periods many hemocytes are sessile and spread out and adhere to the surface of organs and tissues.

1.3.2. Hemopoiesis

Arnold (1974) suggests that hemocyte origins probably vary among insects and that in some species they may be of multiple origins. In some species the MI is sufficiently high to account for replacement of worn-out hemocytes and for periodic increases in THC. There is, however, evidence of hemopoietic organs or tissues in species representing four orders (Hoffmann et al., 1979). The most complex hemopoietic organs occur in crickets and cockroaches (Fig. 1-1). These are composed of hollow, membrane-bound, flattened sacs attached to and opening into the anterior end of the heart. Intermixed with collagenlike fibers, the reticular cells remain in place and are actively phagocytic (Fig. 1-10). Some reticular cells divide and differentiate into hemocyte stem cells. Supposedly, as the stem cells mature, they divide to form isogenic islands of one hemocyte type, which may consist of COs, GRs, or PLs. In *Locusta* and *Melolontha* the hemopoietic tissue stretches along the dorsal diaphragm between the pericardial cells and is not formed into a membrane-limited organ. Reticular cells are present and are said to give rise to differentiated hemocytes which are released into the blood. In

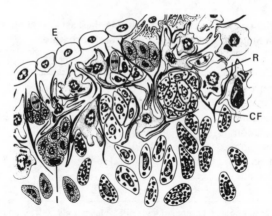

FIGURE 1-10. Postulated hemopoietic organ of *Gryllus*. Some authors term these phagocytic organs (see Fig. 1-1). Epithelial cells (E) form a covering, and some of the cells are very digitated. The cortex is composed of reticular cells (R), collagenlike fibers (CF), and isogenic islets (I) of differentiating hemocytes (note numerous cell divisions). The lumen (lower half) shows accumulations of differentiated hemocytes. The lumen appears to open directly into the heart. (Reprinted with permission from J. A. Hoffmann, D. Zachary, D. Hoffmann, M. Brehelin, and A. Porte, "Postembryonic development and differentiation: Hemopoietic tissues and their functions in some insects," in A. P. Gupta, Ed., *Insect Hemocytes,* Cambridge University Press, 1979, pp. 19–66.)

larval *Calliphora* small clusters of hemopoietic tissues are attached to the heart in the last three abdominal segments, and in *Bombyx* larvae small patches of hemopoietic tissue are located near the imaginal wing buds. Differentiated hemocytes are believed to be released throughout larval life, but in *Calliphora* the hemopoietic tissue breaks down prior to pupation. At that time, differentiating hemocytes appear in the blood, and these circulating cells are the only source of hemocytes in the adult.

Experimental evidence for the function of suggested hemopoietic tissue in *Locusta* is based on X-ray treatment of the dorsal third of the abdomen. After treatment, a 64% decrease in circulating hemocytes, mostly of fragile COs, occurs in 24 hr, and it is 72 hr before new hemocytes appear. On the other hand, X-ray treatment of the ventral third of the abdomen has no effect. Presumably, this treatment inhibits mitosis of the reticular or stem cells.

Although the average MI in insects is usually less than 0.5%, it often varies during development. In larval *Hyalophora* the MI is 0.38% in early instars and peaks at 1.8 during the fourth instar. In *Blatta* the MI decreases before and during apolysis, increasing after each ecdysis. In *Oncopeltus* the MI peaks 30 hr after ecdysis.

In recent years it has been possible to successfully culture PRs and PLs derived from the hemolymph of cockroaches, beetles, and moths (Sohi, 1979). The observed changing appearance of hemocytes in culture supports the theory that all hemocytes are derived from circulating PRs. In addition,

Shapiro (1979) postulates that GRs are the plesiomorphic hemocytes and that they originate as PRs and differentiate through a phagocytic PL stage before becoming distinct GRs. He also believes COs and SPs are further differentiated forms of GRs because he claims to have observed some transformations *in vitro*.

Hemocyte transformations have also been studied by pulse-labeling these cells. Shrivastava and Richards (1965) pulse-labeled 10% of the PRs in *Galleria* larvae, and within 24 hr some PLs had the label. After 48 hr PLs and GRs were labeled, but no PRs. After six days none of the hemocytes were labeled, suggesting a maximum life span of six days for hemocytes. Since colchicine blocks all PR mitoses at metaphase, it has been possible to increase the MI from 0.1 to 10% within 48 hr in *Periplaneta* with this treatment. This indicates a rapid replication rate of PRs. Peake and Crossley (1979) pulse-labeled PRs in larval *Sarcophaga* and estimated that mitosis of PRs took 30 min and that PRs divided every 9 hr. They also found that as PRs age, they increase in size, number of vacuoles, and content of lytic enzymes. They concluded that the PR replication rate was sufficient to account for all PLs in the blood.

It is well established that injury, particularly burning, causes a rapid and large increase in the MI. For example, in *Oncopeltus* a burn trauma can increase the MI from 0.5 to 5.0%, which suggests a humoral control of hemopoiesis. There is also direct evidence of hormonal control of hemopoiesis. If *Rhodnius* is head-ligatured after a blood meal, the percent of PLs remains high instead of showing a normal decline. Removal of the *corpora allata* in *Locusta* reduces the normal increase of COs and GRs during the last larval instar, and supernumerary *corpora allata* cause an increase in the number of PLs. On the other hand, ecdysteroid injections in *Manduca* affect PL motility but not the rate of maturation of PRs into PLs.

1.3.3. Cellular Immunity

Cellular immune reactions in insects have been classified as phagocytosis, encapsulation, nodule formation, and coagulation (Ratcliffe and Rowley, 1979). Which of these reactions occur depends mostly on the size and number of infecting agents. Small doses of single-celled pathogens, such as bacteria, viruses, and protozoans, are phagocytized, and larger doses usually are coalesced into nodules. Metazoan parasites or large inert objects are enclosed in capsules.

Plasmocytes are the principal hemocytes and, in some species, the only phagocytic hemocytes. Other body cells are also phagocytic, for example, the reticular cells found in postulated hemopoietic organs and pericardial cells. Even midgut cells are known to phagocytize inert particles and microorganisms.

There are three steps in phagocytosis: (1) attachment and recognition of foreignness, (2) ingestion, and (3) killing-digestion. Chemotaxis has never

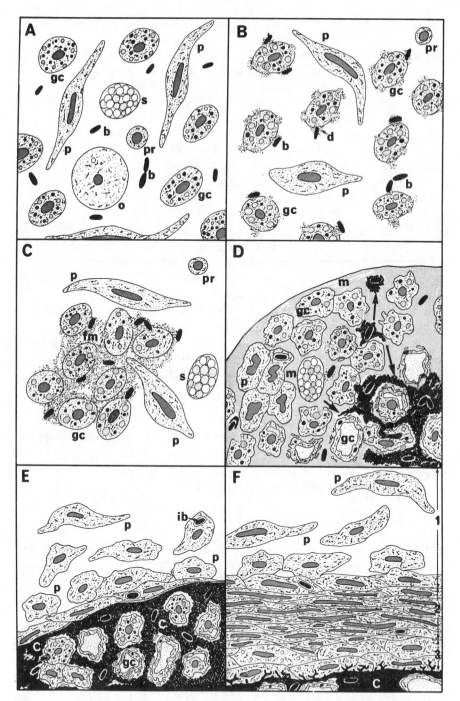

FIGURE 1-11. Diagrammatic representation of main events during nodule formation in *Galleria:* gc, granulocyte (GR) or coagulocyte (CO); pr, prohemocyte (PR); p, plasmocyte (PL); s, spherulocyte (SP). A. Blood immediately after injection of bacteria (b). B. Initial stage of nodule formation. Random contact with bacteria causes granulocytes to rupture and discharge contents (d). C. By 1 min postinjection, localized clot (fm) forms and entraps bacteria. D. At 5–30 min

been demonstrated for phagocytosis, and there is no evidence of opsin proteins that would attract or activate other PLs. Phagocytosis in insects is nonspecific, and while it can be an effective defense against bacteria in some insects, in other species the phagocytic response may be totally ineffectual. For insects, the cuticle is regarded as the main line of defense against pathogens.

The general success of tissue transplants from one insect to another, even between different species, argues against specific cellular recognition. However, damaged tissues always attract more PLs than undamaged tissues. The role of hemocyte phagocytosis during pupal tissue histolysis is not clear. In some species there is no evidence of phagocytosis of lysed tissues; in other species fragments of muscle are seen in phagocytic vacuoles.

Ingestion starts with attachment by fine pseudopodia to the particle. The pseudopodia fuse at their tips or to the cell surface to enclose the particle or bacterium into a phagosome. The fit of the plasma membrane around ingested bacteria is so tight that it often appears as though the bacteria were free in the cytoplasm of the hemocyte. The PLs of *Periplaneta* and *Galleria* can ingest two to four bacteria together at one time. In some species up to 300,000 killed bacteria can be cleared from the plasma in 30 min. It requires 66 million latex beads (1.3 μm) to saturate the phagocytic capacity of the fall armyworm.

Cellular digestion in insect PLs seems to be similar to that in vertebrate neutrophils. The discharge of specific granules, containing alkaline phosphatases and lysozymes, into the phagosome occurs before the discharge of granules containing acid hydrolases.

In *Periplaneta* the speed of amoeboid movements of PLs averages 1.7 μm/min with a top speed of 3.0 μm/min. The general motility in cultured PLs from *Manduca* is increased up to 10 times by the addition of ecdysteroid, but whether such a response occurs *in vivo* is not known. Since cytochalasin B stops both cell movement and phagocytosis, it appears that both activities depend on microfilaments.

The first stage of nodule formation is contact of labile COs or GRs with the bacteria, which results in hemocyte rupture and localized coagulation (Fig. 1-11). The first stage is very rapid, no agglutination of the bacteria occurs, though some phagocytosis by PLs takes place. The second stage involves attachment of PLs to the localized coagulum to form an outer sheath. The attraction of PLs to the coagulum indicates chemotaxis, proba-

---

clumps compact and the matrix (m) starts to melanize (arrows) especially around bacteria. E. Beginning of second stage of nodule formation starts with attachment of large numbers of plasmocytes to inner melanizing core. Some plasmocytes contain phagocytized bacteria (ib). F. By 12–24 hr a mature nodule is formed, and the sheath is divisible into 3 regions: (1) outer region of newly attached plasmocytes, (2) middle region of very flattened cells, and (3) inner region of partially flattened cells with melanized inclusions. [Reprinted with permission from N. A. Ratcliffe and S. J. Gagan, *Tissue Cell* **9**, 73 (1977).]

bly based on factors released from the ruptured COs or GRs. The enveloping PLs become flattened and are joined by desmosomes. The central mass of the nodule becomes melanized, and the degree of melanization seems related to how many of the enclosed bacteria are killed. Inanimate objects are usually less melanized than living cells. Nodule formation is very rapid. In *Galleria, Tenebrio,* and *Pieris* nodules appear within 5 min after injection of bacteria, resulting in an 80% decrease in the total hemocyte count.

Encapsulation is the wrapping of large inert particles or parasites with multiple layers of modified PLs. Like nodule formation, encapsulation starts with contact and lysis of GRs or COs, which results in PLs adhering and flattening out. The PL plasma membrane remains intact with the formation of desmosomes, gap junctions, and possibly septate junctions (Baerwald, 1979). Encapsulation usually occurs within 24 hr of parasite entry, and the parasite is killed when the wall is melanized. Thick-walled capsules may have three layers: an inner layer of necrotic PLs (10 cells thick), a middle layer of very flattened PLs (20 cells thick), and an outer layer of unmodified PLs (5 cells thick). Thin-walled capsules also occur, and in some insects the type of capsule formed depends on the species of parasite. In some insects thick-walled capsules are formed in older instars and thin-walled capsules in early instars, which probably reflects the lower hemocyte concentration in early stars. Humoral encapsulation, involving only plasma proteins, occurs only in Diptera with a low THC and only occurs with fungal or nematode parasites.

Wound healing in insects also appears to occur in two steps. In *Galleria* COs or GRs seal off the wound by rupturing and causing a clot to form at the wound site, usually within 10 min. During the next 2 hr the clot compacts and melanizes. The second stage begins in 6 hr with the massive influx of PLs into the wound site. They flatten down around the clot, and after 12 hr a plaque similar to a disorganized capsule forms at the base of the wound. Thus, nodule formation, encapsulation, and wound healing all have a similar mechanism, the first stage of which involves localized blood clotting.

1.3.4. Humoral Immunity

It has been demonstrated in many insect species that a low dose of pathogenic bacteria or toxins protects the insect against larger doses that would have been fatal without the prior exposure. Some degree of acquired resistance in insects is suggested by the appearance of antimicrobial substances in the plasma after infection, increased resistance with repeated injections, and an increased tolerance to injection of related pathogens. However, all cases of acquired resistance in insects are short-lived, a matter of a few days, and in most instances the acquired resistance is not specific. All cases of increased resistance to bacteria are based on injections and never on the normal (in nature) oral route. Often a similar, though weaker, kind of acquired resistance can be induced by injection of bovine serum albumin

(BSA), enzymes, saline, or India ink. In *Locusta* injections of BSA or amylase induce an antitoxic activity against scorpion venom. There are reports of the passive transfer of increased resistance to bacteria or toxins from one insect to another. However, it was reported that a ^{51}Cr-labeled endotoxin was retained in the blood for several days after injection. Therefore, the transfer of the diluted toxin could in itself induce resistance and not be due to a postulated immune factor from the donor insect.

There is still no proven example of the production of a specific immunoglobulin in insects and virtually no evidence of soluble plasma proteins that constitute a compliment system. However, Rasmusan and Bowman (1979) observed the simultaneous induction of antibiotic activity and the appearance of eight proteins in the plasma of cecropia pupae injected with heat-killed *P. aeruginosa* or live *Escherichia coli*. This provided immunity against normally lethal doses of *P. aeruginosa*. Of the eight bands separated on acrylamide gels, bands P4 and P5 could be isolated and purified. Only when P4 and P5 were injected together was there any increase in antibacterial activity. A rabbit antiserum against P4 could be produced, and it cross-reacted with materials in normal blood and immunized blood. This represented the first indication of an induced immune protein in insects, and though its function is not clear, there are similarities to the vertebrate compliment system.

Much of the natural resistance (not induced) of insects to toxins and certain bacteria is dependent on substances already present in the blood, and increased resistance may be the result of the release of these substances from hemocytes into the plasma. In *Galleria* the lysozyme concentration could increase from 500 to 9000 μg/ml within 24 hr after an injected infection; the lysozyme concentration, however, remained high for three days while the increased resistance declined by the second day.

The polyphenol oxidase system, leading to the production of quinones, has been implicated in nonspecific bacterial activity, but neither the synthesis or actions of these quinones vis-à-vis pathogens are clear. Peroxidases occur in certain hemocytes of *Calliphora*, which may be significant because in vertebrate leucocytes peroxidases are bactericidal and convert halides to bactericidal halogens.

1.3.5. Hemostasis

The need for a blood-clotting mechanism in insects has been questioned because the blood pressure is low, the cuticle protects against most injuries, massive bleeding is generally not fatal, and in some species the blood does not even clot. However, most insects do have an effective hemostatic mechanism, which limits bleeding and the entry of bacteria into the body. In addition, internal localized clotting seems to be the first stage in the defensive processes of nodule formation, encapsulation, and wound healing (Section 1.3.3). One may postulate that the need for cellular defense against bacteria provided the selective pressure for the evolution of the clotting mechanism and not the need for hemostasis.

Supposedly, coagulation starts with the rupture of a specific hemocyte type, usually the coagulocytes (Grégoire and Goffinet, 1979). In many insects coagulocytes constitute half or more of the THC. However, in other species of insects granulocytes are the most numerous hemocytes and are believed to initiate the clotting process. Some investigators who are not convinced that any kind of hemocyte rupture is required to initiate plasma coagulation state that it may be that the plasma coagulum causes hemocyte rupture. Regardless of which comes first, in most insects the speed of clot formation and the density of the coagulum formed is directly related to the number of hemocytes present. The THC is very low in those species in which the blood clots weakly or not at all (see below). Fortunately, with very few exceptions, rapid dilution of the blood 1 : 5 or 1 : 10 with cold Ringer's greatly slows or reduces the amount of clot or in most cases totally prevents clotting. Compared to control *Acheta* larvae, the blood of desiccated larvae clots faster and the blood of rehydrated crickets clots more slowly. The blood of both *Periplaneta* and *Acheta* clots more rapidly after the insects are exercised, presumably because as a consequence of exercise, the concentration of circulating hemocytes nearly doubles. Several days after the injection of iron saccharate into locust larvae the blood does not clot, supposedly because the coagulocytes take up the iron saccharate and die (Table 1-2). Such nonclotting blood, lacking coagulocytes, does clot if normal blood containing COs is added (Brehelin and Hoffmann, 1971).

Over a period of many years, Grégoire and his colleagues examined the morphological patterns of blood clotting in more than 900 species of insects and concluded that there are four patterns of clotting.

Pattern 1. The cytoplasm of the coagulocytes bubbles or froths and the plasma membrane ruptures to release, sometimes explosively, some or much of the cytoplasm into the surrounding plasma. Islands of granular, coagulated plasma protein immediately form along each ruptured CO, which quickly spreads throughout the plasma (Fig. 1-12). With certain preparative techniques, fibrous networks of cytoplasmic threads are observed. This pattern is typical of all orthopteroids, three families of Hemiptera, several families of Homoptera (e.g., Cicadidae), many Coleoptera and Hymenoptera, a few Lepidoptera, and all Neuroptera, Mecoptera, and Trichoptera.

Pattern 2. On contacting a glass slide, the coagulocytes extrude long, straight, threadlike pseudopodia. Collectively these cytoplasmic extensions form a loose mesh within which a transparent, veillike coagulum develops. This pattern is typical of larval Lepidoptera, Scarabaeidae, and a few Diptera.

Pattern 3. This is a combination of patterns 1 and 2. Both a meshwork of pseudopodia and islands of coagulum form. This pattern is typical of most Homoptera and many families of Coleoptera and Hymenoptera.

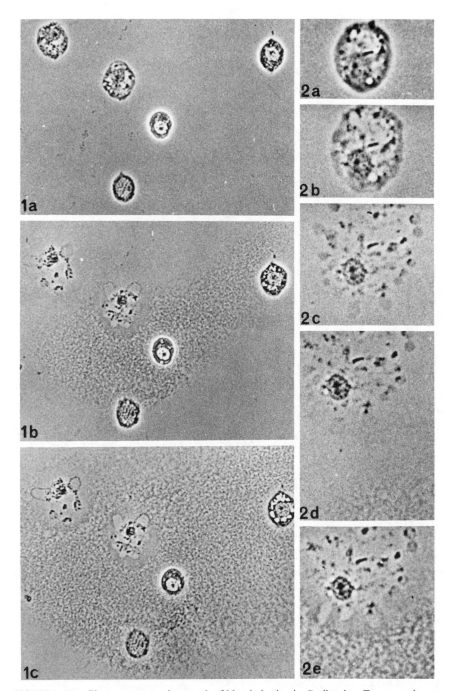

FIGURE 1-12. Phase-contrast micrograph of blood clotting in *Gryllotalpa*. Two coagulocytes (upper left, la) begin to rupture. Cytoplasmic blisters appear on the surface of coagulocytes as a plasma coagulum appears (1b). Coagulum quickly spreads and other cell types are entrapped (1c). Rapid time course of swelling (2b) and rupture (2c) of a single coagulocyte (2a–3). (Reprinted with permission from C. Grégoire, "Hemolymph coagulation," in M. Rockstein, Ed., *The Physiology of Insecta,* Vol. 5, Academic Press, New York, 1974, pp. 309–360.)

Pattern 4. Little or no coagulum is formed. This is what occurs in most families of Hemiptera, some families of Coleoptera, some adult Lepidoptera, a few Hymenoptera (e.g., *Apis*), and in most Diptera.

Seemingly, not all insects have a clottable plasma protein or a biochemical mechanism to induce plasma protein coagulation. However, it is possible that all insects could form a localized cellular clot, thus explaining wound healing and nodule formation in species in which a typical, heavy blood clot does not form. A cellular clot would be formed only from proteins derived from certain hemocytes when such hemocytes clumped or were attracted to certain areas (see Section 1.3.3). In the typical blood clot (pattern 1) involving plasma coagulation, three steps are evident: First, a very rapid change in the morphology of certain hemocytes; second, the release of one or more clot-inducer substances from hemocytes; and third, the selective precipitation of one or more soluble-plasma proteins.

About half of the *in vivo* circulating hemocytes in *Acheta* are spindle shaped, but within seconds of leaving the body almost all hemocytes become round. Within seconds of contacting a glass surface, many hemocytes extend many very fine pseudopodia, and shortly thereafter ruptured edges of plasma membranes are visible. There is a glass factor in which the instability of hemocytes is accelerated.

There is much evidence that hemocytes release one or more clot-inducing factors. In many insects plasma coagulation appears to occur in the immediate vicinity of ruptured hemocytes (Fig. 1-12). If hemocytes can be separated quickly enough from the plasma, presumably before the release of all of the clot-inducing factors, plasma clotting may be retarded for hours or in some cases permanently. There is no effective anticoagulant chemical!

An effective method for hemocyte separation from plasma is rapid dilution and low-speed centrifugation in Ringer's at 4°C. Rupture of separated hemocytes of *Leucophaea* in Ringer's produces a protein gel that, when added to the plasma, induces coagulation of a plasma protein (Barwig and Bohn, 1980). Blood collected from insects that are immersed in 60°C water for 1 min does not clot, no melanization occurs, and the hemocytes do not alter their form. The resultant clear plasma resists microbial fouling for many days at room temperature. Apparently the heat fixation inactivates a number of enzymes or other proteins, specifically the clot-inducing factors and the prophenoloxidases, but does not affect the plasma bacteriocidal factors. Tyrosinases may be involved in plasma clotting in some insects but do not appear to be involved in orthopteroid insects. Addition of phenylthiourea, an inhibitor of tyrosinases, to fresh *Acheta* blood inhibits melanization but has no effect on plasma clotting.

The clotting protein in the plasma of *Locusta* and *Acheta* contains a high-molecular-weight glycolipoprotein, and it appears to constitute 20–30% of the total plasma protein in larval blood. The clot protein has a very high lipid content, and over half of the total plasma lipid content is lost when the coagulum forms. Thus, the clotting plasma proteins in orthopteroid insects include the lipid-transport protein, lipophorin, and in adults the yolk protein vitellogenin. The fluid that is squeezed out of a clot (or is spun out) is serum, and it contains a low titer of lipoprotein.

1.4. PLASMA COMPOSITION

1.4.1. Dynamic Equilibrium

The basic function of plasma is transport, which includes the movement and dispersal of water, salts, sugars, amino acids, lipids, proteins, hormones, and waste products (Fig. 1-13). Under constant environmental conditions, all components of the blood fluctuate in synchrony with the molting and reproductive cycles (Figs. 1-14 and 1-15). Superimposed on the cyclic fluctuations are the fluctuations of plasma composition due to changing environmental conditions, such as food, temperature, light, competition, and disease.

The amount of any plasma component is the product of plasma volume times the concentration. Any feedback control of plasma composition must involve concentration perception because it is difficult to envision a receptor sensitive to amount rather than concentration. The amount of any material transported in the plasma per unit time (turnover) may well exceed the amount of that material in the plasma at any one instant. Therefore, a complete understanding of plasma transport functions requires knowing the concentration, the amount, and the turnover rate.

In terms of being a dynamic tissue, insect blood compares favorably with mammalian blood (Jungreis, 1980). *Homeostasis* simply means that compo-

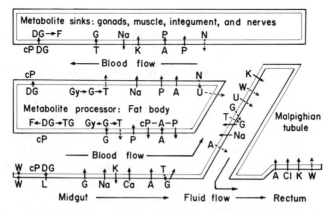

FIGURE 1-13. Schematic representation of dynamic equilibrium. Active processes, solid arrows; passive processes, dashed arrows. See text for details. W, water; Na, sodium ions; K, potassium ions; Cl, chloride ions; Ca, calcium ions; N, ammonia; U, urates; A, free amino acids; P, soluble proteins; CP, lipid carrier protein; L, lipids; DG, diglycerides; TG, triglycerides; F, fatty acids; G, glucose; T, trehalose; GY, glycogen.

sition of the plasma is maintained within certain set limits. The term *dynamic equilibrium* includes homeostasis but emphasizes turnover of plasma components. It implies that though materials may be rapidly entering and leaving the plasma, the composition is carefully controlled. The remarkably stable state of dynamic equilibrium is undoubtedly under neuralhormonal control, which probably brings about equilibrium by controlling cellular membrane permeability and the active transport of materials across membranes. Figure 1-13 gives a schematic overview of dynamic equilibrium. The major movement of materials into and out of the plasma are indicated by arrows. Details of typical plasma concentrations, expected fluctuations, turnover, and regulation of the major components in the plasma are discussed in succeeding sections.

1.4.1.1. Storage Capacity of Plasma.   Because of the large plasma volume in insects, many authors speculate that the plasma may function as a reservoir for readily available metabolites. Plasma metabolites are more quickly available than those in the fat body. The storage capacity of the plasma is more limited than that of the fat body. Larval-specific proteins, thermal-hysteresis proteins, polyglycols, dipeptides, toxins, pigments, and other materials may be temporarily stored in the plasma.

1.4.1.2. Molting Fluid and Plasma.   According to Jungreis (1979), the absence of trehalose, coupled with drastic differences in the concentrations of proteins, polypeptides, labile phosphates, amino acids, and inorganic ions, indicates that unrestricted movement of solutes or water between the blood and the molting fluid is unlikely to occur. Jungreis's model for the secretion of molting fluid requires active potassium transport, and resorption of molting fluid back into the blood requires active bicarbonate transport.

1.4.2. Plasma Volume and Osmoregulation

The typical compartmentalization of an orthopteroid insect shows 30% solids, 20% plasma, 15% gut water, 15% cuticular water, and 20% all other tissue water. Jones (1977) lists the blood volume (or percent) for 54 species. The percent (of wet body weight) plasma is 10% in *Tenebrio,* 15–25% in most orthopteroids, 25–35% in most Endopterygota, and as high as 40% in some dipterous larvae. Larvae generally have a higher percent blood volume than adults, and typically the percent is higher in phytophagous than in other insects. Some examples of blood volumes are: *Aedes* larvae, 0.35 μl; *Acheta* larvae, 50 μl; *Locusta* larvae, 300 μl; and *Hyalophora* larvae, 1200 μl.

Under desiccating conditions many insects can lose up to 50% of their plasma volume over a period of weeks. The loss can be more rapid. For example, *Rhodnius* adults lose 50% of their plasma volume after only 20 min of flight (Gringorten and Friend, 1979). In *Locusta* starved for 5 hr there is a 13% decrease in plasma volume immediately after the next meal due to the entrance of water into the midgut; the plasma osmolality increases for 30 min but returns to normal in 60 min. Rehydration is very rapid in most insects, and it is possible to double the plasma volume in 24 hr. In most insects there is a rapid decrease in blood volume after ecdysis; 30% within 7 hr in *Periplaneta* and a 75% reduction in *Pieris* 2 hr after eclosion (Reynolds, 1980).

Large fluctuations in plasma volume are tolerated because insects are excellent osmoregulators. This they accomplish by delicately controlled removal or addition of salts and/or amino acids from or to the plasma. The overall effect is osmotic stabilization. A stabilized plasma osmolality means that the water concentration in the tissue and plasma remains in equilibrium and body water loss is proportionalized between tissues and plasma. Slight variations in plasma osmolality may trigger feeding; for example, in *Locusta* and *Phormia* a slight decrease in plasma osmolality induces crop emptying, which in turn stimulates feeding. Plasma proteins, lipids, and carbohydrates, which have little osmotic effect compared to plasma ions, tend to become more dilute with rapid increases of plasma volume and more concentrated with decreases of plasma volume.

There is always a slight increase of plasma ion concentration during dehydration, but the increase is far less than would occur if excess plasma ions were not sequestered in an osmotically inactive form. For example, in *Periplaneta* the plasma volume decreases from 16 to 11% after a week of water deprivation, but the plasma osmolality only increases from 410 to 467 mOsM. If all the ions had remained in the plasma, the osmolality would have increased to 600 mOsM [(16%/11%) × 410]. The inorganic ions in *Periplaneta* are sequestered as urates in the fat body, from whence they return to the plasma when the insect is rehydrated. In other insects there is evidence of inorganic ion sequestering by the gut. In many insects the plasma amino acid content varies inversely with the inorganic ion content, thus adding additional stability to plasma osmolality.

Plasma volume and osmolality are regulated by the balance of water and ion input from the gut, urine production by the Malpighian tubules, and resorption in the rectum. Madrell (1977) suggests that plasma volume in *Rhodnius* is hormonally controlled through the regulation of midgut fluid transport. Neurosecretions released from the *corpus cardiacum* in locusts or from thoracic ganglion in *Rhodnius* have diuretic or antidiuretic effects by regulating urine volume and rectal water resorption (Phillips, 1977). In *Calpodes* larvae the Malpighian tubules are permanently in a diuretic state, and the fluid transport increases as the larvae feed and grow (Ryerse, 1978). The tubules are switched off prior to pupation, then turned on again before eclosion. Urine flow in adult *Calpodes* depends on feeding and probably is controlled by a diuretic hormone. In dehydrated *Schistocerca* the urine flow is 1–2 nl/min; in nectar feeding adult *Calpodes,* 150 nl/min; and after a blood meal in *Aedes,* 267 nl/min.

Osmolality: In general, insect plasma osmolality is much higher than that of mammals (290 mOsM), and individual insects tend to tolerate greater individual variations. Larval blood from freshwater mosquitoes can be as

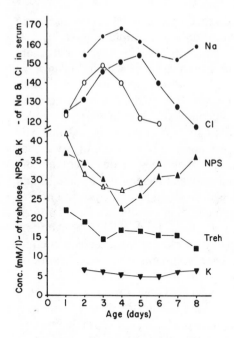

FIGURE 1-14. The normal change in plasma sodium, potassium, ninhydrin-positive substances (NPS), and trehalose concentration during the last two larval instars of *Acheta*. Open symbols, next-to-last instar; closed symbols, last instar. (Reprinted with permission from J. P. Woodring, C. W. Clifford, R. M. Roe, and R. R. Mercier, *J. Insect Physiol.* **23,** 559, Copyright 1977, Pergamon Press, Ltd.)

low as 250 mOs*M* and Odonata larvae, 300 mOs*M*. Litter-inhabiting species such as crickets or some phytophagous caterpillars range from 370 to 400 mOs*M*. Typical terrestrial insects have a plasma osmolality of 420–450 mOs*M*, and species adapted to very dry environments (*Tenebrio*) may exceed 550 mOs*M*.

1.4.3. Plasma Inorganic Ion Composition

The ratio of plasma inorganic ions in insect plasma differs among orders and sometimes among species in the same order (Table 1-3). A high plasma sodium concentration is believed to be the primitive condition (Florkin and Jeuneaux, 1974), and the high potassium (K) and magnesium (Mg) concentration in many families of Hemiptera, Homoptera, Hymenoptera, Coleoptera, and Lepidoptera is believed to have evolved in response to phytophagy. Nonphytophagous Hemiptera have a high sodium (Na) plasma, but nonphytophagous Lepidoptera and Hymenoptera retain the high K–Mg plasma. Phytophagous orthopteroid families, except Phasmidae, have a high sodium plasma (Fig. 1-14). In some endopterygote species both larvae and adults have a high plasma potassium (*Bombyx*), but in other species only the larvae have a high potassium plasma (*Vespula*). The total plasma inorganic ionic concentration is generally higher in exopterygotes than in endopterygotes.

Plasma inorganic anions never balance the cations, and the measured plasma osmolality is always lower than that predicted by summing all inor-

TABLE 1-3. Typical Plasma Inorganic Ion Concentrations in Selected Species and Tabulated Averages for the Major Orders[a]

Order	Species	Na	K	Ca	Mg	Cl	PO$_4$	HCO$_3$
Odonata	Average of 4 species	149	10	7	4	110	4	15
Dictyoptera	*Periplaneta*	157	8	4	5	144	—	—
Orthoptera	*Schistocerca*	108	11	—	—	115	—	—
	Acheta	150	7	8	4	140	—	—
	Average of 5 species	91	20	7	11	97	28	—
Hemiptera	*Rhodnius*	150	4	—	—	130	—	—
	Average of 7 species	106	15	10	6	—	—	—
Homoptera	*Gaeana* (cicada)	18	25	11	23	61	34	—
Coleoptera	*Leptinotarsa* (larvae)	3	60	45	140	—	—	—
	Tenebrio (larvae)	73	39	12	79	—	—	—
	Average of 15 species	77	28	19	48	32	4	—
Lepidoptera	*Bombyx* larvae	15	46	25	101	21	3	—
	Bombyx adults	14	36	15	45	—	—	—
	Manduca 5th instar larvae	1	38	10	50	29	—	—
	Galleria 5th instar larvae	27	31	24	33	—	—	—
	Average of 20 species	11	41	13	61	42	4	—
Diptera	*Drosophila* larvae	56	31	18	—	36	—	—
	Rhynchosciara larvae	34	17	6	91	13	27	7
	Calliphora larvae	140	26	21	34	—	—	—
	Average of 5 species	106	15	10	6	—	—	—
Hymenoptera	*Apis* larvae	11	31	18	20	33	10	—
	Apis adults	47	27	18	1	—	—	—
	Vespula larvae	26	56	19	24	—	—	—
	Vespula adults	93	18	2	3	—	—	—
	Average of 5 species	34	39	6	8	33	—	—

[a] All numbers are in mM/liter.
Source: Adapted from M. Florkin and C. Jeuneaux, "Hemolymph: Composition," in M. Rockstein, Ed., *The Physiology of Insecta,* Academic Press, New York, 1974, pp. 255–307 and others.

ganic ions. This discrepancy is attributable to protein binding; more cations are bound to plasma proteins than are anions. The percents of protein binding in the blood of *Periplaneta* are: Na 22%, Ca 16%, K 0%, Cl 10%, and PO$_4$ 27%. While no potassium is bound in the plasma, much is protein bound within the cytoplasm of cells. In endopterygote insects the plasma chloride (Cl) concentration is low, and citrate is typically high. In *Acheta* and *Rhodnius* the citrate concentration is 2–4 mM, but in *Bombyx* it exceeds 20 mM.

All insects have glial membranes wrapping nerves and ganglia which protect the nervous system from the high levels of potassium and magnesium that occur in the blood of some insects. In spite of the blood–brain barrier, some authors (Lettau et al., 1977) suggest that fluctuating plasma potassium levels may affect neural activity, and that low plasma levels of potassium might stimulate foraging, migration, and so on. Investigators have looked for circadian rhythms of plasma potassium concentration, which might explain the many circadian activity rhythms in insects, but results are inconclusive.

Both monovalent and divalent cations are actively transported across the midgut wall, though the ratio of active and passive flux rates is not always clear. Feeding strongly affects plasma ion concentration and is responsible for the typical fluctuation in the plasma during an instar (Fig. 1-14). In *Periplaneta* sodium is transported from the plasma to the gut lumen, even when the water flow is inhibited. On the other hand, in Lepidoptera and in *Sarcophaga* potassium is transported from the plasma to the gut lumen.

In most insects a K pump is used to drive the fluid secretion from the plasma into the Malpighian tubules. In blood-sucking insects such as *Glossina* a Na pump is used. The osmotic gradient established by the K pump causes the flow of water and other solutes into the tubule lumen. In *Locusta* sodium and glucose are pumped back into the plasma. In addition, most of the essential solutes and urinary water are resorbed back into the plasma from the rectum. By pumping hydrogen ions into the rectal lumen or resorbing bicarbonate ions, the rectal pH is lowered and urates are precipitated. These ion pumps are hormonally controlled. In *Schistocerca* a chloride transport-stimulating hormone (CTSH) from the *corpus cardiacum* controls ion and water flux in the rectum (Spring and Phillips, 1980).

1.4.3.1. Plasma pH and Buffer Capacity. Inexpensive microcombination pH electrodes make the measurement of small volumes (5 μl) of insect blood quick and easy. The normal pH of insect plasma ranges from 6.2 to 7.8, but

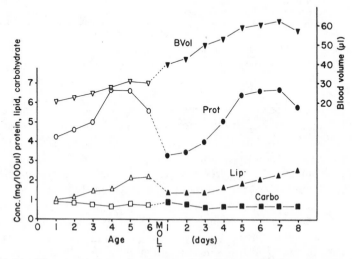

FIGURE 1-15. The normal change in plasma protein, lipid, carbohydrate concentrations, and blood volume during the last two larval instars of *Acheta*. Open symbols, next-to-last instar; closed symbols, last instar. (Reprinted with permission from J. P. Woodring, C. W. Clifford, R. M. Roe, and R. R. Mercier, *J. Insect Physiol.* **23,** 559, Copyright 1977, Pergamon Press, Ltd.)

the blood of most Lepidoptera and Coleoptera is slightly acidic (about 6.6). The blood of most Orthoptera is slightly alkaline, about 7.3. Insects can regulate plasma pH within narrow limits and tolerate induced pH changes far greater than can be tolerated by mammals. After *Acheta* is exposed to CO_2 for 10 sec, the plasma pH declines from 7.6 to 6.5; but in *Trichoplusia* larvae a 1-min CO_2 exposure only depressed the plasma pH from 6.6 to 6.2. During exercise the plasma pH of *Periplaneta* declines from 7.6 to 6.6.

Carbonic anhydrase does not appear to be involved in plasma buffering. On the other hand, this enzyme is present in almost all tissues, including hemocytes, and carbonic acid is quickly formed. Unlike mammals, the region of minimum buffering capacity in insect plasma corresponds to the normal pH. This seems anomalous because adequate concentrations of bicarbonate, phosphate, amino acid, and protein buffers appear to be present.

1.4.3.2. Salines, Ringer's, and Culture Media. A general saline for insect tissues can be quickly made with 7.8 g NaCl in a liter of distilled water (135 mM/liter and 250 mOsM/kg). A Ringer's solution represents a balance of the major cations in the blood. The average composition of the Ringer's listed for 70 species by Jones (1977) is (per liter) 7.7 g (150 mM) NaCl, 0.36 g (5 mM) KCl, and 0.24 (2 mM) $CaCl_2 \cdot 2H_2O$. This ratio probably represents a good "average" insect Ringer's. For some Coleoptera and all Lepidoptera the NaCl content could be reduced to 2 mM and the KCl content raised to 100 mM, a reversal of the "average" Ringer's. The trend of ion ratios in Table 1-3 can provide a useful starting point to prepare a Ringer's. A Ringer's recommended for semi-isolated hearts of both cockroaches and moths by Miller (1979) contains 188 mM NaCl, 18.8 mM KCl, and 9.9 mM $CaCl_2 \cdot 2H_2O$. More complete Ringer's have good buffering capacity, selected osmolality, an energy source (glucose), and a balance of organic-inorganic ions.

Insect hemocytes are among the easiest insect cells to tissue culture, and many lines from many species are established (Hink, 1976). The use of culture media as blood substitute to study embryonic cell, hemocyte, tissue, or organ function has been very successful. On the other hand, most differentiated cells are not bathed in blood and do not grow well in either blood or other media.

1.4.4. Plasma Sugar Composition

Glucose is the plasma sugar of vertebrates and the average concentration is 100 μg/100 μl. In insects the sugar concentration is much higher (Fig. 1-15) and the predominant plasma sugar is trehalose (Table 1-4). However, some species have low concentrations of plasma sugar and some lack trehalose. Although insect blood has been frequently analyzed for the presence of glucose, estimates based on "reducing sugar" tests are not reliable because insect plasma contains other reducing substances such as uric acid, ascorbic

TABLE 1-4. Typical Sugar Composition in Representative Species[a]

Order	Genus	Stage	Plasma Glucose	Plasma Trehalose
Odonata	*Uropelta*	Larvae	90	790
Dermaptera	*Anisolabis*	Adult	30	0
Orthoptera	*Schistocerca*	Larvae	75	1500
	Periplaneta	Adult	30	1700
	Acheta	Larvae	80	600
Neuroptera	Dobsonfly	Larvae?	0	17
Coleoptera	*Popillia*	Larvae	69	280
	Leptinotarsa	Adult	30	0
Lepidoptera	*Bombyx*	Larvae	2	500
	Galleria	Larvae	21	1700
	Manduca	Larvae	90	2500
Diptera	*Phormia*	Larvae	100	0
		Adults	600	600
Hymenoptera	*Apis*	Adult	2000	1000

[a] Glucose values are based on enzyme assays. All numbers are in $\mu g/100\ \mu l$.

acid, amino sugars, and tyrosine. The plasma of some insects contains high titers of fructose, for example, *Apis,* 1500 $\mu g/100\ \mu l$, and *Gastrophilus* larvae, 200 $\mu g/100\ \mu l$.

There is no apparent phylogenetic correlation of high or low plasma sugar, trehalose–glucose ratio, or lack of trehalose. Trehalose is lacking in species that use proline for flight energy, such as *Leptinotarsa* and *Glossina,* but plasma trehalose is just as high in insects that use lipids for flight energy (*Schistocerca*) as in insects that use sugars for flight energy (*Periplaneta*). Actually, locusts use trehalose to start flying, then switch to lipids for cruising during long-distance flight (See Chapters 2 and 12).

Why do insects use trehalose, the synthesis of which expends energy, and why is there a high plasma glycemia in the typical insect? The high glycemia may compensate for poor circulation and slow response to the hyperglycemic hormone (HGH) in the face of periodic high demands for energy. The reason for trehalose utilization may be that it is easier to keep this disaccharide from leaking into the gut or Malpighian tubules than monosaccharides. Also, since trehalose has half the osmotic effect of glucose, rapid plasma trehalose fluctuations would have less effect on the plasma osmolality. Finally, with the reactivity of the aldehydes reduced (1,1 linkage), there is less nonspecific glucosylation with trehalose.

Both trehalose and glycogen are synthesized from absorbed glucose. Glucose is actively absorbed by the midgut and is transported by the plasma to the fat body. In *Phormia* the rate of midgut glucose transport is limited by the maximum release rate of food from the crop (0.015 mol/min). The fat body may make and store glycogen or make and release trehalose into the plasma. The trehalose synthesis rate in *Phormia* is 252 μg/25 mg dry wt. fat body/hr. All tissues use glucose, and all tissues contain one of the membrane-bound or soluble trehalases in order to hydrolyze the plasma trehalose. Glycogen is also made and stored in muscle, gut tissue, and hemocytes but is lacking in the plasma. There is no trehalose pump, and trehalose moves across membranes by diffusion. Trehalose in *Calliphora* was shown to enter Malpighian tubules more slowly than similar-sized molecules. Trehalose is not resorbed in the rectum. It is suggested that the trehalases in the midgut mucosal cells and tubule cells function to hydrolyze any leaking trehalose to glucose, for which there is a steady Na-linked pump to return the glucose to the plasma.

The typical change in the plasma trehalose concentration during each larval instar reflects the feeding and growth pattern in that instar. In *Bombyx* and *Acheta* the plasma trehalose concentration decreases for the first three days and then remains constant or increases. In *Locusta* plasma trehalose increases from 3000 μg/100 μl on day 1 to a midinstar peak of 5000 μg/100 μl, while glucose concentration declines during the same period. Starvation generally reduces the plasma sugar titer, but sometimes much more at one age than another. In *Manduca* larvae 72 hr starvation decreases plasma trehalose from 2500 to 320 μg/100 μl. In *Acheta* 48 hr starvation of two-day-old last-instar larvae decreases plasma trehalose, but starvation of five-day-old last-instar larvae has little effect because they have already reduced feeding preparatory to molting. In *Locusta* starvation of two-day-old last-instar larvae does not decrease plasma trehalose, though the normal increase does not occur, but starvation of four-day-old last-instar larvae greatly decreases plasma trehalose. It requires five days of starvation to deplete plasma sugar in adult *Leptinotarsa,* after which they switch to proline for energy. Flight activity in insects rapidly depletes plasma trehalose, and in some species this limits flight time.

In *Periplaneta* and *Acheta* quick but transient rises in both plasma trehalose and glucose occur in response to any kind of stress. The excitation-induced hypertrehalosemia in *Periplaneta* results in an increase of from 1600 to 2000 μg/100 μl within 10 min of handling. If the insects are not further disturbed, the concentration returns to control levels in 30 min. Since the response occurs in cockroaches with heads ligatured 24 hr prior to handling, the hyperglycemic hormone is probably not involved. Downer (1979) found that injected octopamine doubles plasma trehalose concentration, and he suggested that excitation-induced hypertrehalosemia is neurally induced.

Glucose and trehalose, along with uric acid and serotonin, are the only plasma components for which a circadian rhythm of concentration has been demonstrated (Fig. 1-16). The trehalose rhythm in *Acheta* and *Periplaneta* persists in continuous light or dark. The peak of trehalose concentration occurs 2–3 hr before lights on, but the locomotory and feeding activity peaks during the 6 hr after lights off. Neural regulation appears to be present in *Periplaneta* because the rhythm is abolished with removal of the thoracic

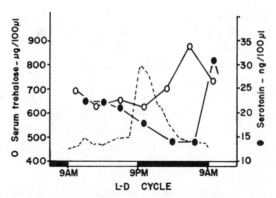

FIGURE 1-16. Rhythm of locomotory activity (dashed line), serum trehalose concentration (open circles), and plasma serotonin concentration (closed circles) in *Acheta* when on a light–dark cycle LD 12 : 12. [Trehalose and activity data unpublished, serotonin data derived from M. Muszynaska-Pytel and B. Cymborowski, *Comp. Biochem. Physiol.* **59A**, 13 (1978).]

ganglia. In *Acheta,* the production and release of serotonin from the brain is circadian and is directly related to the daily fluctuation of serotonin concentration in the blood.

The *corpus cardiacum* (c.c.) is the source of a hyperglycemic hormone (HGH) in all insects examined. Injection of c.c. extracts into normal *Periplaneta, Locusta, Manduca,* and *Phormia* increases plasma trehalose. The response to HGH is slow, trehalose elevation starts about 30 min after injection, and the maximum response occurs 3–5 hr later. *Periplaneta* has the lowest known HGH threshold; 0.002 gland equivalents increase the plasma trehalose titer 300%. HGH is not effective at all times. Before it has any effect, *Phormia* must be starved 24 hr, but *Locusta* must be well fed. *Corpus cardiacum* extracts increase plasma glucose in *Phormia* (200–700 μg/100 μl), but in *Locusta* c.c. extracts decrease plasma glucose. Cardiacectomies in *Calliphora* and *Phormia* result in plasma hypertrehalosemia instead of the expected hypotrehalosemia. According to Chen and Friedman (1977), the brain of *Phormia* may release a hypotrehalosemic hormone (HTH), and they suggest that the hypertrehalosemic condition of cardiacectomized flies results from a lack of HTH.

1.4.5. Plasma Lipid Composition

Diglycerides (DGL) are the predominant lipids in insect plasma, though in some species free fatty acid and triglyceride (TGL) concentrations are very high. The typical range of total plasma lipid concentration is 0.5–2.5%. Some typical values for representative species (in μg/100 μl) are: *Periplaneta,* 350; *Locusta,* 700–2300; *Pyrrhocoris,* 1100; *Popillia,* 650; *Leptinotarsa,* 800; *Hyalophora,* 250; *Manduca,* 140–1040; and *Rhynchosciara,* 520. In *Schistocerca* and *Oncopeltus* steroids comprise about 7% of the plasma lipids, and phospholipids may be as high as 20% in some species. The ratio of free fatty acids, monoglycerides (MGL), DGL, and TGL in the plasma varies among insects (Table 1-5).

The plasma lipid composition changes during development. In addition, plasma lipids vary in response to activities and availability of food. In adult *Pyrrhocoris* the plasma lipid increases from 6 to 1100 μg in 11 days, but most of the increase is due to lipid droplets in the plasma and not soluble lipids. In larval *Acheta* plasma lipids increase steadily during the instar, then decrease during the molt (Fig. 1-15). Starvation of two-day-old last-instar larvae prevents the plasma lipid increase, but plasma lipids increase in starved four-day-old larvae. Five days of starvation of adult *Leptinotarsa* does not affect plasma lipid concentration. In adult male *Locusta* the plasma lipid concentration normally increases from 500 to 700 μg/100 μl, but if starved, the plasma lipids increase to 2000 μg/100 μl (see below).

In insects, lipids are transported in the plasma bound to carrier lipoproteins called lipophorins, which are produced by the fat body. The lipids are loaded and unloaded from the reusable lipophorin. The release of lipophorin into the plasma is independent of diglyceride release from the fat body. These carrier molecules bind and transport fatty acids, MGL, DGL, TGL, phospholipids, and steroids. Freshly absorbed lipids from the midgut of locusts appear in the plasma as DGL bound to lipoproteins, but in cockroaches freshly absorbed lipids appear as TGL bound to lipophorin. The fat body is the major storage site for lipids, and in some species up to 95% of the fat body lipid is in the form of triglycerides. The fat body seems to structurally alter and unsaturate the fatty acids before releasing newly constituted DGLs into the plasma.

An adipokinetic hormone (AKH) is synthesized and released from the glandular portion of the *corpus cardiacum* in *Schistocerca, Periplaneta, Manduca,* and *Tenebrio.* In *Locusta,* AKH causes a large increase in plasma DGL concentration, up to 300% increase in response to only 0.005 parts of an extract of one *corpus cardiacum.* During the first few minutes of rapid flight in locusts, the plasma trehalose is depleted, following which the locusts slow to cruising flight speed and switch to DGL for fuel. At that time the plasma DGL concentration rises from 400 to 1700 μg/100 μl. If AKH is

TABLE 1-5. Typical Plasma Glyceride Ratios in Representative Genera[a]

Genus	Stage	%TGL	%DGL	%MGL	%FFA
Periplaneta	Adult	27	66	3	4
Schistocerca	Larvae	8	78	1	14
Oncopeltus	Adult	23	57	2	19
Hyalophora	Pharate ♂	8	91	1	1
Manduca	Larvae	10	53	10	26

[a] TGL, triglycerides; DGL, diglycerides; MGL, monoglycerides; FFA, free fatty acids.

injected at the start of flight, trehalose is not used up and the switch to DGL occurs sooner. In *Periplaneta,* which is not adapted for sustained flight, AKH decreases plasma lipid concentration. The response to injected AKH seems to be an idiosyncratic species response; cockroach c.c. extracts cause plasma hyperlipemia in the locust, whereas locust c.c. extracts cause hypolipemia in cockroaches.

However, plasma lipid concentration is not entirely regulated by AKH, as illustrated by the fact that starvation of locusts induces a dramatic hyperlipemia (700–2300 μg/100 μl) that is not affected by removal or addition of the *corpus cardiacum.* During starvation the plasma trehalose decreases from 3300 to 400 μg/100 μl. The plasma hyperlipemia of starved locusts is reduced by injections of trehalose. The reduction is dose dependent and lasts for 6 hr (Mwangi and Goldsworthy, 1976). This suggests some kind of inverse relationship of plasma DGL and trehalose. Furthermore, the hyperlipemia is associated with the appearance in the plasma of a specific lipoprotein carrier, which disappears rapidly when feeding resumes and the plasma DGL titer declines.

1.4.6. Plasma Free Amino Acid Composition

One of the most distinctive features of insect plasma is the high concentration of free amino acids. The typical vertebrate plasma concentration is 5 mM/liter, but in insect plasma it ranges from 18 to over 100 mM/liter (Table 1-6). Generally, the plasma concentration is higher in endopterygotes than in exopterygotes. Glutamine, proline, and glycine are predominant in most insect plasma, and aspartate, methionine isoleucine, and phenylalanine are typically present in low concentrations. Although there appear to be no phylogenetic correlations, it has been noted that Lepidoptera often have a higher histidine plasma titer than other insects. Certain species may have unusually high concentrations of a particular amino acid, either lysine, serine, or tyrosine, which sometimes can be related to something specific such as silk production.

The pattern of plasma free amino acids can be modified by development, oogenesis, feeding, cuticular tanning, silk production, or flight activity. The change in total concentration during developmental stages is unpredictable. In *Acheta* and *Locusta* the concentration is lowest at midinstar; in *Chironomus* it is highest at midinstar; and in *Rhodnius* and *Calpodes* there is little change throughout an instar. Often large differences in total aminoacidemia occur between different stages. In *Antheraea* larvae the total concentration is 70 mM/liter, in the pupae 85 mM/liter, and in diapausing pupae the concentration reaches 152 mM/liter, mostly due to large increases in proline and alanine. Two days of starvation of *Calpodes* larvae decreases the aminoacidemia from 103 to 90 mM/liter, but in *Acheta* two days of starvation produces no change in concentration but increases the amount (total plasma content). During the last instar of *Locusta* total amino acids are not reduced

TABLE 1-6. Typical Plasma Amino Acid Concentrations in Representative Genera[a]

Amino Acids	Odonata Aeshna Larvae	Orthoptera Schistocerca Adult	Orthoptera Acheta Larvae	Orthoptera Periplaneta Adult	Hemiptera Rhodnius Larvae	Coleoptera Leptinotarsa Adult	Lepidoptera Cossus Larvae	Lepidoptera Calpodes Larvae	Diptera Chironomus Larvae	Diptera Lucilia Larvae	Hymenoptera Apis Larvae
Aspartic acid	0.61	0.47	0.29	0.04	0.13	1.59	b	b	0.18	0.08	b
Asparagine	0.42	3.40	c	0.02	b	b	3.78	b	b	5.97	2.50
Glutamic acid	1.14	t[d]	0.40	t	t	b	b	0.47	0.15	b	b
Glutamine	2.42	16.13	c	2.83	b	34.91	4.05	b	b	46.00	19.84
Proline	3.58	4.83	5.75	1.93	17.15	55.39	10.66	3.09	0.62	11.80	36.40
Glycine	7.20	19.10	5.20	6.57	3.97	2.23	3.05	10.89	0.41	4.97	11.24
Alanine	5.22	10.47	1.41	0.46	2.52	3.82	4.13	1.85	7.11	5.70	6.56
Valine	2.09	4.07	1.09	0.90	4.94	2.05	3.50	3.18	0.67	5.70	5.05
Methionine	0.92	0.83	0.36	0.71	0.31	t	0.14	b	0.14	0.45	1.60
Isoleucine	1.28	1.77	0.62	0.13	1.48	t	2.76	1.00	0.21	2.67	1.59
Leucine	1.69	1.97	1.45	0.92	2.12	0.92	4.82	1.45	0.30	2.13	2.35
Tyrosine	0.76	2.22	4.85	1.09	2.19	t	6.83	2.24	0.60	3.27	1.71
Phenylalanine	0.69	0.45	0.62	0.19	0.96	0.55	1.52	0.40	0.10	0.81	0.76
Histidine	1.37	1.73	1.89	1.21	2.15	2.76	5.06	8.24	0.83	1.60	1.14
Lysine	0.99	5.70	3.13	1.00	1.42	2.88	10.49	1.77	1.08	2.53	5.08
Arginine	1.10	2.83	1.35	0.61	0.67	1.11	4.95	0.47	0.25	6.23	2.93
Threonine	1.98	3.87	c	b	2.17	1.73	2.24	1.97	0.55	1.78	2.27
Serine	2.33	7.47	c	0.38	1.85	b	b	7.63	1.92	2.33	b
Others	b	2.82	4.04[c]	b	b	b	b	b	2.87	7.10	b
Total	35.85	90.55	32.44	19.00	44.03	109.83	67.98	44.65	18.00	111.14	101.02
Total without glutamine	33.37	74.42	32.44	16.17	44.03	74.92	63.48	44.65	18.00	65.14	81.18

[a] All numbers in mM/liter.
[b] Values not determined.
[c] Added and given as "others."
[d] t = trace.

47

by two days of starvation of two-day-old last instar larvae, but are greatly reduced when four-day-old last instar larvae are starved. In *Bombyx* larvae the essential amino acids quickly disappear from the plasma when excluded from the diet, leading to increases in the nonessential amino acids in the plasma.

The massive production of certain kinds of proteins can alter the plasma amino acid pattern, which is illustrated by silk production in larval Lepidoptera. The principal sericigenous amino acids are glycine, glutamine, serine, threonine, and proline. During silk production these amino acids decline to a lower, stable concentration at a high turnover rate. Removal of the silk glands increases the plasma concentration of the sericigenous amino acids and results in a much lower turnover rate.

Plasma amino acids play a role in osmoregulation in many insects. In *Sialis, Acheta, Philosamia,* and *Calliphora* the total free amino acid concentration in the plasma is inversely proportional to the inorganic ion concentration. The compensation is not perfect because when the insect is denied salt, the plasma osmolality decreases slightly even though the aminoacidemia increases. When starved *Sialis* or *Acheta* are given distilled water to drink, the aminoacidemia increases (plasma salt concentration decreases), but if given saline to drink, the aminoacidemia decreases (plasma salt concentration increases). When *Acheta* larvae are fasting just before molting, all plasma amino acids increase proportionately, but if the larvae are starved during a time when they are normally feeding, then only proline and glycine increase. The use of amino acids for osmoregulation may at times be a passive process in which all amino acids increase or decrease proportionally, but the change in specific amino acid titers in some insects indicates active, regulated processes.

In all insects examined, the plasma tyrosine concentration greatly increases just prior to molting and rapidly declines thereafter. Tyrosine is the primary precursor for the quinones required for tanning the new cuticle. Although tyrosine concentrations increase in the fat body and epidermis at the same time as in the plasma, plasma phenylalanine is not a source of increased tyrosine. A suggested source of tyrosine is small tyrosine-rich peptides in the plasma, which are more soluble than free tyrosine. In *Pieris,* the concentration of such peptides increases 25 times during a larval instar.

Tyrosine is the only plasma amino acid known to be under hormonal control. The suspected site of dopamine and N-acetyldopamine synthesis in *Periplaneta* and *Calliphora* is the hemocytes, and a limiting step in the events leading to cuticular tanning might be the membrane permeability to tyrosine as affected by the hormone bursicon. Ecdysteroids activate the synthesis of phenoloxidase activators, stimulate the increase of dopadecarboxylase, and in *Sarcophaga* promote hydrolysis of an alanine–tyrosine dipeptide in the plasma (Steele, 1976).

A number of insects use proline as an energy source for flight muscle. Proline is transported from the fat body to the flight muscles and alanine is returned to the fat body. In *Glossina* the rate of proline synthesis from lipids and alanine is 1 mol/hr/dry wt. of fat body. In some insects proline is used for a surge of energy to start flight before switching to another fuel, but in *Popillia, Leptinotarsa,* and *Glossina* proline is the sustaining fuel for flight (See Chapter 12).

L-Glutamate is a neurotransmitter in insect muscles, and early reports of high plasma glutamate presented a dilemma. Actually, only traces of free glutamate occur in insect plasma, though large amounts occur in the hemocytes (Irving et al., 1979). Labeled glutamate injected into the plasma is quickly taken up by the hemocytes. Very high levels of glutamine occur in the plasma of some, but not all, insects. The reasons for such differences are not clear. In most insects glutamine is used for transporting nitrogenous degradation products via transamination.

Active uptake of amino acids from the midgut into the plasma has been examined for several species (Nedergaard, 1977). Uptake from isolated *Bombyx* midguts is faster for some amino acids than for others. In locusts much of the absorbed glutamate is metabolized by midgut cells and that not metabolized appears in the plasma as glutamine. Much less alanine is metabolized by the gut cells, and it is absorbed four times faster than glutamate. Almost no plasma amino acids leak back into the midgut lumen.

Amino acids enter the Malpighian tubules by passive, nonspecific processes, and the rate of diffusion is independent of the plasma concentration (Phillips, 1977). Glycine, serine, alanine, threonine, and proline are transported from the rectal lumen of locusts back into the plasma against a large concentration gradient. The transport mechanism is energy dependent and exhibits enzymic characteristics. The relative rates of rectal resorption of amino acids correspond roughly to their relative concentration in the plasma. Phillips (1977) suggests that the pattern of plasma aminoacidemia might be partly based on the relative affinity in the rectal mucosa of a common carrier for amino acids.

1.4.7. Plasma Protein Composition

In insects the total protein concentration rises during each larval instar and declines slightly before apolysis (Wyatt and Pan, 1978). Typical increases during the last larval instar are: *Acheta,* 3–7%; *Pyrrhocoris,* 1–12%; *Leptinotarsa,* 0.4–2.1%; *Pieris,* 0.8–8%; *Phormia,* 5–20%. Also, total-plasma proteins are lowest during early instars (1–2%) and highest during the last instar (6–10%). The pattern of protein concentration in adults depends on sex, diet, and the reproductive cycle. Females generally have a higher plasma titer than males. In species like *Acheta* with a steady oviposition rate, the plasma vitellogenin concentration is stable, but in discontinuous ovipositors, such as locusts or cockroaches, there is a cyclic rise and fall of plasma vitellogenin concentration corresponding to the extent of ovarial maturation of eggs.

Starvation of larvae generally does not result in a decrease in plasma-protein concentration, but the normal increase of storage protein is abolished. In adults that depend on dietary proteins for ovarial growth, starvation greatly reduces plasma-protein concentration. Where ovarial growth is based on proteins accumulated during the larval period, starvation may have little effect.

In most species 20–30 protein bands can be detected with isoelectric focusing or gel electrophoresis, but usually only 4 or 5 bands predominate. There are about eight or nine classes of plasma proteins that have been sufficiently purified to be certain of their identity and for which specific functions are known. These are vitellogenins, lipophorins, JH-binding proteins, plasma storage proteins, plasma enzymes, immune proteins, hemoglobin, and thermal-hysteresis proteins.

1.4.7.1. Vitellogenins. These are lipoglycoproteins that are taken up from the plasma by the ovaries for conversion to the yolk protein vitellin. Their molecular weights are generally in the range of 500–600 kdal, although in Diptera values of 250–300 kdal may be characteristic. Typically there are large subunits of 180–230 kdal and small subunits of about 50 kdal. Vitellogenins contain 5–10% lipid and 2–13% carbohydrate, and the amino acid content is high in terms of aspartic and glutamic acids. In spite of the basic similarities between all insect vitellogenins, the vitellogenins of one species do not appear to be taken up by the ovaries of unrelated species. For example, cockroach vitellogenin injected into silkmoths is not taken up, but it is absorbed by the ovaries of another species of cockroach. In many species JH stimulates fat body synthesis of vitellogenin but only after the fat body acquires competence to respond to JH. On the other hand, the vitellogenin synthesis in *Hyalophora* may not require JH, and in *Aedes* vitellogenin synthesis is induced by ecdysteroids produced in the ovary after a blood meal.

1.4.7.2. Lipophorins. The lipid transport of insect proteins differs in many regards from mammalian high-density lipoproteins. In mammals the core of the complex is of triglyceride cholesterol esters, and the rind is of protein and polar lipids. In insects part of the protein is in the core, and much of the diglyceride (DGL) is on the surface. The molecular weight of the transport lipoprotein in *Philosamia* is 700 kdal (44% lipids) and in *Locusta* 340 kdal (average 31% lipids). Lipid release from the fat body does not require *de novo* lipophorin synthesis. In locusts the transport lipoprotein may contain 16 to 36% lipid, depending on the degree of lipid binding. Lipid loading or unloading is not a simple association reaction. Labeled DGL does not bind to isolated carrier protein, and some specific interaction with a tissue is indicated. In locusts AKH enhances lipid release from the fat body, but there is evidence for a second plasma protein mediating the transfer of DGL to the carrier lipoprotein.

1.4.7.3. JH-binding Proteins. Though JH is sufficiently water soluble to occur in the plasma in an effective concentration, almost all of the JH is carried by specific, low-molecular-weight (28-kdal), high-affinity carrier proteins. These specific carriers are found in species representing five orders. The JH-binding protein protects the JH from the nonspecific esterases present at all times in the plasma but offers no protection against the JH-specific esterases. The JH-specific esterases appear in the plasma during the last half of the last larval instar, at the time when elimination of plasma JH permits expression of adult characters.

1.4.7.4. Storage Proteins. These are high-molecular-weight proteins that accumulate in the plasma during larval growth and are totally resorbed into the fat body as 1–3 μm diameter protein granules prior to pupation. The typical molecular weight is 500 kdal. These proteins are hexameric (85-kdal subunits), have a high content of tyrosine and phenylalanine, and may contain a small amount of lipid (0.5%). Prior to pupation the plasma of *Calliphora* may contain up to 20% storage protein. Similar proteins are found in other Diptera and in a number of Lepidoptera. Presumably, the protein granules taken up by the fat body at pupation are used by the adults as a source of amino acids and energy for protein synthesis. This seems to be an excellent adaptation for insects which have little or no protein in the adult diet. Ecdysteroids may govern plasma storage protein uptake in Lepidoptera, but fat body granule formation in Diptera does not require ecdysteroids. In locusts, cockroaches, and crickets, storage proteins are termed larval-specific proteins (LSP), and these increase in concentration during each instar and decrease at each molt. The LSP are not taken up by the fat body and stored but are presumably utilized directly for cuticular growth during the molt and directly after each ecdysis. The LSP seem homologous to the plasma storage protein of endopterygote insects. Kramer et al. (1980) considered the major function of storage protein in *Manduca* as storage of aromatic amino acids. Fox (1981) showed that parts of the LSP might be directly incorporated into the new cuticle in *Periplaneta*.

1.4.7.5. Enzymes. A large array of diverse enzymes are reported to be present in the blood of insects, but too often there is no proof that such enzymes are free in the plasma rather than bound in the hemocytes. For example, it had been believed that the plasma contained trehalase, but recent studies indicate that all blood trehalase is contained in the hemocytes.

Lysozyme is found in the blood of many insects; the molecular weight in Lepidoptera is 14–16 kdal. This enzyme is probably only present in the plasma when released from certain hemocytes in response to bacterial infection (Section 1.3.3). In other Lepidoptera tyrosinase is present in the plasma as a proenzyme (80 kdal) that is activated by a cuticular factor. On the other hand, this phenoloxidase, or its activator, is bound to hemocytes in the blood of several species of Diptera.

Nonspecific esterases probably occur in the plasma of all insects most of the time. JH-specific esterases are released at specific times during the last larval instar, which helps reduce the JH titer to zero, which in turn is necessary for the metamorphic molt to occur (Sparks et al., 1979).

Many other enzymes have been found in insect blood, especially carbohydrases in Hymenoptera and phosphatases in Lepidoptera, but the sources of these enzymes and their functional roles in the blood are not clear.

1.4.7.6. Immune Proteins. Development of increased resistance to bacteria in cecropia pupae corresponds to the appearance in the plasma of eight new proteins. Two of these have been isolated and characterized. Protein P4 has a molecular weight of 48 kdal and P5, 96 kdal, the latter composed of monomers of 24 kdal. Cross-reactions with other factors in both resistant and nonresistant plasma indicate a possible compliment-type reaction (see Section 1.3.4).

1.4.7.7. Hemoglobin. The plasma of larval Chironomidae contains hemoglobin, which in some species comprises over half the total plasma protein. The monomeric-molecular weight is 16 kdal, and dimeric forms of the molecule occur in some species. Hemoglobin is a larval-specific protein, probably evolved from storage proteins, that increases in concentration throughout larval growth and is completely used up prior to pupation.

1.4.7.8. Thermal-Hysteresis Proteins. Thermal hysteresis is the difference between melting and freezing points. Thermal-hysteresis proteins occur in the plasma of arctic beetles, have a low molecular weight (3 kdal), and function to lower the freezing point of the blood. Similar proteins occur in the blood of arctic fish. The freezing point lowering is much in excess of that expected from the molar concentration of the proteins, and it is believed that they bind ice crystals and prevent other water molecules from joining the crystal lattice.

1.4.7.9. Chromoproteins. Chromoproteins are found in the blood and epidermis of a number of insects, and are responsible along with lipophorin for the green color of many insect larvae such as *Manduca* (Wyatt and Pan, 1978). One chromoprotein, insectocyanin, has a MW of 75 kdal with subunits of 23 kdal.

1.4.8. Plasma Hormone Titers

As mediators of cell function, plasma hormone titers must exhibit temporal patterns in order to regulate the timing of various activities in the insect. As a consequence, hormones have short half-lives in the plasma. The ultimate control of the release of hormones resides in the CNS. The daily concentration of plasma ecdysone is known for species representing five orders, and in several species both the JH and ecdysteroid concentration pattern is known.

Typically, in the next-to-last larval instar and in the pupa there is only one peak of ecdysteroid titer, but a small peak occurs early and a much larger peak late during the last larval instar (Fig. 1-17). The small peak, comprised mostly of ecdysone, is responsible for cellular reprogramming, and the large peak, composed mostly of 20-hydroxyecdysone, is responsible for initiating apolysis. The half-life of ecdysone is very short, so ecdysteroid peaks in the plasma represent greatly elevated synthesis and release from the prothoracic gland. The maximum ecdysteroid titer in the plasma is 5–7 μg/ml.

Plasma JH concentrations are higher in earlier instars than in the last larval instar; during the last larval instar there is typically a midinstar peak and then a complete disappearance of plasma JH. JH suppresses expression of adult characters, and when the JH titer is very low during the last larval instar, 20-hydroxyecdysone causes the molt to the adult (or pupa). Plasma titers of JH increase again several days after the adult molt (or eclosion), which in most insects functions to stimulate synthesis and release of vitellogenin from the fat body. The plasma JH titer at any time during the life of an insect depends on a balance of synthesis and release by the *corpora allata* and on the enzymatic degradation by JH-specific esterases as synthesized by the fat body. The specific esterase titer in the plasma can fluctuate significantly in a few hours.

The hormone bursicon has been found in the plasma of all insects examined, representing five orders. This hormone stimulates quinone synthesis,

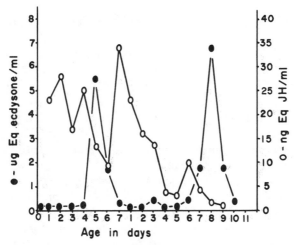

FIGURE 1-17. The normal change in plasma JH (open circles) and ecdysone (closed circles) titers during the last two larval instars of *Locusta*. (Adapted with permission from J. C. Baehr, P. Procheron, M. Papillion, and F. Dray, *J. Insect Physiol.* **25**, 415, Copyright 1979, Pergamon Press, Ltd.)

thus regulating cuticular tanning and having a transient appearance in the plasma. For example, in flies, bursicon is absent at the moment of eclosion, but the titer starts to increase 2–3 min after eclosion and reaches a maximum concentration in 30–60 min. Ten hours after eclosion the plasma bursicon concentration is zero.

The plasma titers of metabolic-regulated hormones and homeostatic hormones have a pattern more related to feeding and specific activities (flight, oogenesis, etc.) than to the molting cycle. Hyper- and hypoglycemic hormones regulate, in part, the plasma sugar concentration. Two adipokinetic hormones (AKH) have been identified in *Schistocerca,* one a decapeptide and the other an octapeptide, which function to mobilize lipid stores in the fat body. In *Locusta* the estimated rate of release of AKH into the plasma is 100 ng/20 min of flight and the half-life is about 20 min. Chloride transport-stimulating hormone (CTSH) and antidiuretic hormones (ADH) regulate water balance. Based on the quick response of urine flow to body water load, the release of ADH is very rapid. Nothing is known about plasma titers or the half-life of these hormones.

There is no evidence for a daily rhythm of plasma hormone titers in insects in the same sense as in vertebrates. In insects, the hormone titers fluctuate more in synchrony with the molting cycle. However, a circadian rhythm of plasma serotonin has been reported in *Acheta,* which peaks at 3 ng/100 μl at lights on (Fig. 1-16). Hormone peaks during an instar (ecdysteroids) or the release of a hormone one time during a life cycle (eclosion hormone) are said to be gated, that is, set by the biological clock but not an actual component of the clock (Beck, 1980).

1.4.9. Plasma Composition of Nitrogenous Degradation Products

There is comparatively little information on plasma concentrations of nitrogenous degradation products. Ammonia, urea, allantoin, allantoic acid, and uric acid have been found in the plasma of various insects. With the exception of hematophagous species, urea is found in very low concentrations. The temporary rise in plasma urea in blood-sucking insects comes from the ingested vertebrate blood; this compound is quickly excreted.

Uric acid is the predominant nitrogenous degradation product in insects, but the concentration in the excreta may not correlate with the plasma concentration. For example, the plasma of *Periplaneta* contains mostly uric acid, but ammonia is the chief nitrogenous excretory product. Similarly, the plasma of *Acheta* contains a nitrogenous degradation product (urea), but it does not occur in the excreta. In most insects uric acid is stored in the fat body, from where it can be mobilized for nucleic acid synthesis or ultimate excretion. It is formed as soluble-plasma urate which diffuses into the Malpighian tubules and is precipitated as the free acid in the tubules, colon, or rectum.

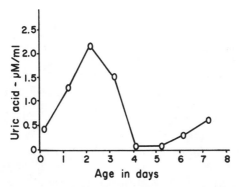

FIGURE 1-18. The normal change in blood uric acid concentration during the last larval instar of *Manduca*. The transition from feeding stage to wandering stage occurs on days 3–4. (Adapted with permission from J. S. Buckner and J. M. Caldwell, *J. Insect Physiol.* **26,** 27, Copyright 1980, Pergamon Press, Ltd.)

The plasma urate concentration in insects ranges from 0.06 to 4.0 m*M*. Some typical concentrations are: *Galleria* larvae, 0.08 m*M*; *Bombyx* adults, 0.8 m*M*; *Vespa* larvae, 0.45 m*M* (adults, 2.5 m*M*); and *Acheta* larvae, 0.75 m*M* (adults, 1.4 m*M*). Plasma uric acid concentration varies with the amount and type of food consumed. It is higher with high protein diets and can vary with the stage of development and with the age in that stage (Fig. 1-18). For example, during the last larval instar of *Manduca,* the plasma uric acid titer peaks on day 3, then falls to zero on day 4 when the larvae cease feeding and start wandering (Buckner and Caldwell, 1980). In *Periplaneta* there is a circadian rhythm of plasma uric acid concentration; the peak (2.0 m*M*) is at lights off and the daily low (1.3 m*M*) is 3 hr later in the scotophase.

Plasma ammonia concentrations can be high in some insects, such as dipterous larvae. When ammonia is excreted, it is secreted directly by the rectum (maggots and cockroaches). The plasma of *Acheta* contains 3–5 m*M* ammonia, a value which is probably typical of insects.

1.5 CONCLUSIONS

Many intriguing problems concerning insect blood and circulation remain. Hemocytopoiesis is still unresolved. Both cellular and humoral immunity constitute very promising areas for both basic and applied research on insect blood. The dynamics of lipid and hormonal transport in the blood require more work with different groups of insects. Understanding the integrated regulation of blood composition in insects would greatly enhance our understanding of how insects function, thereby increasing our ability to control their numbers.

REFERENCES

J. W. Arnold, "The hemocytes of insects," in M. Rockstein, Ed., *The Physiology of Insecta,* Vol. 5, Academic Press, New York, 1974, pp. 201–254.

J. C. Baehr, P. Procheron, M. Papillion, and F. Dray, *J. Insect Physiol.* **25,** 415 (1979).

R. J. Baerwald, "Fine structure of hemocyte membranes and intracellular junctions formed during hemocyte encapsulation," in A. P. Gupta, Ed., *Insect Hemocytes,* Cambridge University Press, 1979, pp. 155–188.

B. Barwig and H. Bohn, *Naturwiss.* **67,** 47 (1980).

S. D. Beck, *Insect Phototperiodism,* Academic Press, New York, 1980.

M. Brehelin and J. A. Hoffmann, *C. R. Acad. Sci., Paris (D)* **272,** 1409 (1971).

J. S. Buckner and J. M. Caldwell, *J. Insect Physiol.* **25,** 27 (1980).

A. C. Chen and S. Friedman, *J. Insect Physiol.* **23,** 1223 (1977).

A. C. Crossley, *Adv. Insect Physiol.* **11,** 177 (1975).

R. G. H. Downer, *J. Insect Physiol.* **25,** 59 (1979).

M. Florkin and C. Jeuneaux, "Hemolymph: Composition," in M. Rockstein, Ed., *The Physiology of Insecta,* Vol. 5, Academic Press, New York, 1974, pp. 255–307.

P. M. Fox, "Circulatory system," in W. J. Bell and K. J. Adiyodi, Eds., *The American Cockroach,* Chapman & Hall, London, 1981, pp. 33–55.

A. E. Glaser and J. F. V. Vincent, *J. Insect Physiol.* **25,** 314 (1979).

C. Grégoire, "Hemolymph coagulation," in M. Rockstein, Ed., *The Physiology of Insecta,* Vol. 5, Academic Press, New York, 1974, pp. 309–360.

C. Grégoire and G. Goffinet, "Controversies about the coagulocytes," in A. P. Gupta, Ed., *Insect Hemocytes,* Cambridge University Press, 1979, pp. 189–230.

J. L. Gringorten and W. G. Friend, *J. Exp. Biol.* **83,** 325 (1979).

A. P. Gupta, Ed., *Insect Hemocytes,* Cambridge University Press, 1979.

B. Heinrich, *J. Exp. Biol.* **64,** 561 (1976).

W. F. Hink, "A compilation of invertebrate cell lines and culture media," in K. Maramorosch, Ed., *Invertebrate Tissue Culture,* Academic Press, New York, 1976, pp. 319–369.

J. A. Hoffmann, D. Zachary, D. Hoffmann, M. Brehelin, and A. Porte, "Postembryonic development and differentiation: Hemopoietic tissues and their functions in some insects," in A. P. Gupta, Ed., *Insect Hemocytes,* Cambridge University Press, 1979, pp. 19–66.

S. N. Irving, R. G. Wilson, and M. P. Osborne, *Physiol. Ent.* **4,** 231 (1979).

J. C. Jones, *The Circulatory System of Insecta,* Charles C. Thomas, Springfield, Ill., 1977.

A. M. Jungreis, *Adv. Insect Physiol.* **14,** 109 (1979).

A. M. Jungreis, "Hemolymph as a dynamic tissue," in *Insect Biology in the Future,* Academic Press, New York, 1980, pp. 273–294.

S. J. Kramer, E. C. Mundall, and J. H. Law, *Insect Biochem.* **10,** 279 (1980).

J. Lettau, W. A. Foster, J. E. Harker, and J. E. Treherne, *J. Exp. Biol.* **71,** 171 (1977).

S. H. P. Madrell, "Insect Malpighian tubules," in A. P. Gupta, Ed., *Transport of Ions and Water in Animals,* Academic Press, New York, 1977, pp. 541–569.

T. Miller, *J. Insect Physiol.* **14,** 1265 (1968).

T. Miller, "Electrophysiology of the insect heart," in M. Rockstein, Ed., *The Physiology of Insecta,* Vol. 5, Academic Press, New York, 1974, pp. 169–200.

T. Miller, *Amer. Zool.* **19,** 77 (1979).

H. Mori, "Embryonic hemocytes: Origin and development," in A. P. Gupta, Ed., *Insect Hemocytes,* Cambridge University Press, 1979, pp. 3–28.

H. Muszynaska-Patel and B. Cymborowski, *Comp. Biochem. Physiol.* **59A,** 13 (1978).

R. W. Mwangi and G. J. Goldsworthy, *Physiol. Ent.* **2,** 37 (1976).

S. Nedergaard, "Amino acid transport," in A. P. Gupta, Ed., *Transport of Ions and Water in Animals,* Academic Press, New York, 1977, pp. 239–264.

H. F. Nijhout, *J. Insect. Physiol.* **25,** 277 (1979).

P. W. Peake and A. C. Crossley, *J. Insect Physiol.* **25,** 789 (1979).

J. E. Phillips, *Fed. Proc.* **36,** 2480 (1977).

C. D. Price and N. A. Ratcliffe, *Z. Zellforsch.* **147,** 537 (1974).

T. Rasmusan and H. G. Bowman, *Insect Biochem.* **9,** 259 (1979).

N. A. Ratcliffe and S. J. Gagen, *Tissue Cell* **9,** 73 (1977).

N. A. Ratcliffe and A. F. Rowley, "Role of hemocytes in defense against biological agents," in A. P. Gupta, Ed., *Insect Hemocytes,* Cambridge University Press, 1979, pp. 331–414.

S. E. Reynolds, *Adv. Insect Physiol.* **15,** 475 (1980).

A. F. Rowley and N. A. Ratcliffe, "Insects," in N. A. Ratcliffe and A. F. Rowley, Eds., *Invertebrate Blood Cells,* Vol. 2, Academic Press, New York, 1981, pp. 421–488.

J. S. Ryerse, *J. Insect Physiol.* **24,** 325 (1978).

M. Shapiro, "Changes in hemocyte populations," in A. P. Gupta, Ed., *Insect Hemocytes,* Cambridge University Press, 1979, pp. 475–523.

S. C. Shrivastava and A. G. Richards, *Biol. Bull.* **128,** 337 (1965).

K. Sláma, N. Baudry-Partiaglou, and A. Provansal-Vaudez, *J. Insect Physiol.* **25,** 825 (1979).

S. S. Sohi, "Hemocyte cultures and insect hemocytology," in A. P. Gupta, Ed., *Insect Hemocytes,* Cambridge University Press, 1979, pp. 259–278.

T. C. Sparks, W. S. Willis, H. H. Shorey, and B. D. Hammock, *J. Insect Physiol.* **25,** 125 (1979).

J. H. Spring and J. E. Phillips, *Can. J. Zool.* **58,** 1933 (1980).

J. E. Steele, *Adv. Insect Physiol.* **12,** 239 (1976).

H. Van der Starre and T. Ringrok, *Physiol. Ent.* **5,** 87 (1980).

J. P. Woodring, C. W. Clifford, R. M. Roe, and R. R. Mercier, *J. Insect Physiol.* **23,** 559 (1977).

G. R. Wyatt and M. L. Pan, *Ann. Rev. Biochem.* **47,** 779 (1978).

NUTRITION AND DIGESTIVE ORGANS

J. E. MCFARLANE
Department of Entomology
Macdonald College of McGill University
Quebec, Canada

CONTENTS

SUMMARY

The basic dietary requirements of insect larvae resemble those of other growing animals, the chief difference being the almost universal insect need for sterol. Carry-over of nutrients from the larval stage may provide protein for oogenesis and occasionally a source of energy for flight, but most adult insects must satisfy these needs by feeding. The ability of the insect to detect the nutritional value of its food appears to be one of the factors affecting the choice of diet. Where the natural diet is inadequate, the insect may harbor symbiotic microorganisms which supply essential nutrients, such as amino acids, B vitamins, and sterols.

The alimentary canal may respond to the nature or quantity of food ingested in a direct way or indirectly via the nervous and endocrine systems. Digestion and absorption occur principally in the midgut but may occur in the foregut and hindgut as well. Digestive enzymes are supplied chiefly by the midgut epithelium and, depending on the species, by the salivary glands and by microorganisms.

2.1. INTRODUCTION

Insects must have in their diet the same basic substances that all higher animals require for growth and development of juvenile forms and for maintenance and reproduction of adults. These include an organic source of nitrogen, an organic source of energy, inorganic salts, water, and a variety of organic chemicals required in rather smaller amounts. In general, these requirements are not in any way unusual, except for the almost universal insect need for sterol. The most significant modification of dietary requirements has been brought about by symbiotic relationships with microorganisms which, in synthesizing essential nutrients, enable many insects to flourish on dietary sources of limited nutritional value.

The variety of substances that insects use as food is reflected in the wealth of physiological adaptations concerned with gaining adequate nutrition. These adaptations are readily observed in the digestive organs, where particular hydrolytic enzymes are usually found where they are needed and are usually absent or weak where they are not. Other adaptations include orientation and feeding mechanisms as well as certain metabolic processes; these will be mentioned only briefly here.

2.2. NUTRITION AND THE CHOICE OF FOOD

Many substances of nutritional value to the insect, such as amino acids, sugars, fatty acids, sterols, vitamins, and so on, stimulate feeding. The question arises as to whether the insect can detect the nutritional adequacy of its food. In experiments with *Agria housei,* a dipterous parasitoid, the quantity and ratio of amino acids and glucose were varied in artificial diets, and the larvae were allowed to choose between diets. Larvae generally chose the diet that promoted the best growth and development when the diets were tested separately (Tables 2-1 and 2-2). It therefore appears that the insect can assess the nutritional quality of its food, although these results may also be interpreted on a strictly chemosensory basis. They do, however, conclusively show the importance of the *balance* of nutrients for the insect, as growth was maximally stimulated on the diet with the higher amino acid level but with the lower glucose level.

Many plant-feeding insects are restricted in their range of host plants, although only rarely is one insect species confined to one plant species. It is usually assumed that all plants are nutritionally adequate for insects. This may be true in a qualitative sense, but plants differ in the proportions of nutrients, and there may be adaptation of the insect to a particular balance of nutrients. Studies of maxillectomized lepidopterous larvae, that is, larvae from which the taste receptors have been removed, show that growth is best on the normal host plant, although other plants may give equally good growth. The problem of host plant specificity therefore seems to resolve itself into one of sensory physiology. Yet nutritional value seems to determine, at least in part, the acceptance or rejection by insects of varieties

TABLE 2-1. Yield of Second-Instar Larvae of *Agria housei* on Diets Containing Different Proportions of Amino Acids and Glucose[a,b]

Diet (%)		Test				
Amino acids	Glucose	I	II	III	IV	Pooled
A 1.125	1.5	88.9	55.5	66.7	81.5	73.1
B 1.125	2.1	65.0	37.0	50.0	64.0	53.3
C 0.75	1.5	7.4	14.8	18.5	20.0	15.1
D 0.75	2.1	3.7	7.4	11.1	0	5.5

[a] Expressed as percentage on four diets/test (and pooled) on the fourth day; four tests of about 25 axenic larvae/diet per test.
[b] See also Table 2-2.
Source: H. L. House, *J. Insect Physiol.* **16,** 2041 (1970).

TABLE 2-2.   **Total Numbers of *Agria housei* Larvae Feeding when Offered Choice of Diets[a,b]**

Diet	Tests								Pooled	Frequency of Performance[c]		
	I		II		III		IV			Best	Equaled	Worst
	1	2	3	4	5	6	7	8				
A	38	33	16	22	15	19	31	28	202	6	1	1
B	26	33	26	12	11	14	22	24	168	1	1	6
A	20	21	29	11	40	20	30	15	186	8	0	0
C	10	9	15	5	24	6	11	11	91	0	0	8
A	30	61	16	5	20	18	39	38	232	7	0	1
D	20	7	4	8	7	15	22	24	107	1	0	7
B	23	18	28	30	33	33	29	18	212	6	0	2
C	12	9	29	27	27	20	16	21	161	2	0	6
B	11	9	32	40	12	33	29	22	188	6	0	2
D	15	6	10	13	15	4	8	12	83	2	0	6
C	20	24	25	28	31	26	26	35	215	8	0	0
D	19	18	11	18	9	12	6	17	110	0	0	8

[a] Shown by diets in pairs per each and all of eight tests and a tally of frequency of performance of one diet excelling the other.

[b] Diets A, B, C, and D are the same as in Table 2-1.

[c] A, Excelled another 21 times, equaled once, and was worst twice; B, excelled another 13 times, equaled once, and was worst 10 times; C, excelled another 10 times and was worst 10 times; D, excelled another 3 times and was worst 21 times.

Source: H. L. House, *J. Insect Physiol.* **16,** 2041 (1970).

within a plant species. Extracts of peas susceptible to the pea aphid, *Acyrthosiphon pisum,* show a concentration of total water-soluble amino acids that is generally almost twice as high as extracts from resistant varieties. They are two to four times higher in terminal growth, which is susceptible, than in middle growth, which is resistant. The nutritional adequacy of the susceptible plant for the pea aphid may depend, however, on the balance of amino acids, on differences in phagostimulation, and on other nutritional factors as well as on the concentration of amino acids.

Nutritional chemicals are not the only ones that affect feeding. Plant allomones such as glycosides, alkaloids, and so on, are of no nutritional value but act as "token" stimuli. (An allomone is a compound produced by individuals of one species which affects the behavior or physiology of individuals of another species in a way that is adaptive for the former.) They sometimes stimulate feeding but are more often deterrents to feeding. With a few insects, the allomones appear to be crucial to monophagy or oligophagy. For example, the diamondback moth, *Plutella maculipennis,* feeds only on cruciferous and other noncruciferous plants containing mustard oil glycosides. In the majority of insects host specificity is determined by the total pattern of sensory stimulation, both physical and chemical, provided to the central nervous system (CNS).

Nutritional substances may also act as token stimuli. Most blood-sucking insects engorge when offered 0.15 *M* NaCl solutions (isoosmotic with blood) containing one of a variety of nucleotides, including adenosine mono-, di-, and triphosphates. ATP, which is present in red blood cells and platelets, is probably the natural gorging stimulant. The reduviid bug, *Rhodnius prolixus,* continues to engorge when transferred to an NaCl solution without ATP after feeding on an NaCl solution with ATP for only 2 min. It appears that ATP stimulation triggers CNS activity, and engorging is completed without continuous stimulation being required.

2.3. DIGESTIVE ORGANS

2.3.1. Functional Morphology of the Alimentary Canal

A generalized diagram of the insect gut is presented in Fig. 2-1. The foregut is lined with cuticle, as it is derived from an epidermal invagination. The salivary glands originate from separate epidermal invaginations on the mandibular, maxillary, and labial segments and may later gain connection with the buccal cavity. The most important glands for digestion are generally the labial glands. The crop is a storage organ in many insects and also serves as a site for digestion, particularly in Orthoptera; digestive enzymes are secreted

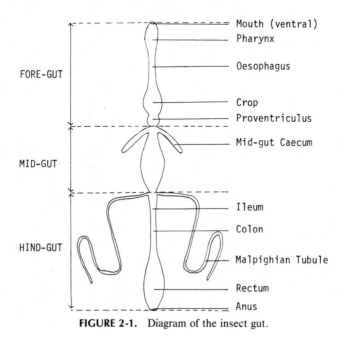

FIGURE 2-1.   Diagram of the insect gut.

by the salivary glands or passed anteriorly from the midgut or derived from microorganisms. (The crop bacteria in locusts have an additional role. In *Locusta migratoria* lignin from the locust's plant food is degraded by bacteria to guiacol (*o*-methoxyphenol) from which the gregarization pheromone, 2-methoxy-5-ethylphenol or locustol, is synthesized. Locustol is passed out of the insect with the feces, becomes airborne, and is taken up by the hemolymph of receiving larvae where it produces the morphological and physiological changes associated with gregarization.) The proventriculus is a triturating organ in some insects, and in most it provides a valve controlling the entry of food into the midgut.

The midgut or intestine with its ceca is the chief locus for digestion, for secretion of the digestive enzymes, and for absorption. The cells are typically columnar with a brush border when viewed with the light microscope. With the electron microscope, they are microvillate (Fig. 2-2) with basal infoldings (Fig. 2-3). The increase in area provided by the folding of these surfaces facilitates passage across the epithelium. Small replacement cells also occur for renewal of lost epithelial cells, and there may be other cell types such as the goblet cell of larval Lepidoptera and certain other insects. The goblet cell appears to function in the active transport of potassium ions in the direction hemolymph to lumen side. The goblet cells of lepidopterous larvae may not be evenly distributed along the length of the midgut. Physiological specialization of the regions of the midgut is, in fact, not uncommon. Digestion of protein in *Glossina morsitans* takes place only in the posterior section of the midgut. Lipolysis in the lumen of the southwestern corn borer, *Diatraea grandiosella,* also occurs only in the posterior section of the midgut, whereas absorption and transport into the hemolymph occur mainly in the anterior region; the epithelia of the mid- and posterior regions act as temporary storage sites for absorbed lipid. Digestive enzymes in insects may be secreted without change in the histology of the midgut. Where changes in the midgut epithelium are seen, it is not always clear how the changes are associated with secretion.

Food in the midgut is enclosed in the peritrophic membrane, which is secreted by cells at the anterior end of the midgut in some insects or formed by the midgut epithelium in most. It is secreted continuously or in response to a distended midgut, as in biting flies. It is likely that the peritrophic membrane has several functions, although the evidence is not conclusive. It may protect the midgut epithelium from abrasion by food or from attack by microorganism or it may be involved in ionic interactions within the lumen. It has a curious function in some coleopterous larvae, where, in various ways, it is used to make the cocoon. In *Gibbium psylloides* the peritrophic membrane collapses in the gut shortly before the cocoon forms, leaving the anus as a flattened thread which is used to form the larval cocoon. The peritrophic membrane consists of chitin and protein which form fibrils that are often organized into various meshwork arrangements (Fig. 2-4).

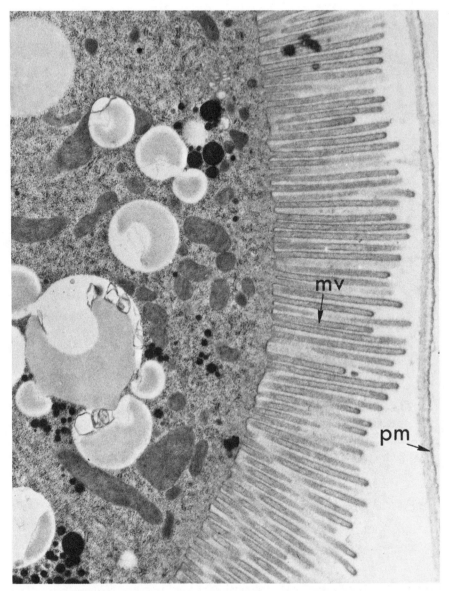

FIGURE 2-2. The apical portion of an absorptive cell from the midgut of a fed *Stomoxys calcitrans*. The surface area of the cell is greatly increased by the long microvilli (mv). The microvilli are separated from the food by the overlying multilayered peritrophic membrane (pm). ×12,600. (Original photograph courtesy of Dr. M. J. Lehane.)

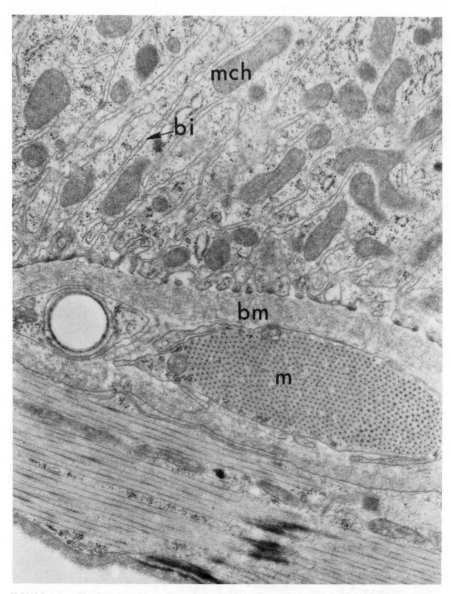

FIGURE 2-3. The basal portion of an absorptive cell from the midgut of *Stomoxys calcitrans* and part of the musculature (m) surrounding the gut. The basal infoldings (bi) and their association with mitochondria (mch) can be clearly seen, as can the narrow openings of the basal infoldings at the inner edge of the cell's basement membrane (bm). ×25,500. (Original photograph courtesy of Dr. M. J. Lehane.)

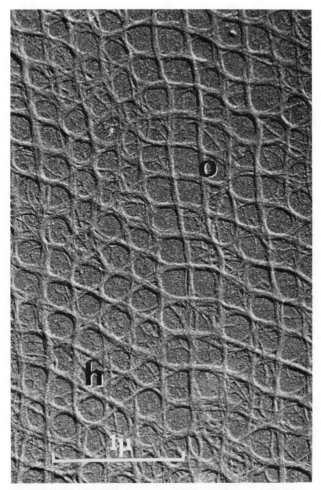

FIGURE 2-4. Freely merging orthogonal (o) and hexagonal (h) network patterns in the peritrophic membrane of *Gibbium psylloides*. Scale = μm. (Reprinted with permission from J. N. Tristram, *J. Insect Physiol.* **23**, 79, Copyright 1977, Pergamon Press, Ltd.)

The hindgut, like the foregut, is derived from an epidermal invagination and is therefore lined with cuticle, but the cuticle is much thinner than that of the foregut. This is probably because the hindgut, and especially the rectum, functions in absorption from the lumen. The Malpighian tubules form, with the rectum, an excretory system; secretion of substances from the hemolymph takes place in the Malpighian tubules and reabsorption from the rectum. Digestion in the hindgut may be brought about by enzymes passed down from the midgut or by microorganisms. In termites the hindgut is the principal site of cellulose digestion by microorganisms.

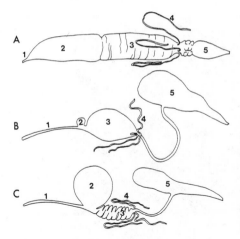

FIGURE 2-5. Changes in the structure of the gut during metamorphosis in *Malacosoma americanum*. A, larva; B, pupa; C, imago. 1, esophagus; 2, crop; 3, midgut; 4, Malpighian tubule; 5, rectum. (From *Principles of Insect Morphology* by R. E. Snodgrass. Copyright © 1935, McGraw-Hill Book Company. Used with permission of McGraw-Hill Book Company.)

The dramatic changes that occur in external body form during the metamorphosis of endopterygote insects are paralleled by changes in the form of the gut. Figure 2-5 shows the changes in the form of the alimentary system in the moth *Malacosoma americanum,* changes which also have their functional counterparts as the larva is a leaf eater whereas the adult is a nectar feeder.

Extraintestinal digestion may take place in some insects: it is found in the larvae of Myrmeliontidae and Dytiscidae, which inject digestive juices into their prey. It also occurs in the larvae of higher Diptera, such as blowflies, which live in a semiaquatic environment, passing their gut secretions out through the anus and reingesting them.

2.3.2.   Control of Gut Activities

There do not appear to be any continuously feeding insects, although some aphids in the laboratory may maintain a continuous rate of excretion, and so probably of feeding, for several days. Insects may be regarded as either frequent or occasional feeders. The efficiency of digestion is affected by the rate at which food passes through the gut. The gut, which has layers of circular and longitudinal muscle, shows rhythmic movements that are under nervous, and possibly endocrine, control. The contractions of the three regions of the gut seem to be independent of one another. The rate of movement of food from the crop to the midgut in *Periplaneta americana* fed a liquid meal declines exponentially with time. This movement is regulated by the frequency and extent of opening of the proventricular valve. The frequency of opening also decreases exponentially and is related to the osmotic pressure of the meal (Fig. 2-6). Crop-emptying in this insect is mediated entirely by the retrocerebral nervous system, sensory input coming from osmotic pressure receptors in the pharynx, and possibly from viscosity and

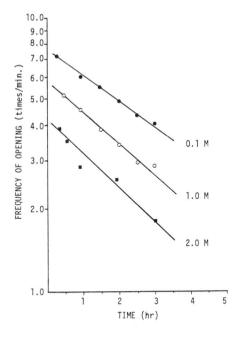

FIGURE 2-6. The decline in frequency of opening of the proventriculus in *Periplaneta americana* fed with 0.1, 1.0, and 2.0 *M* glucose. [From K. G. Davey and J. E. Treherne, *J. Exp. Biol.* **40,** 763 (1963).]

stretch receptors. In another cockroach, *Leucophaea maderae,* crop-emptying seems to depend simply on distension when solid or semisolid food is fed.

The stimulus for digestive enzyme synthesis has been investigated, particularly in hematophagous insects in which engorgement makes them most suitable for this kind of study. In the tsetse fly, *G. morsitans,* in *R. prolixus,* and in *Aedes aegypti,* protease synthesis appears to be due to a secretogogue mechanism, that is, the presence of protein in the gut stimulates protease synthesis (Fig. 2-7). In the stable fly, *Stomoxys calcitrans,* it has been suggested that the midgut cells produce and accumulate proteases until they are "fully loaded." Following a blood meal, this store of digestive enzymes is immediately released.

However, in *A. aegypti* the neurosecretory system and ovaries are required for a substantial part of the protease activity normally found in the midgut after a blood meal. This endocrine-dependent activity represents secretion of enzyme in excess of the proteolytic requirements.

In other insects the presence of a meal in the foregut stimulates secretion of digestive enzymes. The mechanism in *Locusta* appears to involve stretch receptors that record distension of the foregut; the information is then transmitted to the brain via the frontal ganglion; and a hormone controlling secretion is released. There is immunochemical evidence for a gastrinlike peptide in the neuroendocrine system of *Manduca sexta,* but it is absent from the guts of *M. sexta, Dermestes maculatus, Musca domestica,* and *M. autum-*

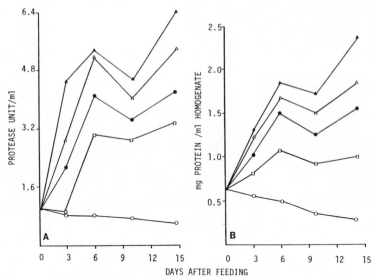

FIGURE 2-7. Protease activity (A) and protein content (B) in extracts of the midgut of fifth-instar *Rhodnius prolixus* larvae fed on: saline (○——○) and saline diluted blood (25% □——□; 50% ●——●; 75% △——△ and whole blood ▲——▲). (Reprinted with permission from E. de S. Garcia and M. L. M. Garcia, *J. Insect Physiol.* **23,** 247, Copyright 1977, Pergamon Press Ltd.)

nalis and so is believed to have a neurotransmittive, and not a digestive, function.

There may be a nutrient class specificity in the control of inducible midgut enzymes. The dermestid beetle, *Attagenus megatoma,* shows a consistently higher level of sucrase activity on casein–sucrose diets than on all-casein diets.

2.3.3. Functions of the Salivary Glands

The salivary glands may secrete digestive enzymes, the occurrence of which correlates well with the nature of the food. The saliva of hematophagous insects contains very few, if any, digestive enzymes. The saliva of nectar feeders contains sucrase. Silkworm saliva, on the other hand, can hydrolyze protein, starch, sucrose, maltose, trehalose, cellobiose, and mellibiose. A lipase is present in the saliva of insects which, like the seed-eating lygaeid bugs, feed on substances rich in fat. These enzymes appear to function within the alimentary canal.

In certain insects the salivary glands may secrete digestive enzymes which function externally as well as a variety of other components, some of them important in feeding. An anticoagulin is present in the saliva of some hematophagous insects, such as tsetse flies, but it is absent from the yellow fever mosquito, *A. aegypti.*

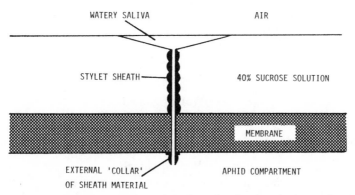

FIGURE 2-8. Diagram showing the formation of a stylet sheath and release of watery saliva when *Myzus persicae* adults were fed through a membrane on a 40% sucrose solution. (Reprinted with permission from P. W. Miles, *J. Insect Physiol.* **11,** 1261, Copyright 1965, Pergamon Press, Ltd.)

In Homoptera and some Heteroptera, the salivary glands secrete a lipoprotein sheath around the stylets which also forms a collar around the rostrum at the point of entry into the epidermis (Fig. 2-8). Hardening of the lipoprotein sheath involves the formation of disulfide and hydrogen bonds, which is prevented in the glands by reducing conditions and the dielectric effects of other materials. Additional hardening of the lipoprotein sheath may result from quinone tanning. The sheath appears to prevent the loss of plant sap and saliva through the wound in the epidermis.

In aphids a second, more watery salivary secretion produced independently of the sheath contains pectinase, which may digest the pectin of the middle lamella, thereby helping penetration of the stylets. Some hemipteran saliva is also rich in amino acids, which may cause feeding damage to the plant or serve as precursors of the plant auxin, 3-indoleacetic acid. This auxin, present in saliva, may cause the formation of galls. It is probably of plant origin, although it is possible that it may be synthesized by the salivary glands.

Adults of the blowfly *Calliphora* salivate when the labellar taste papillae are stimulated by crystalline sugar, sugar in solution, or water. The saliva always contains digestive enzymes, whether required or not. The production of saliva in isolated glands is stimulated by a factor present in the blood of salivating flies: the endocrine system as well as the nervous system is therefore involved in salivation in this insect.

2.4. EFFICIENCY OF FOOD UTILIZATION

To determine the overall efficiency with which insects utilize their food, food intake and growth must be measured. Such determinations, which have been made mainly with phytophagous insects feeding on natural plant food-stuffs, have usually involved the calculation of one or more of the following coefficients:

1. Efficiency of conversion of ingested food (ECI):

$$\text{ECI} = \frac{\text{Weight gained}}{\text{Weight of food ingested}} \times 100$$

2. Efficiency of conversion of digested food (ECD):

$$\text{ECD} = \frac{\text{Weight gained}}{\text{Weight of food ingested} - \text{weight of feces}} \times 100$$

ECD does not correct for losses due to energy metabolism.

3. Approximate digestibility (AD):

$$\text{AD} = \frac{\text{Weight of food ingested} - \text{weight of feces}}{\text{Weight of food ingested}} \times 100$$

AD estimates the assimilation of digested food, and does not correct for urine, peritrophic membrane, or exuvium. The "feces" of insects includes the urine as well as the egesta.

These coefficients vary widely with the insect species. The ECI and ECD of lepidopterous larvae are about double those of orthopterous larvae, the AD being about the same. In a comparison of the utilization of mulberry leaves by penultimate-stage larvae of the oligophagous *Bombyx mori* and the polyphagous *Prodenia eridania,* the following coefficients were calculated:

	AD	ECD	ECI
Bombyx	44	47	21
Prodenia	50	42	21

The higher ECD of *Bombyx* was interpreted to mean a specific adaptation to the nutrient balance of the natural food. The higher AD of *Prodenia* was seen as an adaptation that compensates for its low ECD. Yet nutrients may not be the only factors involved. For example, plant allomones may markedly affect these coefficients (Table 2-3). The mechanism is obscure, and it is clear that an understanding of these relationships must be sought in other ways.

The coefficients of food utilization also vary with age (both within and between instars) and sex and with environmental factors such as temperature, humidity, and degree of crowding. Coefficients of utilization for energy and for individual components of the diet, such as nitrogen, lipid, and various carbohydrates, have also been determined.

TABLE 2-3. Effects of Some Allomones (3.75×10^{-2} M) on Larval Weight Gain, Ingestion, and Nutritional Indices in the Black Cutworm, *Agrotis ipsilon*[a]

Allelochemic	Amount Ingested	Weight Gain	AD	ECD	ECI
p-Benzoquinone	71.3[c]	71.7[c]	74.7[c]	139.8[c]	102.5
Duroquinone[b]	26.6[c]	5.5[c]	210.3[c]	10.4[c]	18.1[c]
Catechol	95.9	75.3[c]	97.1	82.2[c]	79.4[c]
L-Dopa	111.2	64.3[c]	77.0[c]	74.2[c]	56.3[c]
Resorcinol	43.8[c]	51.4[c]	108.7	114.9[c]	100.4
Phloroglucinol	80.7	61.3[c]	88.0	85.1[d]	73.1[c]
Cinnamic Acid	90.7	88.2	99.7	98.3	98.5
Benzyl Alcohol	123.9	115.7	108.3	82.4[d]	88.0

[a] All values expressed as percent of control and all determinations were made on a dry weight basis [% = (experimental value)/(control value) × 100].
[b] A synthetic compound not reported in plants.
[c] Difference from the control significant at the 0.01 level of probability.
[d] Difference from the control significant at the 0.05 level of probability.
Source: S. D. Beck and J. C. Reese, *Recent Adv. Phytochem.* **10**, 41 (1976).

2.5. DIETARY REQUIREMENTS, NATURE OF THE DIGESTIVE ENZYMES, AND ABSORPTION

2.5.1. Determination of Dietary Requirements

The nutritional requirements of insects have been largely determined by rearing insects on artificial diets. It is important to rear the insect axenically, that is, in the absence of other organisms that might influence its nutrition. It is particularly necessary to rid the insect of the hereditary microbial symbiotes in its gut lumen or in its tissues. Gut symbiotes may usually be removed by surface sterilization of eggs and intracellular symbiotes, by feeding antibiotics. A novel method of producing aposymbiotic individuals, used with the tsetse fly, *G. morsitans,* involves feeding the flies on rabbits which have been immunized with the symbiotes. Intracellular symbiotes (Fig. 2-9), variously called yeasts, bacteroids, and so on, are often present in mycetocytes (specialized cells of the midgut or fat body). (The mycetocytes may be concentrated in a "mycetome.")

Axenic rearing requires sterile conditions, and it is not always practicable to take the elaborate precautions necessary to prevent contamination by other species. Xenic rearing is therefore always subject to the limitation that the precise contribution of microorganisms to the nutrition of the insect cannot be known. However, with insects that do not have hereditary symbiotes, much useful information has been gained with considerably less effort by xenic rather than axenic rearing.

It is desirable to rear the insect on a diet that consists of the simplest

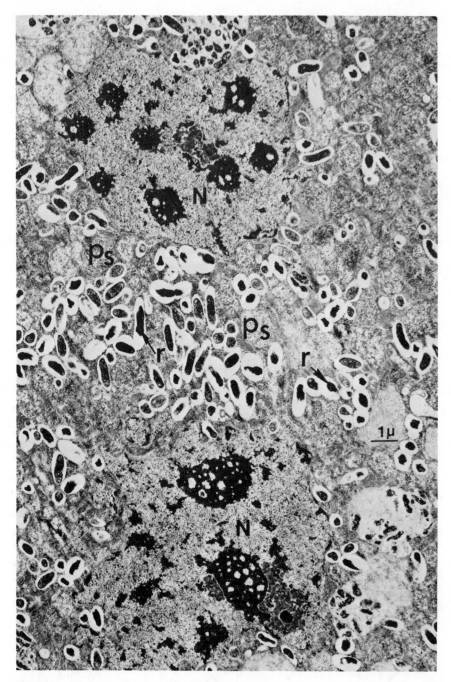

FIGURE 2-9. Portion of a mycetome of a male *Cimex lectularius* showing the occurrence of rod-shaped symbiotes (r) and pelomorphic symbiotes (Ps) in syncitial cytoplasm. The clear spaces surrounding the symbiotes are probably fixation artifacts. The absence of an obvious membrane boundary between the two nuclei (N) is suggestive of the syncitial nature of the mycetome. (Reproduced by permission of the National Research Council of Canada from the *Canadian Journal of Microbiology*, Vol. 19, pp. 1075–1081, 1973.)

chemicals which can support life, maintain optimal growth, and provide for reproduction. When the chemicals required may be obtained in a relatively pure form, the diet is said to be chemically defined, or holidic. Many insects have been reared successfully on holidic diets, principally the plant-sucking insects that obtain their nitrogen in nature in the form of amino acids. It is, however, being increasingly realized that unknown or undefined impurities may be important in insect nutrition, and it is perhaps best simply to refer to all specially compounded diets as artificial.

Because of their small size, which makes rearing in large numbers very efficient, insects are good experimental animals for nutritional research. But there are problems. Most temperate species have a diapause, which is often a period of arrested morphogenesis. Many economically important insects grow slowly and have a long life cycle. Some insects require a liquid or semiliquid diet that must be kept sterile. Others must obtain the liquid diet by piercing through a membrane. Finally, insects may require special physical or chemical stimulants to induce them to feed.

2.5.2. Proteins, Peptides, and Amino Acids

2.5.2.1. Requirements. New protein is synthesized during growth and reproduction, and the protein of the adult is constantly being replaced. The source of the required amino acids is dietary protein in most insects. Certain plant-sucking species, feeding in nature on a mixture of amino acids rather than protein, can be reared successfully on artificial diets. Most insects, however, like many other animals, cannot be reared on artificial diets in which the amino acid mixture resembles the composition of any natural protein, even though growth is satisfactory with the protein as the sole amino acid source. Work with mammals indicates that this is because many of the amino acids are actually absorbed as peptides. In mammals, the rates of transport of individual amino acids from the intestinal lumen have been found to vary widely, partly because of their differences in solubility. Peptides, on the other hand, are generally more soluble, and their transport appears to provide the tissues with a pattern of amino acids more favorable to growth. This finding may also apply to insects, in which peptides also appear to be absorbed, even though absorption of amino acids has been found to be passive, rather than active, as in mammals. Amino acid mixtures have been prepared that will support good growth of both insects and mammals, which normally require protein, but these mixtures bear no relation to the composition of natural protein.

In spite of generally poor growth on a diet containing a mixture of amino acids, deletion of individual amino acids from mixtures has been the usual method of determining amino acid requirements. The method works because omission of a required amino acid provides very little growth or no growth or survival, a result that contrasts sufficiently wiht that of an inadequate "complete" mixture. Other methods may be used with insects. By feeding or injecting glucose-U-^{14}C and then recovering tissue amino acids in order to determine their radioactivity, the amino acids synthesized by the animal may be identified. This method, useful when the deletion method is not applicable, often gives different results from the deletion method when the two are compared, and it is not considered to be entirely reliable. Nitrogen balance studies, which are used in human nutrition, have not been carried out with insects but are feasible with insects like the house cricket, *Acheta domesticus,* which will feed on dry diets and excrete a dry feces.

There are 20 amino acids coded for in the genetic material of living organisms, and, generally speaking, 10 of these are required for growth in insects. These are the same 10 that are essential for growth of the rat and other animals (Table 2-4). Much of this work has been done with xenically reared insects not having hereditary symbiotes, which suggests that the contribution of "accidental" gut lumen symbiotes to amino acid nutrition in these insects is negligible. While several insects require 1 or 2 amino acids in addition to the usual 10, radical departures from the norm occur only in insects with hereditary symbiotes, such as the German cockroach, *Blattella germanica.* Here the intracellular symbiotes synthesize various amino acids, including methionine by using inorganic S as a precursor, and are able to convert stored urates to amino acids.

What is essential or required depends on the criteria for growth. An amino acid or other nutrient may not be essential for survival or for limited growth, yet may improve growth. In fact, a mixture of the nonessential and essential amino acids usually provides for better growth than a mixture of essential amino acids. Clearly it would be desirable to have optimal growth as a reference point, optimal growth usually, but not always, being defined as growth on the best natural diet. (Certain insects, such as the nitrogen-deficient wood feeders, grow better on artificial diets than on their natural diets.) The growth of most insects on artificial diets usually falls short of the optimum. This is likely due in large part to a lack of the proper balance of nutrients rather than to unknown qualitative requirements. For these reasons, a precise definition of the nutritional requirements of most insects seems a long way off.

A number of attempts have been made to demonstrate whether insects that feed on diets low in nitrogen, like termites and aphids, are able to obtain organic nitrogen by the fixation of atmospheric nitrogen. Recent studies

TABLE 2-4. The 10 Essential
Amino Acids

Arginine	Methionine
Histidine	Phenylalanine
Isoleucine	Threonine
Leucine	Tryptophan
Lysine	Valine

have shown that nitrogen fixation does occur in termites. It is believed, however, that this is accomplished by symbiotic bacteria, and in any case proceeds too slowly to provide for the nitrogen needs of the insect.

The amino acid requirements of the adult insect are not nearly as well known as the requirements for growth and metamorphosis. Adult male insects do not usually require amino acids at all: spermatogenesis occurs in immature stages, and the adults may have sufficient reserves for metabolic needs and for synthesis of accessory gland proteins. Adult female insects, on the other hand, require amino acids for oogenesis. Carry-over from the larval stage is obviously necessary where the adult does not feed. Until recently, it was thought that carry-over supplied the nitrogenous requirements of all nectar-feeding adult Lepidoptera. However, analysis of nectar shows considerable amounts of amino acids to be present, although the variety is limited. Flowers with nectars richest in amino acids are visited by females with no alternative source of nitrogen (Table 2-5). These observations suggest that nectar is an important source of amino acids for insects, as well as being an essential source of carbohydrate for flight.

Carry-over is also seen in autogenous species of mosquitoes and black-flies, that is, species that can produce at least one cycle of eggs without a blood (i.e., protein) meal. Anautogenous species are infertile without a blood meal. Where there is insufficient carry-over and inadequate amounts of amino acids in the diet, insects such as queen ants, water beetles, and *Dysdercus* (Pyrrhocoridae) may break down their flight muscles to produce the amino acids required for oogenesis.

2.5.2.2. Digestion and Absorption. Those endopeptidases of the insect gut that have been adequately characterized resemble the trypsin and chymotrypsin of mammals, though they are not identical. The trypsinlike enzyme of *Tenebrio molitor* has a relatively low number of —S—S— bonds

TABLE 2-5. Amino Acid Concentration of Nectar in Flowers of Different Pollination Types

Flower Pollination	No. of Species	Mean (on histidine[a] scale)	Range	Coefficient of Variation (%)
Bee	95	4.76	0–9	47
Butterfly and bee	44	6.02	3–8	23
Butterfly	41	6.68	4–10	21
Moth	30	5.60	0–9	45
Bird	49	5.22	1–9	37
Fly (unspecialized)	34	3.77	0–9	57
Fly (specialized)	8	9.25	7–10	—

[a] Nectar treated with ninhydrin and color compared with standard histidine solutions. 1 = 49 μM; 10 = 25 mM.
Source: H. G. Baker and I. Baker, *Nature*, **241**, 544 (1973).

compared with bovine trypsin and so differs in conformation. The endopeptidases have been identified from the gut lumen, although work on digestive enzymes tends to be done on whole guts or tissues and lumen contents together because of the small size of most insects. There may be more than one species of these enzymes in a single insect. For example, in the filter-feeding larvae of *A. aegypti* where whole guts have been examined, at least 12 different electrophoretic species of endopeptidases exist. Most are trypsin- and chymotrypsinlike, but some are neither. In the honeybee, there is a change in the proteinases with metamorphosis. Whereas the midguts of both larval and adult workers contain two species of chymotrypsin, shown by immunological techniques to be identical in both stages, a tryptic protease was found only in adults.

The pH of the midgut is usually in the range 6–8, which favors the activity of enzymes of this kind. Buffering agents vary with the insect species, but phosphates, amino acids, and proteins may be involved, as well as other organic acids and salts. Low pH values have been recorded for a few insects. Fly larvae have enzymes active at an acid pH; in the absence of studies on substrate specificity, one cannot conclude that a pepsinlike enzyme is present. The pH optima of gut proteinases ranges from very acid to very alkaline in hematophagous insects (Fig. 2-10).

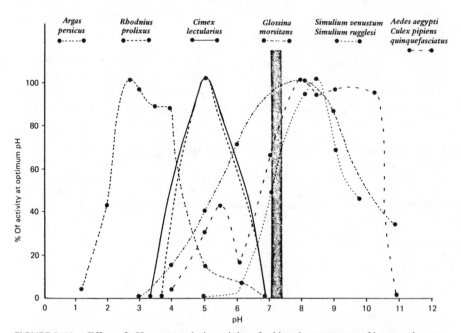

FIGURE 2-10. Effect of pH on proteolytic activity of midgut homogenates of hematophagous arthropods. The shaded region indicates the pH values reported for vertebrate blood. [From R. H. Gooding, *Acta Trop.* **32**, 96 (1975)].

Among the exopeptidases purified electrophoretically, carboxypeptidases A and B, aminopeptidases, and dipeptidases occur in the gut lumen.

Amino acids appear to be absorbed by passive diffusion. Although the hemolymph titer of amino acids is characteristically high, water is rapidly absorbed from the midgut lumen following the active absorption of Na^+, so that diffusion will account for the movement from gut to hemolymph.

Of special interest to the nutritionist is the digestion by certain insects of keratin and collagen, structural animal proteins that are not digested by other animals. Keratin, the protein of hair, skin, and nails, has its polypeptide chains cross-linked by disulfide bonds, which must usually be broken by reduction before the peptide chains become vulnerable to enzymic attack. Mallophaga, clothes moths (Tineidae) and their allies, and certain carpet beetles (Dermestidae) are able to digest keratin. The clothes moth, *Tineola,* has a keratinase which when purified can digest *in vitro* considerable amounts (up to 30%) of the native protein. The midgut cells contain cystine reductase, which converts the cystine produced on hydrolysis of the keratin, and subsequently absorbed, to cysteine. Cysteine desulfhydrase, which is also found in the midgut epithelium, can degrade cysteine, releasing hydrogen sulfide. The cysteine and hydrogen sulfide passing into the lumen both reduce the disulfide bonds of keratin and thereby promote digestion.

Collagen, the protein of connective tissue, has three polypeptide chains so tightly bound by hydrogen bonding that they are usually resistant to enzyme attack. By boiling collagen, the hydrogen bonding is broken and gelatin, which is a readily digestible protein, is formed. Collagenase, an enzyme that can attack the native protein, is found in blowfly larvae. Collagenase is found in mammals, but not in the gut lumen.

Silk consists of fibroin, a structural protein, and sericin, a globular protein binding the fibroin chains. In larval Lepidoptera silk is secreted by the labial glands. The maxillary galeae of the adults of *Antheraea pernyi* and *A. polyphemus* secrete a nonspecific tryptic proteinase called cocoonase, which is an adaptation for escaping from the cocoon. Cocoonase secreted in an almost pure form is activated at the appropriate time by a secretion of the labial glands, a solution buffered at a pH of 8.3–8.7, a pH range favorable for the activity of the enzyme. The enzyme digests sericin, thus softening the cocoon and permitting escape of the adult insect.

2.5.3. Carbohydrates

Carbohydrate is the form in which energy is most accessible to an animal. It is therefore not surprising that most insects require an exogenous source of carbohydrate for flight, the most energy-demanding of all life activities. This is not to say that carbohydrate is the only fuel for flight. Lipid, which is used to a considerable degree by Hemiptera, Lepidoptera, and locusts, is probably used to some extent by all insects. In locusts, carbohydrate is the fuel used to initiate flight, whereas fatty acids are metabolized during sustained flight. The tsetse flies, which feed exclusively on mammalian blood, a food low in carbohydrate, use proline, an amino acid, as their principal energy source. Other insects, such as blowflies, can also metabolize proline rapidly. (see Chapter 12.)

For larvae the need for carbohydrates is less as metabolism is generally slower. Carbohydrate is usually abundant in plant food, but carnivores and parasitoids encounter relatively little and carrion feeders virtually none. Table 2-6 presents the carbohydrate requirements of selected species. Some insects do not need carbohydrate, although in nature they may feed on it; and carbohydrate may actually be inhibitory to growth in some, such as blowfly larvae. Where carbohydrate is required for optimal growth, there is a wide variation in optimal dietary proportion. Certain insects, such as *A. domesticus* and *Tenebrio molitor,* do not grow at all without dietary carbohydrate.

In dietary experiments it is usually found that glucose will supply the carbohydrate requirement of the insect. Glucose therefore usually serves as the standard against which the utilization of other carbohydrates is measured. A great number of carbohydrates has been tested for their utilization by insects. Much of this work is of purely biochemical interest, indicating only the possibilities of metabolic interconversions, as the compounds either do not occur in nature or the insect does not normally feed on them.

Among the polysaccharides, starch is hydrolyzed by insects that feed on it and is therefore utilized; the enzyme(s) involved is an α-amylase. Polysaccharides like cellulose, which have β-linked units, are utilized by few insects. Lower termites and wood roaches depend on symbiotic gut protozoa to break down the cellulose anaerobically; the principal product utilized is

TABLE 2-6. **Carbohydrate Requirements of Various Insects**

Do not require carbohydrate:
    Attagenus sp.
    Lasioderma serricorne
    Tribolium confusum
    Ptinus tectus
Require carbohydrate for optimal growth:

	% in diet
Acheta domesticus	20
Anagasta kuheniella	80
Blattella germanica	30
Cadra cautella	50
Locusta migratoria	26
Macrocentrus ancylivorus	5
Myzus persicae	10–20
Ostrinia nubilalis	25
Phalera bucephala	50–80
Tenebrio molitor	80–85

acetic acid. Higher termites and scarab beetles apparently use cellulolytic bacteria; only silverfish have a C_1^- cellulase. Inulin, a β-fructofuranoside, is digested by the silkworm and several cockroaches. Chitinase must be present in the molting fluid of all insects as the old cuticle is digested during the molting cycle, but it is absent from the gut lumen, except for the midgut of *P. americana* and perhaps a few other species of insects.

Polysaccharides grouped in the class of hemicelluloses, which includes galactans, mannans, arabams, and so on, are utilized by a number of wood-boring beetles.

With oligosaccharides, enzymic studies have usually confirmed the results of feeding trials. Generally speaking, insects have α-glucosidases, which hydrolyze a variety of oligosaccharides—sucrose, maltose, trehalose, and melezitose. Other glycosidases are present in various species: these include β-glucosidase, α-galactosidase, β-galactosidase, and β-fructofuranosidase.

An unusual aspect of carbohydrate digestion occurs in the plant-sucking aphids and scale insects. These insects ingest great quantities of plant sap, the solids of which are largely sugar, and excrete the excess sugar as honeydew. The amount of sap ingested depends to a large extent on the turgor pressure of the plant, not, as was once thought, because plant sap is low in nitrogen. In fact, amino acids are excreted in the honeydew. Turgor pressure is not the only or even a necessary factor, however, as aphids may feed at a greater rate than turgor pressure can account for and will feed through artificial membranes in the absence of positive pressure.

Honeydew may contain one or both of two types of trisaccharide, melezitose or glucosucrose (Fig. 2-11). These trisaccharides are produced by the action of two distinct α-glucosidases or sucrases on sucrose. The first adds a glucose unit to C-3 of the fructose moiety producing melezitose; the second adds glucose units to C-4 of the glucose moiety of sucrose to produce the trisaccharide glucosucrose, followed by glucotetrasucrose, glucopentasucrose, and so on. It has been suggested that the purpose of oligosaccharide synthesis is to prevent absorption of unwanted carbohydrate. This sounds reasonable, as some insects excreting honeydew also have "filter chambers" whereby much of the sugar and water ingested bypasses the midgut.

Carbohydrates are generally absorbed as monosaccharides, which would normally be glucose and fructose. Absorption of oligosaccharides also appears to occur to some extent in insects as it does in mammals.

Absorption of glucose is passive, but facilitated diffusion of glucose has been shown to occur in a few insects and may be widespread. The principal hemolymph sugar of most insects is trehalose (glucose-1-glucoside) which is synthesized in the fat body. The fat body in many insects surrounds the gut. Glucose absorbed from the gut moves into the hemolymph and then quickly into the fat body where it is converted to trehalose. The rapid removal of glucose by the fat body therefore creates a steep concentration gradient across the gut wall. In dipterous and other insects in which glucose and not

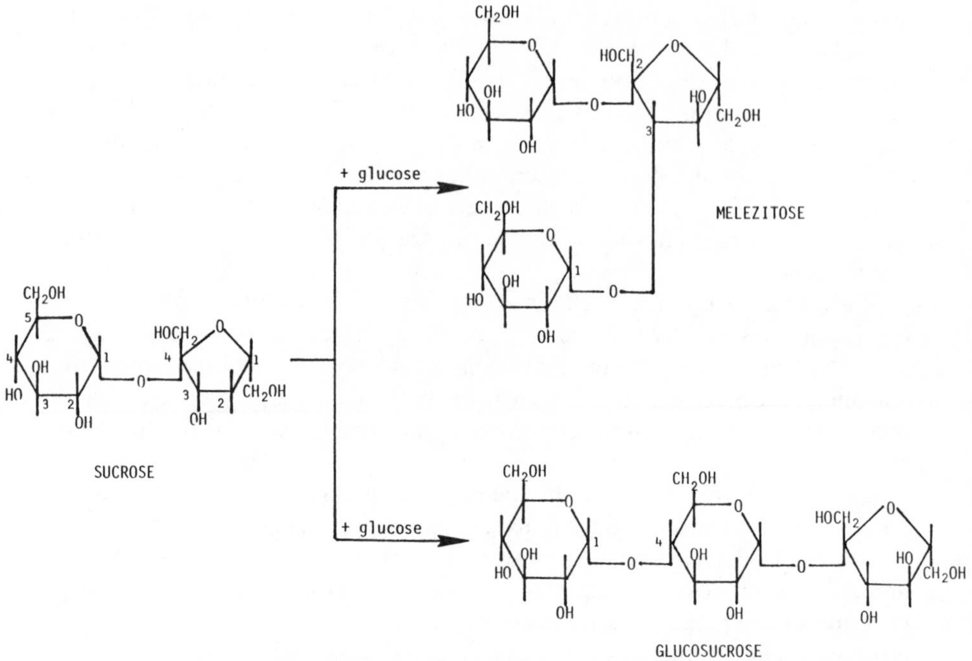

FIGURE 2-11. Oligosaccharide synthesis from sucrose to form melezitose and glucosucrose.

trehalose is the principal hemolymph sugar, glucose may be actively transported across the gut wall as it is in mammals.

2.5.4. Lipids

2.5.4.1. Fatty Acids, Fats, and Waxes.   Fatty acids form a structural part of the cell membrane and serve as a source of energy. Although normally a part of the insect diet, fatty acids are synthesized by many insects in the process of accumulating fat as a storage form of energy.

Many insects require polyunsaturated fatty acids, and in most of these species linoleate and linolenate are equally effective as, for example, in growth and wing development of *Diatraea grandiosella*. In other insects, they may not be nutritionally equivalent, or equivalent for all functions. Generally they are equivalent for growth, but linolenate is required for metamorphosis. Thus, in three species of Lepidoptera, *Trichoplusia ni*, *Autographa californica*, and *Heliothis zea*, linolenate was essential for the development of normal wings; linoleate could substitute at a high level, but only in *H. zea*. For larval development of these species, linolenate was more effective than linoleate.

A third polyunsaturated fatty acid, arachidonic acid, is required for flight and survival of newly emerged adults of *Culex pipiens*. It is believed that this requirement is due to the loss of the biosynthetic pathway for converting linoleic acid to arachidonic acid.

Other fatty acids may promote growth although they are not essential. Oleic acid improves growth of *Agria housei* and of three lepidopterans, *Argyrotaenia velutinana, B. mori,* and *Sitotroga cerealella.*

Some insects, like *Galleria mellonella,* are influenced very little by the relative composition of fatty acids in the diet. Other insects, such as *Pieris brassicae,* alter the composition of dietary fatty acids through selective incorporation and synthesis.

Dietary fatty acids are usually present as esters of glycerol, that is, triglycerides. These esters may possibly be absorbed without alteration, but most insects have a gut lipase which hydrolyzes triglycerides. Emulsification of the fat appears to be brought about by fatty acylamino complexes in *Gryllus bimaculatus.* Bile salts have not been reported from insects.

The midgut is the chief site of fat digestion and absorption. In *Pieris brassicae* labeled triolein is hydrolyzed to free fatty acids, diglycerides, and monoglycerides (Fig. 2-12). Diglycerides are absorbed rapidly, free fatty acids less readily. In the midgut wall, label was found in phospholipids, triglycerides, diglycerides, and free fatty acids. In the hemolymph most of the label was found in diglycerides and phospholipids. This suggests that diglycerides are released from the gut wall and carried by the hemolymph to the fat body where fatty acids are stored as triglycerides. They are released from the fat body as diglycerides, triglycerides, and sterol esters.

A study of the fine structure of fat-absorbing midgut cells of *Stomoxys calcitrans* suggests that free fatty acids pass from the lumen across the microvilli to the rough endoplasmic reticulum and are then transported to the Golgi bodies, where they are converted to triglycerides and phospho-

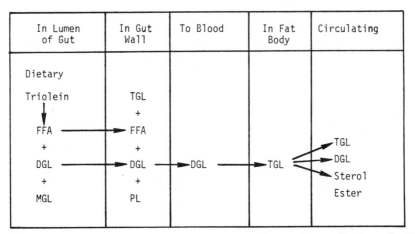

FIGURE 2-12. The fate of labeled triolein fed to larvae of the cabbage white butterfly. For explanation see text. [From data of S. Turunen, *J. Insect Physiol.* **21,** 1521 (1975)].

lipids. They may continue to the basal infoldings or be stored in the form of lipoid spheres.

Esters of fatty acids other than with glycerol are commonly found in insects' cuticular wax or as special wax secretions. These are not normally digested. An exception is myricyl (C_{30}) palmitate, the most important constituent of beeswax. Up to 50% of beeswax may be utilized by the wax moth, *Galleria mellonella,* but when the insect is reared axenically, relatively little wax is utilized. It appears that hydrolysis of wax is accomplished by both larval and microbial enzymes.

2.5.4.2. Sterols. Unlike vertebrates, insects and other arthropods cannot synthesize sterols. Nearly all insects require sterol in the diet, the only exceptions being certain members of the plant-sucking Homoptera and Heteroptera, in which intracellular symbiotes provide the sterol.

The sterol requirement may be met in all cases studied by cholesterol, apart from two species, *Drosophila pachea* and *Xyleborus ferrugineus,* which require a $\triangle^7$-sterol (Fig. 2-13). With these two exceptions, insects appear to need a small amount of cholesterol itself to serve an essential though unknown function. To produce the steroid hormone ecdysone, small amounts of cholesterol or a $\triangle^7$-sterol are mandatory. Much larger amounts of sterol are required as structural components of the cell membrane and cuticular wax and to function in fatty acid transport. Plant and yeast sterols are often more effective in the diet than cholesterol, depending on the natural diet to which the particular insect is adapted. Yet cholesterol is still the main body sterol in these species.

Without prior esterification, sterols are absorbed chiefly from the crop in some insects and from the midgut in others.

2.5.5. Accessory Growth Factors

The accessory growth factors include organic substances which, while essential, are required in relatively small amounts. They may be divided into

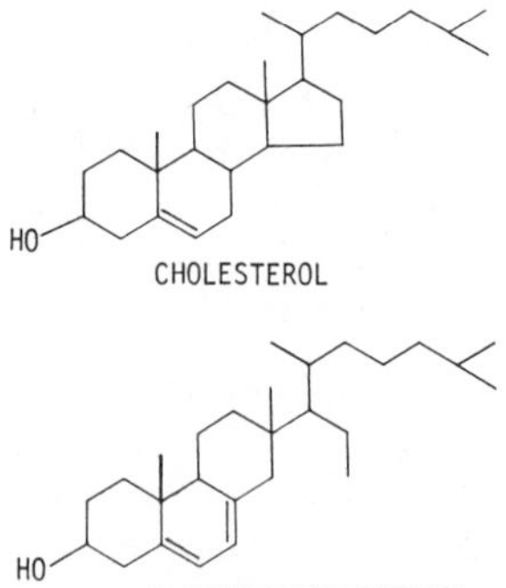

CHOLESTEROL

7-DEHYDROCHOLESTEROL

FIGURE 2-13. The structure of cholesterol and 7-dehydrocholesterol.

water-soluble and lipid-soluble classes. A listing of the factors generally required by insects, as well as an indication of their function, is presented in Table 2-7.

Water-soluble factors include the B vitamins, the lipogenic factors choline and inositol, and vitamin C. The insect requirement for B vitamins is similar to that of other animals. Seven vitamins are required: thiamine, riboflavin, nicotinic acid, pantothenic acid, pyridoxine, folic acid, and biotin. As these function as part of coenzymes, they are essential, although adequate amounts may be supplied in the egg for development through one genera-

TABLE 2-7. Growth Factors in Insect Nutrition and
Their Physiological Function(s)

	Principal Physiological Function(s)
Water-soluble factors	
B vitamins	
Thiamin	Component of coenzyme thiamine pyrophosphate (TPP) or cocarboxylase
Riboflavin	Component of coenzymes flavine mononucleotide (FMN) and flavine adenine dinucleotide (FAD)
Nicotinic acid (nicotinamide)	Component of coenzymes nicotinamide adenine dinucleotide (NAD) and nicotinamide adenine dinucleotide phosphate (NADP)
Pyridoxine (vitamin B$_6$)	Component of coenzyme pyridoxal phosphate
Pantothenic acid	Component of coenzyme A
Biotin	Coenzyme for enzymatic reactions involving addition of CO$_2$ to other units
Folic acid	Involved in the transfer of single carbon units
Vitamin C	Incompletely understood. Reversible oxidation and reduction capacity suggests role as hydrogen transporter in cellular respiration
Lipogenic factors	
Choline	Component of the phospholipids lecithin and sphingomyelin and of acetylcholine, source of biologically labile methyl groups
myo-Inositol	Incompletely understood. Component of phospholipid phosphatidylinositol
Lipid-soluble factors	
Vitamin A	Component of rhodopsin. Other functions, such as effect on growth, incompletely understood
Vitamin E	Antioxidant, especially for fats. Other functions, such as role in cellular respiration, incompletely understood
Vitamin K	Involved in prothrombin synthesis in mammals and in electron transport and oxidative phosphorylation in microorganisms

tion. Vitamin B_{12} improves the growth of a few insects, and vitamin B_T or carnitine is essential for a number of tenebrionid beetles. Carry-over from the larval stage may supply the vitamin needs of many, though certainly not all, endopterygote insects. The hymenopterous parasitoid *Exeristes comstockii* has been shown to require B vitamins for oogenesis.

Hereditary symbiotes may supply the B vitamins in blood-sucking and sap-sucking species, blood and plant sap being low in these factors. Species in which all stages are restricted to blood as a source of food, for example, lice and bedbugs, have symbiotes. Those insects (fleas, mosquitoes, etc.) with other sources of food in the larval stages with which they are able to accumulate adequate B vitamin supplies for the entire life cycle lack symbiotes.

The lipogenic factors, choline and *myo*-inositol, are frequently classified with the B vitamins, although they are not coenzymes and are required in somewhat larger amounts. They are components of phospholipids and are therefore an integral part of cell membranes. Choline functions in methyl group transfer as well.

Vitamin C is required for good growth and survival of a number of plant-feeding insects including *Schistocerca gregaria, Ostrinia nubilalis, B. mori, D. grandiosella, Anthonomus grandis,* and *Myzus persicae.*

Of the lipid-soluble factors, vitamin A is probably required by all insects, as it is a component of rhodopsin, the visual pigment of animals. Rhodopsin has been demonstrated to occur in a number of insects. Since it is needed in very small amounts, carry-over in the egg is sufficient to meet the need through at least one generation. Consequently, the requirement is not readily demonstrable; it has been shown only for *Drosophila, M. domestica,* and *A. aegypti.* A precursor of vitamin A, β-carotene, is also required for normal pigmentation of *Schistocerca gregaria, Melanoplus bivittatus,* and *Plodia interpunctella.* In the last-named insect, there is an effect on development as well.

Other fat-soluble vitamins have been shown to be required by certain insects. Vitamin E or α-tocopherol is necessary for reproduction in the beetle *Crypotlaemus montrouzieri,* for growth and spermatogenesis in *A. domesticus,* and for growth as well as for reproduction in the female of *Agria housei.* Vitamin K_1, the form of the vitamin found in plants, improves growth of *A. domesticus;* synthetic vitamins K are without effect.

The growth of divers, though not all, Diptera reared on artificial diets is improved by RNA or its components, as is reproduction in *M. domestica.* The larval requirement for RNA is absolute in *Drosophila melanogaster, Cochliomyia hominivorax,* and *Culex pipiens.*

2.5.6. Water

Water is a dietary requirement for most terrestrial insects, and although the majority of insects will drink free water, the usual source of water is in the food. The water content of insects ranges from about 60 to 80%, and it is more or less maintained even under dry conditions. The efficiency of the rectum in absorbing water has been known for a long time, and it has been experimentally shown that the rectum absorbs water against a concentration gradient. Stored insects like the confused flour beetle, *Tribolium confusum,* and the Mediterranean flour moth. *Anagasta kuheniella,* growing at low humidities cannot obtain their water requirements directly from their desiccated food but conserve metabolic water to maintain their water content.

A number of insects can absorb water from subsaturated atmospheres. In larvae of the mealworm, *T. molitor,* and in the firebrat, *Thermobia domestica* (Fig. 2-14), water absorption occurs through the anus, and it appears that the rectum is the site of uptake by a mechanism that is not yet understood.

2.5.7. Minerals

The mineral requirements of insects are little known, although one would expect them to be qualitatively similar to other animals. Obtaining the necessary minerals other than trace elements seems to be a problem only for

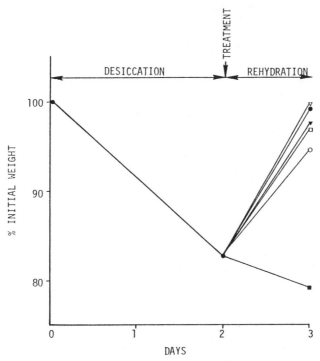

FIGURE 2-14. Effects of various treatments on subsequent rehydration of previously desiccated *Thermobia domestica.* ●, insects before operation and unoperated controls; ▼, all sham-operated combined; ▽, mid-dorsal abdomen treated with wax; □, mouth blocked with wax; ○, posterior ventral abdomen treated with wax; ■, anus blocked with wax. [From J. Noble-Nesbitt, *J. Exp. Biol.* **52,** 193 (1970)].

certain species living in fresh water. Thus mosquito larvae have been shown to actively absorb chloride ions through the anal papillae.

Mineral requirements are best studied in insects that can be reared on amino acid mixtures, as complex molecules of natural origin, such as proteins, are always contaminated with small amounts of inorganic ions. Studies of aphids have shown that iron, zinc, manganese, copper, potassium, phosphate, sulfate, calcium, and probably sodium chloride are required. When trace elements are studied, contamination may be crucial. Trace element deprivation has been shown to affect the intracellular symbiotic microorganisms of aphids and cockroaches.

In spite of the restriction mentioned above, less pure artificial diets have often been used successfully to investigate mineral requirements. Thus, where casein has provided dietary protein, zinc and potassium have been shown to influence the expression of vitamin B_T deficiency in *T. molitor,* and copper and zinc have been shown to be required and to interact in *A. domesticus* (Fig. 2-15). In the latter insect, an interaction which affects testis development has also been shown between copper and vitamin E.

The absorption of ions by the midgut is less well known than rectal absorption, which has been studied in relation to the excretory system. Much work has, however, been done on the midgut epithelium of *Hyalophora*

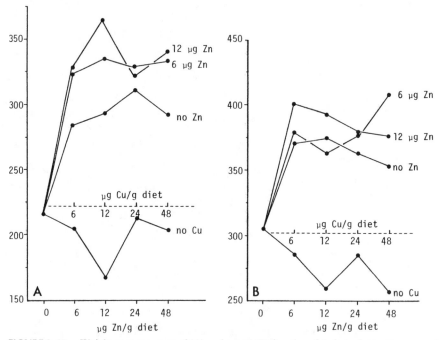

FIGURE 2-15. Weight at emergence of (A) males and (B) females of *Acheta domesticus* reared on diets containing varying amounts of added copper and zinc. [From J. E. McFarlane, *Can. Entomol.* **108,** 387 (1976). Courtesy of the Entomological Society of Canada.]

cercropia as an ion transport system. Large amounts of potassium ions are found in the midgut, derived from the leaves on which the insect feeds, but the hemolymph level is about 10 times lower. Potassium ions are actively transported from the hemolymph to the lumen. In Lepidoptera, the rate of fluid secretion by the Malpighian tubules is too low to permit cation regulation of the hemolymph.

2.6. CONCLUSIONS

A considerable amount of basic information is available on the nutritional requirements of insects which permits broad generalizations to be made. Because of the great diversity of the insect group, many exceptions are found: to the comparative nutritionist, a dietary peculiarity may provide the opportunity to examine the physiological action of a compound in depth. Economic entomologists may find applications for basic nutritional information in the mass rearing of insects for biological control procedures. They may also use insights into an insect's adaptation to its source of food to develop more effective methods of control.

ACKNOWLEDGMENTS

The author is grateful to Dr. M. J. Lehane, University College, Bangor, Wales, for permission to use the original electron micrographs of Figs. 2-2 and 2-3. Dr. J. N. Tristram and Dr. K. P. Chang kindly made available original copies of their published electron micrographs. The author is also grateful to Mr. Pierre Langlois for preparing the graphs and for photographic work.

SUGGESTED READING

G. Bhaskaran, S. Friedman, and J. G. Rodriguez, Eds., *Current Topics in Insect Endocrinology and Nutrition,* Plenum Press, New York, 1981, p. 362.

R. H. Dadd, *Ann. Rev. Ent.* **18,** 381 (1973).

R. H. Dadd, "Insect Nutrition," in M. Florkin and B. T. Scheer, Eds., *Chemical Zoology,* Vol. 5, Academic Press, New York, 1970, p. 35.

R. H. Dadd, "Digestion in insects," in M. Florkin and B. T. Scheer, Eds., *Chemical Zoology,* Vol. 5, Academic Press, New York, 1970, p. 117.

H. L. House, "Nutrition," in M. Rockstein, Ed., *The Physiology of Insecta,* 3rd ed., Vol. 5, Academic Press, New York, 1973, p. 1.

H. L. House, "Digestion," in M. Rockstein, Ed., *The Physiology of Insecta,* 3rd ed., Vol. 5, Academic Press, New York, 1973, p. 63.

J. Rodriguez, Ed., *International Conference on the Significance of Insect and Mite Nutrition,* North Holland Publishing Co., Amsterdam, 1973, p. 702.

J. M. Scriber and F. Slansky, *Ann. Rev. Ent.* **26,** 183 (1981).

3

EXCRETORY SYSTEMS

DONALD G. COCHRAN
Department of Entomology
Virginia Polytechnic Institute and State University
Blacksburg, Virginia

CONTENTS

SUMMARY

Excretion in insects is carried out mainly by the Malpighian tubule–rectum complex. The Malpighian tubules produce the primary urine, which is essentially isoosmotic to the hemolymph and resembles it in chemical composition. The driving force for urine production is apparently the active transport of potassium and other ions rather than a hydrostatic-pressure filtration system. The Malpighian tubules are extremely diverse structures among the insects which have been studied. A unifying feature is that they all appear to have cells with elaborately developed apical and basal plasma membranes in their principal secretory regions. Several hypotheses have been advanced to explain the mechanism of transport, taking the available facts into account. Perhaps the best known of these is the *standing gradient hypothesis.* It emphasizes the active transport of ions but neglects *vesicular transport,* which also appears to play a role in urine production. Diuretic and antidiuretic hormones regulate the rate of urine flow.

The primary urine is modified somewhat by the anterior parts of the hindgut, but it is the rectum that plays the major role in completing the excretory process. It functions by reabsorbing water, ions, and many other small-sized urinary molecules including nutrients like amino acids and sugars. Reabsorption occurs in response to the physiological needs of the insect. Because of this, it is here that the processes of excretion and osmoregulation and water balance become intertwined. From an excretory point of view, the result is that the nitrogenous waste products and other substances present in excess of metabolic needs remain in the rectal lumen and are voided to the exterior. In most insects the rectum is characterized by specialized regions called pads or papillae. The cells in these structures have highly developed apical and especially lateral plasma membranes. Rectal reabsorption appears to be dependent on active processes carried out by these specialized cells. Hormonal regulation of rectal function is known to occur and may involve the same hormones that act on Malpighian tubules.

Other excretory functions in insects include sequestration of certain substances in cells. Mineral concretions, uric acid spherules, and pigment granules housed within cells are often mentioned as possible examples of storage excretion. Other cells, such as nephrocytes and pericardial cells, may degrade large waste molecules from the blood. Labial or cephalic glands are sometimes considered to be alternative excretory systems, particularly in primitive insects that do not have Malpighian tubules.

3.1. INTRODUCTION

The most obvious function of the excretory system of animals is to remove chemical substances from the body which are either ametabolic or are present in quantities in excess of those required to carry out normal metabolic processes. For most animals the compounds of primary concern are those containing nitrogen. An extreme example is ammonia, a nitrogenous compound that is quite toxic at low concentrations and must be dealt with quickly before it causes damage to the tissues. The problem is not confined to nitrogenous materials, however, and the urine of animals contains a sizable number of chemical compounds and ions in an aqueous matrix. While small quantities of some of them may be inadvertently lost with the urine, the assumption is that most are voided because they are present in excess of needs.

In viewing the excretory system of insects, one finds it is not a hydrostatic pressure filtering system as it is in many vertebrates. Rather it consists of the physically discontinuous Malpighian tubule–rectum complex. The Malpighian tubules produce a primary urine which generally reflects the chemical constituents present in hemolymph. This fluid is passed via the digestive tract to the rectum where it may undergo extensive modification. Examination of the final urine produced by the rectum is rendered difficult because by this time it is usually mixed with fecal wastes for voiding via the common anal opening. Nevertheless, it is generally recognized that the rectum modifies the primary urine by selectively reabsorbing substances from it.

The system described above is a very efficient excretory apparatus for insects. With few exceptions it is present throughout the class Insecta. What may not be obvious at first glance is that it also plays a vital role in osmoregulation and water balance. The selective reabsorption of water, ions, and certain chemicals by the rectum is carried out in response to the physiological needs of the insects. Thus, in a dry environment the rectum will reabsorb water from the urine and void a dry excreta. In a freshwater habitat, the problem is to conserve ions and void excess water. This function of the excretory system, in conjunction with other components of the osmoregulatory system, has played a major role in allowing insects to invade diverse habitats.

In the pages that follow an attempt will be made to discuss the insect excretory system in reasonable detail. Emphasis will be placed on integrating the structure and function of the system and on recognizing instances where adaptive changes have occurred. A limiting factor is that detailed analyses are available for only a rather meager sample of the class Insecta. This is an undesirable situation, but the temptation to overgeneralize is, in this instance, dampened by the great diversity in both structure and function which has already been described.

The pioneering work of Wigglesworth (1931) and Ramsay (1953, 1954) in the field of insect excretion deserves special mention. Their efforts laid the groundwork for much of what we know today. The current body of knowledge has been contributed to by a large number of people (Cochran, 1975). In addition, a number of review articles have been published which are pertinent to excretion in insects. Among them are Maddrell (1971, 1977a, 1977b), Reigel (1972), Blankemeyer and Harvey (1977), Gee (1977), Phillips (1977), Phillips and Bradley (1977), Wall (1977), Zerahn (1977), and Wessing and Eichelberg (1978). These articles plus the current literature have been used as the primary sources of information in the preparation of this chapter.

3.2. MALPIGHIAN TUBULES

3.2.1. General

The Malpighian tubules of insects are blind tubes that arise embryologically as evaginations of the gut, usually at the junction of the mid- and hindgut. They are extremely variable in number, size, and shape. In *Drosophila* larvae, for example, there are only four tubules (Fig. 3-1), while *Periplaneta* adults have 150 or more. Often where the number of tubules is reduced, as in *Drosophila,* they are rather large in relation to the size of the gut. Contrarily, in *Periplaneta* the tubules are long and threadlike. Usually Malpighian tubules lie free in the body cavity where they have ready access to the hemolymph. However, in some groups of insects the tubule blind ends are embedded in the surface of the hindgut, producing what is called a cryptonephridial system. This arrangement is believed to facilitate the reabsorption of water from the hindgut and make it available to recycle through the excretory system.

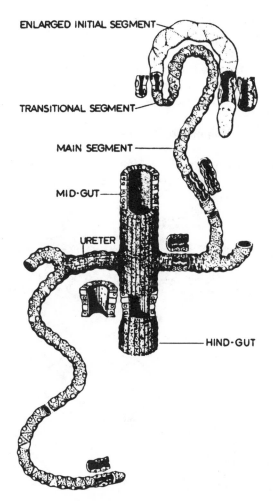

ENLARGED INITIAL SEGMENT

TRANSITIONAL SEGMENT

MAIN SEGMENT

MID·GUT

URETER

HIND·GUT

FIGURE 3-1. Diagram of larval Malpighian tubules of *Drosophila melanogaster*. Different segments of the tubule and gut are cut open to show structural details. [With permission from A. Wessing and D. Eichelberg, in *The Genetics and Biology of Drosophila*, M. Ashburner and T. R. F. Wright, Eds., Vol. 2C, Academic Press, London, 1978. Copyright by Academic Press Inc. (London) Ltd.]

Where there are many tubules, some of them may be closely associated with fat body. This apparently facilitates the transfer of wastes from the fat body just as being bathed by hemolymph probably facilitates waste transfer from it. Malpighian tubules move about in the body cavity either passively in response to body or gut movements of the insect or actively due to the presence of muscles on the tubule surface. The Malpighian tubules of *Periplaneta* have helical muscles which allow them to move with a corkscrew type of motion. These muscles are innervated, which should indicate controlled contractions. In addition, the Malpighian tubules are richly endowed with tracheae. Sometimes tracheation is largely restricted to the proximal part of the tubule, but it is often more extensive. In such cases the tracheal system may lend structural support to an integrated Malpighian tubule, gut, fat body complex. Where this occurs, it may indicate a closer interrelationship than one might otherwise expect.

3.2.2. Structure of Malpighian Tubules

3.2.2.1. Gross Morphology. The Malpighian tubules studied to date offer evidence of great variability within the insect world. They range from the

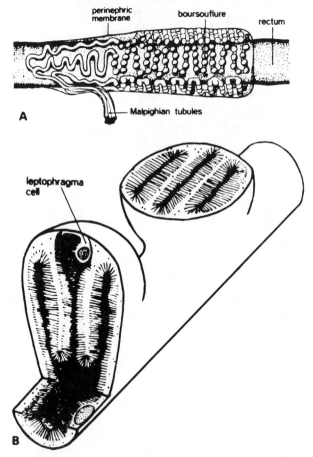

FIGURE 3-2. A. Simplified diagram of the rectal complex of *Tenebrio molitor* showing some of the Malpighian tubules and a very reduced number of boursouflures. B. Enlargement of a short tubule section cut open to show two boursouflures in different perspectives. Note the location of the leptophragma cell. [With permission from Grimstone et al., *Phil. Trans. Roy Soc. London Ser. B* **253,** 343 (1968).]

rather simple tubules of dipteran larvae, through the cryptonephridial system (Fig. 3-2) of certain coleopterans and lepidopterans, to the rather complex filter chamber arrangement of the highly specialized homopterans (Fig. 3-3) that consume large amounts of a liquid diet. In some species whole tubules may present an entirely different appearance from their neighbors. For example, in *Gryllotalpa* there are three types of tubules designated as white, yellow, and embryonic. Each of these types is reported to have a different function. In addition, Malpighian tubules frequently have more than one region within a single tubule. In *Periplaneta* (Fig. 3-4), Wall et al. (1975) describe four distinct regions. Three of them (distal, middle, and proximal) are part of the Malpighian tubule proper, and a further subdivision of the middle region has been reported for larvae. The fourth is the ampulla, into which several individual tubules empty. In this case the middle region is again characterized by a yellowish color which has been attributed to the presence of riboflavin. No satisfactory explanation for the occurrence of the vitamin in the tubules has been presented.

At the cellular level Malpighian tubules consist of a single layer of two to

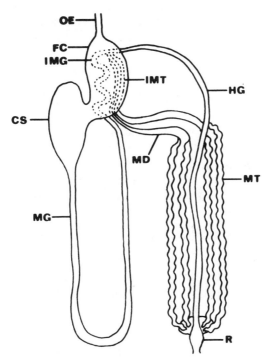

FIGURE 3-3. Diagram of the general organization of the cicadoid and cercopoid gut. Esophagus, OE; filter chamber, FC; conical segment, CS; midgut, MG; internal midgut, IMG; Malpighian tubules, MT; Malpighian ducts, MD; internal Malpighian tubules, IMT; hindgut, HG; rectum, R. [With permission from A. T. Marshall and W. W. K. Cheung, *Tissue Cell* **6,** 153 (1974).]

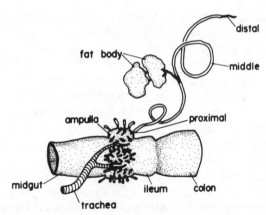

FIGURE 3-4. General arrangement of the gut, Malpighian tubules, trachea, and fat body of *Periplaneta americana*. The full length of only one tubule is shown, although each ampulla may drain about 30 tubules. Similarly, only one trachea and a small piece of fat body bound to the tubule by tracheoles are shown. [With permission from Wall et al., *J. Morph.* **146,** 265 (1975).]

six cells when viewed in cross section (Fig. 3-5). The tubules have a hollow center, referred to as the *lumen,* which may be more or less visible depending on the physiological state of the insect and the point along the tubule where it is observed. The cells surrounding the lumen are complex. A tubule may have two or more cell types interspersed along the entire length of the tubule or there may be distinct regional specialization, with each region having one or more cell types. It cannot be stated with certainty how many cell types exist in insect Malpighian tubules. Currently, there are three or four, with others probably still undiscovered. To complicate matters further, accumulating evidence indicates that specific cell types may change their appearance in response to the function being performed at any given time. Thus, we see that morphological specialization in Malpighian tubules may be achieved not only by variations due to life stage but also by tubule variation, by regionalization, or by mixing cell types in a tubule or region thereof in response to functionality.

3.2.2.2. Fine Structure. In view of the great specilization found in the gross morphology of Malpighian tubules, it is surprising to find unifying features. Nevertheless, at this level of organization, most of the variations just described can be understood easily in terms of the fine structure of the cells present in tubules. Since the production of primary urine is essentially a secretory process, cells of the Malpighian tubules should show the properties of secretory cells. This was demonstrated in *Rhodnius* (Wigglesworth and Salpeter, 1962). The Malpighian tubules of this blood-sucking insect are divided into a distal and a proximal region. Cells in the distal region are quite complex (Fig. 3-6). On the hemolymph or basal side there is, first, a discrete basement membrane. Moving upward in the cell, this is followed by deep

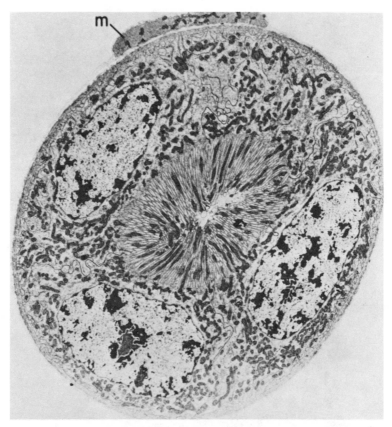

FIGURE 3-5. Cross-sectional view of the short distal Malpighian tubule region of *Periplaneta americana*. A single strip of muscle (M) is embedded in the connective tissue. Original photo ×4600. [With permission from Wall et al., *J. Morph.* **146,** 265 (1975).]

and intricate infoldings of the basal cell membrane, which create a series of complex cavities. Numerous mitochondria are associated with these basal infoldings, indicating a possible site of energy-requiring processes. Internal to the basal foldings is a less specialized region of the cell sometimes termed *intermediate zone*. Few mitochondria, but Golgi bodies, smooth and rough endoplasmic reticulum, the nucleus and often vacuoles and other inclusion bodies, such as mineralized spheres, occur in this zone. Finally, the apical or lumenal border of the cell is very complex. Its striated border consists of a large number of long, closely packed microvilli. Many are invaded by long, slender mitochondria that extend to their tips. Other microvilli may have vesicles in them. Here, too, the structure indicates active processes on the cell apical boundary.

Cells in the proximal region of *Rhodnius* Malpighian tubules are easily distinguishable from those just described. The most obvious differences in-

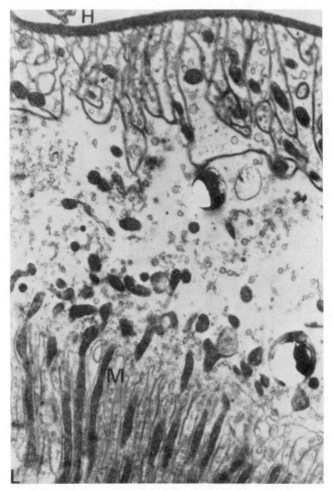

FIGURE 3-6. Transverse section of a Malpighian tubule distal region cell from *Rhodnius prolixus*. Hemolymph, H; lumen, L; mitochondria, M. Original photo ×4800. (With permission from V. B. Wigglesworth and M. M. Salpeter, *J. Insect Physiology* **8**, 299, Copyright 1962, Pergamon Press, Ltd.)

clude a more prominent mitochondrial concentration among the basal invaginations, an even less elaborate intermediate zone, and shorter, less densely packed microvilli along the apical border. Mitochondria are also much less conspicuous among the microvilli. These cells are obviously equipped to function differently than those of the distal region.

Cells from Malpighian tubules of other species of insects have also been examined at the ultrastructural level. In *Periplaneta* the distal region of the Malpighian tubule is short and highly contractile (Fig. 3-5). A dense display of microvilli, many of which have filamentous mitochondria, largely obscures the lumen. Some microvilli contain extensions of smooth endoplasmic reticulum (SER) (Fig. 3-7). The intermediate zone of these cells lacks vacuoles but has numerous mitochondria. The basal zone has infoldings of the basal cell membrane with few mitochondria.

The middle region constitutes most of the Malpighian tubules of *Periplaneta*. It has a distinct lumen, and cells in its proximal end are somewhat less differentiated than those located distally. The two cell types in this region are referred to as *primary* and *stellate* cells. Primary cells are reminiscent of the proximal cells of *Rhodnius* tubules (Fig. 3-8). Their microvilli are

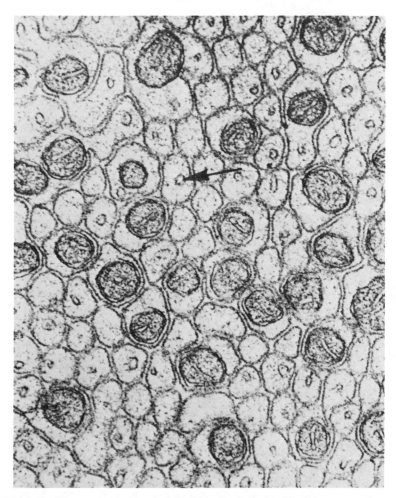

FIGURE 3-7. Transverse section of microvilli in distal tubule of *Periplaneta americana*. Two kinds of microvilli are present: those containing mitochondria and those containing extensions of smooth endoplasmic reticulum which appears as a circle (arrow) in profile. Original photo ×32,500. [With permission from Wall et al., *J. Morph.* **146**, 265 (1975).]

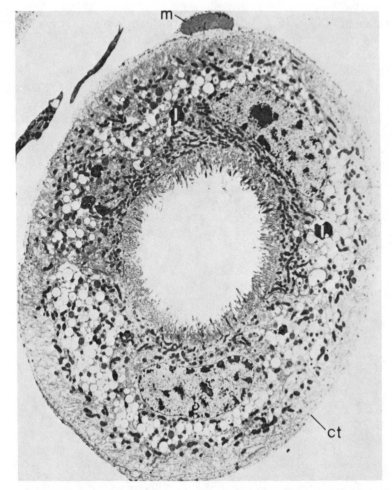

FIGURE 3-8. Survey micrograph of *Periplaneta americana* Malpighian tubule middle region. Note the relative size of the lumen and the apical microvilli. Connective tissue, ct; muscle, m. Original photo ×4000. [With permission from Wall et al., *J. Morph.* **146**, 265 (1975).]

short and mostly without mitochondria. However, they do contain fingers of SER, and dense concretions are sometimes seen at their tips. The intermediate zone has numerous clear vacuoles and mineralized concretions. It also contains some larger structures that resemble lysosomes as well as a moderate number of mitochondria. Microtubules are also in evidence in this region. The basal zone has extensive invaginations with mitochondria. The basement membrane is laminated. Embedded in a layer of connective tissue outside of it are the Malpighian tubule muscle and the tracheoles, when present.

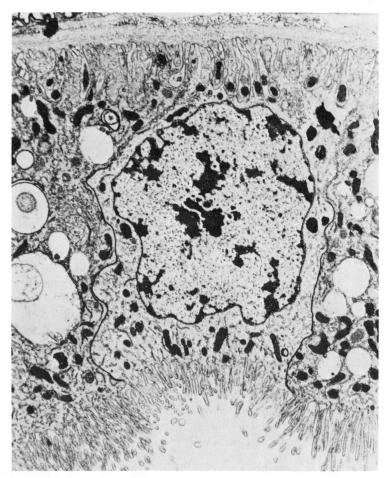

FIGURE 3-9.   Stellate cell from *Periplaneta americana* Malpighian tubule in a region where it extends across the whole tubule. Original photo ×11,500. [With permission from Wall et al., *J. Morph.* **146,** 265 (1975).]

Stellate cells of the middle region (Fig. 3-9) are rare and can easily be missed. As the name implies, they are often star-shaped. They are quite small, approximating the size of a primary cell nucleus. Their basal zone is similar to that of the primary cells, but with fewer mitochondria. The intermediate zone has little cytoplasm, no vacuoles, and contains a rather large nucleus. The apical zone has short microvilli which contain no mitochondria or SER. *Carausius* tubules have "mucocytes" which appear to be similar to stellate cells except for their numerous Golgi bodies and extensive granular endoplasmic reticulum (GER).

The two remaining portions of the *Periplaneta* tubule—the short proximal region and the ampulla into which numerous individual tubules empty—are similar at the ultrastructural level. They are more like the midgut than other regions of the Malpighian tubules. The sparse microvilli of their cells are devoid of mitochondria; however, mitochondria seem to be concentrated in the cytoplasm near the base of the microvilli. The intermediate zone lacks vacuoles but contains large dense bodies with lamellar interiors. The basal zone has rather extravagant and distended infoldings with some mitochondria. Outside of a basement membrane is a rather prominent layer of connective tissue. The ampulla, in particular, has extensive muscles and tracheae embedded in this layer.

Several other variations in fine structure should be mentioned. In *Musca* the main part of the tubule has three intermingled cell types. Type I is the primary excretory cell. It has deep basal infolds extending more than halfway through the thickness of the cell (Fig. 3-10). Many of the infolded membranes have elongated mitochondria in close association with them. The cytoplasm has numerous vacuoles, many of which contain dense material. The microvilli are closely packed and contain mitochondria. Type II cells, which are considerably smaller, occur throughout the main part of the tubule, but much less frequently than Type I cells. Although they have extensive and dilated basal infolds, they are probably similar to the stellate cells of *Periplaneta*. Type III cells are similar to Type I and are characterized primarily by having an extensive GER which occupies much of the cytoplasm. The GER cisternae form an extensive network of tubuli and spaces in the intermediate zone of the cell. Numerous dense bodies also occur in the cytoplasm. Type III cells are unusual in Malpighian tubules. While they may represent a new cell type, one must remember that cells may present different appearances depending on what function they are performing at any given time. Type IV cells can be found in *Musca* where the tubules join the intestine. These cells are similar to the proximal and ampulla cells of *Periplaneta*.

The cryptonephridial and filter chamber modifications of Malpighian tubule structure were mentioned earlier. While they are complex structures in the gross sense, the cellular basis for their functioning is rather simple. Cells of the primary tubule type are present in most cryptic situations. In addition, the *Tenebrio* cryptonephridial system has small cells called leptophragma which may be highly modified forms of stellate cells. They have a thin sheet of cytoplasm at the periphery of the tubule with a cell body which hangs freely into the tubule lumen (Fig. 3-11).

Accumulating evidence indicates that Malpighian tubule cells have membrane-lined channels traversing the entire cytoplasm. Since the mechanism of transcellular transport is not clearly understood, the existence and function of such channels need further study, as does the formation of membrane-bounded vesicles of various types, which has been observed from several species.

Before leaving the fine structure of Malpighian tubule cells, the cell

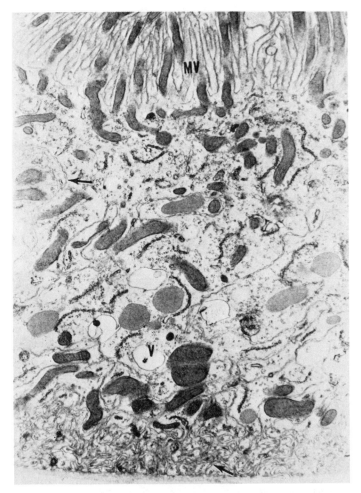

FIGURE 3-10.   Transverse section through the Malpighian tubule of *Musca domestica* showing Type I cell. Inflections of the basal plasma membrane (arrows) anastomose to form a network of canaliculi. Microvilli, MV; vacuoles, V. Original photo ×21,000. [With permission from R. S. Sohal, *Tissue Cell* **6**, 719 (1974).]

boundaries must be discussed. An important point is that cell boundaries are very irregular. Indeed, in many instances adjoining cells have extensive interdigitations (Fig. 3-12). The reason for this is not entirely clear, but it may facilitate the intercellular transfer of substances. It may also impede the passage of materials from the hemolymph to the lumen between adjoining cells often bound together by structures such as desmosomes. The cells of the distal region of the *Rhodnius* tubule have septate desmosomes, implying that the epithelial layer is tightly bound and not subject to excessive leakage.

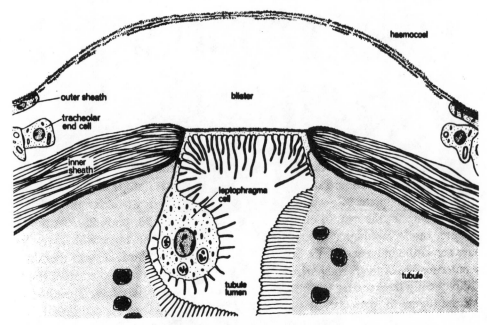

FIGURE 3-11. Diagram showing a leptophragma cell from *Tenebrio molitor* and its relation to the inner and outer sheaths of the perinephric membrane. [With permission from Grimstone et al., *Phil. Trans. Roy. Soc. London Ser. B* **253**, 343 (1968).]

Contrarily, except for cells of the ureter, the cells of *Drosophila* tubules do not possess desmosomes and may allow passage of material between them. The cell junctions of Malpighian tubules need further study.

3.2.3. Function of Malpighian Tubules

3.2.3.1. Production of Primary Urine. The Malpighian tubules may be thought of as the initiators of excretion. They accomplish this by production of the primary urine, the fluid that accumulates in the lumen of tubule distal regions. In insects this is primarily a secretory process not dependent on a pressure-filtration system. Although excretion in insects has been studied extensively, the process is still not well understood. Some facts seem central to the functioning of Malpighian tubules. However, here also there is considerable variation among species. There is also an extensive body of theoretical interpretations explaining how the excretory function operates, taking into account the facts which are available.

The most obvious fact about the excretory function is that the Malpighian tubules produce a fluid that flows from the distal to the proximal end of the tubule where it is eventually emptied into the hindgut. This primary urine contains a high concentration of K^+ and a low concentration of Na^+ in relation to their concentrations in the hemolymph or to an artificial medium

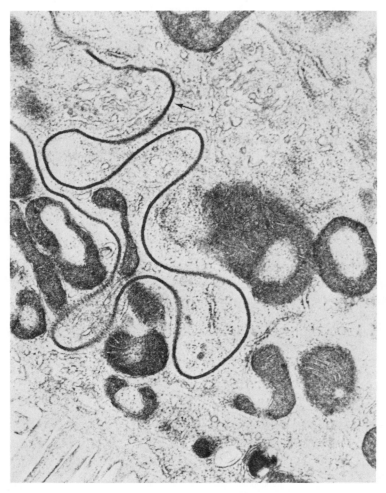

FIGURE 3-12. Micrograph from middle region of *Periplaneta americana* Malpighian tubule showing irregular boundary (arrow) between adjacent cells. Original photo ×21,000. [With permission from Wall et al., *J. Morph.* **146,** 265 (1975).]

surrounding the tubules (Berridge and Oschman, 1969). Other cations are also present in the urine at low concentrations. From numerous species, it is known that the urine is nearly isoosmotic with the hemolymph and that its osmolality is essentially independent of the rate of urine flow. Furthermore, there is an electrical potential difference across the tubule wall with the lumenal side usually being 20–30 mV positive to the hemolymph side. These facts are interpreted to mean that the transfer of K^+ from the blood into the tubule lumen occurs against a significant concentration gradient and requires energy. In other words, K^+ are actively transported. Significantly, the cells of the fluid-secreting regions of Malpighian tubules are well equipped to carry out active processes, and ion-specific ATPases have been reported to be present in apical and basal plasma membranes.

The movement of K^+ alone is not likely to bring about fluid flow into the tubule lumen. However, anions also accompany this movement, and together they foster water flow because of solute drag. The anions that appear to be most important normally are chloride and phosphate. Their action is independent, but phosphate, which supports the best rate of fluid flow, is present in the urine in concentrations higher than in the blood. Other cations like Na^+ and NH_4^+ may also support fluid flow at a very reduced rate in the absence of K^+. Na^+ greatly stimulate flow when K^+ levels are low. Thus, fluid flow appears to be dependent on and coupled with ion movement.

This description of fluid production was developed mainly from work on *Carausius* and *Calliphora* Malpighian tubules (Maddrell, 1971). When other species are examined, it is obvious that variations occur. For example, in *Calpodes* the secretory rate is slowed if Na^+ are omitted from the medium surrounding the tubules, and Na^+ alone are more effective in fostering flow than in *Carausius* and *Calliphora*. Likewise, major differences also occur in *Rhodnius*. Here, even in the absence of K^+ in the surrounding medium, secretion occurs, and Na^+ have a pronounced effect on flowrate. The active transport of Na^+ is indicated. In addition, Cl^- is the main anion present, with phosphate supporting a rate of flow about 1% that of Cl^-. The transwall electrical potential difference is around 30 mV, with the lumenal side negative to the hemolymph side. This is interpreted as evidence that Cl^- are actively transported and probably provide the main driving force for urine production. The adaptive significance of this situation should be noted. *Rhodnius* is a blood-sucking insect. Vertebrate blood is high in Cl^- and Na^+ and relatively low in K^+ concentrations. The modifications described above take advantage of this situation and allow *Rhodnius* to utilize this food source. Other species' variations also exist, but those described illustrate the point.

3.2.3.2. Mechanisms of Transport. The general concept of urine production just outlined reflects many of the pertinent facts currently available and indicates the beginnings of an interpretation. However, it falls far short of explaining the mechanism of excretion. Several hypotheses have been advanced. Perhaps the best known is the *standing gradient hypothesis* depicted in Fig. 3-13. It attempts to explain fluid movement, in the absence of a significant osmotic pressure, on the basis of localized compartments within an epithelium. In this hypothesis the epithelial cells of the Malpighian tubule secretory regions, with their deep basal infolds and apical microvilli, are adequately endowed for this purpose. The long narrow channels across which active solute transport can take place, offer the opportunity for local gradients to exist. Thus, isotonic fluid enters the basal channels but becomes hypotonic as the result of active solute transfer. On the inner side of the membrane, solute molecules accumulate in that channel producing a hyper-

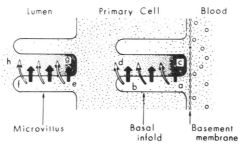

FIGURE 3-13. Diagram of the *standing gradient hypothesis* of solute-linked water transport to provide a model for urine production by Malpighian tubules. Local osmotic gradients are set up by the basal infolds (a–b) and microvilli (e–f). These gradients are indicated by a variation in density of dots (representing solute molecules). The basement membrane probably functions as a molecular sieve, preventing the channels from being clogged with large molecules such as proteins (open circles). Fluid entering the open ends of the channels (a to e) is isotonic but becomes hypotonic toward the closed ends (b and f) because solutes are pumped into the cell or lumen (closed arrows). Since solutes enter the long narrow compartments (c–d and g–h), these regions also have standing gradients. Fluid in the closed ends (c and g) is hypertonic but becomes isotonic at the open ends (d and h) because water enters the cell and lumen (open arrows) from the hypotonic fluid in the basal infolds and apical microvilli, respectively. The overall result is an isotonic secretion in the absence of an osmotic gradient between the lumen and the blood. [With permission from M. J. Berridge and J. L. Oschman, *Tissue Cell* **1**, 247 (1969).]

tonic situation. Solute molecules here begin to diffuse into the cytoplasm, and water moves inward in response to the local osmotic gradient, bringing about a return to isotonicity in the cytoplasm. A similar circumstance is depicted as occurring in the apical channels of the microvilli. The result is the production of an isotonic primary urine containing an excess of the actively transported ion(s).

This is an attractive hypothesis, which was evolved to explain the key role apparently played by the active transport of potassium and other ions. However, it does not account for all of the known facts and observations on excretion, and other hypotheses have been developed. One of them has been called the *simple osmotic model*. It demands that there be an increase in osmolality from the hemolymph to the cytoplasm and from the cytoplasm to the lumenal fluid. This hypothesis requires a hyperosmotic primary urine. As previously mentioned, most primary urines are isoosmotic or nearly so. There is some variation: for example, *Periplaneta* has a somewhat hyperosmotic urine, and other insects, such as *Rhodnius,* produce a slightly hypoosmotic urine, although perhaps because of proximal tubule reabsorptive activities. In no case is the deviation from isoosmosity sufficient to make this a really viable hypothesis.

Two related hypotheses involve the formation and discharge of vesicles. The first, the *formed-body hypothesis* of Riegel (1966), envisions that lysosomelike vesicles are formed in the secretory cells and contain both proteins and proteases. These are subsequently discharged into the lumen, whereupon the proteases become activated and hydrolyze the proteins, thus increasing both the molecular species inside the formed bodies and their osmotic pressure. Water enters the formed bodies from the lumen, causing them to swell. Solutes in the lumen fluid, unable to enter the formed bodies, become concentrated and draw water into the lumen from the tubule cell cytoplasm by osmosis. This hypothesis is supported by the fact that formed bodies can be observed in urine samples under certain conditions. While it has the advantage of providing a mechanism for voiding large molecules, it fails to account for the apparent dependence of urine production on active ion transport.

The remaining mechanism, that of Wessing and Eichelberg (1975), may be called the *vesicular transport hypothesis*. It is based on the fact that careful analysis of Malpighian tubule cellular fine structure (including the use of histochemical techniques) has revealed the presence of numerous vesicles, vacuoles, microtubules, membrane-lined channels, and other cytoplasmic entities. This suggests that the ground cytoplasm is highly organized and that the standing gradient hypothesis is too simple to account for transcellular transport in its entirety. Specifically, the *vesicular transport hypothesis* envisions the involvement of the extracellular mucopolysaccharide layer, which lies between the basement membrane and the basal plasmalemma. It suggests that this layer binds positively charged substances from the blood, enclosing them in vesicles created by infoldings of the basal cell membrane. These vesicles pass to the apical side of the cell and release their contents into the lumen by extrusion. Various changes in the vesicles are found as they pass to the apical side. Among them are changing stainability of the mucopolysaccharides, release of protons from the vesicles through the action of carbonic anhydrase, alkalinization of the vesicle contents, and the appearance of concretions in the vesicles at the apical cell boundary. These concretions are released into the lumen where they may become enlarged by the addition of layers of materials. Many of the events described in this paragraph can be observed in electron micrographs (Fig. 3-14).

Mainly using histochemical techniques, Wessing and Eichelberg (1975) have also demonstrated the association of various ions and organic molecules with specific cytological entities, such as the cisternae of the endoplasmic reticulum. They have proposed that as many as six pathways may exist for the passage of substances through the Malpighian tubule epithelium (Fig. 3-15). Of these pathways, the first is vesicular transport. The second involves what are, in effect, vesicles formed by the ER, sometimes enclosing specific substances for extrusion into the lumen. The third proposed pathway is closest to that described in the standing gradient hypothesis. The remaining pathways simply take into account that transport events may be occurring at the lateral cell boundaries including the intercellular channels. Perhaps the most significant point to this diagram is that the *standing gradient hypothesis* and the *vesicular transport hypothesis* are not mutually ex-

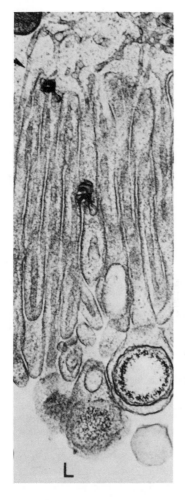

L

FIGURE 3-14. Apical surface of a Malpighian tubule middle region principal cell from *Periplaneta americana*. Note the dense concretion that appears to be pinching off into the lumen (L). Original photo ×40,000. [With permission from Wall et al., *J. Morph.* **146,** 265 (1975).]

clusive. Indeed Fig. 3-15 serves to emphasize the complexity of transcellular transport.

3.2.3.3. Composition of Primary Urine. The primary urine of insects is a complex, variable substance which, at least in a general way, reflects the composition of the hemolymph. Thus, the primary urine contains not only substances normally considered to be excretory products but also substances, such as sugars and amino acids, normally thought to be useful metabolites. It is apparent that insects have evolved a system in which it is more desirable to selectively reabsorb useful substances later on than to prevent their loss from the blood via the Malpighian tubules. However, this description must be qualified because several factors impact on the initial composition of the primary urine.

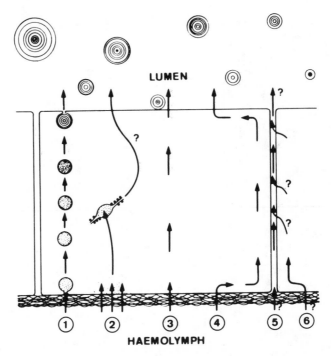

FIGURE 3-15. Schematic drawing of different pathways of transepithelial passage of substances in the cells of the initial Malpighian tubule segment of larval *Drosophila melanogaster*. 1. Vesicular transport. 2. Accumulation of invaded substances in the cisternae of the endoplasmic reticulum. 3. Transport through the cytoplasm of substances not bound to membranes. 4. Free transport through the cytoplasm along the outer plasmalemma. 5. Passage through the intercellular space. 6. Passage through the intercellular space after passing the cell membranes. [With permission from A. Wessing and D. Eichelberg, *Fortsch. Zool.* **23**, 148 (1975).]

One factor is that the basement membrane is selectively permeable. There are indications that large molecules, such as proteins, do not pass this structure. Even here some species' variation exists. In some species, horseradish peroxidase, added to the medium bathing isolated tubules, may reach the epithelial cytoplasm or even the Malpighian tubule lumen. Nevertheless, most naturally occurring large molecules appear to be prevented from entering the cells by the basement membrane.

Another factor is the rate at which ions and small molecules pass the epithelial cell cytoplasm. From some of the early work on Malpighian tubules, it appears that differences exist in the rate at which K^+ and Na^+ are transported. Obviously, this may be related to the active transport process, but other substances also show differing rates of passage. The point is that the cells themselves may influence the rate at which certain substances traverse their internal space.

Finally, it is becoming increasingly clear that many substances are transported in the blood bound to carrier proteins and possibly other large molecules. It is known that hormones may be transported in this manner. Certain ions and compounds like pteridines and uric acid may also be bound to macromolecules. The mechanisms by which such molecules and ions are bound and released are not well established at present, but it is clear that while they are bound to a macromolecule, they will not be lost via the Malpighian tubules. In this regard, it is worth noting that the *vesicular transport hypothesis* suggests the binding of positively charged substances from the blood by the basal mucopolysaccharide layer. Perhaps this is a mechanism for unbinding substances in the blood which are attached to carrier molecules.

With these considerations in mind, we can discuss the composition of the urine. Obviously, various ions and water are present as conspicuous components. In addition to K^+, Na^+, and Cl^-, ions like Ca^{2+}, Mg^{2+}, Mn^{2+}, PO_4^{3-}, SO_4^{2-}, and several other trace ions have been found in the primary urine of insects. As with most urinary components, the amount of each ion present may vary greatly, depending on concentration in the blood, active transport, species, sex, stage of development, diet, habitat, and so on. Meaningful compilations are rendered difficult because of these variations. In addition, all components of insect urine are not necessarily present in the urine of any given species.

Those components of urine which are closest to the classical excretory function are the compounds containing nitrogen. They provide the vehicle for voiding excessive amounts of this element. *Uric acid is considered to be the classical nitrogenous excretory product of insects.* It is well suited for this purpose as it is highly oxidized, has low solubility in water, and can be voided essentially dry. However, uric acid is not the only nitrogenous compound found in the urine of insects, and in some cases it may be absent. For example, in some species uric acid may be converted to allantoin and/or allantoic acid prior to voiding. In other cases ammonia seems to be the principal nitrogenous excretory product as exemplified by aquatic immatures, blowfly and other dipteran larvae, and in certain cockroaches. Still other compounds including amino acids, other amino compounds, pteridines, other purines, tryptophan derivatives, urea, creatine, creatinine, and even proteins are present in insect urines in greater or lesser amounts.

The urine of many insects also contains carbohydrates of various types. Simple mono- or disaccharides are perhaps the most common, but the urine of homopterous insects with a filter chamber type of excretory system may contain an extensive array of carbohydrates in the honeydew. Simple sugars again predominate, but higher-level saccharides may also be present.

An excretory function often minimized or even overlooked is that of voiding xenobiotics. This function is usually, but mistakenly, considered primarily in relation to the excretion of man-made compounds (e.g., insecticides). It was probably evolved as a mechanism for insects to deal with toxic compounds present in plants such as cardiac glycosides or alkaloids like nicotine. By studying the excretion of organic dyes, it has been established that diverse compounds like amaranth, *p*-aminohippuric acid, and other acidic or basic dyes are excreted by the Malpighian tubules. Other complex foreign compounds, including insecticides, are also excreted, but detailed studies of these are not abundant.

Among the compounds reported from insect urine are certain lipids and fatty acids, calcium oxalate, calcium carbonate, oxalic acid, juvenile hormone, and the exogenous protein inulin. In addition, the Malpighian tubules of some species secrete proteins and mucoid substances. The proteins may be used by certain aquatic insects for building dwelling tubes or by some homopterans to produce foams. The role of the mucoid substances is not well understood. It has been suggested that they may keep solid particles in the primary urine from precipitating out in the tubules, thereby facilitating urine flow. A newly discovered structure called an axepod, which has been reported from the proximal tubule lumen of *Rhodnius,* may have a similar function.

How specific urinary components find their way into the tubule lumen has not been adequately focused on thus far. Active transport of various ions apparently provides the driving force for urine flow. Other urinary constituents which can become ionized under physiological conditions might serve as the oppositely charged ion to that which is actively transported. In addition, small molecules may be voided as solutes in water via solvent drag. Larger molecules may be included in vesicles, transported across the cell cytoplasm, and extruded into the lumen. The pathways depicted in Fig. 3-15 are obviously pertinent here, but more detailed information is needed with reference to specific excretory components.

3.2.3.4. Reabsorption by Malpighian Tubules.   As mentioned above, metabolically useful compounds are found as components of the primary urine. Certain of these materials are selectively reabsorbed as the primary urine traverses its normal route to the exterior. At least in some species, the Malpighian tubules play a role in this process. For example, the proximal region of *Rhodnius* tubules reabsorbs K^+ and Cl^-; this occurs particularly in the proximal most one-third of the proximal tubule. Tubules of certain other species function similarly. However, while tubule reabsorption is characteristic of some species, it does not seem to be critical to insects in general indicating that it is of special adaptive significance for those species in which it occurs.

3.2.3.5. Role of Hormones in Urine Production.   The Malpighian tubules of many species function at dramatically different rates, depending on life stage or in relation to feeding activities. It has been shown that these changes are under hormonal control. Blood-sucking insects such as *Glossina, Anopheles,* and *Rhodnius* have been best studied. After a blood meal, which may be very large in relation to the size of the insect, a rapid and major-magnitude diuresis occurs. At least in *Rhodnius,* it appears to be a response to nervous stimulation from abdominal stretching. The response is the release of a diuretic hormone into the blood within 15 sec after feeding begins. The hormone, which is present in the mesothoracic ganglionic mass, is released into the hemolymph at neurosecretory axon endings located outside this mass. In other species the brain, *corpus cardiacum,* and *corpus allatum* may be involved. The Malpighian tubules respond rapidly to hormone release and produce urine at a high rate. Maximal response is reached quickly as hormone titer increases. The Malpighian tubules destroy the hormones so that the length of diuresis is controlled by the duration of hormone secretion. This in turn may be controlled by stretch or other receptors in the abdomen.

Maddrell (1976) has shown that other fine tuning of the system occurs in *Rhodnius.* Reabsorption of KCl from the proximal region of the tubule is also under hormonal control, and absorption of blood from the midgut into the hemolymph may be so regulated. All of these events are intimately involved in the process of diuresis. It seems logical that each must be closely regulated so that the process can function as an integrated whole without disrupting blood volume or content. Probably several hormones are involved. In addition, 5-hydroxytryptamine stimulates rapid urine flow in *Rhodnius,* while cAMP has a similar effect in other species. Each may be acting as a "second messenger."

Some efforts have been made to isolate, purify, and identify the diuretic hormones. They appear to be peptide hormones which may contain other constituents as well. In *Glossina* it is heat stable, nondialyzable, alcohol soluble, and contains amino acids, glucose, and sialic acid residues.

An antidiuretic hormone may also exist and function in water conservation. Still other hormones may be involved in the control of excretion. Indeed, recent reports have implicated ecdysone-type compounds in Malpighian tubule functioning.

3.3   RECTUM

3.3.1.   General

The hindgut of insects receives the discharge from the Malpighian tubules, that is, the primary urine. While some modifications of the primary urine occur in the hindgut anterior to the rectum (Section 3.4.2), the rectum carries out the major modifications. In this sense, it is usually spoken of as the second part of the discontinuous excretory system of insects. It also plays a major role in water balance and osmoregulation.

3.3.2.   Structure of Rectum

3.3.2.1.   Gross Morphology.   The rectum of insects is the enlarged, posteriormost section of the hindgut. It is an extremely variable structure among

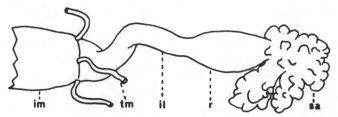

FIGURE 3-16. Hindgut of the firebrat, *Thermobia domestica*. Midgut, im; Malpighian tubule, tm; ileum, il; rectum, r; anal sac, sa. [With permission from C. Noirot and C. Noirot-Timothée, *J. Ultrastr. Res.* **37**, 119 (1971).]

insects but is usually readily distinguishable from the rest of the hindgut. In some instances the rectum itself may be differentiated by the posterior part being more highly developed than the anterior part. An extreme example of this is the anal sac of *Thermobia* (Fig. 3-16), which is a region specialized for the uptake of water from the atmosphere. More typically the rectum is characterized by a series of specialized regions referred to as rectal pads, cones, or papillae (Fig. 3-17). They range from a relatively simple, thickened epithelium, as in *Tomocerus* (Fig. 3-17A), to the highly complex, multilayered papillae of some adult dipterans such as *Calliphora* (Fig. 3-17F). There may be only one specialized region (Fig. 3-17B) or a large series of small pads (Fig. 3-17E). While no particular arrangement is typical, six elongated pads are often found distributed around the rectum (Fig. 3-17C).

The rectum is, of course, a part of the proctodaeum. As such it has a cuticular lining on the lumenal side which is shed at molting. On the hemolymph side it is usually covered by a prominent layer of muscles, including both circular and longitudinal fibers. In addition, the rectum is richly endowed with tracheae which penetrate the muscular layer and ramify extensively among the cells of the rectum itself. In some of the more highly specialized types of pads and papillae, components of the neurosecretory system are also present.

At the cellular level the rectal pads may be a simple layer of enlarged, specialized epithelial cells. Contrarily, the highly developed rectal papillae are characterized by two or more layers of cells arranged in a complex manner. The cells are of three or more types, each with its own special characteristics. Intercellular spaces, subepithelial sinuses, and other features, such as the intimate involvement of large tracheae, are prominent aspects of rectal papillae. Intergradations in complexity of structure between these two extremes also occur (Fig. 3-17). The cryptonephric system fits into this picture since the ends of the Malpighian tubules are embedded in the rectum (Fig. 3-18). They lie close to the rectal pad basal region, and the entire structure is surrounded by the perinephric membrane.

3.3.2.2. Fine Structure. A reasonable amount of detailed information is now available on the fine structure of the insect rectum. While considerable

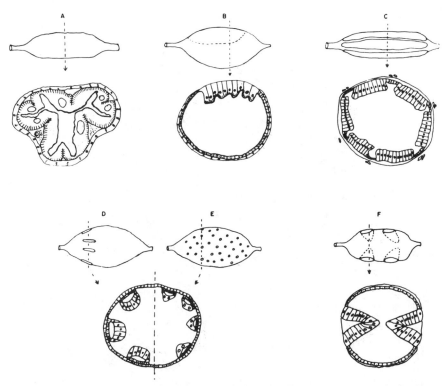

FIGURE 3-17. Diagram of "rectal glands" in various orders of insects. A. Relatively unspecialized epithelium of *Tomocerus vulgaris* (Collembola). B. Thickened region comprised of
columnar cells found in Hemiptera, based on *Nezara viridula*. C. Cushion-shaped rectal pads as
in Dictyoptera and Orthoptera, based on *Periplaneta americana*. D. Rectal pads in Hymenoptera, based on *Vespa vulgaris*. E. Multiple round rectal pads found in some Lepidoptera, such
as *Hyalophora cecropia*. F. Rectal papillae in adult Diptera, such as *Calliphora erythrocephala*.
[With permission from B. J. Wall and J. L. Oschman, *Fortsch. Zool.* **23,** 193 (1975).]

variation exists even at this level of organization, there are unifying features
which aid in understanding the function of the rectum. Perhaps the best
studied group of insects are the cockroaches. Accordingly, they will be
discussed first.

Somewhat surprisingly, there are at least two variations in the structure of
the cockroach rectum. The first is the *Periplaneta* type (Fig. 3-19) which has
four different kinds of cells. The rectal epithelial cells line the general wall of
the rectum exclusive of the rectal pads. These are relatively simple epithelial
cells, which probably play only a minor part in rectal functioning. They are
joined to a series of long, thin, junctional cells at the boundary of the pads.
These very attenuated cells provide the lateral boundaries of the pads with a
complex series of cell membranes that, in total, are undoubtedly quite impermeable. At the lumenal side the junctional cells are closely joined to the
cuticular lining; at the basal side they are joined to basal cells by septate

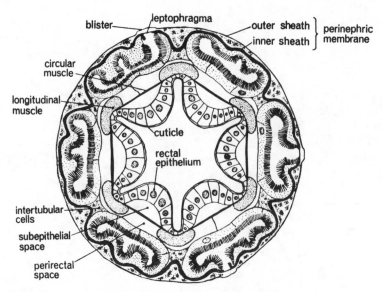

FIGURE 3-18. Transverse section of the cryptonephric complex from *Tenebrio molitor* showing the distal ends of the Malpighian tubules enclosed by the perinephric membrane (also see Fig. 3-11 for details of the leptophragma cell). [With permission from Grimstone et al., *Phil. Trans. Roy. Soc. London Ser. B* **253**, 343 (1968).]

junctions. Thus, the lateral boundary of the pads appears to be well sealed all the way around.

The remaining two cell types are presumed to be the functional components of the pads. They are the principal cells and the basal cells. The basal cells are a layer of flattened cells facing the hemolymph. Their basal surface is simple, but evidence exists for pinocytic activity there. The cytoplasm has mitochondria and microtubules but is relatively simple. The apical membrane is in contact with the basal surface of the principal cells. The two cell types are joined by macular desmosomes, gap junctions, and mitochondria–scalariform junctional complexes (MS complexes) which also characterize the lateral boundaries of the principal cells (Fig. 3-19). Basal cells are joined by septate junctions and sometimes by gap junctions and small desmosomes.

In addition to the elaborate lateral MS complexes, the principal cells are characterized by an extensively folded apical border with mitochondria intimately associated with the infolds. There is a small subcuticular space separating the convoluted apical membrane from the overlying cuticle. At the lumenal side the principal cells are joined by septate junctions which lead into the MS complexes that in turn feed into intercellular spaces. The intercellular spaces sometimes lead to the basal cell boundary, while in other instances they are invaded by tracheae. Whether there are openings around the tracheae that could allow fluid movement remains unsettled, but in any case the tracheae pass through the muscular layer and the subepithelial sinus

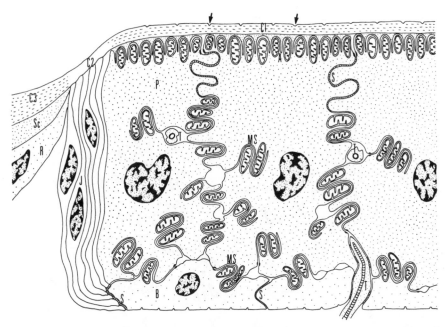

FIGURE 3-19. General organization of the rectum of *Periplaneta americana*. The rectal papillae are composed of two layers of cells; principal cells (P) and basal cells (B), covered with lamellate cuticle (C1) bearing epicuticular depressions (arrows). Each papilla is limited along its edge by very narrow junctional cells (J), supporting the sclerotized cuticle of the frame (C2). Between the papillae, the rectal epithelium (R) supports a lamellate cuticle (C3) including a thick subcuticle (Sc). On the principal cells, the apical complex (A) is provided by folds of the plasma membrane, associated with elongated mitochondria; on the lateral and basal faces, mitochondrial scalariform junction complexes (MS) are very well developed. MS complexes also occur between principal and basal cells. Tracheae and tracheoles (T) are insinuated between the basal cells and ramify between the principal cells. The intercellular junctions are very varied; note the presence of septate junctions (S), both at the apical and basal poles of the papilla. [With permission from C. Noirot and C. Noirot-Timothée, *Tissue Cell* **8,** 345 (1976).]

prior to their insinuation between the basal and principal cells. In the epithelium they ramify extensively, thus ensuring an adequate supply of oxygen.

The second kind of organization found in cockroach rectal pads is the *Supella–Blattella* type (Fig. 3-20). It is quite similar to the *Periplaneta* type except on the basal or hemolymph side. The main differences are that the basal cells are reduced in number, are discontinuous, and that the junctional cells have long cytoplasmic leaflets underlying the entire hemolymph side except where the basal cells exist. The junctional cell leaflets are also interrupted where tracheae penetrate the principal cell intercellular space. The basal cells are horseshoe shaped and are again relatively simple. There is evidence of pinocytic activity on their basal side, and tracheae and neurosecretory components are present in the basal concavity.

Other species of insects can be used to show more or less specialization of

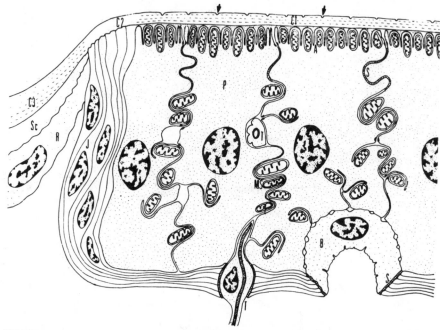

FIGURE 3-20. General organization of the rectum of *Supella longipalpa*. The basal cells (B) form a discontinuous layer and the junctional cells (J) are prolonged beneath the principal cells (P). The tracheae traverse the multiple layers of the junctional cells. For other abbreviations see Fig. 3-19. [With permission from C. Noirot and C. Noirot-Timothée, *Tissue Cell* **8**, 345 (1976).]

the rectum. On the one hand, the freshwater mosquito, *A. aegypti,* has relatively simple rectal epithelial cells. They have basal and apical infolds, but the infolds are not extensive. There are few mitochondria associated with the apical convolutions, but mitochondria are present in the cytoplasm. On the functional side it is known that this mosquito does not regulate well in brackish water. The relatively simple epithelial cells of the rectum correlate well with this finding. By contrast, in saltwater mosquitoes, such as *A. campestris*, the rectum is divided into distinct anterior and posterior regions. The anterior rectum is similar to that just described, but the posterior rectum is highly developed. Here the single-cell type is characterized by elaborate apical infolds with associated mitochondria and no special lateral-boundary developments.

The cone-shaped papillae of the hymenopteran *Nasonia* represent another variation of rectal structure (Fig. 3-21). Exclusive of the rectal epithelial cells, there are four cell types. They are cone cells, collar cells, junction cells, and basal cells which surround a closed central cavity. The cone cells are most similar to the principal cells of *Periplaneta*. They have "membrane stacks," which are reminiscent of the MS complexes and which lead into intercellular spaces. The latter are open to the central cavity. The collar cells

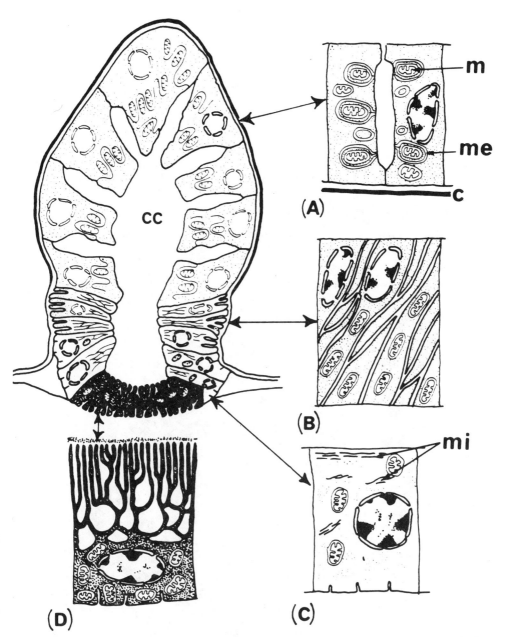

FIGURE 3-21. Diagram of the rectal papilla cells of *Nasonia vitripennis*. A. Cone cells form the body of the papilla and are separated from the rectal lumen by a layer of cuticle (c). The cytoplasm is packed with mitochondria (m) found within "membrane stacks" (me) at the lateral plasma membrane. B. Collar cells form a distinct collar around the base of the rectal papilla. C. Junction cells are found at the junction of the rectal papilla and the rectal epithelium. They contain large numbers of microtubules (mi). D. Basal cells separate the central cavity of the cone from the hemolymph. Both the apical and basal plasma membranes are well developed. [With permission from I. Davies and P. E. King, *Cell Tissue Res.* **161,** 413 (1975).]

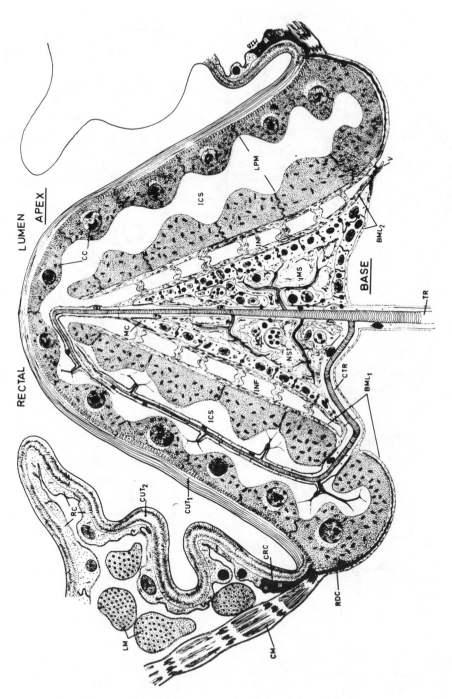

and junction cells are less complex and may play a mainly structural role. The basal cells separate the central cavity from the hemolymph. They are quite well developed, particularly the apical plasma membrane which is thrown into deep and extensive folds. Mitochondria are closely associated with the base of this region. The basal plasma membrane also has infolds which tend to form dilated channels.

Perhaps the most complex rectal structure studied thus far is the papilla of the blowfly *Calliphora* (Fig. 3-22). It consists of a cortex and a medulla. The cortex has two layers of cortical cells separated by a large intercellular space. The cell layers, which are continuous at the base of the papilla, can be thought of as one layer folded back on itself. The medulla fills the hollow cone produced by the folded cortex and has a layer of medullary cells on its outer surface. They are separated from the inner cortical cell layer by an incompletely septated infundibular space. This space is continuous with the intercellular space at the apical end of the papilla and is separated from the hemolymph by a valvelike structure at the base. The medulla also has connective tissue, neurosecretory elements, and a large trachea which branches and forms loops that traverse the intercellular space and ramify between the cortical cells. The edge of the papilla has junctional cells which isolate the papilla from the relatively unspecialized rectal epithelial cells.

At the fine structural level the cortical cells are of primary interest. They appear to be similar to the principal cells of the cockroach rectum. On the outer cell layer, their apical border faces the lumen or cuticle side, while it faces the hemolymph at the base of the papilla and the infundibular space on the inner layer. The apical plasma membrane is highly folded, but mitochondria do not appear to be concentrated here. The cytoplasmic surface of the folds is coated with particles, as is not uncommon with transport epithelia including those of other insects. The most outstanding feature of these cells is their elaborate series of lateral membrane stacks closely associated with mitochondria and leading into the intercellular spaces. They are, of course, similar to the MS complexes of cockroach principal cells. The basal plasma membrane does not appear to be elaborately folded, and basal cells have not been reported from this species.

FIGURE 3-22. Diagrammatic representation of a rectal papilla from *Calliphora erythrocephala*. Note the relationship of the intercellular space (ICS), the infundibulum (INF), and the hemocoel. The bridges traversing the infundibulum are bar-shaped and not complete septa. The tracheal branches in the intercellular space are now shown in the right half of the papillary complex. The relative dimensions of the structures are significantly altered for the sake of diagrammatic clarity. BML_1–BML_2, basement lamellae of the cortex and medulla, respectively; CC, cortical epithelial cells; CM, circular muscles; CRC, junctional cell; CTR, cortical trachea; Cut_1–Cut_2, cuticular intima of the papilla and rectal pouch, respectively; LM, longitudinal muscle; LPM, lateral plasma membranes of the cortical epithelial cells; MC, medullary cells; MS, medullary space; NST, neurosecretory terminals; RDC, radial cells; RC, rectal cells; TR, tracheal trunk; V, medullary or infundibular valve. [With permission from B. L. Gupta and M. J. Berridge, *J. Morph.* **120**, 23 (1966).]

Still other variations in structure of the insect rectum have been reported, but those discussed illustrate the important details. These details are that the apical plasma membrane of the principal cells is complex, that their lateral cell boundaries are elaborately developed, that intercellular spaces are prominent features of the insect rectum, that basal cells often interface with the hemolymph, and that distinctive and varied junctional arrangements exist between cells. How these structural details correlate with function is discussed in the next section.

3.3.3. Function of Rectum in Excretion

3.3.3.1. Modification of Primary Urine. In the overall sense, the excretory function of the rectum is to transform the primary urine that enters it into the final urine voided via the anus. This simple statement translates into a series of actions at the operational level. Central to this functioning of the rectum is its ability to reabsorb specific components from the primary urine. Before discussing reabsorption, however, it is necessary to consider two other factors. One is the cuticular lining of the rectum. This lining probably has several functions, but of major concern in this discussion is that it acts as a molecular sieve. In effect, it severely restricts the molecular species to which the cells of the rectal pads or papillae are exposed and on which they can exert their reabsorptive powers. In particular, larger molecules are sieved out and must remain in the rectal lumen. More detailed information is needed with respect to the exclusion of specific molecules, but most ions and small molecules apparently are not impeded by this lining.

The other factor of significance is that the pH of the rectum is often rather acid. This is usually quite different from the pH of the gut sections anterior to the rectum. Acidification of the rectum may occur because of the release of protons into the lumen, perhaps in exchange for certain other cations that are being reabsorbed by the rectum. In any case, acidification of the rectum has important consequences. For example, in an acidic medium uric acid tends to precipitate out of solution and thereby remove itself as an osmotic affector. For insects that void uric acid to the exterior, this may be crucial because uric acid is a relatively small molecule and might otherwise be reabsorbed as a result of solvent drag.

It is evident that the substance presented to the rectal pad or papilla cells may already be quite different from the urine that enters the rectum. In this connection, it must also be remembered that by this stage the urine will usually be mixed with fecal materials, products of gut microbial metabolism, digestive enzymes and their breakdown products, and other constituents present in the gut. Although the primary focus here is on urinary products, the reabsorptive powers of the rectal cells will act on the total substance which reaches them irrespective of the origin of specific molecules.

The urinary modifications associated with rectal cells revolve primarily around the removal of Na^+, K^+, Cl^-, and water from the rectal lumen and their transfer to the hemolymph. As occurs with Malpighian tubules, there is an electrical potential difference of 30–35 mV across the rectal wall with the lumen side being positive. This is interpreted to mean that Cl^- are actively transported against a considerable concentration gradient and that Cl^- transfer may be the driving force for fluid movement. Other evidence indicates that Na^+, K^+, and certain amino acids are also actively transported. It is assumed that water movement normally follows the movement of these ions and molecules and is coupled with their transfer. Carried along with water movement by solvent drag will be other ions and molecules dissolved in the water, including sugars, amino acids, and other nutritionally useful substances.

One apparent inconsistency in our understanding of function is that, in certain instances, water movement has been shown to occur in the apparent absence of solute reabsorption. Rectal function is dependent on energy derived from oxidative metabolism as can be shown by the inhibitory effect of anoxia, dinitrophenol, and cyanide on rectal function. ATPases are also known to occur in cells of rectal pads and papillae.

3.3.2.2. Mechanism of Transport. Perhaps even less is known about the mechanism of transport in the rectum than is true of the Malpighian tubules. Ultrastructural details clearly indicate that the two are quite different. In the rectum it is the apical and lateral cell boundaries which are elaborately developed. The hypothesis offered to explain rectal function suggests that there is active ion transport from the lumen into the cell cytoplasm along the apical border, at least in some species. This sets up water movement into the cytoplasm. The function of the very complex lateral cell boundaries is suggested to be the pumping of high concentrations of solutes into the intercellular spaces. In *Apis* it has been shown that the rectal lumen has an osmotic pressure of 160 mOsM, while the rectal pad lumen is at 720 mOsM and the hemolymph at 470 mOsM. The rectal pad of *Apis* is similar to that of *Nasonia* (Fig. 3-21). The numbers reported show a very high osmotic pressure within the pad and tend to support the hypothesis.

The intercellular spaces, sealed by septate junctions at the apical cell border and sometimes also at the basal cell border (Fig. 3-19), are quite inextensible. As a result, water movement from the rectal lumen via the cytoplasm into the intercellular spaces, occurring in response to the high internal osmotic pressure, sets up a considerable hydrostatic pressure in the spaces. A question not satisfactorily resolved at present is how transfer from the intercellular spaces into the hemolymph is accomplished. Wall and Oschman (1975) have suggested that in *Calliphora* it occurs by means of valves at the basal side of the infundibular space (Fig. 3-22). They proposed a recycling of ions from the infundibular space back into the intercellular space brought about by action of the inner layer of cortical cells. This would explain how the osmolality of the hemolymph is maintained at a lower level than that of the intercellular space. The recycling of ions back into the intercellular space is also used to explain how water can be transported out of the rectal lumen without active ion transport from the lumen. In this instance, water movement into the intercellular space would occur passively.

While this explanation may be satisfactory for *Calliphora,* it has problems when applied to other species. Wall and Oschman (1975) suggested that in *Periplaneta* the transfer of fluid into the hemolymph occurs through narrow, basement-membrane-lined spaces between the rectal cells and the tracheae which insinuate into the intercellular spaces. However, Noirot and Noirot-Timothée (1976) have restudied the *Periplaneta* rectal pad and found that these spaces are probably closed by septate junctions and may not be capable of transporting fluid, as is also the case for *Suppella.* They also discovered the existence of a layer of thin basal cells lying between the principal cells and the hemolymph in *Periplaneta* (Fig. 3-19) and suggested that these cells may play a role in fluid transport into the hemolymph.

In the hymenopteran *Nasonia* (Fig. 3-21), the central cavity of the rectal papilla is closed to the hemolymph. Here a layer of basal cells is also present. The only logical conclusion is that they must play an important part in fluid transport into the hemolymph and its regulation. As was true with the cockroaches, the basal membrane of these cells shows evidence of pinocytic activity. Whether this is related to fluid transport is not known.

Before leaving mechanisms of rectal function, a few comments about the cryptonephric complex should be made. In *Tenebrio* (Fig. 3-18) the rectum produces a very dry fecal pellet. The cryptonephric system presumably assists in this process. It is known that leptophragma cells transfer ions from hemolymph into the buried ends of Malpighian tubules. By setting up a hyperosmotic situation in the Malpighian tubules, the movement of water from subepithelial and perirectal spaces should be facilitated. This, in turn, should bring about further water removal from the rectal lumen and may also aid in nutrient reabsorption. The tandem operation of the rectal epithelial and Malpighian tubule cells appears to be a very efficient system.

3.3.3.3. Relationship of Excretion to Water and Ion Balance (Osmoregulation).

The process of excretion has been discussed up to this point, more or less, as a separate and distinct physiological system. Concern has been focused largely on the voiding of excretory products such as uric acid and other nitrogenous waste materials. One problem confronted by insects is, in the excretory process, not to lose physiologically useful products. It has been estimated that a very high percentage of the primary urinary constituents are reabsorbed prior to voiding. Thus, for most insects the Malpighian tubule–rectum complex works rapidly and well as an excretory apparatus. The useful components of the primary urine are reclaimed, and the excretory products are expelled to the exterior.

For insects, the other major problem related to excretion is water conservation. While numerous methods of water conservation have been evolved by insects, the one of interest here is that of the rectum removing water from the urine and transferring it back to the hemolymph. It is here that excretion and osmoregulation became intertwined. Osmoregulation is primarily concerned with water and ion balance. The rectum is involved in osmoregulation in a major way but is not the only structure which contributes to it. Osmoregulation has been reviewed in detail by Stobbart and Shaw (1974) and Arlian and Veselica (1979) and will be treated only briefly here to clarify its relationship to excretion.

The main relationship between excretion and osmoregulation is that the same processes, namely, ion and water movement, are used in both. The most significant difference is that in osmoregulation those movements occur in response to the physiological needs of the insect, that is, maintaining the hemolymph and body tissues in a steady state. After all, from the standpoint of excretory function, it makes little difference whether the excreta, including the final urine, is liquid, semisolid, or solid as long as the excretory products are voided. However, from the standpoint of osmoregulation it is crucial that water and ions either be retained or voided in accordance with the physiological needs of the insect. These needs are influenced greatly by habitat and type of food consumed.

For most terrestrial insects the major problems are conserving water and keeping the ionic concentrations of the hemolymph correctly regulated or balanced. For an insect like *Periplaneta* the rectum is adequately equipped to carry out these functions. It can reabsorb water in the absence of ion transport, and it can selectively reabsorb ions. In addition, this insect can react well to stresses, like a lack of water, by reducing its blood volume. The fluid apparently goes into the tissues to keep them hydrated and the excess osmotic constituents are sequestered within the body. When water again becomes available, the situation is reversed, but throughout, the osmotic pressure of the hemolymph changes little.

The situation is vastly different for insects that live in an aquatic environment or for terrestrial insects that consume a liquid diet. The problem for them is to void excessive liquids and retain ions. Freshwater mosquito larvae and blood-sucking insects are good examples. In both cases large amounts of water pass through the system as a consequence of feeding. The rectum produces a dilute urine by efficiently removing ions from the lumen. Particularly in the blood-sucking insects, ion removal may be selective because vertebrate blood is quite different in ionic constitution from the insects' blood. Like the terrestrial insects discussed previously, these insects osmoregulate well in relation to their food and habitat.

Another situation regarding osmoregulation involves insects living in brackish or salt water. Successful osmoregulation is accomplished in brackish water by many species; their task is to retain water and remove excessive ions. These insects void a hyperosmotic urine largely through rectal action. However, few species of insects are capable of osmoregulating in very brackish or sea or salt water. It has been suggested that this is an important reason why insects do not inhabit the seas in large numbers. Those that do have evolved some special capabilities. In saltwater mosquitoes, such as *A. campestris,* the posterior rectum has highly developed cells. It has been shown that these cells actively secrete Na^+, K^+, Mg^{2+}, and Cl^- into the lumen. This ensures production of an extremely hyperosmotic final urine. The anterior rectum is thought to be reabsorptive in function and apparently is responsible for transfer of water into the hemolymph. Other components of the total osmoregulation system also help insects to deal with varied environments. While they will not be detailed here, it is obvious from this brief discussion that the ability to osmoregulate has played an extremely important role in the capacity of insects to adapt to diverse habitats.

3.3.3.4. Role of Hormones in Rectal Function. Very little detailed information is available about the hormonal control of rectal function. It has been shown that diuretic factors from the *corpora cardiaca* of *Schistocerca* increase fluid production by Malpighian tubules and decrease reabsorption of water by the rectum. Presumably, a similar situation occurs in those blood-sucking insects which show a rapid diuresis. Also in *Schistocerca* homogenates of the *corpora cardiaca*, cAMP, theophylline, and blood from recently fed insects all stimulate rectal transport of Cl^-. In addition, an antidiuretic hormone concerned with water conservation has been reported from several insect species and is presumed to act on the rectum.

In view of the many and varied functions carried out by the rectum, it would not be surprising if several additional hormones are eventually shown to be involved. For example, the selective reabsorption of ions by the rectum could be mediated by specific hormones. The existence of neurosecretory axons in the rectum supports this conjecture, since they would presumably release their hormones within the rectum where these differential functions occur. The stimulus for release of specific hormones could come from receptors located elsewhere in the body. Obviously, much additional research is needed in this area. At present it is not even known with certainty whether the diuretic hormone affecting the Malpighian tubules is the same as that which acts on the rectum.

3.4. OTHER PARTS OF GUT

3.4.1. General

As indicated, the Malpighian tubules and the rectum are the principal excretory organs of insects. However, mounting evidence indicates that other parts of the gut are involved in or carry out activities related to excretion. Consideration of these functions will follow morphological distinctions.

3.4.2. Hindgut

The hindgut of insects is variously developed. Perhaps a typical arrangement consists of an ileum and a rectum (Fig. 3-16). In some instances the anterior part of the hindgut is further subdivided into an ileum and a colon. In others the entire hindgut may be a rather simple, straight tube with little if any apparent specialization. Finally, in cases where the rectum is simple or carries out other than its usual functions, the ileum may show a surprising degree of cytological specialization. Where this occurs, the ileum may, at least to some extent, take over the function of the rectum. A variation of this situation occurs in *Blattella,* which has a functioning rectum, but whose ileum cells have apical microvilli and basal infoldings more or less typical of transport organs.

This description of the hindgut supports the contention that its anterior parts play a role in excretion. It has been shown in *Periplaneta* and *Pieris* that rapid reabsorption of ions, particularly K^+, occurs in the ileum. In *Dytiscus* and *Acilius,* with the rectum consisting of only a simple layer of epithelial cells, ion uptake is almost exclusively by the ileum. Other modifications of the primary filtrate suggested for the ileum consist of a drying effect (fluid reabsorption) and a continuation of the acidification of the primary filtrate as it moves rearward. One of the most striking capabilities of the ileum occurs in *Sarcophaga* larvae. Here the entire hindgut is a rather simple tube, the middle portion of which actively secretes K^+, NH_4^+, and Cl^- into the lumen. The result is a hyperosmotic urine produced by secretion of ions rather than absorption of water as occurs in the adults. This species voids a significant amount of nitrogen with its urine via this route.

Two other events occurring in the ileum have an impact on excretion. First, the ileal epithelial cells of some species, such as *Blattella,* have mineral concretions in their cytoplasm. These cells are the same as those described earlier as having membrane elaborations typical of transport organs. The concretions contain at least P, Cl, Ca, Mg, K, and Fe in a glycoprotein matrix. It is not clear whether the concretions represent a permanent sequestration of these minerals or are involved in voiding them into the lumen. Perhaps both occur.

The impact of gut microorganisms on the primary filtrate needs also to be mentioned. These organisms, mainly located in the anterior hindgut, almost certainly possess the enzymes necessary to bring about important changes. Unfortunately, little specific information on their actions is available, but they probably modify the primary urine in some ways.

Clearly, the anterior parts of the hindgut can perform important excretory functions; nevertheless, the rectum typically plays the major hindgut role in excretion.

3.4.3. Midgut

The insect midgut functions mainly to digest food and absorb nutrients. The midgut is usually a more or less straight tube with no distinctive subdivisions. Sometimes regional distinctions can be made at the cellular level as in *Manduca* where cell morphology changes along the length of the midgut.

From its anatomical features one would not expect the midgut to play a major role in excretion—and it probably does not. However, certain aspects of its operations bear on excretion. The most overt of these occurs in *Gilpi-*

nia where it has been reported that uric acid is excreted from special structures, called gastric crepts, located in the posterior part of the midgut. While the site of urate production here is unusual, in actual practice this situation differs little from those where Malpighian tubule content may be transported forward into the posterior midgut by antiperistaltic contractions of the gut. In both instances, the excretory products are probably soon passed along into the hindgut.

The midgut has also been implicated in the voiding of hemoglobin in certain blood-sucking insects and in ridding the insect of injected acidic dyes. Both of these processes suggest a capability for handling relatively large molecules which may be related to the structure of certain highly specialized cells in the midgut epithelium. In addition, the midgut epithelial cells of numerous species have been shown to contain mineral concretions in their cytoplasm as have hindgut cells. These concretions consist of several minerals plus chlorides, carbonates, and phosphates. In one instance the entire mixture was in an organic matrix of polysaccharides. In some species the number and size of the concretions increase with age of the insect, which suggests a more-or-less permanent sequestration. In the collembolan *Tomocercus* the midgut epithelium is shed at molting, possibly causing the loss of its stored concretions.

Another function of insect midgut which relates to excretion is its ability to transport certain ions across the epithelial cells. This ability is perhaps most highly developed in certain lepidopterous larvae whose food is quite high in K^+ content. Nonetheless, the hemolymph maintains a very moderate level of K^+ because the goblet cells of the midgut wall have a K^+ pump which actively transports K^+ into the midgut lumen. The pump, located in the apical plasma membrane of the goblet cells, will transfer K^+ from either the hemolymph or from those ions entering the epithelium via the midgut lumen. There is a potential difference across the cell wall with the lumen side being positive to the hemolymph. In other instances ions may be transported from the lumen into the hemolymph. For example, Mg^{2+} and Ca^{2+} are transported in *Manduca* and Na^+ in *Sarcophaga*. In the latter case there is a potential difference across the wall, with the lumen being negative. While this transport has been shown to be energy dependent, it is probably more related to absorption or ion balance than to excretion.

The midgut has also been implicated as a restricter of fluid movement. This is an active process sensitive to the action of DNP and hormones (see Section 3.2.3.5).

3.5. STORAGE EXCRETION

3.5.1. General

The concept of storage excretion has been in existence for a relatively long period of time. What is implied by this concept is that insects are capable of sequestering certain substances present in their bodies in excess of requirements. On the surface it appears that this process could play a role in excretion. However, for it to do so, it is necessary to specify the types of substances involved and the interval over which such sequestration occurs. If the substances in question are waste products and are permanently removed from the metabolic pool over the remainder of the insect's life, there is little question that it can be considered a form of excretion. This would also be true when waste materials are stored for a shorter period and then voided to the exterior at some specific point such as at a molt or at hatching.

Opposing these examples of storage excretion are those in which insects sequester useful substances in the form of dynamic, mobile reserves. It is well known, for example, that the fat body normally has deposits of proteins, carbohydrates, and lipids which can be stored (even for long periods of time) or mobilized, depending on the physiological needs of the insect at any particular time. At the crux of the concept of storage excretion is the question of what are useful and what are waste products. A prime illustration is recent evidence that uric acid deposits in the fat body of cockroaches are, in fact, a useful mobile reserve rather than the permanent sequestration of a major nitrogenous excretory product. In the paragraphs that follow, this and other questions related to storage excretion will be explored.

3.5.2. Cuticular Substances

The cuticle of insects is a complex structure consisting of the cuticular epithelial cells and the products elaborated mainly by these cells. In many species, pigments are a prominent feature of the cuticle. They may occur either in the main body of the cuticle or in elaborations thereof such as the wing scales of Lepidoptera. Of interest here is that certain pigments are pteridines or purines, particularly uric acid. Since these two types of compounds are known nitrogenous excretory products of insects, the question of storage excretion is again raised. There can be little question that, once deposited in the cuticular scales or even the cuticle proper of an adult insect, these substances are more or less permanently removed from the metabolic pool of the insect. In this sense the definition of storage excretion seems to be satisfied. However, one must also consider that pigmentation has adaptive significance for the species involved. Therefore, the sequestration of both pteridines and purines in the cuticle cannot be thought of solely as cases of storage excretion. It is perhaps more appropriate to view this situation as an example of metabolic parsimony, where the utilization of a given compound helps to accomplish two or more functions simultaneously.

3.5.3. Intracellular Storage

3.5.3.1. Mineral Concretions. The ability of insects to form complex intracellular mineral concretions has been mentioned in discussions of the hindgut (Section 3.4.2) and midgut (Section 3.4.3). These discussions must be broadened to include other parts of the insects' body. Indeed it appears that the formation of mineral concretions in insect cells is quite widespread. In addition to the midgut and hindgut, they occur in the fat body, cuticular epithelium, Malpighian tubules, eyes, and eggs. The variable composition of the concretions may be influenced by species, sex, and diet. Most commonly they contain Ca, Mg, K, and Fe. Varying amounts of Mn, Zn, Cl, Cu, and Na may also be present. The concretions are often combined as phosphates or urates and may be present in a more complex organic matrix. Other minor constituents such as oxalates, calcite, and xanthine have also been reported.

The function of these complex concretions is not fully understood. Clearly, one possible function is their involvement in excretion. In the case of Malpighian tubule cells, the concretions could represent a stage in vacuolar transcellular transport. On the other hand, since concretions may also be voided at the molt if in gut cells, they could represent an example of storage excretion. The apparent permanent retention of concretions in some cells also supports the storage excretion hypothesis. In *Musca* it has been reported that mineralized concretions show an age-related increase in adults. Contrarily, it is possible that the concretions act as metabolic reserves in the event of a mineral stress. Indeed, their presence in the cytoplasm of the various metabolically active cells suggests that this is likely to occur. Here, too, the difficulties of distinguishing between permanent sequestration versus mobile reserves and useful versus waste products are all too evident.

3.5.3.2. Urate Concretions.   The ability of cockroaches to accumulate uric acid in their fat body has been known for many years. If one places a cockroach on a diet high in nitrogen, urate storage may become excessive. These findings correlate well with the cellular constitution of the cockroach fat body. It consists of trophocytes (typical fat-body cells), mycetocytes which contain microbial symbiotes, and urate cells that house the stored urates (Fig. 3-23). The latter two cell types increase in number significantly as urate storage accelerates. The urocytes become packed with discrete urate spherules, each of which is associated with a unique cytoplasmic organelle. The entire structure, including the urate spherule, has been called a urate structural unit (Fig. 3-24).

Because analysis has shown that the urate spherules contain K and sometimes Na, but not significant levels of other mineral elements, the spherules are referred to as urate concretions instead of mineral concretions. Nevertheless, they sequester what is typically an insect excretory product, and the phenomenon, consequently, has often been referred to as storage excretion.

But is it? Cockroaches present an unusual picture in this context because it is known that, with a few exceptions, they do not void urates to the exterior. However, the situation is more complex than it appears because when cockroaches are placed on a diet deficient in nitrogen, they deplete whatever stored reserves of urate they possess over time—but none are voided to the exterior. Thus, it is evident that the deposits of urate in cock-

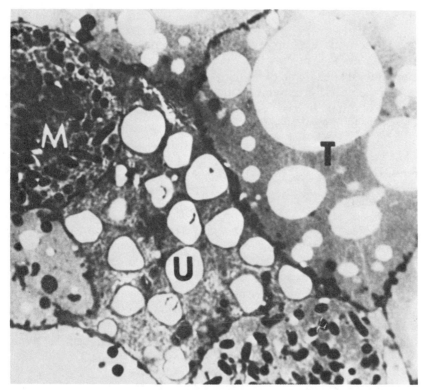

FIGURE 3-23. Fat body from adult *Periplaneta americana* showing the three cell types. M, mycetocyte; T, trophocyte; U, urocyte. Original photo ×800. [With permission from Cochran et al., *Ann. Entomol. Soc. Amer.* **72,** 197 (1979).]

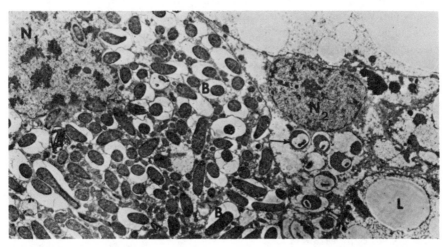

FIGURE 3-24. Fat body from adult *Periplaneta americana* showing the urate structural unit with small, dark-centered urate spherules (arrows). B, bacteroids in mycetocyte; L, trophocyte lipid vacuole; N_1–N_2, nuclei of mycetocyte and urocyte, respectively. Original photo ×800. [With permission from Cochran et al., *Ann. Entomol. Soc. Amer.* **72,** 197 (1979).]

roach fat body do not represent a case of storage excretion. Rather they serve as a metabolic reserve that can be mobilized in times of dietary stress. At any rate, this example of fat body storage of urates in cockroaches emphasizes the need for extreme caution before labeling a given phenomenon storage excretion. Indeed, unequivocal examples of storage excretion in insects may be unusual if not rare.

3.6. OTHER TYPES OF EXCRETORY SYSTEMS

3.6.1. General

The Malpighian tubule–rectum complex serves as the primary excretory organ for the vast majority of insect species. However, for completeness it must be mentioned that certain other structures and cells are often considered to have an ancillary excretory role. Besides the midgut (as described), these include labial glands, pericardial cells and nephrocytes, and cockroach male accessory glands. While some phenomena associated with these cells and structures impinge on excretion, there is little evidence to support a vital role for any of them, with the possible exception of the midgut K^+ pump.

3.6.2. Labial Glands

In the head or anterior thorax of certain apterygotes and adult saturniid moths are paired structures called labial glands, antennal glands, cephalic glands, and other names as well. They usually have a single external opening somewhere on the lower part of the head. In the collembolan *Orchesella* the glands can be divided into several sections. In *Orchesella* and in the saturniids some of the gland epithelial cells have elaborate basal and/or apical plasma membrane invaginations with mitochondria, indicating the possibility of active transport. Beyond this, cells in the *Orchesella* glands seem to undergo cyclic changes in structure which suggest variable metabolic capabilities.

   The function of these glands is not well understood. In several species they secrete fluid which is isoosmotic to the blood but which contains high concentrations of potassium bicarbonate. This could be a mechanism to bring about a rapid reduction in blood volume. Or, in the case of the moths, the fluid may be important at adult ecdysis as a buffered medium for the enzyme cocoonase. While these proposed functions appear plausible, they do not show any overwhelming excretory implications. Yet because of the apparent active transport capabilities of these glands, it is difficult to ignore their possible importance.

3.6.3. Pericardial and Other Cells

In the immediate area of the heart and aorta there are a series of cells of mesodermal origin called pericardial cells. These cells, as well as the nephrocytes, which are similar cells located in other parts of the body, are often considered to have an excretory function. In structure they are rather complex, with peripheral cytoplasmic digitations and an internal layer rich in mitochondria. The cell interior is highly organized with readily discernible coated vesicles, tubules, dense bodies, vacuoles, and lysosomes. Each cell appears to function individually, and there are no ducts or ductlike structures leading to its exterior.

While there is no clear indication of the importance of these cells in excretion, several of their activities can be described. For example, they have pinocytic activities which may explain their ability to pick up dyes, hemoglobin, and other large molecules. In fact, a recurring suggestion is that one of their principal roles is the removal of colloidal particles from the blood. Proteins are prominent in this connection, and pericardial cells can apparently digest them and return their constituent amino acids to the blood. Furthermore, they produce lysozyme, and this activity is stimulated by the introduction of bacteria into the hemolymph. This, of course, implies that pericardial cells may play a role in defense against invasion by extraneous organisms. In *Periplaneta* it has been shown that pericardial cells respond to a feeding stimulant by producing indolalkylamine, which acts on the heart to make it beat faster. Finally, there is evidence that the pericardial cells exhibit considerable selectivity in their actions. This may be correlated with the presence of desmosomelike slits in the peipheral structure which limit what gets into the cells and possibly what emerges as well.

Clearly, pericardial cells are active metabolically and can perform several functions. Regarding excretion, removal of colloidal particles has the most relevance. Regardless of whether such substances are stored in the pericardial cells or digested and returned to the hemolymph in another form, they are removed from the general circulation.

3.6.4. Utriculi Majores

The *utriculi majores* are part of the male accessory gland apparatus of cockroaches, and in some species these glands are filled with uric acid in adults. At the time of mating the male deposits a spermatophore in the female genital opening and covers it with a mass of uric acid. After sperm transfer, the female drops this entire mass and may eat it.

Since uric acid is involved, it has been suggested that this behavior represents a form of excretion, but other possible explanations are likely. First, it occurs only in adult males of some species, but not others. For it to have important excretory implications, the phenomenon should be widespread, if not universal, in this group. Secondly, adult males prevented from voiding uric acid by this route suffer no apparent ill effects. Finally, the close association of urate voiding with mating suggests a function related to reproduction.

Some recent work supports the latter conjecture. Mullins and Keil (1980) showed that *Blattella* females consume radiolabeled urates and deposit them in their fat body and in developing egg cases. Furthermore, females on a nitrogen-restricted diet have higher levels of labeled urates in the fat body and especially in egg cases. This finding clearly indicates that urates deposited by the males at mating may be utilized by females in providing nutrients to the developing eggs. In view of these results, it is difficult to support an excretory implication for voiding urates by this route even though uric acid is an important normal insect excretory product.

3.7. CONCLUSIONS

From the preceding sections, it is evident that excretion in insects has been studied intensively for many years. Indeed, when viewed cumulatively the literature in this area of study is quite overwhelming. While little attempt has been made here to detail that literature, the references listed mainly in the introduction will serve as a means to enter it.

The state of the art on excretion in insects is that a good understanding of the morphology of the excretory organs currently exists at both the gross and fine structural levels. While important details are sure to be uncovered in the future, it is abundantly clear that the cells are elaborately developed to carry out active secretory processes. Information on the mechanisms of ion and water transport is in a less satisfactory state of development, but several theories have been advanced which take into account most of the available information. A more thorough understanding of the phenomenon should occur in the years ahead.

Finally, the close interrelationship between the processes of excretion and osmoregulation should be reemphasized. Most review articles tend to stress one of these areas and minimize the other, so that the interrelationship may be lost or obscured. Stated simply, the two processes overlap by using some of the same structures and mechanisms. While in some respects they may be operationally identical, two separate physiological needs are satisfied.

ACKNOWLEDGMENTS

I wish to express my appreciation to a number of people who were most helpful in the preparation of this chapter. They are: Drs. D. E. Mullins and J. L. Eaton for critically reviewing the manuscript, Miss B. J. Waller for typing it, Mrs. N. F. Boles for preparation of the figures, and various authors and publishers who generously allowed the use of photographs and figures from their published works.

REFERENCES

L. G. Arlian and M. M. Veselica, *Comp. Biochem. Physiol.* **64A,** 191 (1979).

M. J. Berridge and J. L. Oschman, *Tissue Cell* **1,** 247 (1969).

J. T. Blankemeyer and W. R. Harvey, "Insect midgut as a model epithelium," in A. M. Jungreis, T. K. Hodges, A. Kleinzeller, and S. G. Schultz, Eds., *Water Relations in Membrane Transport in Plants and Animals,* Academic Press, New York, San Francisco and London, 1977, p. 161.

D. G. Cochran, "Excretion in insects," in D. J. Candy and B. A. Kilby, Eds., *Insect Biochemistry and Function,* Chapman and Hall, London, 1975, p. 177.

D. G. Cochran, D. E. Mullins, and K. J. Mullins, *Ann. Ent. Soc. Amer.* **72,** 197 (1979).

I. Davies and P. E. King, *Cell Tissue Res.* **161,** 413 (1975).

J. D. Gee, "The hormonal control of excretion," in B. L. Gupta, R. B. Moreton, J. L. Oschman, and B. J. Wall, Eds., *Transport of Ions and Water in Animals,* Academic Press, London, New York, and San Francisco, 1977, pp. 265–281.

A. V. Grimstone, A. M. Mullinger, and J. A. Ramsay, *Phil. Trans. Roy. Soc. London Ser. B* **253,** 343 (1968).

B. L. Gupta and M. J. Berridge, *J. Morph.* **120,** 23 (1966).

S. H. P. Maddrell, "The mechanisms of insect excretory systems," in J. W. L. Beamont, J. E. Treherne, and V. B. Wigglesworth, Eds., *Advances in Insect Physiology* Vol. 8, Academic Press, London and New York, 1971, p. 199.

S. H. P. Maddrell, *Amer. Zool.* **16,** 131 (1976).

S. H. P. Maddrell, "Insect Malpighian tubules," in B. L. Gupta, R. B. Moreton, J. L. Oschman, and B. J. Wall, Eds., *Transport of Ions and Water in Animals,* Academic Press, London, New York, and San Francisco, 1977a, pp. 541–569.

S. H. P. Maddrell, "Hormonal action in the control of fluid and salt transporting epithelia," in A. M. Jungreis, T. K. Hodges, A. Kleinzeller, and S. G. Schultz, Eds., *Water Relations in Membrane Transport in Plants and Animals,* Academic Press, New York, San Francisco, and London, 1977b, p. 303.

A. T. Marshall and W. W. K. Cheung, *Tissue Cell* **6,** 153 (1974).

D. E. Mullins and C. B. Keil, *Nature* **283,** 567 (1980).

C. Noirot and C. Noirot-Timothée, *J. Ultrastr. Res.* **37,** 119 (1971).

C. Noirot and C. Noirot-Timothée, *Tissue Cell* **8,** 345 (1976).

J. E. Phillips, "Problems of water transport in insects," in A. M. Jungreis, T. K. Hodges, A. Kleinzeller, and S. G. Schultz, Eds., *Water Relations in Membrane Transport in Plants and Animals,* Academic Press, New York, San Francisco, and London, 1977, p. 333.

J. E. Phillips and T. J. Bradley, "Osmotic and ionic regulation in saline-water mosquito larvae," in B. L. Gupta, R. B. Moreton, J. L. Oschman, and B. J. Wall, Eds., *Transport of Ions and Water in Animals,* Academic Press, London, New York, and San Francisco, 1977, pp. 709–734.

J. A. Ramsay, *J. Exp. Biol.* **30,** 358 (1953).

J. A. Ramsay, *J. Exp. Biol.* **31,** 104 (1954).

J. A. Riegel, *J. Exp. Biol.* **44,** 379 (1966).

J. A. Riegel, *Comparative Physiology of Renal Excretion,* Oliver and Boyd, Edinburgh, 1972, pp. 1–204.

R. S. Sohal, *Tissue Cell* **6,** 719 (1974).

R. H. Stobbart and J. Shaw, "Salt and water balance; Excretion," in M. Rockstein, Ed., *The Physiology of Insecta,* Vol. 5, Academic Press, New York and London, 1974, pp. 361–446.

B. J. Wall, "Fluid transport in the cockroach rectum," in B. L. Gupta, R. B. Moreton, J. L. Oschman, and B. J. Wall, Eds., *Transport of Ions and Water in Animals,* Academic Press, London, New York, and San Francisco, 1977, pp. 599–612.

B. J. Wall and J. L. Oschman, *Fortsch. Zool.* **23,** 193 (1975).

B. J. Wall, J. L. Oschman, and B. A. Schmidt, *J. Morph.* **146,** 265 (1975).

A. Wessing and D. Eichelberg, *Fortsch. Zool.* **23,** 148 (1975).

A. Wessing and D. Eichelberg, "Malpighian tubules, rectal papillae and excretion." in M. Ashburner and T. R. F. Wright, Eds., *The Genetics and Biology of Drosophila*, Vol. 2c, Academic Press, London, New York, and San Francisco, 1978, p. 1.

V. B. Wigglesworth, *J. Exp. Biol.* **8,** 411 (1931).

V. B. Wigglesworth and M. M. Salpeter, *J. Insect Physiol.* **8,** 299 (1962).

K. Zerahn, "Potassium transport in insect midgut," in B. L. Gutpa, R. B. Moreton, J. L. Oschman, and B. J. Wall, Eds., *Transport of Ions and Water in Animals,* Academic Press, London, New York, and San Francisco, 1977, pp. 381–401.

4

THE INTEGUMENT

H. R. HEPBURN
Department of Physiology
University of the Witwatersrand
Johannesburg, Republic of South Africa

CONTENTS

139

SUMMARY

The insect integument is a sensorially rich and metabolically dynamic organ consisting of cuticle and epidermis. The former is divisable into a very thin set of chemically complex layers, the epicuticle, which is subtended by a relatively gross set of layers comprising the procuticle. The epicuticle may contain an apparently hydrophobic cement layer, a predominantly lipid wax layer that is important in waterproofing, a lipoproteinaceous outer epicuticle that probably defines the shape and size of an instar, and a polyphenol protein-complexed inner epicuticle of possible importance to integumental wound repair.

The procuticle is a composite material of which the principal components are a proteinaceous matrix (but with some lipid) and fiber phase of chitin microfibrils. The physical and chemical properties of this region are in a constant state of flux, and this allows for the differentiation of procuticle into the substituent fractions of exocuticle, mesocuticle, and endocuticle. Chitin, a cellulosic derivative, is a viscoelastic polymer that is crystallographically polymorphic. The great stability of cuticle mainly depends on the extent to which matrix proteins are stabilized by tanning and sclerotization, the latter having several routes.

Thus, the material properties of cuticles vary enormously, and the range of observable behavior extends from the highly distensible and rubbery resilin-bearing cuticle to the heavily tanned, solid, tough or brittle, and inextensible sclerites. These properties vary vertically and horizontally within the same animal and are constantly changing during the course of development.

Although some parts of an insect's exoskeleton may increase in size within an instar, major increases in surface area are usually accomplished through a graded and polarized process, metamorphosis. This consists of hormonally mediated changes beginning with apolysis and extending through and beyond ecdysis into the new instar. So, it is possible for the head of an insect to be at pharate pupal stage while the abdomen is still in the larval condition. These macroscopic changes move in sequence with the cyclical activities of the relevant subcellular particles: plasma membrane plaque formation and Golgi complex activities alternate in controlling the synthesis and degradation of various cuticular components.

Aside from performing normal cellular activities, the epidermis is a major regulator both of cuticular properties and of itself. Epidermal cells are functionally syncytial, and the activities of "neighborhoods" of such cells seem to depend on an anisotropic ionic coupling in the junctional membrane region. This results in a very directional distribution of chemical information between neighboring cells. It is reasonably clear that segmentally arranged physiological gradients are just as real as are skeletal segments.

4.1. INTRODUCTION

The insect integument is the interface between a living animal and an environment which is at the same time both hostile to protoplasm and necessary for its existence. It is a sensorially rich organ, capable of perceiving stimuli that may impinge on it in chemical, electrical, mechanical, photic, and thermal forms. The integument consists of cuticle and epidermis, neither of which has any physiological integrity without the other. Both must act together to function as an organ. Cuticles in all of their complexity and variation are collections of layers that serially represent a temporal record of the synthetic activities of the epidermal cells from which they are secreted. Such a record is by no means static; the numbers and kinds of layers are in a constant state of flux as the epidermis wends its way from embryonic blastula through senescence to death in some larval cul-de-sac or aged adult.

The cuticle consists of two principal fractions, protein and α-chitin, which form a composite material the final properties of which derive from the intrinsic characteristics of each material and the kinds of interactions that exist between them. Hence, there is the solid cuticle of sclerites which has a remarkable strength or stiffness compared to its mass. This kind of cuticle has great structural rigidity, which makes it ideal for the protection of internal organs, but is bioenergetically costly because molting is its only solution to the limited distensibility of α-chitin and covalently bonded matrix. The cuticle fraction of the integument also serves as a site for the attachment of muscles and to define the shape and size of insects. All of these properties are modulated by the epidermis which can also produce arthrodial membrane, resilin, and transitional type cuticles to give locomotory life to these arthropodal creatures.

The epidermis is the principal site of synthesis of cuticular materials and of their general maintenance and turnover as well as the seemingly endless variation encountered in the exoskeletal phenotype. These nuances of skeletal structure and function relate to the fact that epidermal cells are ionically coupled when in membrane contact. This allows for the functional integration of epidermal cells and, owing to the polarized transmission of electrical

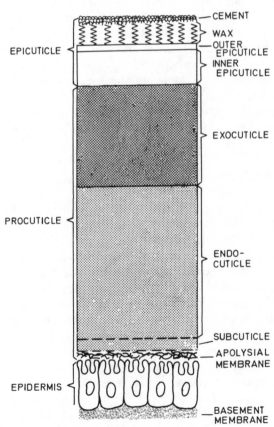

FIGURE 4-1. The general arrangement of the components of the integument between molts. The cuticle fraction is represented as a cross section of solid cuticle.

(and material?) information among them, accounts for the occurrence of discrete neighborhoods in which cells communicate different kinds of information: cells of the developing pulvillus are clearly exchanging information of a different sort from those of the median ocellus.

The general arrangement of the components of the integument are shown diagrammatically in Fig. 4-1.

4.2. EPICUTICLE

4.2.1. General

The epicuticle is a complex of minute layers that enjoys wide acclaim among integumental biologists. Indeed, it is credited (perhaps rightly) with so many functions as to verge on the magical. It is generally held that the epicuticle is the ultimate morphogenetic determinant of the shape of an insect down to the Byzantine intricacies of surface sculpturing. It is definitely known to control the surface properties of the integument with regard to permeability, hence, water economy; accommodate and limit intrastadial exoskeletal expansion in the growth of endopterygote larvae (but not termite queens) and in blood and plant sap-sucking insects as a consequence of engorgement; and facilitate the transport of molting fluid activator(s) into the apolysial space and the return of digests of the old cuticle, at the same time preventing the dissolution of the presumptive exocuticle of the pharate instar beneath it (see Section 4.5). Moreover, the epicuticle can withstand wide variations in diel and in seasonal and habitat temperature and humidity; it also appears reasonably adept at resisting pathogenic attack. It may also serve as a metabolic dustbin as well as a reservoir for contact pheromones and even juvenile hormones. Mechanically, it is usually a brittle lacquer stronger in compression than in tension.

Unfortunately, the epicuticle is steeped in terminological controversy, and problems of interspecific homology abound. Nonetheless, a generalized epicuticle consists of four more or less distinct regions each with some unique attributes. Progressing from the outside of an insect inward, these are cement and wax layers, an outer epicuticle, and an inner epicuticle. Unlike the subdivisions of the procuticle, those of the epicuticle are formed in nearly the reverse sequence in which they occur *in situ:* the outer epicuticle is first formed, followed in turn by the inner epicuticle, wax layer, and cement layer.

4.2.2. Cement Layer

The cement layer is believed to be the product of two different groups of dermal glands and to consist of protein and lipid stabilized by polyphenols. The formation of the cement layer is analogous to that of the ootheca of

Periplaneta, which is stabilized by a mixing of separate secretions from the right and left colleterial glands (Section 4.6.2). The composition, extent, and thickness of the cement layer varies interspecifically, and the layer may occasionally be absent (*Apis*). Its properties are based on inferences from lac: shellac is thought to be combined with wax, the stabilized product forming the cement layer of *Laccifer.* The probable function of the cement layer is to protect the subtending wax layer, an inference based partially on events associated with apolysis and ecdysis (see Section 4.5).

At ecdysis, the new cuticle is hydrophobe, but shortly after eclosion (*Rhodnius*) dermal gland secretions penetrate or cover the cement layer and this newly coated or impregnated surface becomes hydrophil. A day later the epicuticle becomes hydrophobe again for either or both of the following reasons: (1) because the cement layer itself becomes chemically stabilized by a process analogous to tanning and (2) because it is impregnated with waxes that are intrinsically hydrophobic.

4.2.3. Wax Layer

Just below the cement layer is the wax layer(s), a mixture of odd- and even-numbered hydrocarbons in the C_{25}–C_{31} range, esters of even-numbered fatty acids, and alcohols in the C_{24}–C_{34} range. Saturated and esterified fatty acids lead to wax hardness; unsaturated analogues are soft and greasy. The waxes are mainly long-chain saturated alcohols esterified with acids. But because lipids are labile and most analyses have been made on derivatized compounds (as in beeswax), the precise composition of epicuticular wax ought to be regarded as tentative pending further analysis by solid-state techniques.

Wax layer lipids are probably secreted by the epidermis just prior to ecdysis and transported via pore canals to even finer wax canals and thence to the surface. On the basis of histochemically defined esterases, it has also been suggested that wax is synthesized in the wax canals from epidermally derived precursors (*Apis, Calpodes, Galleria, Tenebrio*). Generally, lipid is in a mobile phase, but at ecdysis it changes to cuticular wax. In contrast, free diphenols in the cuticle apparently inhibit oxidation of unsaturated hydrocarbons, thus keeping the lipid in a mobile greasy state (*Periplaneta*). Other waxes occur external to the cement layer in the form of blooms secreted from distinct tubercles (*Cryptoglossa, Calpodes*) or from dermal glands as scales (*Apis, Epiptera, Eriosoma*).

The functional importance of the wax layer in insects is probably best documented with respect to waterproofing, cuticular transpiration vis-à-vis changes in temperature. As an example, there are marked discontinuities in the permeability of cuticle related to rising temperature (Fig. 4-2). There is an initial critical temperature at which there is a steep rise in permeability followed by a plateau region and yet another steep rise beyond environmental temperatures (not shown in Fig. 4-2). Traditionally, these results have been interpreted as related to the dissolution of a mobile lipid phase followed

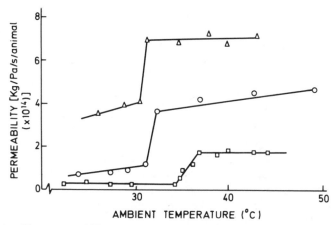

FIGURE 4-2. Water permeability as a function of temperature. Triangles, *Pieris* pupa; circles, *Rhodnius* larva; squares, *Tenebrio* pupa. (After Beament.)

by the dissolution of saturated hardened waxes. However, it is not entirely clear as to whether the failure of either lipid phase is related to melting points and/or shifts in structure toward liquid crystal phases. Nor is it clear precisely how branched lipids (*Eleodes, Cryptoglossa, Centrioptera*) affect permeability in the wax layer.

Additional evidence further suggests that changes in epicuticular waxes may be induced by the neuroendocrine system. It has also been observed that diapausing pupae produce three times more epicuticular wax than nondiapausing pupae (*Manduca*). Composition may also change seasonally (*Eleodes*).

As a final point we can note that while the wax layers ultimately provide the basis for waterproofing in most insects, the situation in maggots could well differ; waterproofing in these insects is not disrupted by steeping in chloroform, a result which likely precludes the wax layer's role in waterproofing in these larvae.

4.2.4. Outer Epicuticle

The outer epicuticle (=cuticulin) appears to be the only region of the epicuticle universal among insects. Besides being a component of the exoskeleton, it also lines tracheoles and serves as an insertion point for the tonofibrillae of muscles. It is absent from some sensory receptor areas of the cuticle. It can be visualized as a trilaminar membrane of about 12–18 nm thick and is formed from the accretion of plaques into continuous layers.

Because the outer epicuticle is the first formed layer of the pharate cuticle, it is to this region that the properties of defining the shape and size of an insect are ascribed. Shortly after formation of the outer epicuticle, pores of 3 nm diameter are discernible in the layer (*Rhodnius, Calpodes*). These are thought to allow the resorption of lysed endocuticle and to prevent the inward flow of molting fluid enzymes, thus protecting the presumptive exocuticle of the pharate insect below. It is of considerable interest that the pharate outer epicuticle is itself resistant to dissolution by molting fluid enzymes.

Notwithstanding the fact that a complete chemical characterization of the lipoprotein outer epicuticle continues to elude us, most authors agree that this is the most important of epicuticular layers. Owing to its position within the cuticle, it must (and has been shown to) have certain properties. It functions as a selectively permeable barrier, which is permeable to whichever factor(s) activates the molting gel and allows resorption of old cuticle digests. In some cases it may be permeable to waxes (*Rhodnius*) or to water where the latter is actively taken up from the environment (*Thermobia, Tenebrio, Sialis*).

The outer epicuticle also imposes constraints on intrastadial growth be it intussusceptive* (*Calpodes*) or by stretching (*Bombyx*); it also defines the size of each successive instar (Brooks' rule = Dyar's rule). Similarly, the ultimate distensions achieved by insects which feed by engorgement depend on the distensibility of the outer epicuticle. The elaboration of surface patterns and sculpturing, of diffraction gratings and plastron geometry, all occur at the time of formation of the outer epicuticle. In assigning all of these important roles to the outer epicuticle, a major and unresolved question arises: to what extent is the outer epicuticle the "prime mover" vis-à-vis the microvilli of the apical plasma membrane of the crenulate epidermis, which actually secretes the outer epicuticle?

4.2.5. Inner Epicuticle

The inner epicuticle is laminar (*Rhodnius, Tenebrio*) and thought to be secreted by the epidermis with possible contributions from oenocytes. Physical and chemical evidence to date suggests that this layer is a polyphenol–protein complex (tanned?). The detection of polyphenol oxidase in the inner epicuticle is consistent with a polyphenol composition. It has been suggested that this layer is the likely source of surface-wound repair materials which tan damaged cuticles (*Calpodes, Musca, Sarcophaga*).

4.3. PROCUTICLE

4.3.1. General

The procuticle constitutes the bulk of the integument. Its properties depend on the interactions of its components. Any assessment of its constituents is

---

* Intussusceptive growth refers to the actual incorporation of new and additional material within a previously formed structure be it cuticle, cell walls, or membranes. Such growth has been documented for both plants and animals.

made difficult by the fact that often drastic physical and chemical treatments are necessary to isolate them. Thus, there is no guarantee that the properties of procuticular derivatives are the same as those occurring *in situ*. For this reason it is desirable to consider relatively unaltered forms of chitin and protein and then to construct a picture of procuticle by extrapolation and circumspection.

4.3.2. Chitins

4.3.2.1. Crystallographic Forms.

The structural polymer chitin, poly-β-(1,4)-N-acetyl-D-glucosamine, is a cellulosic derivative the biosynthesis of which is ancient cell tradition. Chitin is produced by protozoans, fungi, and most protostomian invertebrates. It is crystallographically heterogeneous and three forms are known: α-, β-, and γ-chitin.

α-Chitin, the ubiquitous exoskeletal polymorph of insects, has an orthorhombic unit cell with dimensions 1032 pm in the c or fiber axis, 474 pm in the a axis, and 1886 pm in the b axis. The unit cell contains disaccharide sections of two antiparallel chains (Fig. 4-3). The chain is bonded in two ways. There is an intramolecular hydrogen bond (03'—H $\cdots$ 05), and successive chains along the a axis are joined by C=O $\cdots$ H—N hydrogen bonds to form sheets. The CH$_2$OH side chains, consequently, have different configurations on the two chains in the unit cell: the CH$_2$OH is hydrogen bonded intramolecularly to the carbonyl group of the next residue (06—H $\cdots$ 07'), while on the second chain the hydrogen bond (06—H $\cdots$ 07') is an intersheet bond. All hydroxyl groups are hydrogen bonded, which accounts for α-chitin's resistance to hydrate formation. The amide groups are also hydrogen bonded.

β-Chitin is rare among insects and is known from a few beetle cocoons (*Cionus, Cleopus, Murmidius*). Its monoclinic unit cell has the dimensions 1038 pm in the c axis, 485 pm in the a axis, and 926 pm in the b axis. Since the unit cell contains only one repeating disaccharide, the chains are aligned in parallel, just the opposite of α-chitin (Fig. 4-4). β-Chitin forms hydrates, a property that is facilitated by the absence of intersheet hydrogen bonding.

γ-Chitin, less rare than β-chitin, occurs in the cocoons of beetles (*Rhynchaenus, Prionomerus, Gibbium, Ptinus*) and in peritrophic membranes (*Blaberus, Schistocerca, Antheraea, Phymatocera*). Its suggested cell dimensions are $\sim$1030 pm in the c axis, $\sim$470 pm in the a axis, and $\sim$2900 pm in the b axis. While the crystal structures of α- and β-chitin have been confirmed by several workers, that of γ-chitin awaits further confirmation (Fig. 4-4). Like β-chitin, γ-chitin is capable of considerable deformation on solvation. The significant lack of swelling in water by α-chitin makes it useful for skeletal purposes but might also be one of the reasons why most insects must molt in order to greatly increase their external dimensions.

Mechanically, all of the chitins are strain rate dependent (wet or dry) on stretching. Thus, they are viscoelastic polymers for which stiffness, strength, and extensibility are time dependent, that is, vary with the rate of

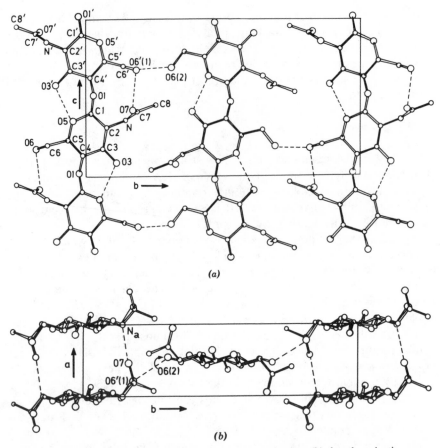

FIGURE 4-3.   Structure of α-chitin. (a) The *bc* projection; (b) the *ab* projection.

deformation. These characteristics vary within and among chitins, differences which are related to crystal structure, microfibrillar packing within macrofibrils (the samples actually used for analysis), and degree of orientation. The mechanical properties of all chitins are sensitive to the effects of solvation in various media. Mechanochemical studies indicate two important properties of chitins: it would appear that α-chitin deforms viscoelasti-

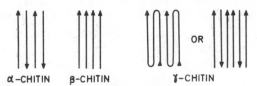

FIGURE 4-4.   Schematic representation of chitin chains in the unit cells of three polymorphs of chitin. (After Rudall.)

cally by chain slipping but that viscoelastic deformation of β- and γ-chitin can largely be explained in terms of molecular conformational changes. Such studies further suggest that the ↶-shaped unit cell rather than the ↑↓↑-shaped cell is the likely configuration of γ-chitin (analogous to the X-β conformation of *Chrysopa* silk). Finally, one can induce a β to α and a γ to α transconformational change but not vice versa.

Within the insect, chitin crystals form aggregates of microfibrils of about 3 nm diameter and are always associated with a protein matrix be it in a cuticle, cocoon, or peritrophic membrane. We are principally concerned with skeletal α-chitin, and by averaging various estimates, the microfibril consists of about 18–21 chains passing through a crystallite at any one level. This is consistent with having three sheets of 6 or 7 chains. By analogy to synthetic fibers, chitin microfibril diameter is possibly independent of direct cellular control.

4.3.2.2. Architecture. The arrangement of chitin microfibrils in the three-dimensional procuticle has long been the subject of intense controversy. The problem is how to interpret the familiar parabolic patterns that occur in thin sections of cuticle viewed electron microscopically (Fig. 4-5). How do they arise and by what model can they be explained? A helicoidal model (Fig. 4-6a) has been proposed in which all of the microfibrils in any one sheet are parallel to one another and to the surface of the cuticle. Moreover, neighboring sheets (in each of which all of the microfibrils are also parallel to one

FIGURE 4-5. Transverse section of the pupal cuticle of the cotton leaf worm, *Spodoptera littoralis,* showing the parabolic patterns of chitin microfibrils within lamellae.

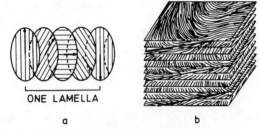

ONE LAMELLA

a   b

FIGURE 4-6. Models for the interpretation of parabolic arcs (as in Fig. 4-5) proposed by Bouligand. (a) A number of sheets all with parallel chitin microfibrils gradually rotated with respect to one another and passing (from left to right) through 180° to form one lamella. (After Neville.) (b) Variation on the above wherein the chitin microfibrils of any one sheet are progressively curved. (After Bouligand.)

another) are rotated by some small angle with respect to one another and always in the same direction of rotation. Any collection of sheets that rotates through 180° is defined as a lamella. Alternatively, the microfibrils of a layer may be curved rather than straight (Fig. 4-6b).

A deceptively similar model has also been proposed. In this the microfibrils are continuously secreted onto those previously formed so that a given microfibril gradually changes its position vertically within the lamella, thus contributing to more than one sheet of microfibrils. This screw-carpet model is shown in Fig. 4-7. These models are geometrically pleasing; whether they are correct in all cases is another matter. Among the unresolved difficulties are that (1) both models preclude the existence of lamellae as distinct entities, but there are cases where this is so (*Periplaneta, Limulus, Scylla, Carcinus*); (2) specimens of cuticle prepared as truncated pyramids should show intersecting lamellae out of register on at least one corner, but there are cuticles where this is not so (*Carcinus, Cancer, Homarus*); (3) the helicoidal model fails to account for the occurrence of vertical microfibrils (*Pachynoda, Locusta, Limulus, Pieris*); and (4) both fail to account for ring patterns normal to the cuticle surface (*Xyleborus, Photuris, Boreus*).

Nonetheless, it has been proposed that all insect chitin architecture can be explained as follows: layers of microfibrils can be helicoidally arranged in

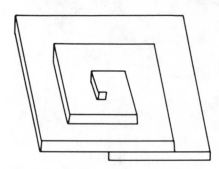

FIGURE 4-7. Screw-carpet model of Weis-Fogh proposed for the arrangement of chitin microfibrils in which fibers are continuous and rotated as shown. This is analogous to crystal growth of certain plastics.

the protein matrix and/or neighboring layers of microfibrils can all be oriented unidirectionally to form a "preferred" layer. Either may occur alone or in various permutations. Alternatively, it has been suggested that microfibril architectures can be defined essentially as fabrics of variable weave embedded in a protein matrix (analogous to a range of fiberglass/resin composites) and/or sheets of microfibrils are completely separated by discrete layers of matrix protein. All of the models discussed are of heuristic value; however, for the moment we must assess them as unproven.

4.3.2.3. Synthesis and Degradation. Substantial evidence from electron microscopic and radiochemical studies indicates that chitin is a product of the epidermis. It may be secreted as monomers, and the extracellular enzyme chitin synthetase attached to the plasma membrane is responsible for its polymerization and for the elaboration of microfibrils. Chitin synthetase is produced as a zymogen in the endoplasmic reticulum of the epidermis and transferred in chitosomes to the plasma membrane. The zymogen is activated by proteases and the synthetase molecules assemble the chitin microfibrils.* The chitin synthesis pathway as currently understood is outlined in Fig. 4-8.

Owing to the cyclical nature of molting in insects, degradation of chitin should be considered along with its synthesis. Insects degrade accessible chitin (probably endocuticular in origin) during molting, and some of this material may be reused in the deposition of the pharate cuticle. It is likely that "chitinase" and "chitobiase" are actually a potpourri of enzymes derived from proenzymes in the cuticle which become activated by some calcium ion–dependent molting fluid factor (*Hyalophora, Manduca, Bombyx*). Because there may be nonacetylated sites on the chitin molecule (possible points of bonding to matrix proteins), the most plausible hypothesis would be that several different enzymes degrade chitin, certain enzymes to split acetal bonds and others to split putative chitin–protein bonds. The chitin microfibril is always associated with protein to which it might be bound directly or indirectly. Given that the microfibril has a diameter of about 3 nm, only some of the peripherally located chains could participate in chitin–protein bonding.

4.3.3. Protein

4.3.3.1. Heterogeneity. No primary structural analyses of cuticular proteins have been reported, but there is a large literature on protein bands

---

* Although this account is true of certain fungi, it has not actually been shown that chitin in insects is secreted as monomers. N-Acetylglucosamine residues are presumably transferred, directly or indirectly, from UDPAG to the growing chain. If the enzyme sits in the membrane, UDPAG could well come from the inside while the chitin chains grow out in the extracellular space. We hazard the guess that the enzymic synthesis of chitin will be found to parallel that of fungi.

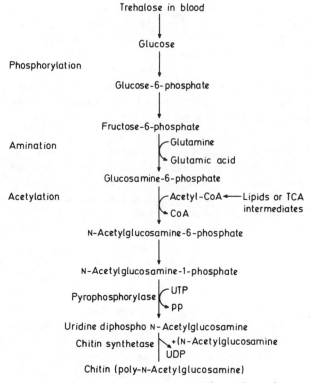

Trehalose in blood

Glucose

Phosphorylation

Glucose-6-phosphate

Fructose-6-phosphate

Amination ⟨ Glutamine

Glutamic acid

Glucosamine-6-phosphate

Acetylation ⟨ Acetyl-CoA ← Lipids or TCA intermediates

CoA

N-Acetylglucosamine-6-phosphate

N-Acetylglucosamine-1-phosphate

Pyrophosphorylase ⟨ UTP

PP

Uridine diphospho N-Acetylglucosamine

Chitin synthetase ⟨ +(N-Acetylglucosamine

UDP

Chitin (poly-N-Acetylglucosamine)

FIGURE 4-8. Pathway for the biosynthesis of chitin. (After various sources.)

obtained from cuticle using various solvents and separated with various electrophoretic and chromatographic techniques. Numerous protein bands have been obtained from a single species (up to 55 in *Agrianome* alone), and this kind of heterogeneity occurs within an individual, between sexes, and interspecifically. Because of the extreme sensitivity of cuticular proteins to variations in analytical technique, it is likely that many of the bands are actually fragments of larger proteins or collections of oligomers. Variation in amino acid composition of protein digests is equally marked and occurs horizontally (between sclerites and arthrodial membranes) as well as vertically. Similarly, protein polymorphism may occur interstadially (*Galleria*) or may not (*Locusta*).

The functional significance of this heterogeneity is therefore rather moot. It has been suggested that natural populations of insects are polymorphic at thousands of structural gene loci and that each individual in a large outbreeding population is likely to have a unique protein profile, as exemplified by *Drosophila* species. The argument follows that any phenotypic character need only be selectively nonlethal or neutral to be retained in a population. Nevertheless, by comparing the amino acid residues of untanned cuticle and grouping them according to the second letter of their genetic code origins, the general trend for many different protein bands becomes very similar, especially in contrast to residue compositions of other kinds of cuticle. Thus, untanned cuticles contain a "family" of similar amino acid composition that is markedly different from that of resilin and other cuticles. It may be that kinds of protein rather than particular proteins are of importance, a generalization consistent with the effects of residue composition on molecular behavior and versatility (see next section).

Before turning to the attributes of amino acids and certain proteins, it is worth mentioning their possible origins. While it is tacitly assumed that cuticular proteins are synthesized in the epidermis, this is probably not the sole source of such proteins. Proteins found in the hemolymph can be recovered from cuticle, indicating that the epidermis exercises some discretion in the transport of hemolymphal proteins to the cuticle. While this only indirectly implicates some other organ system, it has also been shown that some specific proteins are synthesized at the time the epidermis has a relatively low rough endoplasmic reticulum content.

4.3.3.2. Amino Acid Properties.

Proteins extracted from untanned cuticles of endopterygote larvae are rich in residues with bulky side chains. These residues are associated with loose molecular packing as well as flexibility. Mechanically, untanned cuticles containing high concentrations of such residues are of relatively low stiffness and strength but are often tough. These same properties extend to resilin in which there are dimers and trimers of tyrosine.

On balance, the smaller amino acid residues dominate the hydrolysates of sclerites, and where they occur these residues naturally allow closer molecular packing of peptide regions. In addition, there is a correlation between the content of nonpolar residues and the relative hardness of cuticle (*Schistocerca*). It has also been demonstrated that the number of free amino groups available for substitution decreases as sclerotization progresses.

The distribution of groups of amino acids according to types of cuticle is shown in Table 4-1. Further functional relationships between kinds of amino acid residues and properties apply to cysteine, which is known to contribute to the formation of disulfide cross-links (*Lucilia, Machilis, Tenebrio*).

Amino acid composition has been related to metamorphosis in the following way: In the Exopterygota there are only small interstadial changes in amino acid composition, while in the Endopterygota there are large and dramatic changes in residue composition associated with the pupal/adult metamorphosis. There is indirect evidence that juvenile hormone controls protein synthesis (hence, residue composition) because the amino acid profiles of juvenile hormone–treated *Tenebrio* pupae, allowed to develop into supernumerary pupae, are similar to those of normal pupae.

4.3.3.3. Egg and Oothecal Proteins.

Proteinaceous oothecae are well known from Cursoria, Mantodea, and Orthoptera but are also made by

TABLE 4-1. Amino Acid Composition of Various Insect Cuticles[a]

Species[b]	Residues with Small Side Chains	Highly Polar Residues	Proline	Alanine	Cuticle Type
Coleoptera					
Agrianome spinicollis (1)	32	43	10	—	Arthrodial/caterpillar
Tenebrio molitor (1)	40	36	9	?	Solid
T. molitor (2)	42	36	7	?	Solid
T. molitor (3)	42	27	9	?	Solid
Pachynoda eppipiata (1)	37	45	8	—	Arthodial/caterpillar
P. eppipiata (2)	38	42	10	Trace	Solid
P. eppipiata (3)	50	29	7	Trace	Solid
Xylotrupes gideon (3)	47	25	8	1	Solid
Cursoria					
Periplaneta americana (3)	43	17	10	25	Solid
Diptera					
Lucilia cuprina (1)	29	43	9	—	Arthrodial/caterpillar
L. cuprina (4)	28	46	9	9	Anthrodial/caterpillar
Calliphora augur (4)	28	47	7	11	Arthrodial/caterpillar
Drosophila melanogaster (4)	26	47	5	1	Arthrodial/caterpillar
Musca domestica (4)	29	45	6	8	Arthrodial/caterpillar
Hemiptera					
Rhodnius prolixus (1)	53	9	8	—	Arthrodial/caterpillar
Lepidoptera					
Bombyx mori (1)	31	47	10	—	Arthrodial/caterpillar
B. mori (2)	21	38	10	14	Arthrodial/caterpillar
Sphinx ligustri (1)	28	51	9	?	Arthrodial/caterpillar
S. ligustri (2)	38	41	12	?	Solid
S. ligustri (3)	44	39	8	?	Solid
Hyalophora cecropia (1)	29	50	10	?	Arthrodial/caterpillar
Aglais urticae (1)	29	52	9	?	Arthrodial/caterpillar
Xylophasia monoglypha (3)	40	38	7	+	Solid
Odonata					
Aeshna juncea (3)	64	20	7	—	Resilin
Orthoptera					
Schistocerca gregaria (3)	63	20	7	—	Transitional

[a] The data are given as residues of amino acids/100 residues.
[b] (1), larvae; (2), pupae; (3), adults; (4), puparia. Compiled from various sources and where more than one analysis is available for a given case, averaged values are used.

beetles and moths. They (Blatta) are of historical importance to the study of cuticle as source materials of the first modern analysis of the nature of tanning. Oothecal proteins are usually secreted as liquids but quickly become solid, crystalline, and are often tanned. Those of mantids consist of ribbons (Mantis), and those of beetles, tactoids (Aspidomorpha). The latter are crystallographically and mechanically similar to feather keratin, and both are only slightly distensible on hydration. These proteins are generally rich in glycine, tyrosine, and polar residues, and measures of their (vis-à-vis solid cuticle) comparative hardness values would be extremely interesting.

The insect egg shell is a multilaminar product of the follicular epithelium and, like cuticle, is a physical record of the synthetic activities of the cells that produced it. The principal structural protein of the chorion, chorionin, has been analyzed crystallographically, chemically, and microscopically for most orders of insects. The combined analyses from many different species suggest that chorionin is by no means a unique species but rather a genuine menagerie of proteins. A general physical model visualizes the chorion as consisting of protein microfibrils combined into larger fibrils that form layers. The fibrils of a layer are all parallel to one another, while adjacent layers cross orthogonally. The vitelline membrane and serosal proteins of eggs are poorly known.

While little more than the amino acid compositions of some egg proteins are known, egg proteins must perform unequivocal physiological functions. They must accommodate the respiratory needs of the embryo and control water and ion ingress and egress. When an egg takes up water prior to hatching (*Bittacus*), the proteins must be capable of distension. Similarly, eggs must be elastic or plastic to be physically squeezed through an ovipositor. In some wasps (*Pimpla*) this is actually required to trigger embryogenesis. Eggs excised prior to oviposition fail to develop, while excised eggs forced through a capillary tube develop in the normal way.

In addition, the chorion of eggs must resist physical insult and predator attack. In view of the numerous families of wasps that parasitize eggs, it would be interesting to compare the mechanical properties of eggs that are susceptible to a high frequency of attack with those that are less so. In the end, the larva must hatch, and it has been suggested that, in addition to having lines of weakness in the chorion, some species secrete proteolytic enzymes to digest portions of the shell. How any of these unquestionable functions are related to the amino acids and proteins of eggs in molecular terms remains to be seen.

4.3.3.4.   Resilin.   That resilin is the best characterized of cuticular proteins owes something to mechanics: it occurs in pure form in the tendons of dragonflies (*Aeshna*). The principal physical properties of this protein are its long-range elasticity, elastic efficiency (an index of energy recovered when deformation is discontinued), and its notable lack of creep. When a dragonfly tendon was held stretched at 200% of its initial length for several months, it immediately returned to its original length when released. The material is mechanically isotropic as well.

Our understanding of resilin derives from independent physical and chemical studies which complement one another. The need for long-chain molecules with a weak secondary structure and few cross-links between peptides was predicted for resilin. This was subsequently confirmed chemically, and it was established that the tertiary structure consists of peptides cross-linked by dimers and trimers of tyrosine, di-, and tertyrosine respectively, which are formed of tyrosine.

While resilin has not been sequenced nor its biosynthesis determined, the median neurosecretory cells of the brain (*Calliphora*) are required for its deposition. These same cells secrete bursicon, which is associated with the deposition of endocuticle, and could be essential for the synthesis of resilin as well.

Functionally, resilin is an insect rubber (analogous to mollusc abductin and vertebrate elastin) with diverse mechanical roles. It serves as a tension spring in dragonfly and wasp tendons; a compression spring in the jumping of fleas, flipping of click beetles, and proboscis recoil in certain Lepidoptera; and as a bending spring (tension and compression) in the wing hinges of locusts and the tympani of cicadas. Resilin also occurs in mixed form in transitional cuticles.

4.3.3.5. Silk Proteins

4.3.3.5.1. Structure. The protein silks of insects are of great importance to the study of integuments for the insight they provide into the relationships between molecular structure and behavior. Aside from their obvious uses in the cocoons of Endopterygota, silk functions are many and bizarre: symbolic prenuptial offerings in Ephydridae, mooring ropes on the eggs of certain Odonata laid in fast-running streams, the stalks of eggs and spermatophores, the snares of Trichoptera, and the tunnels of Embioptera and Psocoptera not to mention leaf-weaving by ants (*Oecophylla*). There is another reason to compel our interest in these proteins. It has been suggested that the low extensibility of α-chitin constrains the sizes of insects. By the same token, the sclerotized protein is cast with an exuvium, and judging from the properties of amorphous proteins (resilin) and even the more crystalline ones (silks), sclerotization is probably required to check the high order of extensibility and lack of structural integrity of cuticular proteins in general.

The discovery that the same protein could exist in at least two different configurations, folded and unfolded forms, is a landmark in molecular biology. This work was extended by the discovery that yet a third form could be obtained from thermally supercontracted preparations. These three forms, which are crystallographically distinct, are the α, $\|\beta$, and $X\beta$ configurations. Moreover, transconformational changes (analogous to those of β- and γ-chitin) can be induced experimentally, thus $\alpha \rightarrow \|\beta \rightarrow \alpha$ and $X\beta \rightarrow \|\beta \rightarrow X\beta$. Naturally occurring examples of all three configurations were eventually found: the α-helical form in bee silk (*Apis*), the $\|\beta$ form in *Antheraea*, and the $X\beta$ form in the eggstalk of *Chrysopa* and the hanging threads of glowworms (*Arachnocampa*). The distinctness and behavior of these configurations greatly depend on the secondary structures, that is, their hydrogen-bonding topologies. Transconformational changes require disruption of the secondary structure and changes of position by the crystallites with reference to their fiber axes.

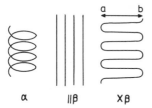

FIGURE 4-9. Diagrammatic representations of chain folding for α-helical, parallel-β, and cross-β fibroins. (After Rudall.)

The structure of these silks are as follows. The Xβ consists of well-oriented fibers in which the c axis of the crystallites is parallel to the fiber axis and the ends of the crystallites are joined by short bends forming continuously folded chains held together by secondary forces. The α-helical silk is thought to consist of four helices coiled on one another as fibers in a thread and held together by hydrogen bonding. The $\|\beta$ crystallites of *Galleria*, *Anaphe*, *Bombyx*, and *Antheraea* are in pleated sheets with the b axis parallel to the fiber axis (Fig. 4-9).

4.3.3.5.2. Crystallinity. The degree of crystallinity (the ratio of sharply defined to poorly defined spacing on X-ray diffraction) greatly affects the material properties—chemical, mechanical, thermal, electrical, optical—of any polymer. Interestingly, crystalline content tends to increase with decreasing fiber diameter for many materials. For fibrous proteins, those amino acids with bulky or long side chains (LC) tend to predominate in the amorphous regions, while those with short side chains (SC) tend to occur in the more crystalline regions when the unit cell in the b axis is small. Since only these latter residues (glycine, alanine, threonine, and serine) occur in highly crystalline segments of *Bombyx* fibroin, their abundance relative to bulky residues in various fibroins has been used to estimate the degree of crystallinity for well-known examples of the three configurations, α, $\|\beta$, and Xβ (Table 4-2). There is relatively good agreement between the SC–LC ratios of the $\|\beta$ fibroins and diffraction measurements as well as tensile mechanical measurements; however *Apis* silk is anomalous because it is highly crystalline on diffraction but has a very low glycine content. This can be partially explained by the fact that LC residues can occur in crystalline regions (Braconidae).

Relative crystallinity can also be assessed by the ratio of stiffness in the dry state to that of the wet fiber. The rationale for employing this index of crystallinity, which is common to all structural biopolymers, is based on the observation that water can penetrate amorphous regions of polymers and then diminish crystallite interactions and/or compete for potential hydrogen-bonding sites within the fiber. In both cases fiber stiffness will be reduced. Plotted in this way (Table 4-2), results agree well with actual diffraction data for the fibroins concerned.

4.3.3.5.3. Extensibility. All of these fibroins are related transconformationally. When specimens of *Apis* and *Chrysopa* silk are stretched a suf-

TABLE 4-2. Relative Crystallinity of α, Xβ, and $\|\beta$ Fibroins Assessed in Terms of Elastic Moduli and Amino Acid Residues

Specimens	(SC/LC)^a	(E_{wet}/E_{dry})^b
$\|\beta$		
Anaphe	23.4	0.81
Bombyx	6.6	0.64
Antheraea	3.9	0.32
Galleria	3.0	0.18
Xβ		
Chrysopa	10.9	0.29
α		
Apis	1.5	0.53

^{a} Based on number of amino acid residues estimated per 1000 residues. LC, amino acid residues with long side chains. SC, amino acid residues with short side chains. (Compiled from various sources.)
^{b} E, elastic modulus in MPa.

ficient amount, the main features of their diffraction patterns are indistinguishable from those of *Bombyx* or *Galleria* fibroin. The transconformation X$\beta \rightarrow \|\beta$ fibroins requires some 700% stretching and that of $\alpha \rightarrow \|\beta$ some 200%; both can be achieved after solvation in water. Considering only the $\|\beta$ fibroins, reversible extensibility is related to (1) relative crystallinity and (2) the fact that fiber deformation is limited to the extent to which the peptide backbone can stretch and the content of amino acid residues having bulky side chains. When any of the α and Xβ fibroins are stretched, there is an initial extending of, and then breaking of, hydrogen bonds within the crystals followed by backbone unfolding. The most important point to be made relative to cuticular proteins in this context is that structural integrity depends not only on the covalent bonding of the primary structure but also on the extent to which peptides are free to interact at the levels of secondary and tertiary structure. These are great indeed. By extrapolation we can conclude that sclerotization is indispensable for holding the highly distensible nature of proteins in check within the whole cuticle.

4.3.4. Other Constituents

4.3.4.1. Pigments. Most of the proper pigmentation of the cuticle is black or brown and results from sclerotization and/or melanization. This dark coloration is usually distributed uniformly through a particular layer as in the exocuticle. The extent of melanization can vary with incident light (*Pieris*), temperature (*Sarcophaga*), population density (*Eudia, Schistocerca, Lo-*

custa), and genetically (*Drosophila, Schistocerca*) but in all cases is likely mediated or influenced by bursicon. In addition, black pigments also occur in discrete granules in certain larvae (*Celerio, Papilio*) and in toad bugs (*Gelastocoris*). Most of the other pigments that give rise to color in insects occur as cytoplasmic granules in the epidermis, the hemolymph, fat body, or other organs. Significantly, most of the spectacular, scintillating colors of insects (*Morpho*) derive from optical effects.

While limited space precludes an in-depth treatment of color chemistry and physics, the physiological significance of coloration cannot be overemphasized. Coloration has myriad functions: aposematism, mimicry, cryptic coloration, mate recognition, facilitation of body warming through sun-basking (*Polistes*), dissipation of heat, and storing of detoxified metabolic waste products.

4.3.4.2. Calcification. Calcification is rare in insects. Except for odd deposits of calcium here and there, calcium seems to function principally as an adjunct to hardening puparia (*Rhagoletis, Musca*) and oothecae (*Periplaneta, Blattella, Orthodera*). By inference from Crustacea, calcification imparts compressive strengthening, a property which is desirable for puparia since they so often occur in shifting soils and other unstable substrates. Crystals of calcium oxalate have been reported from the procuticles of Lepidoptera and Hymenoptera, but their functions (if any) are unknown. Such salts would certainly heavily tax flight economy.

4.3.4.3. Lipids. Procuticular lipids have been only slightly studied vis-à-vis those of the epicuticle. Thus far, about all we can say is that they might contribute to the stabilization of the matrix (*Rhodnius*).

4.4. COMPOSITE CUTICLES

4.4.1. General

The physical and chemical properties of cuticles vary enormously both vertically and horizontally in the insect skeleton, ranging from the dragonfly tendon of pure resilin to the sclerites of beetles and wasps. These cuticles are by no means inert: both the electrical and permeability properties of cuticles are modulated by the epidermis, and it is likely that the thermal properties are similarly affected. Nor is cuticle metabolically inert as is evidenced by the complex changes during molting and sclerotization and by the fact that it can store metabolic wastes as well as the pigments that make the animal cryptic or aposematic long after ecdysis. Similarly, the properties of cuticle may change during development in response to feeding and oviposition and even to invasions by parasites and parasitoids.

Clearly then, there is no "the cuticle." Rather, cuticles possess a vast spectrum of properties that vary in time and space. The end points are relatively discrete, but there are innumerable nuances that make any generalizations as susceptible to exception as is English pronunciation. Nevertheless, on functional mechanical grounds cuticles can be grouped into four major spectra, each with its own variations: the solid cuticles of sclerites, podites, and apodemes; the resilin-based, rubberlike cuticles; arthrodial membranes and caterpillarlike cuticles; and transitional types which are mixtures of solid and rubberlike cuticles as in compound eyes and the caudal edges of abdominal tergites.

4.4.2. Classification Schemes

Most schemes for classifying cuticles have been primarily concerned with properties of the procuticle. (This is somewhat surprising in view of a perhaps as great a variability in the properties of the epicuticle as well as of epidermal cell populations which produce all integumental variability). Basically there are three schemes for classifying procuticular variation in solid cuticle: structural, histological, and mechanical. Each of these stresses different aspects of variation and, hence, has certain advantages and limitations.

The structural classification, as recently summarized by Neville (1975), contains the following descriptive terms. The exocuticle is that part of the procuticle which is sclerotized and hard, amber to dark in color, resistant to enzymic lysis by the molting fluid, and contains chitin crystallites arranged in helicoidal fashion. Endocuticle is unsclerotized and consists of well-defined lamellae that are lysed and resorbed during molting. It is in the endocuticle that permutations of chitin architecture and daily growth layers (when present) occur (Section 4.3.2.2). Mesocuticle is vaguely considered to be impregnated with "stabilized" lipid but not tanned and to resemble endocuticle in "texture."

The histological classification of Richards (1951) is based on cuticular reaction to Mallory's Triple Connective Tissue stains. Thus, exocuticle is refractory to staining and ranges in color from amber to black (exocuticles I, II, and III), mesocuticle stains red, and endocuticle stains blue. Corollary to this is that exocuticle is hard, relatively nondeformable, and not digested by molting fluid; mesocuticle is "elastic" and endocuticle "soft and pliable," and both are digested by the molting fluid.

While both of the above schemes contain much information, they are qualitative and thus discrimination between somewhat similar kinds of procuticle cannot be made. These limitations led to the development of a mechanical classification scheme which, while also imperfect, reflects some of the functional properties of cuticles. Moreover, mechanical data (though more difficult to obtain) are numerical in form so that direct quantitative comparisons of vastly different or very similar kinds of cuticle can be made.

Similarly the temporal changes in cuticle that accompany sclerotization or endocuticular deposition can be considered. This approach recognizes a wide spectrum but particularly allows discrimination of solid cuticular subtypes.

The end points are sharply defined as follows: on the one hand, there are A-type solid cuticles with a heavily tanned brittle matrix, a high relative stiffness, elastic modulus and breaking stress, and low breaking strain. Fractographicaly these cuticles are brittle and matrix-dominated. On the other hand, there are B-type solid cuticles in which there is a highly plastically deformable matrix, a low relative stiffness, modulus, and breaking stress, and a correspondingly high breaking strain. Fractographically these cuticles show the influence of the chitin fiber phase. Either type may be planar isotropic or anisotropic. In pure form the former is typical of sclerites like that of *Dytiscus* or chrysidid wasps, the latter, of butterfly wing membranes, not wing veins (*Danaus*). The vast majority of cuticles fall somewhere between these end points and are usually blends of both types. As determined histologically, type A cuticle is analogous to exocuticle and type B to endocuticle (Fig. 4-10).

4.4.3. Composite Cuticles

4.4.3.1. Solid Cuticle. The most complete pictures that we have for solid cuticles fortunately include exopterygote insects (*Schistocerca, Locusta*) and an endopterygote one (*Apis*). Following the arguments for a functional mechanical classification, we can monitor the development of a sclerite from the time of apolysis of the pharate adult honeybee, through ecdysis, and into the foraging life, and in the locusts from the final larval instar through ecdysis to egg-laying age. These solid cuticles have been monitored mechanically, chemically, and histologically.

Relative changes in the amounts of NaOH nonextractable protein and in the stiffness (elastic modules), as well as reactions to histological changes as a function of time during the development of the third abdominal tergite of the pharate and postpharate adult honeybee, are shown in Fig. 4-11. Similar parameters for the larval and adult locust hind femur are shown in Fig. 4-12. Obviously, the trends are the same in both cases. Stiffness is highly correlated with NaOH nonextractable protein (an index of cross-linking in the matrix) so that, as the amount of nonextractable protein increases (relative to the total mass), the stiffness increases. The introduction of additional cross-links in the matrix renders it increasingly insoluble (and the recovery of cross-link derivatives following drastic hydrolysis is correlated with the amounts of nonextractable protein). These relationships hold even though different tanning systems occur in the examples: the locust is β-sclerotized and the honeybee mainly quinone tanned. Therefore, the generalization emerges that the precise nature of the cross-link is not so important as the extent of cross-linking.

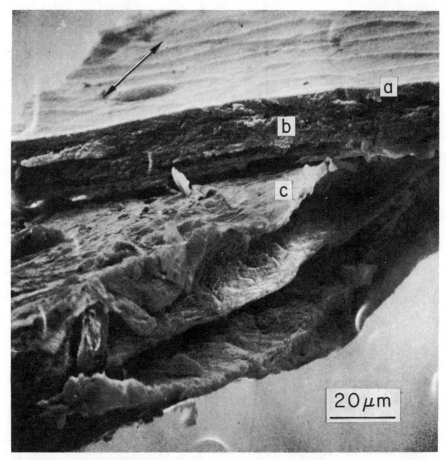

FIGURE 4-10. Scanning electron photomicrograph of locust (*Locusta*) femoral cuticle fractured in tension. Tensile axis is indicated by arrow on surface of cuticle. a, epicuticle; b, exocuticle; c, endocuticle.

That the locust and honeybee are less stiff as fully formed than in earlier stages reflects the formalisms of physical measurements, but to the biologist this appears "intuitively" wrong. However, this occurs because of the deposition of *more* endocuticle in both locust and honeybee *after* the exocuticle has been tanned, and the stiffness depends on the *total* cross-sectional area of the specimen. However, by compensating for thickness and replotting the data as *relative* stiffness, the appropriate "intuitively" correct results are obtained (Figs. 4-13 and 4-14).

The relative stiffness is also highly correlated with NaOH nonextractable protein. A dip in the real stiffness curves will occur when the rate of sclerotization falls below that of endocuticle deposition (as in the one-week-old adult locust and postecdysial honeybee). But note, it is only the *relative*

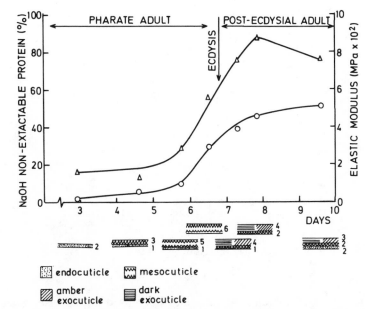

FIGURE 4-11. Relative changes in NaOH nonextractable protein (circles) and elastic modulus (triangles) over the course of pharate and postecdysial development of the third abdominal tergite of the worker honeybee, *Apis mellifera adansonii*. Histological staining patterns are given in figure key and associated numbers indicate cuticular thickness in micrometers.

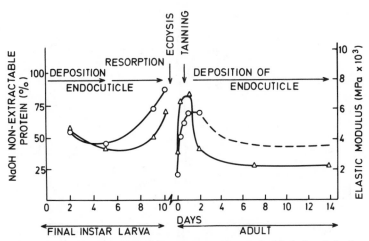

FIGURE 4-12. Relative changes in NaOH nonextractable protein (circles) and elastic modulus (triangles) over the course of larval and adult development of locust (*Locusta migratoria migratorioides*) hindleg femoral cuticle.

163

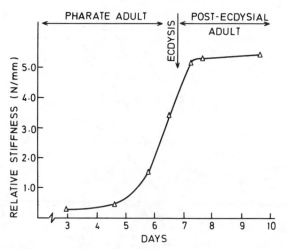

FIGURE 4-13. *Relative* stiffness of honeybee cuticle (cf. Fig. 4-1).

amount of nonextractable protein that has decreased in this case, since the *absolute* amount cannot decrease. The *relative* stiffness (N/mm) is directly related to the absolute amount of insoluble (tanned) matrix and, hence, to the *total* number of cross-links. For real stiffness (N/mm²) it is the *relative number* of cross-links that is reflected when two different pieces of cuticle are compared. The trends for ultimate tensile strength are the same as those

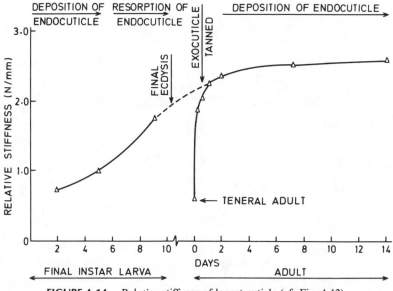

FIGURE 4-14. *Relative* stiffness of locust cuticle (cf. Fig. 4-12).

of stiffness, while the breaking force (force to break the material irrespective of its thickness) follows that of the *relative stiffness*.

The total extensibility of the cuticles (breaking strain) has a high negative correlation with the stiffness and nonextractability. This is because, as the extent of sclerotization proceeds, the ability of proteins to slip past one another or to be stretched (Section 4.3.3.5) is greatly reduced owing to the cross-linking of the matrix. Where the matrix is not cross-linked to any extent, the material can flow plastically; where it is well tanned, it fails in a brittle manner (Fig. 4-15).

Returning briefly to cuticular classification, it can be seen that locust and honeybee solid cuticle vary enormously over time, and we can ask how these changes relate to histological staining reactions. Figure 4-11 shows that there is, indeed, a *qualitative* relationship between staining and the physicochemical properties: day-3 honeybee cuticle is blue-staining endocuticle, highly extractable, and soft and flexible, but neither stiff nor strong, suggesting little (if any) covalent cross-linking. This is further supported by the fall in stiffness and extractability beginning on day 8 when blue endocuticle appears again. At $6\frac{1}{2}$ days the cuticle is entirely red-staining mesocuticle, and its extractability has decreased and stiffness increased (cross-links have been introduced). The appearance of nonstaining exocuticle leads to further rises in stiffness and even less extractability, so that it appears that meso-

FIGURE 4-15. Tensile fracture of mesosternum of the beetle, *Pachynoda sinuata,* showing brittle failure of the exocuticle (a) and plastic failure of endocuticle (b). (Note orthogonal arrangement of endocuticular lamellae; compare with Fig. 4-10; scale bar = 100 μm).

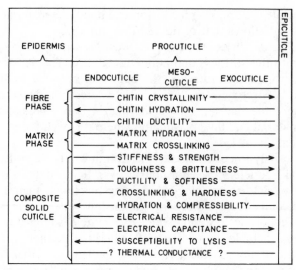

FIGURE 4-16. Summary of the physical properties of insect solid cuticle.

cuticle does represent an intermediate stage on its way to becoming exocuticle. But note that histological exocuticle can vary greatly in stiffness (an order of magnitude difference exists between the elytron of the meloid, *Ecapatoma,* and the mesosternum of the scarab, *Pachynoda*) so that, although a qualitative relationship exists between stiffness, nonextractability, and staining reactions, molecular rationales for the relationship remain equivocal. The physical properties of solid cuticles are summarized graphically in Fig. 4-16.

4.4.3.2. Other Cuticles

4.4.3.2.1. Arthrodial Membrane and Caterpillarlike Cuticles. These cuticles stain blue with Mallory's stain and range in relative stiffness over an order of magnitude from the larval abdominal tergites of *Bombyx* and *Uresiphita*. Their tensile behavior on extraction with solvents like formamide also varies. These cuticles are far more extensible than solid cuticles, and their bulk properties on deformation are to a large extent dominated by the chitin microfibril phase since the matrix is not stabilized through covalent bonding. They are both structurally and mechanically anisotropic.

4.4.3.2.2. Resilin and Transitional Cuticles. Resilin, which is red-staining with Mallory's dyes, is unique in being mechanically isotropic. It can sustain recoverable deformations of about 300% and when hydrated is of quite low modulus, 2 MPa.*

---

* Bulk (resilin) modulus = stress (minus the increment in pressure)/strain (ratio $\Delta v/v$). $m = 10^6$ or kilogram meters/second². Pa = Pascal or newton/meter².

Transitional cuticle is best known from the prealar arms of locusts (*Schistocerca*) which consist of a laminate of chitin microfibrils embedded in a resilin matrix. Like pure resilin, this composite is of low stiffness (2–8 MPa), which makes it ideal for use in oscillating structures such as the flight system where it minimizes the expenditure of power in deforming the structure. Although a systematic investigation of transitional cuticles remains to be undertaken, it is very likely that this kind of cuticle is very widespread in the insect exoskeleton.

4.5. APOLYSIS AND ECDYSIS

Molting consists of both intra- and interstadial changes and can be characterized temporally as a series of events associated with apolysis and ecdysis which terminate in the postecdysial instar. Apolysis specifically refers to the gradual anteroposterior detachment of the epidermis from the cuticle, a process mediated by 20-hydroxyecdysone. As such, it defines the beginning of an instar, and the animal within the loosened cuticle is the pharate stage, be it larva, pupa, or adult. Once detached from the cuticle, the epidermis undergoes mitoses, which eventually lead to a greater template for pharate cuticle production. Ecdysis refers to the actual shedding of the old or exuvial cuticle and is mediated by the eclosion hormone. (See Chapter 13, Section 13.5.5.)

A generalized sequence of events associated with molting are (1) apolysis along an anteroposterior gradient; (2) epidermal mitoses (increasing cell numbers) and folding of the epidermis; (3) secretion of the molting fluids; (4) formation of the pharate outer epicuticle at the surface of the apolysed and crenulated epidermis, resulting in the definition of the surface pattern of the pharate cuticle; (5) secretion of the pharate inner epicuticle; (6) activation of molting fluid enzymes and lysis and resorption of old endocuticle; (7) deposition of presumptive pharate exocuticle; (8) ecdysis; (9) expansion of the new cuticle; (10) onset of tanning; (11) secretion of endocuticle; (12) wax secretion; (13) continued deposition and tanning of endocuticle; and (14) formation of the apolysial membrane for the next molt.

Epidermal mitosis may precede the completion of apolysis; for example, extensive mitotic activity in the head may be near completion before that of the abdomen has commenced. Apolysis usually commences with the discharge of a foamlike secretion into the forming apolysial space whose boundaries are the plasma membrane of the epidermis and the old cuticle. The secretion is contained in small, membrane-bound vesicles, and these apolysial (= ecdysial) droplets reach the apolysial space by exocytosis of the plasma membrane. The release of apolysial droplets may precede apolysis (*Lucilia, Calpodes*) or follow it (*Cecropia, Galleria, Ctenicera*). The apolysial space becomes greater after apolysis probably owing to the accumulation of molting fluid. Since the apolysial droplets are so closely associated in timing with the dissolution of the old endocuticle, they might well be collections of cuticulolytic enzymes or proenzymes. In any case, apolysis and lysis of the old cuticle are two separate processes.

The thin apolysial (= molting, exuvial, ecdysial) membrane is appressed, but not connected, to the pharate cuticle. It seems to occur in most insects where there is a definite lysis of the old endocuticle, but it is absent otherwise (*Drosophila, Lucilia*). Evidence from *Ctenicera* and *Calpodes* suggests that this membrane arises from the innermost lamella of the old endocuticle and becomes "tanned" at the same time that tanning of the pharate outer epicuticle occurs. The functional significance of this membrane is equivocal. Following apolysis and the release of endocuticulolytic enzymes into the apolysial space, lysis of the old cuticle proceeds in varying degrees or not at all. The fate of the digested products is uncertain. Possibly they are resorbed by the pharate insect before ecdysis and reutilized in further cuticular synthesis.

The secretions of the numerous components of the cuticle can be accommodated in terms of the alternation of two forms of secretory activity. These activities involve either the plasma membrane plaques, which are concerned with the outer epicuticle and the procuticle, or the Golgi complex, whose secretory vesicles give rise to the inner epicuticle and ecdysial droplets. The Golgi complex also supplies lytic enzymes in the turnover of apical plasma membranes and other cuticular components (Fig. 4-17).

The fate of muscle attachments during molting is now well known; they remain attached to the old cuticle just up to the time of ecdysis. This accounts for the nonquiescent behavior of supposedly quiescent pupae, since in reality such "pupae" are pharate adults using muscle attachments to the outer pupal cuticle for locomotion. Obviously, the tonofibrillae are remarkably resistant to digestion, and these connections are only severed (by molting fluid enzymes?) very near to the time of ecdysis. As an aside, it is worth noting that bacterial inclusions are sometimes seen in cuticle (*Ctenicera, Boreus*). When they occur in the apolysial space, components of both apolysial membrane and procuticle can be digested by the bacterial enzymes, but the epicuticle remains apparently unaffected. These observations very indirectly support the interpretation that the apolysial membrane may be related to old tanned procuticle and not tanned pharate outer epicuticle.

Occasionally the presence of a subcuticle has been observed (*Ctenicera, Lucilia, Tenebrio*) lying between the epidermis and the procuticle and consisting of parallel microfibrils (chitin?) extending from the tips of the microvilli into the innermost developing layer of the endocuticle. Its composition is unknown (mucoprotein?). Among the several suggested, but untested, functions are that it binds the procuticle to the epidermis between molts, maintaining a space and a link during apolysis or that it is the space where endocuticle formation takes place.

No major changes are known to occur in the intrastadial epicuticle, but there is probably continued deposition of lipids (*Ctenicera*) as well as secretion of greases (*Periplaneta*) and waxes generally. The continuous inclusion

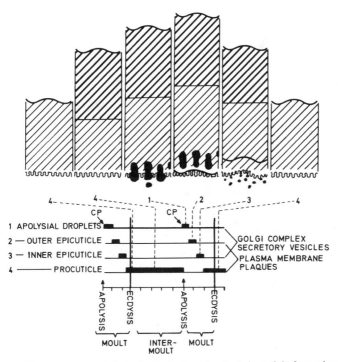

FIGURE 4-17.   The events associated with apolysis and ecdysis in cuticle formation and dissolution. There is a cyclical alternation of plasma membrane plaques and Golgi complex activity.
CP is the critical period for prothoracic gland activity. (After Locke.)

of lipoidal material into the procuticle has also been noted (*Hypoderma,
Ctenicera*). The greater part of procuticular differentiation, growth, and tanning in both exopterygote and endopterygote adults and immature stages
occurs well after ecdysis (Section 4.4.3.1).

4.6.   SCLEROTIZATION

4.6.1.   Major Pathways

Sclerotization (= tanning, hardening) encompasses the processes that lead to
matrix stabilization. The chemical pathways responsible for cuticular hardening have been extensively studied and, while not every detail has been
confirmed, most cuticles are tanned by either or both β-sclerotization and
quinone tanning. Since both quinone tanning and β-sclerotization utilize N-
acetyldopamine as a substrate, it is likely that intermediate reactions are
common to both pathways (Fig. 4-18). However, it has recently been established that N-β-alanyldopamine is the preferred tanning substrate for pupal
cuticular *o*-diphenol oxidase for insects in several orders. This finding may

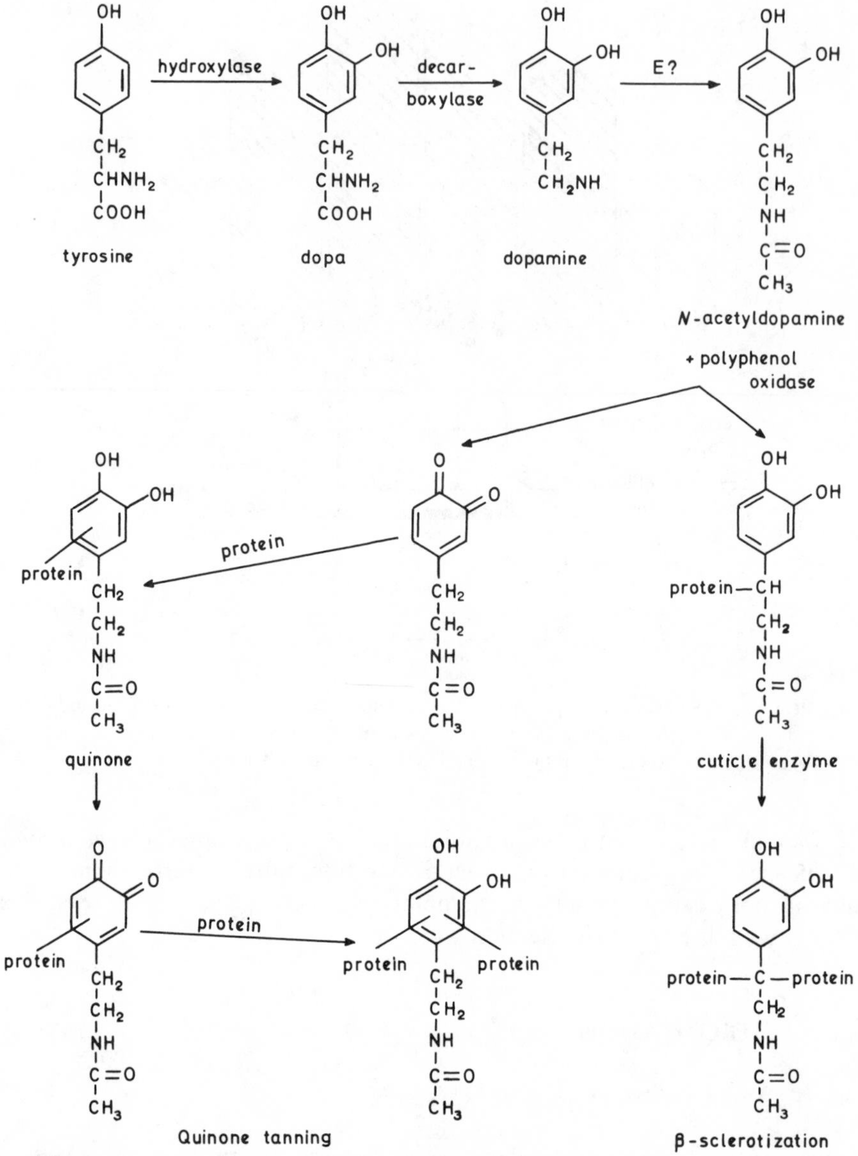

FIGURE 4-18. Pathways for cuticular hardening: quinone tanning and β-sclerotization.

indicate that N-acetyldopamine will have to share its paramountcy as the major catecholamine metabolite in sclerotization with N-β-alanyldopamine.

In quinone tanning (*Ephestia, Calliphora, Pachynoda, Schistocerca, Romalea,* and many others) the general scheme is as follows: Tyrosine is metabolized to 3,4-dihydroxyphenylalanine, which in turn is converted to N-acetyldopamine, the presumed cross-link precursory molecule in this pathway (Fig. 4-18). N-Acetyldopamine is secreted by the epidermis and possibly by hemocytes (*Periplaneta*) into the pharate and/or postecdysial cuticle where it enters the pore canals and percolates to the outer limit of the presumptive exocuticle. Here, the substrate comes into contact with the enzymes (polyphenoloxidases?) situated in or just below the epicuticle and is oxidized to form the corresponding quinone. This now reactive molecule can then combine with sulfhydryl, amino, or phenolic groups or other quinone molecules to form quinone polymers. As the groups immediately around the pore canals become occupied, the quinones diffuse further inward so that interprotein cross-linking (matrix stabilization) is effected from the outer procuticle inwardly toward the epidermis (an interpretation consistent with histological observations and chemical and mechanical measurements). So the cuticle becomes hardened and darkened, *but* darkening is not causally related to hardening.

The term β-sclerotization (*Schistocerca, Apis, Tenebrio, Hyalophora*) was proposed after it was demonstrated that a number of *o*-diphenols can be released from hardened cuticles and that the β-carbon atom rather than the ring of N-acetyldopamine is involved in the reaction; hence, "β-sclerotization." There is additional evidence that free amino and phenolic groups of the matrix proteins are involved in the incorporation of N-acetyldopamine and that noncovalently bound by-products may also be trapped within the matrix. That N-acetyldopamine derivatives act as cross-links between proteins is suggested by the fact that ketocatechols can be recovered from sclerotized but not from teneral cuticle. Using double-labeling techniques (tritium at either the β-position or in the ring position), lightly colored and sclerotized cuticle showed β-position-directed activity while darker cuticles showed activity to both ring and β-position, indicating that both quinone tanning and β-sclerotization can simultaneously occur in a sclerite. It is likely that the same enzyme is involved in both pathways.

4.6.2. Minor Pathways

In the oothecae of *Blatta* and *Periplaneta*, the left colleterial gland secretes protein, phenoloxidase, and the β-glucoside of 3,4-dihydroxybenzoic acid and the right gland β-glucosidase. When the secretions mix, the β-glucosidase cleaves the glucoside releasing 3,4-dihydroxybenzoic acid which is then oxidized by phenoloxidase to form the *o*-quinone. The latter reacts with the free amino groups of the oothecal proteins and so cross-links them. (See Chapter 11, Section 11.4.2.)

While the enzyme-dependent quinone tanning and β-sclerotization modes of matrix stabilization are reasonably well established pathways, we should still consider the nonenzymatic *p*-quinone hypothesis, which is based on the following observations: When *Calliphora* cuticle is incubated with tyrosine or phenylalanine, a *p*-hydroxyphenol and *o*-dihydroxyphenols are obtained by nonenzymatic hydroxylation. The hydroquinone reacts nonenzymatically with an o-quinone to form the putative precursor, p-benzoquinone, and it is from these molecules that cross-links are forged. However, it has been reported that labeled hydroquinone is not incorporated into *Calliphora* cuticle. This and other evidence cast doubts on the reality of the p-quinone hypothesis. Contrarily, there are cuticular enzymes active toward p-dihydroxyphenols (*Drosophila*), and when the epicuticle is removed (*Sarcophaga*), there is a darkening of the underlying cuticle which proceeds even after the preparation has been heated to 100°C. Thus, p-quinone tanning (enzymatically or nonenzymatically) should be regarded as metabolically feasible and worthy of investigation with modern analytical techniques.

β-Alanine cross-linking has been implicated in the hardening of fly puparia (*Calliphora, Drosophila, Sarcophaga*). The dipeptide β-alanyltyrosine (= sarcophagine) present in the hemolymph disappears just prior to pupariation. Its disappearance is correlated with the incorporation of β-alanine into the puparium and seems to require cAMP. Three possible cross-linking possibilities have been mooted: β-Alanyl residues may (1) form new amino terminals on cuticular proteins, (2) attach to chitin residues, or (3) incorporate schemes in pupariation: tyrosine-o-phosphate (*Drosophila*) and γ-glutamylphenylalanine (*Musca*). In both cases the evidence is indirect and similar to observations on the disappearance of the compounds from the hemolymph and their reappearance in the cuticle. The extent of covalent cross-linking is small relative to noncovalent cross-links in the hardened puparium of *Calliphora*, and this is likely true for puparia in general.

Though not tanning in the strict sense (protein–cross-link–protein), cross-linking between chitin and protein, peptidochitodextrins, has been suggested (*Agrianome, Lucilia, Sarcophaga*) on the basis of cuticular stability toward chitinases and on the resistance of aspartyl and histidyl residues to alkaline hydrolysis. The existence of molecular cross-links between chitin and protein remains an unproven possibility. Similarly, it has been postulated that sterols conjugated to protein are incorporated into cuticle and that the presence of aromatic bonding derived from the oxidation of tyrosine provides a basis for cross-linking (*Rhodnius, Samia, Periplaneta*). The presence of procuticular lipids is beyond doubt; the roles they may play remain uncertain.

Tanning in silks has been investigated and 3-hydroxyanthranilic acid (a tryptophan metabolite) and gentisic acid have been identified as the cross-link precursors. Both are secreted as glucosides and the phenols are released by a glucosidase. The phenols are then oxidized to form cross-links in the sericin fraction of silk (*Hyalophora, Samia, Bombyx, Antheraea*). In addition, di- and tertyrosine have been recovered from silk (*Antheraea*) and the vitelline membrane of eggs (*Sympetrum*), and they probably serve as cross-links in these proteins.

Of the tanning schemes discussed, quinone tanning and β-sclerotization are most firmly rooted experimentally. Both may occur in the same sclerite and are likely directed by the same enzyme(s). Indeed, the exocuticle of

Pachynoda is quinone tanned as well as being β-sclerotized. It has also been shown that it is the amount of substrate rather than distribution of enzyme which is rate limiting in tanning (*Schistocerca*). There is good evidence that this is under epidermal control.

The epidermis is, of course, affected by juvenile hormone concentration and in conjunction with ecdysone is controlled with respect to chitin deposition, cross-link formation, the elaboration of surface features, and the differentiation of cuticular types. Ecdysone greatly influences the epidermis from apolysis to sclerotization, including the induction of dihydroxyphenol metabolism for tanning procuticle and all of the events associated with epicuticular tanning as well as for the calcification of puparia (see Chapter 13, Section 13.5.3).

The hormone which actually controls hardening following ecdysis is a separate hormone, bursicon, that is probably ubiquitous in insects (*Periplaneta, Calliphora, Pieris, Tenebrio,* and others). Bursicon also mediates melanization (*Sarcophaga*) and possibly affects endocuticular deposition (*Sarcophaga, Locusta*) (see chapter 13, Section 13.5.5).

4.7. EPIDERMIS

4.7.1. General

The epidermis usually consists of a single layer of cells (*Oncopeltus* is an exception) subtending the cuticle; covering apodemes, fore- and hindguts; and lining the inner surfaces of wings, gills, and styli. Though a specialized tissue, the epidermis performs many functions in common with other kinds of cells: the elaboration of organelles, the uptake of chemicals for synthesis and secretion, general self-maintenance and metabolism, as well as the production of a cuticle.

Epidermal cells have other interesting features. They regulate the number of cells in their layer through controlled mitoses, possibly assisted by pre-programmed cell death. They recognize their position and orientation within and between segments, possibly through the electrochemically polarized transfer of information through junctional membrane contacts anisotropically distributed in the planar direction and piezoelectrically normal to the cuticle surface. In addition, the epidermis engages in tracheole-fetching behavior under conditions of low oxygen tension. Epidermal cells are capable of amoeboid movement in the repair of wounds, of changing shape so as to control lens aperture in vision, and of producing endless nuances of structure for sensory perception. They also control the movement of water and ions and the production of waxes and other exocrine secretions such as pheromones and defensive compounds.

The great versatility of the epidermis is, of course, modulated by numerous intrinsic and extrinsic factors. Among the former, membrane permeability, nutritional status, quantity and kinds of ions, and cell health speak for themselves. In the case of extrinsic factors, epidermal cells are affected by humoral factors (hormones), the relative density of cells, death and disease of neighbors, cell–cell interactions (polarized signal passing), wounding, and parasitism. There are also the effects of mechanical loading during postecdysial expansion, engorgement on feeding, and locomotion, not to mention the effects of temperature, water, and humidity.

4.7.2. Cellular Conversations

4.7.2.1. Ionic Coupling. For many years biologists have been trying to unravel precisely how functional coordination is achieved in the epidermis. The problem is the more complex because these cells live in small neighborhoods (recognizable by organelle density and staining intensity) and as such make separate "decisions." For example, given excesses of enzyme and substrate, the mandibular epithelium of *Schistocerca* secretes a very thick and heavily tanned cuticle, while those cells underlying an abdominal tergite secrete a thinner product, considerably less tanned and stiff than that of the mandibles. Similarly, some cells in the gills of mayflies specialize in chloride movement while others do not. While we still cannot explain how these differences in behavior arise, recent advances in electrophysiology have shed much light on the problem.

Each epidermal cell is surrounded by a plasma membrane, and each cell is attached to its neighbors by a specialized junctional membrane region situated between the apical plasma membrane and a much larger noncontacting region of the plasma membrane (Fig. 4-19). The junctional contact region contains both septate and gap junctions, and it is through the latter organelle that neighboring cells make contact. Recent measurements indicate that groups of epidermal cells behave as functional syncytia in that cells are ionically coupled when in membrane contact. Because of low differences in electrical resistance between contact regions of the membranes, there is an anisotropic movement of molecules from cell to cell through cytoplasmic channels in the junctional plasma membrane, but there is little leaking into

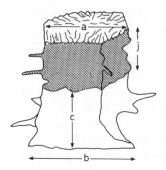

FIGURE 4-19. External morphology of an epidermal cell. The stippled area (j) is junctional membrane. a, apical plasma membrane; c, lateral noncontact membrane; b, basal plasma membrane (a, b, and c are nonjunctional areas). (After Caveney and Podgorski.)

extracellular spaces because of the high resistance between noncontact regions of adjacent membranes.

The magnitude of the electrotonic potential is related to the distance between stimulating and recording electrodes, hence to distance between cells. The *resistance* of the intercellular pathway is very low at all stages of development, and the movement of ions over a distance of about 25 cells has been recorded. Ionic conduction does, however, fluctuate during development, especially at the pupal–adult metamorphosis (Fig. 4-20). Both ecdy-

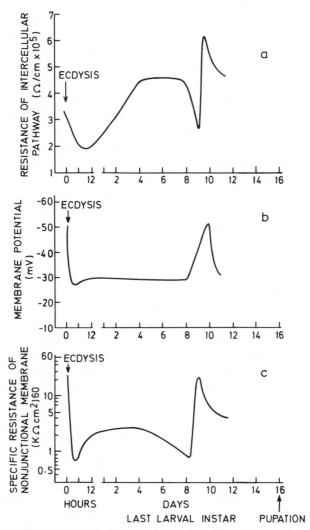

FIGURE 4-20. Changes in the ionic properties of the epidermis of the final larval instar of *Tenebrio molitor*. (a) Resistance of the intercellular pathway; (b) membrane potentials; (c) specific resistance of nonjunctional membrane.

sone and cAMP increase contact membrane resistances, and juvenile hormone lowers the resistance of nonjunctional membrane regions. While hormones clearly affect the resistivity of cells, not all cells are equally competent to respond. The reasons for this are unknown. With respect to K ionic concentrations, the electrical resistance of nonjunctional membrane regions is highly susceptible, whereas that of junctional regions is highly stable. The permeability of junctional membranes to ionic flow can be reduced by increasing calcium concentrations in the epidermal cytoplasm.

Because epidermal cells form functional syncytia, an interesting problem arises: If one member in a group of connected cells dies, there is the possibility of all of the others literally leaking to death. This problem is apparently overcome by the shutting down of junctional membrane connections, the mechanism for which is at least partially mediated by cytosolic calcium derived from mitochondrial and endoplasmic reticular stores.

The diffusion of charged molecules through junctional membranes implies the transfer of electrical signals, and this is one way in which cells communicate. The content of these conversations awaits translation. However, the significance of ionic coupling is pregnant with suggestion: It is related to the regulation of growth and the formation of spatial patterns, and it may well be that the signals passing through these junctional channels are morphogenetic in nature. Similarly, the polarity of cuticular patterns, including the arrangement of microfibrils in the procuticle, is probably ordered by a segmental gradient in which individual cells behave in a manner that suggests that they are "aware" of their position within the epidermis.

In addition to the planar transfer of electrical and chemical signals, cuticle exhibits piezoelectricity. This means that mechanical deformation of the cuticle generates potential differences across the integument so that electrical signals also impinge on the epidermis perpendicular to the ionically coupled junctional membrane flow route. Moreover, detachment of the epidermis from cuticle reduces conductance properties of the cuticle. At present it is difficult to assess the relative significance and possible interactions of these two sources of electrical information. One can envisage piezoelectric transduction as one element in a complex feedback system as, for example, in differential tanning. Resolution of this three-dimensional problem lies in the future.

4.7.2.2. Segmental Gradients. The transfer of information between cells has also been investigated with transplant techniques. Such studies are greatly facilitated by the fact that transplanted cuticle—allo-, homo-, and heterografts—is only very slightly associated with interspecific incompatibility and rejection problems. On the other hand, it is immunologically curious that there is a "positional incompatibility" problem so that grafts are comfortably accepted *laterally* but not *axially*. Ultimately, this finding provides a powerful tool for understanding cellular communication.

Transplants of pieces of larval sternites (*Rhodnius*) put in exactly the same place in the recipient as they occurred in the donor result in complete acceptability and normality of cuticular pattern. But if the donor cuticle is rotated through 180° in the recipient, there are discontinuities in the surface pattern. These observations imply that continuity of surface pattern depends on continuity of epidermal contacts. Similarly, intersegmental transplants are associated with discontinuities, which in turn implies an axial gradient in the control of growth and surface pattern regulation. As a consequence, such experiments imply that cells receive information as to their whereabouts and the general topography of the growth pattern. On the other hand, if pattern recognition depended to any extent on humoral factors, one would not expect such results. The transfer of information between cells is polarized and consistent with the anisotropic distribution of junctional membrane complexes. The possible contributions of piezoelectric information in transplantation experiments has not yet been investigated.

The extensive experiments of various investigators essentially tell us that the epidermal cells are arranged in a segmentally repeating gradient of some kind. Cells are alike side-to-side but not axially. The characteristic that varies quantitatively is the recognition of position in a linear axial order. When pieces of cuticle with their attached epidermis are displaced from the normal axial order, the host and graft patterns interact in a way that is predictable on the basis of their relative positions. This again implies an anisotropic distribution of junctional complexes and again confirms that epidermal cells have neighborhood characteristics and each neighborhood its own electrical grid system.

4.7.2.3. Tracheal Growth. There are some unique aspects of tracheal growth. It is known that epithelial continuity is necessary for the quantitative control of growth of tracheae; but the epidermis appears to behave in a polarized manner with respect to tracheal development and metamorphosis. In an exopterygote insect like *Rhodnius,* tracheal diameter increases at each molt. The increase in the cross-sectional diameter of the main tracheae is equal to the rate of increase in diameter of the final terminal tracheae. Diameter control is usually attributed to the epithelium that secretes the tracheae.

The question here is a teleological one: How do the cells "decide" on how big tracheal diameter ought to be? Perhaps tracheal diameter is simply determined by the number of cells in the neighborhood so that diametrical growth is closely analogous to that of general exoskeletal growth. Significantly, the growth of tracheae begins with the secretion of cuticle at the spiracles and proceeds inward to the terminal tracheoles. Therefore, if tracheal epidermal cells are essentially like other epidermal cells, then it would not prove surprising if the control of tracheal growth lies in the transport of electrical signals from cell to cell by means of the junctional complex route.

4.7.3. Growth and Size

During metamorphosis cuticle changes in form and size are related to major changes in the epidermis. Changes in the latter include the recruitment and proliferative growth of anlagen (primordia), major variations in rates and kinds of synthetic activities (differential rates of cuticle deposition and tanning as in *Schistocerca* and *Apis*), and, of course, cell aging and death. All of these possible changes are further affected by the hormonal milieu; for example, juvenile hormone both directs the kinds of syntheses of certain cells and affects the rate of cell death (low titers, more death; high titers, less death). Moreover, there is an inverse relationship between molting hormone levels and cAMP levels in the epidermis and a positive correlation between bursicon activity in the hemolymph and epidermal cAMP, implying that cAMP could be a mediator of bursicon production and release* (see Chapter 13, Section 13.5.1).

The phenomenon of cell death is more than just of thanatological interest because of the relationship between the numbers and sizes of cells and the extent of cuticle synthesis. Since the ultimate size of a new instar depends on the population of cells available for and engaged in cuticle secretion, the ratio of functional cells to dying and dead cells ought to have a bearing on new instar size, if not form. It might be instructive to plot the standing crop of dead cells against size increases in cuticle. Perhaps this would provide a solution to the variation in Brooks' rule between taxa.

Although the size of an insect is related to feeding, how feeding relates to epidermal cell growth is somewhat obscure because growth can be measured as post hoc postecdysial expansion or it may be intrastadial in nature. There is the additional problem of epidermal mitosis: mitoses are initially associated with a relative reduction in the amounts of nuclear material, that is, the cells actually increase in size before completion of division. For the presumptive solid cuticle of sclerites, increases in size are largely restricted to an unfolding of a slightly pleated new cuticle; but in the soft cuticle of endopterygote larvae, where there is an intrastadial increase in size, this may be due to creep (*Bombyx*), or to intussusceptive growth (*Calpodes, Galleria*), or to a combination of both modes. The events associated with growth and size are summarized in Fig. 4-21.

Recently some of the intracellular changes that occur during growth have been documented, mainly for *Calpodes*. Cuticle deposition and composition are controlled by dynamic changes in the secretion of synthesized products as well as by the uptake of various molecules. Synthesis is cyclical in nature, and this is related to the turnover of the apical plasma membranes. The turnover of these membrane plaques is apparently related to molt–intermolt

---

* Alternatively, it could also be that molting hormone regulates the reduction in cAMP titers whereas bursicon regulates increases in cAMP, so that it is not cAMP which is a mediator of bursicon production but rather that bursicon is a mediator of cAMP production. The distinction remains to be resolved.

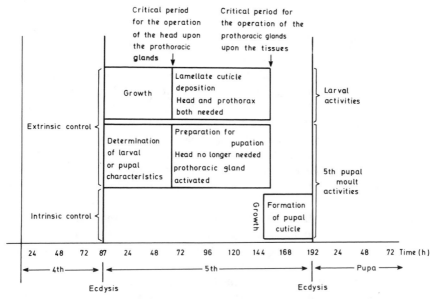

FIGURE 4-21. Activities of the epidermis of *Calpodes ethlius* during the molt–intermolt cycle. (After Locke.)

cycles and not to the kind of cuticle being secreted. Membrane plaques are lost to the membrane when ecdysial droplets are secreted. The cyclical formation and involution of membrane plaques is synchronous with the hormonal control of molting. The dissolution of plaques is effected by lytic enzymes elaborated in the Golgi complex and carried in lysosomes to multivesicular bodies. The latter digest fragments of the plasma membrane. The Golgi complex is also the source of apolysial droplets which digest portions of the old cuticle after apolysis. Thus, the apical plasma membrane is in a dynamic state of cyclical formation through the addition of new plaques and loss of them through endocytosis.

The cyclical behavior of plaque formation and dissolution are aspects of epidermal function affected by hormones. Indeed, both epidermal cell morphology and cuticle production are influenced by both ecdysteroids and juvenile hormones. In *Manduca* apolysis is associated with high ecdysteroid titers and ecdysis with high juvenile hormone titers. These observations can be extended to the molecular synthetic level in that ecdysone also seems to induce chromosomal puffing in certain regions of the epidermis and to suppress puffing in other regions (*Sarcophaga*). Similarly, polytene chromosomes have been found in the pupal epidermis of *Sarcophaga* specifically related to the synthesis of the entire dorsum of the pulvillus. These observations are quite encouraging as they hold the possibility of locating the precise region of the genome responsible for the synthesis of individual constituents of the cuticle.

4.7.4. Special Functions

4.7.4.1. Water Uptake.   In view of surface-to-volume considerations, insects clearly need control mechanisms for the conservation of water. Despite water-conserving mechanisms like spiracular closure and the resorption of water by the Malpighian tubules, an impermeable cuticle would also appear to be required. Impermeability is a property of the cuticle, while the active uptake of water is a property of the epidermis.

The active uptake of water by terrestrial insects has been demonstrated using deuterium oxide or tritiated water. Water vapor may be taken up through the anus (*Thermobia, Tenebrio*) or generally through the cuticle (*Periplaneta*). Similarly, active uptake of water has been demonstrated for aquatic insects (*Sialis, Corixa, Notonecta*). Indeed, some insects can actively take up water from air of lower saturation than the hemolymph.

Evidence for the epidermal control of water uptake comes from very different kinds of observations. If, for example, epicuticular lipids are removed or damaged, water uptake ceases until wound repairs have been made. Moreover, the epidermis must be in contact with the cuticle since uptake processes stop during apolysis. Anesthesia abolishes the ability of the epidermis to take up water. Finally, it can be noted that water may be taken from the environment in either gaseous or liquid phases and that the epidermal cells of different regions of the body may be specialized in the uptake of one or the other phases.

4.7.4.2. Daily Growth Layers.   Daily growth layers in cuticle appear as pairs of bands and are analogous to the annual growth rings of trees. Each pair consists of a nonlamellate band secreted during the day and a lamellate band secreted at night. These layers were proven to be secreted daily simply by selectively removing locust hind legs at known intervals following ecdysis (*Schistocerca, Locusta*) and correlating the number of band pairs with time. Layers have been recorded from the larvae and adults of at least nine orders of exopterygote and endopterygote insects in structures as diverse as ocelli, apodemes, and tibiae. Daily growth layers are, however, not always exactly correlated with age and may vary depending on the availability of food (*Belostoma, Rhodnius*). Such growth layers may also be temperature and light dependent, the effects of which vary among different populations of epidermal cells. Given constant low temperature and fixed illumination conditions (light or dark), some cells produce only lamellate cuticle. The same conditions can result in the uncoupling of the circadian clock and the production of only nonlamellate cuticle by regions of the epidermis. However, other regions of the epidermis apparently cannot be uncoupled from their circadian clocks. Thus, some epidermal cells are apparently obligatorily and others only facultatively coupled to circadian clocks, leading to the production of both the lamellate and nonlamellate cuticle. Recent experimental evidence suggests that the epidermis itself is capable of light perception, but this does not exclude the possibility of the transfer of light information by neural and/or endocrine routes.

4.7.4.3. Pore Canals and Tonofibrillae.   Pore canals are extensions of the epidermis and occur in most insects. They are typically twisted, flattened ribbons, helical in nature, and in the epicuticle may remain as single fine canals or be arborescent. Aside from functioning as conduits for the passage of compounds utilized for tanning and waterproofing, they may also function in assisting in the fusion of several small pore canals into larger ones because of the presence of inclusive cytoplasmic (contractile?) filaments.

Muscle attachments are made to cuticle in the following way: microtubules originating in the muscle pass through the epidermal cells and attach to sockets of the subcuticular plasma membrane of the epidermal cells from which tonofibrillae arise and extend to epicuticle. They appear to continue to grow in length as the procuticle becomes thicker between molts.

4.8. CONCLUSIONS

The phenomenal success of insects in occupying almost every conceivable nook and cranny on our earth reflects the great adaptability of the exoskeleton which in turn derives from the monumental versatility of the epidermis. Hopefully, the previous pages provide adequate testimony that we have learned much as to how these integuments are constructed and the ways in which they work; we know something of their versatility and a bit about their limitations. Despite the existence of about 20,000 articles dealing with various aspects of the integument, one has the distinct feeling that we are at the threshold of a future pregnant with exciting possibilities.

Ultimately, we shall want to solve the presently intractable physical problems of chitin microfibrillar arrangements in cuticle. In addition, we should determine the sequences of both epicuticular and procuticular proteins, but more importantly, to establish their conformational states. And certainly we will want to characterize the precise compositions and functions of epicuticle and subcuticle. Additionally, for those of Darwinian bent it will be of great interest to further investigate the electrical and thermal conductance properties of the integument in the context of niche adaptation and diversity. We shall also want to know how it is that different neighborhoods of epidermal cells behave differently in an ostensibly similar hemolymphal milieu. Perhaps most interesting of all, we shall want to find the Rosetta stone that will allow us to translate the conversations, chitchat, and neighborhood gossip of epidermal cells.

Although nothing has been said of insecticides vis-à-vis the cuticle, this omission certainly does not reflect an unawareness of the great importance of this topic. Indeed, it is common knowledge that the vast majority of the world's entomologists are principally concerned with killing insect pests with topical insecticides. In exploring the elegance and mystery of the insect integument in this chapter, a major effort has been made to present a contemporary picture of it as a remarkable evolutionary development that has permitted insects to flourish in all corners of the globe. As our comprehension of the insect integument has increased, so have the practical benefits.

The relevance of an understanding of the basic physics, chemistry, and biology of the integument to pest control can be readily seen in the recently developed chitin synthesis inhibiting polyoxins and substituted urea compounds, powerful and effective fungicides and insecticides, respectively, whose development is a direct outgrowth of basic science. Armed with more knowledge, tomorrow's chemists will undoubtedly be able to develop more sophisticated and specialized insecticides that will control those insects we do not want without devastating the whole of that very thin crust of earth which sustains life.

ACKNOWLEDGMENTS

There is more involved in the preparation of a textbook than the drudgery of library journal research and the time necessary to sift and sort published data. One attempts to both interpret data and to formulate a general synthesis of a vast literature. In my view this last aspect is most important, and the student should know that much of my approach has been influenced by discussions through the years with many scientists, particularly Professors S. O. Andersen, R. Dennell, A. G. Richards, and Dr. K. M. Rudall. This is not to say that they endorse what I have written in the preceding pages. It is also necessary to note that Professor S. O. Andersen has kindly reviewed this chapter in manuscript form and saved me once again from publishing more than one howler. He is, of course, not responsible for any errors that might still remain.

REFERENCES

R. M. Alexander, *Animal Mechanics,* Sidwick and Jackson, London, 1968.

J. W. L. Beament, J. E. Treherne, and V. B. Wigglesworth, *Advances in Insect Physiology, and Onwards.* Academic Press, London, 1963.

E. B. Edney, *Water Balance in Land Arthropods,* Springer, Berlin, 1977.

M. Florkin and E. H. Stotz, *Extracellular and Supporting Structures,* Elsevier, Amsterdam, 1968.

R. D. B. Fraser and T. B. MacRae, *Conformation in Fibrous Proteins,* Academic Press, New York, 1973.

H. R. Hepburn, *The Insect Integument,* Elsevier, Amsterdam, 1976.

H. E. Hinton, *Biology of Insect Eggs,* Pergamon, Oxford, 1980.

J. S. Huxley, *Problems of Relative Growth,* Dial Press, New York, 1932.

C. Jeuniaux, *Chitine et Chitinolyse,* Masson, Paris, 1963.

G. A. Kerkut and L. I. Gilbert, *Comprehensive Insect Physiology, Biochemistry and Pharma-cology,* Pergamon, Oxford, 1985.

J. T. Martin and B. E. Juniper, *The Cuticles of Plants,* Edward Arnold, London, 1970.

T. A. Miller, *Cuticle Techniques in Arthropods,* Springer, New York, 1980.

R. R. A. Muzzarelli and E. R. Pariser, *Proceedings of the First International Conference on Chitin/Chitosan,* Massachusetts Institute of Technology, Cambridge, 1978.

A. C. Neville, *Biology of Arthropod Cuticle,* Springer, Berlin, 1975.

L. Picken, *The Organization of Cells and Other Organisms,* Clarendon Press, Oxford, 1960.

H. Przibram, *Connecting Laws in Animal Morphology,* University of London Press, London, 1931.

A. G. Richards, *The Integument of Arthropods,* University of Minnesota Press, Minneapolis, 1951.

D. W. Thompson, *On Growth and Form,* The University Press, Cambridge, 1963.

S. A. Wainwright, W. D. Biggs, J. D. Currey, and J. M. Gosline, *Mechanical Design in Organisms,* Edward Arnold, London, 1976.

A. G. Walton and J. Blackwell, *Biopolymers,* Academic Press, New York, 1973.

V. B. Wigglesworth, *The Control of Growth and Form,* Cornell University Press, Ithaca, N.Y. 1959.

5

RESPIRATORY SYSTEMS

JAMES L. NATION
Department of Entomology and Nematology
University of Florida
Gainesville, Florida

CONTENTS

SUMMARY

Gas exchange is facilitated in insects by a vast network of air-filled tubes, the tracheae, that penetrate the body and tissues. The tracheae open to the outside of the body through segmentally arranged spiracles. The primitive number of spiracles is 11, but most insects have fewer. Some, such as certain aquatic species, have no functional spiracles. The tracheal supply to wing muscles is especially adapted to meet the very high requirements for O_2 by working wing muscles.

The tracheae are lined by a cuticular intima that is continuous with the cuticle of the exoskeleton, and in larger tracheae the intima is molted. Thickened spirals of the intima, the taenidia, strengthen the tracheae and help resist collapse. The tracheae branch, repeatedly, and the smallest branches, called tracheoles, have a diameter of 0.2–1 μm and make intimate contact with cells. New tracheoles develop within outgrowths of tracheal epithelial cells.

Diffusion alone is adequate for exchange of gases in very small insects, but larger insects ventilate the system by abdominal movements and wing movements when flying. The hole fraction concept introduced by Weis-Fogh relates the cross-sectional area of the tracheae in a plane through a tissue to the cross-sectional area of nontracheal tissue. The high efficiency of the tracheal system is related to large hole fraction values that vary from 10^{-1} to 10^{-3}.

Some insects, particularly pupae, have evolved a pattern of discontinuous CO_2 release, or CO_2 bursts to keep the spiracles closed much of the time and conserve water that would otherwise be lost by evaporation.

Although most aquatic insects have an open tracheal system and breathe air in the same manner as terrestrial insects, some have a closed tracheal system and regulate gas exchange cutaneously. Many dependent on cutaneous exchange have flaplike cuticular gills that are richly tracheated. Most aquatic insects have a hydrofuge region, often surrounding a spiracle, that enables them to break the water surface more easily. Some aquatic insects submerge with a bubble of air that must be periodically renewed. A few aquatic insects have an incompressible gas space, called a plastron, that is guarded by closely packed nonwettable hairs. The plastron can take O_2 from well-aerated water continuously, and insects thus equipped can stay submerged indefinitely. Some insect eggs also have a plastron, usually composed of micro-air spaces in the chorion, that serves as an O_2 source during temporary wetting. Relatively few species of insects produce the respiratory pigment hemoglobin, and its role in most of them is uncertain or poorly defined.

5.1. INTRODUCTION

Respiration has been widely used to refer to two separate processes. It may be used, as in this chapter, to refer to the mechanisms and organ systems involved in transport of gases between the environment and tissues. It has also been used to describe the metabolic processes for energy production occurring at the cellular level.

Insects and a few other arthropods have evolved a respiratory system that directly delivers oxygen to the cellular site and removes carbon dioxide through a series of air-filled tubes, the tracheae. This system is remarkably adaptive for insects. Indeed, it is so effective that it can deliver all the oxygen needed by flight muscles to sustain long periods of activity at the highest rate of aerobic metabolism/unit volume of tissue known. For example, a blowfly adult, *Lucilia sericata,* uses 33–50 μl O_2/min/g body weight at rest, but it can increase this to 1625 μl/min/g almost instantly upon initiation of flight. This rate of oxygen use is some 30–50 times the maximum use of leg or heart muscle/unit volume in man. The tracheal system is capable of delivering O_2 at such a high rate for long periods of flight. Even in the fastest flying insects it generally has been possible to demonstrate no, or only a slight, oxygen debt.

This vast network of air-filled tubes presents an astonishing surface area over which water may be lost by evaporation. Unsurprisingly, many examples of adaptations for water conservation have been discovered in insects. Of course, not all the water-conserving mechanisms occur within the tracheal system, but in it we can find intricate examples of structural, metabolic, and behavioral adaptations for reducing water loss. The purpose of this chapter is to explore the functional morphology and physiology of this marvelous respiratory system.

5.2. BASIC STRUCTURE OF THE RESPIRATORY SYSTEM

5.2.1. General Morphology and Body Plan

Although there are many variations in the arrangement of the tracheal system, basically it consists of paired, segmentally repeated, external openings—the spiracles—connected to a meshwork of ramifying tubes—the tra-

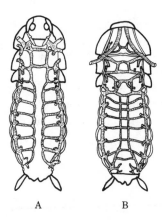

FIGURE 5-1. Tracheal system of a cockroach, *Periplaneta*. A. Dorsal tracheae. B. Ventral tracheae. (From O. W. Richards and R. G. Davies, "The Respiratory System," in *Imm's General Textbook of Entomology*, Vol. 1, *Structure and Development*, 10th ed., Chapman and Hall, London, 1977.)

A   B

cheae and the tracheoles. While in a few less well-developed systems the tracheae from neighboring segments remain independent and unconnected, the most common feature of more highly developed systems is the anastomosis of tracheae from each segment to form continuous laterodorsal and lateroventral longitudinal trunks (Fig. 5-1) (Richards and Davies, 1977). Transverse connectives unite the two sides. The heart and dorsal musculature usually are supplied by branches in each segment from the dorsal longitudinal trunks. The viscera and internal reproductive organs receive tracheal branches from the lateral trunks. The ventral nerve cord and ventral musculature are supplied by branches from the ventral trunks or ventral transverse connectives. The legs and wings are supplied by tracheae from the thoracic spiracles, while the head and associated structures are generally supplied with branches from the first spiracle and the dorsal longitudinal trunk.

5.2.2. Adaptations in Very Active Muscles

The wing muscles of insects make great demands on the tracheal system for delivery of oxygen, so it is not surprising to find special adaptations in the tracheal supply to muscles. Weis-Fogh (1964a) claims that all large flying insects have basically the same type of supply, which he divides into primary tracheae, secondary tracheae, tertiary tracheae, and tracheoles. The primary tracheae originate at a thoracic spiracle and pass through the core of a muscle (centroradial) or run along its surface (lateroradial) (Fig. 5-2). In both cases the tracheae may balloon into an air sac beyond the muscle. A third arrangement (laterolinear), which is more common in small insects, is characterized by the presence of an air sac that lies on the surface of the muscle and provides the primary oxygen supply. Some insects, such as *Schistocerca gregaria*, exhibit each of the three arrangements in some wing muscles.

Secondary tracheae branch from the primary tracheae and radiate into the spaces between muscle fibers. The secondary tracheae give rise, in turn, to

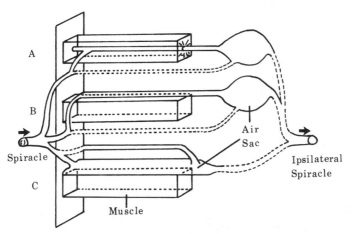

FIGURE 5-2. Three basic types of primary tracheal supply to wing muscles. A. Centroradial. B. Lateroradial. C. Laterolinear. The arrows indicate the direction of air flow. A muscle shunt, typical of many muscles, is shown by a broken line. [From T. Weis-Fogh, *J. Exp. Biol.* **41,** 207 (1964a).]

the tertiary tracheae. In the tergosternal muscle of dragonflies, *Aeshna* spp., tertiary tracheae branch from the secondary tracheae at about 20-μm intervals for a total of about 24 tertiaries. The tertiary tracheae eventually narrow and branch into tracheoles; in *Aeshna* each produces about 20–30 tracheoles.

A further specialized adaptation occurring in many active tissues, and frequently in wing muscles, is the growth of a tracheoblast against the limiting membrane of another tissue cell, so that the tracheoblast indents the cell, much like a finger pushed into a balloon. The tracheole that develops in the indented portion of the tracheoblast (Fig. 5-3) can thus deliver air within a few tenths of a micrometer to the actual site of use, the mitochondria (Smith, 1964). These tracheoles have been called "intracellular" tracheoles, but in reality they are not inside the cell membrane of the cell being served. Such a tracheole is still surrounded by its tracheoblast membrane, and the whole is separated from direct contact with the recipient cell cytoplasm by the membrane of the cell. The regulation of growth in such interesting tracheoblast–cell interactions awaits detailed study.

5.2.3. Tracheae and Tracheole Structure

Tracheae and tracheoles are lined with a cuticulin intima that is similar to, and continuous with, the cuticulin layer of the integument. Ectodermal tissue in the embryo gives rise to the tracheal system as it does to the integument. Sometimes, but not universally, an endocuticular layer containing chitin lies beneath the cuticulin layer in larger tracheal trunks. A hydrophobic substance, probably lipoidal in nature, is secreted on the lumen surface

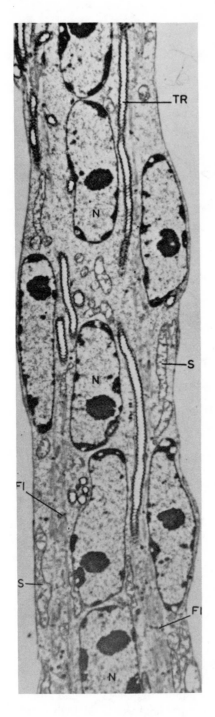

FIGURE 5-3. Indentation of flight muscle fiber of an adult carabid, *Nebria brevicollis,* a fully winged but flightless beetle. The fiber is about 5 μm in diameter. N, muscle; TR, tracheoles inside the muscle fiber; S, filamentous sarcosomes with few cristae; FI, extremely narrow fibrils in the richly particulate sarcoplasmic matrix. ×7000. [From D. S. Smith, *J. MOrph.* **114,** 107 (1964).]

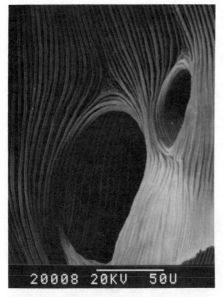

FIGURE 5-4. A. Scanning electron micrograph of the interior of a trachea from a mole cricket, *Scapteriscus acletus,* showing the origin of two smaller tracheae and taenidial windings. B. Enlargement of the photograph in A to show taenidia.

of tracheae and reduces water loss by evaporation. Thickened, tight spirals of the intima, the taenidia, strengthen tracheae and tracheoles, provide elasticity, and help the tubes resist compression and collapse (Figs. 5-4 and 5-5). In larger tracheae the taenidia are up to 450 nm in width and are spaced about 300 nm apart, but in tracheoles the taenidia are smaller (50–80 nm width) and are spaced further apart than their width. Within the taenidial folds there is a component, probably similar to procuticle (Fig. 5-5), that adds strength to the taenidia (Bordereau, 1975). The cuticular intima is as thick as 200 nm at the taenidial spirals and as thin as 10–40 nm between spirals. The micelles of the cuticulin layer are oriented so that their long axis is parallel to the long axis of a trachea or tracheole in intertaenidial areas, but perpendicular to the long axis in the taenidial thickenings. This orientation strengthens the tubes.

When tracheae first form, they have smooth walls. The taenidia soon appear, however, and their formation has been the subject of several investigations. Locke (1958) proposed that taenidial formation is the result of expansion and buckling of the tracheal wall (Fig. 5-6). With a mathematical model, he accounted for the frequency of taenidia, tube-wall thickness, and orientation of the cuticulin micelles within the taenidia and intertaenidial regions. The buckling hypothesis is not universally accepted, however, and further investigations are warranted.

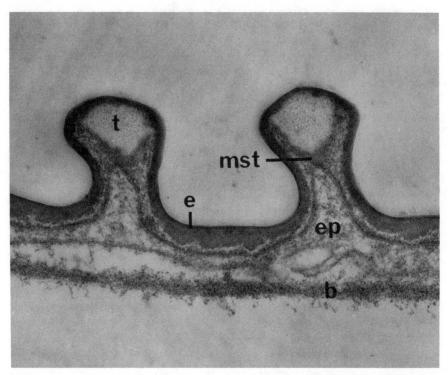

FIGURE 5-5. A. Electron micrograph of taenidia from a female termite, *Cubitermes fungifaber*. e, epicuticle intima; t, taenidia; mst, granular material under taenidia; ep, epithelium; b, basement membrane. Approx. ×80,000. (Reprinted with permission from C. Bordereau, *Int. J. Insect Morph. Embryol.* **4,** 431, Copyright 1975, Pergamon Press, Ltd.)

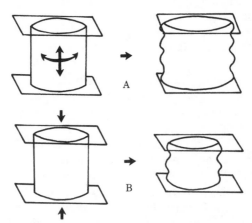

FIGURE 5-6. Buckling hypothesis of taenidial formation. Buckles in an expanding cylinder restrained in the long axis (A) are viewed similar to effect caused by compression in long axis (B). [From M. Locke, *Quart. J. Micr. Sci.* **99,** 29 (1958). With permission of Oxford University Press.]

The major distinction between tracheae and tracheoles is one of size. Tubes down to about 1 μm in size usually are called tracheae, while those smaller than 1 μm are tracheoles. The smallest tracheoles have a diameter less than 0.2 μm. Because tracheoles usually have an irregular cross-sectional shape, Pickard (1974) has recommended measuring the greatest and least dimensions and calculating the cross-sectional area of this ellipse as the effective tracheolar diameter. The effective tracheolar diameter continues to decrease until it approaches the mean free path of diffusing oxygen molecules. According to Pickard (1974), the mean free path (diameter) for air at 300°K and 1 atm is about 0.072 μm. Published electron micrographs show that the smallest tracheoles generally are about 2–3 times this diameter.

5.2.4. Development of New Tracheoles

A tracheole develops as an intracellular canal within a tracheoblast (also called a stellate cell, transition cell, and tracheal end cell). A tracheoblast develops when a tracheal epithelial cell, associated with a larger tracheal trunk, begins to grow out toward a region deficient in tracheoles (Keister, 1948) (Fig. 5-7). As it grows, the tracheoblast tends to become spindle shaped, and a single unbranched tracheole may form within it at this stage. More often, however, the cell becomes stellate before a tracheole forms; then a tracheole forms with branches in several or all of the extended processes of the tracheoblast. By maintaining at least a thin process in contact with the tracheal epithelium from which it originally grew, the tracheoblast provides a pathway for the tracheoles to grow toward the larger tracheal

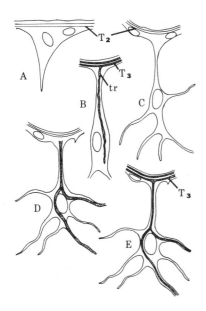

FIGURE 5-7. Stages in the development of new tracheoles in *Sciara coprophila*. A. Development of tracheoblast by enlargement of tracheal epithelium cell; ×1000. B. Fully formed, fluid-filled tracheole in a spindle-shaped tracheoblast that is little changed from that shown in A. ×1000. C. More common situation with tracheoblast in stellate form before formation of new tracheoles, ×650. D. Cell similar to C, but with partially formed, fluid-filled tracheoles not yet making connection with larger trachea, ×650. E. Similar to D, but with connection made to newly formed third-instar tracheal trunk, ×650. T_2, T_3 refer to second-instar or third-instar tracheal trunk, respectively. tr, tracheole formed prior to molt into third instar. [From M. L. Keister, *J. Morph.* **83,** 373 (1948).]

trunk. The tracheole eventually will make contact with the tracheal trunk and become air filled and functional at the next molt. The tracheoblast may continue to enlarge and develop more processes, and additional tracheoles may form in later molts.

5.2.5. Molting of Tracheae

Although tracheae are shed at each molt, tracheoles usually are not molted (Fig. 5-8). Prior to molt, a tube of pale, homogeneous-appearing cytoplasm forms around each old trachea. Slowly a darkening and thickening occurs at the boundaries until a definite tube filled with fluid has formed around the old trachea. At molting, the old, air-filled tracheae are pulled out of the newly formed, fluid-filled ones. Wigglesworth (1959) found that tracheoles, which are never shed in *Rhodnius,* become cemented to the new tracheae at each molt by a ring of adhesive material. Pneumatization, or the filling of the system with air, occurs and the new tracheae become functional. The new tracheae do not have to fill with air through the spiracle. An air bubble may appear at any point within a new trachea and quickly enlarge and spread to fill the system. Although details are scarce, it seems likely that the fluid in the new system is actively reabsorbed, thus making way for the air.

5.2.6. Migration of Tracheoles

Tracheoles in some insects move or can be pulled into tissue regions where oxygen tension is low. Wigglesworth (1959, 1977) described thin contractile strands from epidermal cells that attached to tracheoles in the epidermis of *Rhodnius* and pulled them to distances greater than 700 μm toward regions that had been experimentally detracheated (Fig. 5-9). The movement of tracheoles has also been observed in cultured *Galleria mellonella* wing discs and in organ transplants in Diptera. Tracheole migration is ecdysteroid dependent in *Galleria* and may be related to the presence of large numbers of

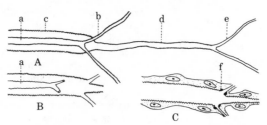

FIGURE 5-8. Molting of tracheae. A. Trachea and tracheoles just before a molt of *Rhodnius prolixus*. a, old air-filled trachea; b, old tracheoles; c, new fluid-filled trachea; d, newly formed terminal trachea; e, newly formed tracheoles. B. Old trachea with short segments of old tracheoles being pulled from the new trachea. C. The new trachea with covering of epithelial cells and attached old tracheoles. Rings of cement (f) secure old tracheoles to new trachea. [Modified from V. B. Wigglesworth. *J. Exp. Biol.* **36**, 632 (1959).]

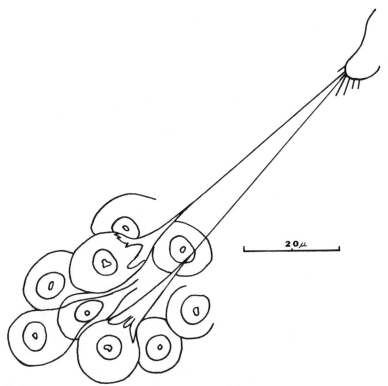

FIGURE 5-9.   Contractile strands from epidermal cells pulling tracheoles into oxygen deficient region in epidermis of *Rhodnius prolixus*. [Modified from V. B. Wigglesworth, *J. Exp. Biol.* **36,** 632 (1959).]

microtubules in cultured wing discs (Oberlander, 1980). Oberlander showed that microtubule formation was disrupted and tracheole migration in the discs was prevented by treatment with 10^{-8} M vinblastin or 10^{-5} M cholchicine. In *Galleria* wing discs it appears that the tracheoles uncoiled themselves.

5.2.7. Air Sacs

Dilations of both primary and secondary tracheae, commonly known as air sacs, occur in many insects and especially in flying insects. These sacs vary in size, but frequently are large in flying insects such as honeybees, some Diptera, cicadas, and some scarabaeid and buprestid beetles. The intima of air sacs may contain typical taenidia, but these may be reduced or irregular. An air sac frequently provides the primary air supply to flight muscles (Fig. 5-2), and this is especially common in small fast-flying insects such as *Drosophila* (Weis-Fogh, 1964a). Since these air sacs provide a large surface in contact with the muscle, diffusion exchange of gases is very effective. Due to their generally oval shape and reduced wall strength, even slight pressure causes the air sacs to collapse, and the rhythmical squeezing action of working flight muscles acts like a pump on the air sacs to increase the total flow of air.

The air sacs serve other functions as well as respiration. For example, in young adults they reduce body weight to the degree that they take up space, but by collapsing they allow growing tissue to fill the space. Significantly, young adult insects often have air sacs in the abdomen that become displaced by the developing ovaries and testes, and the general body shape remains little changed in spite of this growth. The air sacs also serve a hydrostatic function in some aquatic insects and in some sound-producing insects allow the tympanic membrane more freedom to vibrate. They may increase hemolymph concentration of solutes without necessarily increasing total solute by restricting the space that the volume of circulating hemolymph must serve.

5.2.8. Spiracle Structure and Function

The openings of the tracheal trunks to the outside of the insect are called spiracles. Many variations exist in spiracular structure. The spiracle may be as simple as an unguarded opening of the tracheal trunk, but frequently there is an atrium just inside the opening. When an atrium is present, usually a cluster of tracheae arise from it and radiate inward. If an atrium is present, there also may be external valves, internal valves, and/or dust-catching setae or hairs, or it may be covered by a sieve plate containing many small pores. A sieve plate is present in some terrestrial Diptera, Coleoptera, Lepidoptera, as well as in some aquatic insects where it assists in keeping water out of the trachea. Usually the spiracle is surrounded by a ring of sclerotized cuticle, the peritreme.

5.2.9. Orientation and Number of Spiracles

Some Diplura have 11 spiracles, but other insects have 10 or fewer segmentally arranged along the pleural region of the body. Insects from many orders have 8–10 spiracles on each side and are classified (Hoyle, 1961) as having a polypneustic system. Further subcategories within the polypneustic system are holopneustic with 10 spiracles, peripneustic with 9 spiracles, and hemipneustic with 8 spiracles. Holopneustic insects have one mesothoracic and one metathoracic spiracle, while insects in the remaining two subcategories have only one mesothoracic spiracle, with all other spiracles located on the abdomen.

Insects with only one or two functional spiracles are classified as having an oligopneustic system. This category, too, is further divided into amphipneustic with two spiracles (one mesothoracic and one postabdominal), which is common in larval Diptera; metapneustic with only one postabdominal spiracle, occurring in larvae of Dytiscidae, Culicidae, and Tipulidae; and propneustic with only one mesothoracic spiracle, found in some pupae of Diptera. Some insects, particularly aquatic species, have no functional spiracles at all, and the tracheae, although present, do not open to the outside. Gas exchange between these tracheae and the environment occurs by diffusion through a thin cuticle.

5.2.10. Spiracular Muscles and their Physiology

Spiracles may or may not have a closing mechanism. Lack of a closing mechanism occurs in Apterygota, many aquatic larvae, and some aquatic adults. Most terrestrial insects can close the spiracles, an adaptation evolved for water conservation. The simplest closing mechanism consists of folds of the integument that can be pulled together over the spiracular opening by a closer muscle. Alternately, there are valves (Fig. 5-10) (Whitehead, 1973) that rotate and close over the spiracle when the closer muscle contracts. Opening of the spiracle is generally produced by the natural elasticity of the surrounding integument. Spiracle 1 on the thorax of Orthoptera has both closer and opener muscles. Although the spiracle opens partly by natural elasticity after relaxation of the closer muscle, the opener muscle can cause the spiracle to open more widely during increased ventilation (Miller, 1960).

The muscles associated with the spiracle are innervated by a branch of the median nerve from the ganglion in the same segment or from the ganglion in the anterior segment (Fig. 5-11) (Case, 1957). Regularly occurring action potentials from the median nerve cause contracture of the closer muscle and close the spiracle. It might be expected that sectioning the median nerve supply to a closer muscle would allow it to relax and allow the spiracle open. The observed effect, however, of sectioning the median nerve varies in different insects and in response to different physiological conditions.

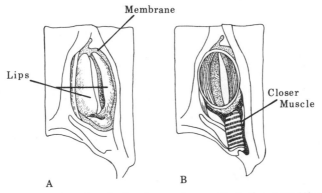

FIGURE 5-10. Second thoracic spiracle of a grasshopper illustrating outer (A) and inner (B) views of valves. (Modified from *Principles of Insect Morphology* by R. E. Snodgrass. Copyright © 1935, McGraw-Hill Book Company. Used with permission of McGraw-Hill Book Company.)

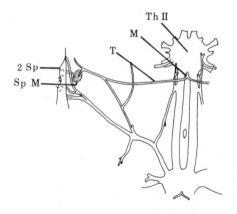

FIGURE 5-11. Innervation of the second thoracic spiracle, 2 SP, of *Periplaneta americana*. The median nerve, M, arises from the second thoracic spiracle, ThII, and divides into a left and a right transverse nerve T, that innervates the spiracle closer muscle, SpM. [Modified from J. F. Case, *J. Insect Physiol.* **1**, 85 (1957).]

Beckel and Schneiderman (1957) found that the spiracles of *Hyalophora cecropia* pupae closed and remained closed for days following closer muscle denervation. The muscle has a neural-independent myogenic rhythm leading to slow pacemaker potentials and subsequently to spike discharges (Fig. 5-12) (Van Der Kloot, 1963). The rate of spike generation was related to the membrane potential reached following each repolarization. The fastest rate of firing occurred when the repolarization membrane potential fell to -54 mV, whereupon the pacemaker potential caused the membrane potential to fall at a constant rate of 21 mV/sec until spikes were generated at a membrane potential of -20 mV and at intervals of 1.8 sec. At a membrane repolarization potential of -81 mV, the pacemaker potentials developed more slowly (at a rate of 19 mV/sec) until the spike threshold was again reached and spikes occurred every 2.2 sec. At a repolarization potential of -85 mV, no pacemaker potentials developed and the muscle relaxed, opening the spiracle. This action could be precipitated by long exposure to CO_2, which hyperpolarized the membrane. Exposure of the muscle to 10% CO_2

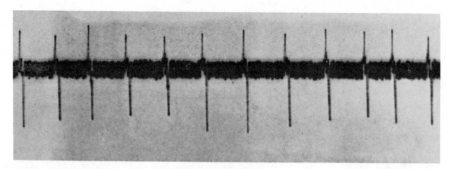

FIGURE 5-12. Spontaneously generated action potentials in fibers of a denervated spiracular muscle from a cecropia pupa. Calibrations indicate 100 μV and 1 sec. [From W. G. Van Der Kloot, *Comp. Biochem. Physiol.* **9**, 317 (1963).]

slowed firing from 34 spikes/min to 27 spikes/min. The manner in which the myogenic rhythm is integrated with the normal neural control from the median nerve has not been clarified.

Miller (1960) found that while denervation of the mesothoracic spiracle of *S. gregaria* sometimes produced immediate closure, at other times closure occurred only after 1–5 days. Hoyle (1961) failed to get any closure of spiracles after sectioning the second thoracic median nerve of bran-fed locusts, but immediate closure occurred in 12 of 14 grass-fed locusts treated similarly. He attributed these results to the high level of K^+ in the hemolymph of grass-fed locusts and locusts about to molt, which also showed immediate closure. He experimentally showed that raising the K^+ concentration caused contracture in the closer muscle (Fig. 5-13). This can be viewed as an adaptive mechanism that operates normally to cut down on water loss during temporary water deprivation or under very dry conditions that lead to changes in concentration of the hemolymph K^+ as a result of natural, unavoidable water loss. In addition to closing the spiracle under these conditions and thus further reducing the loss of water, the same process will close the spiracles at the time of molting when water loss is again critical.

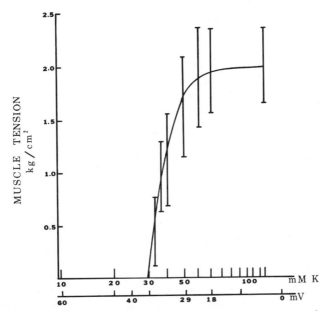

FIGURE 5-13. Relationship between the tension developed by *Schistocerca* spiracular muscle and the potassium concentration in the bathing fluid. Potassium was raised by replacing sodium chloride in the bathing saline with potassium chloride. The vertical bars show the range and the curve the mean of values from different experiments. The lower horizontal scale shows the mean membrane potential of muscle fibers. (Reprinted with permission from G. Hoyle, *J. Insect Physiol.* **7,** 305, Copyright 1961, Pergamon Press Ltd.)

5.3. MOVEMENT OF O_2 AND CO_2 IN TRACHEAE

Movement of O_2 and CO_2 through tracheal tubes is promoted by diffusion and by active ventilation. Diffusion is influenced by the molecular weight of the diffusing molecules, the partial pressure difference along the diffusion path, and the permeability of the medium through which the molecules must pass. Diffusion of O_2 is about 15% faster than CO_2 in an air path, other factors being equal, because of its smaller molecular weight. In a tissue or aqueous medium, however, CO_2 is much more soluble than O_2 and its transport is greatly increased over that of O_2. Generally, then, a system that will meet the tissue need for O_2 will also be adequate for CO_2 removal. Most of the experimental work on insects has been related to supply of O_2 to the tissues, and especially to the wing muscles, which are capable of the greatest known rate of O_2 use/unit volume of tissue.

The highly successful adaptation of a system of tubes carrying air deep into tissue masses can be appreciated by understanding that O_2 diffuses about 10^6 times faster in air than in an aqueous tissue medium. The permeability constant for diffusion of O_2 in an air path at 20°C is 11 ml min^{-1} cm^{-2} atm^{-1} cm^{-1}, whereas in water it is 3.4×10^{-5} ml min^{-1} cm^{-2} atm^{-1} cm^{-1}. Although the permeability constant for O_2 in insect tissue has not been measured, we may reasonably assume it to be similar to that in frog muscle, 1.4×10^{-5} ml min^{-1} cm^{-2} atm^{-1} cm^{-1} (Chapman, 1969).

The hole fraction concept introduced by Weis-Fogh (1964b) mathematically relates the cross-sectional area of all the air tubes in a plane through a tissue to the cross-sectional area of nontracheal tissue in that plane. A larger hole fraction value indicates proportionately more air tubes per unit of tissue served and, consequently, better aeration of the tissue. Therefore, for a hole fraction of 10^{-1}, one-tenth of the area is air space through which diffusion can occur very rapidly. The very high efficiency of the tracheal system of insects is due to the large hole fraction value that varies from 10^{-1} to 10^{-2} for the secondary tracheal supply, to the smaller values of 10^{-2} to 10^{-3} for the tertiary and tracheolar supply.

Some insects, usually small ones such as *Drosophila,* show no ventilatory movements, and it has generally been believed that diffusion is sufficient to support the metabolic demand for O_2. Krogh in fact came to this conclusion after careful study and mathematical treatment of the process of diffusion in the tracheal system. Some of his assumptions, however, have been questioned, and Weis-Fogh reexamined the problem. In brief, Weis-Fogh confirmed Krogh's earlier conclusions that diffusion is, indeed, adequate in small insects. The following discussion of diffusion is based on Weis-Fogh's calculations (1964b) in which he assumed that an insect uses about 25% (0.05 atm) of the O_2 it breathes. In order to calculate whether diffusion can supply the O_2 needed, one must know the rate at which O_2 is used. Table 5-1 shows some representative data for O_2 use by active wing muscle.

TABLE 5-1. Utilization of O_2 by Wing Muscle Tissue during Steady-State Flight

	ml O_2/g muscle/min
Schistocerca	1.4–2.8
Aeshna	1.8
Butterflies, moths	1.4–3.5
Aphis	1.4–1.8
Drosophila	2.0–2.3
Vespa crabro	2.6–3.3
Lucilia	5.6
Apis mellifera	7.3

Source: T. Weis-Fogh, *J. Exp. Biol.* **41,** 207 (1964a).

5.3.1. Diffusion in Tracheal Tubes

The rate of diffusion is proportional to the area of the medium through which diffusion must occur and to the concentration gradient as follows:

$$\frac{\delta s}{\delta t} \propto A \frac{\delta c}{\delta x} \tag{1}$$

where A is area, c is concentration of molecules, and x is the distance of movement along the x axis. The differential $\delta s/\delta t$ expresses the rate of transport in moles per unit of time. The differential $\delta c/\delta x$ expresses the concentration gradient.

Since the rate of diffusion is influenced very greatly by the nature of the medium through which it occurs, a value representing the effect of the medium upon diffusion, the diffusion coefficient, D, must be introduced into equation (1). The equation then becomes an equality expressed as

$$-\frac{\delta s}{\delta t} = -DA \frac{\delta c}{\delta x}. \tag{2}$$

The negative sign in equation (2) denotes that diffusion occurs from a region of higher concentration to a region of lower concentration. Before equation (2) can be used for practical calculations of diffusion in the tracheal system, additional mathematical transformations must be made and the rate of tissue use of oxygen incorporated.

A single formula describing diffusion in tracheal systems cannot be used because diffusion varies with the nature and diameter of tracheal branching patterns (i.e., whether there is a constant area as in *Cossus* larvae and *Rhodnius* systems or a decreasing area as in *Schistocerca* and *Aeshna*) and with the type of primary tracheal supply to muscles (Fig. 5-2). Each case must be considered separately, and Weis-Fogh developed 19 expressions to describe known variations in tracheal systems. For detailed treatment of diffusion in insect tracheae, the reader is referred to his excellent paper (Weis-Fogh, 1964b).

One of the simplest cases occurs when the summed cross-sectional area of all tubes is constant at all distances from the last point of branching, as it is in the main branches of *Cossus* larvae, *Aphelocheirus,* and *Rhodnius.* The equation for calculating the pressure drop, p, necessary to cause O_2 to diffuse over the distance L (in cm), to satisfy the metabolic rate, m, in ml O_2/g muscle/min, is given by the expression

$$\Delta p = \frac{mL^2}{2aP} \tag{3}$$

where a is the hole fraction for the tissue in question and P is the permeability constant for O_2 diffusion in an air path, that is, 11 ml min^{-1} cm^{-2} atm^{-1} cm^{-1}.

Tracheae may branch and then taper in such a way that either areas of tubes or tube diameters decrease linearly with increasing distance. Both types of situations occur in the dragonfly *Aeshna cyanea.* If the tracheae taper so that area decreases linearly from the point of branching, then the pressure drop is calculated as

$$\Delta p = \frac{mL^2}{aP}. \tag{4}$$

If the secondary or tertiary tube diameters decrease linearly with distance from the point of branching, then the formula is

$$\Delta p = \frac{mL^2}{aP} \ln \frac{L}{L - x} \tag{5}$$

where x is the distance from the reference point. Equation (5) can be used with data from Weis-Fogh to illustrate a typical calculation showing that diffusion is sufficient to account for movement in the secondary and tertiary tracheae. For example, in one rectangular lobe of a wing muscle from *Aeshna,* the diameters of secondary tubes decrease linearly with distance after branching from the primary tracheae. The length of the lobe of muscle, and hence of the path over which oxygen must diffuse, is 407×10^{-4} cm, and the hole fraction, a, is 1.5×10^{-2}. The secondary tracheae give rise to tertiary tracheae at a distance into the lobe of 342×10^{-4} cm. Thus, 342×10^{-4} cm becomes the value of x in equation (5). The metabolic rate in *Aeshna* (from Table 5-1) is about 2 ml O_2/g/hr. Substitution of the above data into equation (5) allows one to make the following calculation:

$$\Delta p = \frac{(2)(0.0407)^2}{(1.5 \times 10^{-2})(11)} \ln \frac{0.0407}{0.0407 - 0.0342} \tag{6}$$
$$= 0.037 \text{ atm.}$$

The calculation shows that a pressure drop of only 0.037 atm between the origin of the secondary trachea and the beginning of the tertiary tracheae is necessary for diffusion to supply 2 ml O_2/g/hr over the distance involved. If an insect uses about 25% of the inspired O_2, then a pressure drop of about 0.05 atm between the spiracles and the tissue cells can be expected, or more than enough to suffice. By similar reasoning and calculations, Weis-Fogh showed that diffusion in primary tracheae is not adequate in large insects to supply O_2 at the higher rates characteristic of flight, but it might be sufficient for resting metabolism. The primary tracheae must be ventilated.

5.3.2. Diffusion from Tracheoles to Mitochondria

The final path for O_2 movement is from the tracheole, across the cell membrane, and into the mitochondria. Diffusion is the only mechanism at work. Since the diffusion pathway is through an aqueous medium, we may expect slower movement than in the air path of the tracheae and, consequently, the tracheoles must come very close to the mitochondria in order to meet the rates at which O_2 is used by working muscles. Although the calculated distance for diffusion depends on several characteristics of tracheole structure, the site of use can be only about 8 μm away from a tracheole in the cases of insects with lower metabolic rates (dragonflies and locusts), and only about 4 μm away in insects with higher rates (honeybee) (Chapman, 1969). Although there may be little or no difference in the permeability of larger tracheae and tracheoles to O_2, it is clear that only tracheoles usually will be close enough to the site of use for O_2 to diffuse to the mitochondria in the quantities needed.

Since many muscle fibers are several times the diameter of the 4–8-μm distance that O_2 can diffuse in the quantity needed, tracheoles that only touch the surface of a muscle fiber will not supply the necessary O_2 if the fiber is larger than about 20 μm. This then is a limitation on the size of muscle fibers in some insects. In some groups, such as Diptera and Hymenoptera, the tracheoles regularly indent muscle fibers and cells of other active tissues. The tracheoles may even indent the membrane of the mitochondria. In these and similar systems, tracheoles are often only 2–3 μm apart, and there is a safety factor of two- to threefold in the system. Clearly, the terminal ends of the tracheal system are highly adapted to the very high oxygen demands made by working flight muscles.

5.3.3. Ventilation of Tracheae

Based on the calculations of Weis-Fogh, the primary tracheae must be well ventilated to meet the demands of working wing muscles. Two mechanisms of active ventilation described from flying insects (Weis-Fogh, 1967) follow:

1. Abdominal pumping, either by dorsoventral compression of the abdomen or by telescoping abdominal segments.
2. Thoracic pumping due to wing movements.

Abdominal pumping, common in large insects at rest, typically continues during flight. In some fast-flying insects, such as the wasp *Vespa crabro* and possibly other hymenopterans, abdominal pumping alone is sufficient to supply the wing muscles during flight. In many other large insects, such as dragonflies and locusts, thoracic pumping is important for ventilation of wing muscles. Indeed, in both dragonflies and locusts, thoracic pumping is sufficient in the absence of abdominal mechanisms.

Thoracic pumping tends to create a tidal flow that moves air in and out of the same spiracles. While abdominal pumping may also cause a tidal flow in some insects, in others there is a directed flow into the thoracic spiracles and out through the abdominal spiracles. The opening and closing of the spiracles must, of course, be correlated with the pumping action for effective unidirectional flow. Unidirectional flow is more effective than tidal flow, because it does not leave a dead air space unventilated as a tidal flow does.

At rest, abdominal pumping of *Schistocerca gregaria* can move 40 liters air/kg/hr in through the thoracic spiracles and out through the abdominal ones. During the first 5 min of flight the abdomen can pump 180 liters/kg/hr, but this then falls to 150 liters/kg/hr. The 180 liters/kg/hr pumped on initiation of flight is partitioned as follows:

30 liters to the head and ventral trachea of the thorax and abdomen (i.e., it serves the central nervous system),

70 liters to the pterothorax, and

80 liters to the other parts of the body (Weis-Fogh, 1967).

Abdominal pumping in the locust is not necessary for flight, however, as selective injury to the nervous system blocks or greatly reduces abdominal pumping without altering tethered flight. As a result of wing action during flight, thoracic pumping ventilates the locust thorax with 250 liters air/kg/hr. This is more than enough for wing muscles.

Each abdominal ganglion of *Schistocerca* is capable of initiating ventilatory movements for its segment, but the rhythm that synchronizes the overall movements probably originates in the metathoracic ganglion since its rate of firing is faster than that of the abdominal ganglia. The basic rhythm of movements can be altered by CO_2 and, to a lesser extent, by hypoxia. Local perfusion of the head or thoracic ganglia with gas mixtures indicates that each can alter the rhythm of ventilation.

A nonflying locust can utilize ventilatory mechanisms in addition to the regular and continuous abdominal movements of raising and lowering the sterna. These are

1. longitudinal telescoping of abdomen,
2. protraction and retraction of head ("neck ventilation"), and
3. protraction and retraction of prothorax ("prothoracic ventilation") (Miller, 1960).

While these mechanisms are not used under normal resting conditions, they do operate for short intervals following periods of great activity such as after flight. Table 5-2 summarizes the data on ventilation in the locust.

Dragonflies (*Aeshna* spp.) and wasps (*Vespa crabro*) represent two additional groups of fast-flying insects that strongly ventilate in order to supply the O_2 needed during flight. In sustained horizontal flight dragonflies and wasps use about 20 liters/kg/hr of oxygen (Weis-Fogh, 1967), a rate about 40–50% higher than that of flying locusts. Dragonflies supply the wing muscles entirely from thoracic pumping. Wasps, however, have a very hard, rigid pterothorax that allows little thoracic pumping. Abdominal ventilatory movements increase in amplitude and frequency (to 180/min) during flight, and Weis-Fogh concluded that ventilation from these abdominal movements was the primary mechanism for O_2 supply.

5.3.4. Water Balance during Flight

It might seem logical to expect a flying insect to lose large amounts of water due to evaporation from the tracheal surfaces as large volumes of air are ventilated through the system. We may expect to find physiological and behavioral adaptations, particularly in migratory and other long-distance flyers, to prevent excessive water loss and desiccation. One such adaptation is the use of fat, characteristic of Lepidoptera, Orthoptera, and possibly other groups, as fuel for sustained flight. In *Schistocerca gregaria* about 7 g fat/kg/hr are burned during flight, resulting in the production of 8.1 g/kg/hr of water (Weis-Fogh, 1967). This gain of metabolic water may be sufficient to offset the loss from evaporation.

Water regulation in insects will clearly be related to the prevailing relative humidity and temperature of the air. Figure 5-14 illustrates the combinations

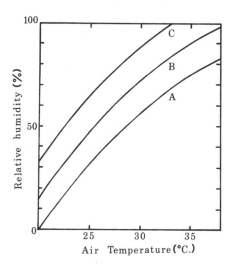

FIGURE 5-14. The interaction between relative humidity and air temperature on water balance in the thorax of a flying desert locust, *Schistocerca gregaria*. Locusts gain water from metabolism for points above the curves, while they lose water by evaporation for points below the curves. A. No net radiation heating. B. Thoracic temperature 2°C higher than in A due to net absorption of radiation heating of 0.55 cal/cm²/min. C. Thoracic temperature 4°C higher than in A caused by radiant heating of 1.25 cal/cm²/min. [From T. Weis-Fogh, *J. Exp. Biol.* **47,** 561 (1967).]

TABLE 5-2. Measurements of the Flight Ventilation System in _Schistocerca gregaria_

Measurement	Method	Values
Metabolic rate	Measurement of heat production, of oxygen consumption, etc.	65 kcal/kg/hr or 13.4 liters O_2/kg/hr
Volume ventilated by the abdomen in steady flight (after first 10 min)	Abdomen enclosed in a small plethysmograph joined to a spirometer	144 (39–245) liters air/kg/hr
Volume of air ventilated in unidirectional anterior-to-posterior flow during steady flight	Spiracles 4–9 waxed; spiracle 10 intubated and output led to capillary with kerosene droplet	33 liters/kg/hr
Volume change in pterothorax caused by flight movements	Dead locusts (HCN); thorax in sealed vessel; thoracic spiracles and soft cuticle waxed; abdomen in separate vessel and air movement through abdominal spiracles measured when wings were moved by lever	25 μl/stroke = 760 liters/kg/hr
Pressure changes in pterothorax resulting from abdominal ventilation during flight	Hypodermic needle in hemolymph adjacent to air sacs; pressure transducer	2.22–3.7 mm Hg (peak-to-peak)
Pressure changes in pterothorax resulting from flight movements		1.48–1.85 mm Hg (during abdominal expiration) 0.75–1.11 mm Hg (during abdominal inspiration)
Pressure changes in air sacs resulting from abdominal ventilation during recovery from CO_2		+14.8 mm Hg (expiration); −3.7 mm Hg (inspiration)
Volume of pterothoracic tracheal system in a standard locust	Injection	100–150 μl
Actual volume of air ventilating flight-muscle tracheae in flight resulting from flight movements		250 liters/kg/hr
Total volume of air ventilating flight muscle tracheae in flight		320 liters/kg/hr (250 from thoracic movements and 70 from abdominal movements)
Gas content of pterothoracic air sacs in flight	Samples extracted from spiracle 2 and analyzed	10–15% oxygen; 4.8–7.9% CO_2

Source: P. L. Miller, "Respiration: Aerial gas transport," in M. Rockstein, Ed., _The Physiology of Insecta,_ Vol. 6, 2nd ed., Academic Press, New York, 1974.

of temperature and relative humidity at which there is a balance in flying *S. gregaria* between water production and water loss from the thorax. Clearly, while a desert locust can stay in water balance during sustained flight at moderate to high relative humidity and at temperatures between 25° and 30°C, at very low relative humidity at even 25°C, water will be lost and this loss will be intensified at higher temperatures. Although detailed data are not available for other insects, similar relationships undoubtedly apply. Other adaptive mechanisms that insects use to regulate water loss may relate to behaviors such as failing to take flight or to sustain flight for long periods; flight at night when temperature is usually lower and relative humidity usually higher; and flight at higher altitudes where temperatures are lower.

5.3.5. Discontinuous Release of CO_2

A beautiful example of an adaptive mechanism for conserving water is the process of discontinuous release of CO_2, also referred to as passive suction ventilation. Its occurrence was first described in the 1940s, but a comprehensive physiological explanation and related details were not elucidated until the 1960s. The process is known to occur in *Periplaneta,* in *Schistocerca* at certain ages, in prepupae of *Orthosoma* (Coleoptera), and in several diapausing pupae of Lepidoptera. It has been most thoroughly studied in the diapausing pupae of the cecropia silkmoth, *Hyalophora cecropia.*

The prepupa of cecropia releases CO_2 continuously, and discontinuous release begins in the pupa. Inspection of Fig. 5-15 (Schneiderman and Williams, 1955) reveals that CO_2 is released discontinuously in large "bursts" at intervals that vary with the stage of development. Oxygen uptake is continuous, though the level decreases with pupal age because the pupa normally enters diapause, a stage of very low metabolic activity. A dilemma arising from the early experiments was to explain how O_2 uptake could be continuous when all spiracles appeared to be closed between CO_2 bursts. With careful observations and recording of spiracular valve movements with a thread attached to the valves and thence to a lever arm writing on a kymograph drum, Schneiderman (1960) found that the valves are not tightly closed all the time between bursts. Following an initial period of tight closure, a valve begins to flutter open and close rapidly, with the amplitude of the flutter movements increasing with time. Finally the burst occurs and the spiracle stays open for a few minutes, then beginning a "decline phase" in which it alternately opens and closes. Finally it closes tightly again and the cycle starts over. Table 5-3 presents mean values for the spiracular movements, but it should be noted that there is variation correlated with ages of pupae and variation between individuals.

It proved to be possible to cannulate the spiracles of cecropia pupae (Fig. 5-16) and to withdraw gas samples at intervals during a burst cycle for measurement of the O_2 and CO_2 concentrations (Levy and Schneiderman, 1966a). These analyses showed that at a burst the inward movement of O_2

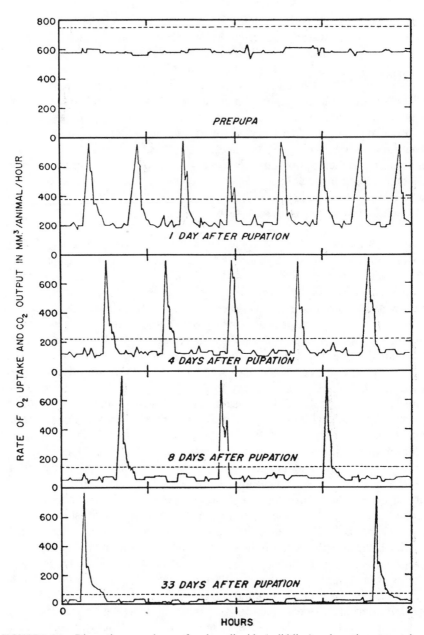

FIGURE 5-15. Discontinuous release of carbon dioxide (solid line) and continuous uptake of oxygen (broken line) in a cecropia silkworm during larval–pupal transformation. [From H. A. Schneiderman and C. M. Williams, *Biol. Bull.* **109**, 123 (1955). With permission of Marine Biological Laboratory.]

TABLE 5-3. Mean Duration of Spiracular Valve Movements during Cyclical CO_2 Bursts in Lepidopterous Pupae[a]

Species	Total Cycle	Closed	Fluttering	Open	Opening and Closing
H. cecropia	56.35	15.08	36.12	1.28	3.35
	±23.25	±5.71	±21.96	±1.28	±1.82
A. polyphemus	8.04	0	5.52	1.70	0.78
	±3.98		±3.84		

[a] Minutes ± SD.
Source: Adapted from H. A. Schneiderman and C. M. Williams, *Biol. Bull.* **109,** 123 (1955).

raises the intratracheal O_2 tension to about 18%, only slightly reduced from the 21% O_2 in outside air due to some mixing with the large dead air space of the tracheae. At the same time CO_2 moves rapidly outward and the intratracheal CO_2 falls to about 3%. On closing of the valve following a burst, O_2 tension falls due to use by the tissues and CO_2 tension gradually begins to rise. Inspection of Fig. 5-17, which summarizes the event, shows that O_2 tension falls much more rapidly than CO_2 tension rises (Levy and Schneiderman, 1966b). The explanation for this is that CO_2 accumulates largely as the bicarbonate ion, HCO_3^-, in solution in the tissues and hemolymph, and only a small percentage of the CO_2 exists in the gaseous form. The falling O_2 tension creates a small vacuum in the tracheal system (about -4 mm Hg, Fig. 5-17), and when O_2 tension falls below about 5%, the spiracular valves begin to flutter. Each flutter allows a little air to be sucked in, hence the alternative name of this process—passive suction ventilation. The inward movement of air gradually relieves the vacuum in the system and maintains O_2 tension at

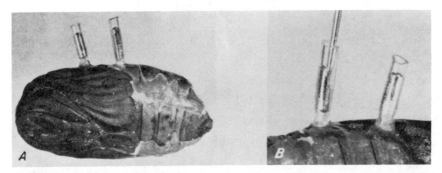

FIGURE 5-16. A. Photograph of cecropia pupa with cannulae inserted into first and third abdominal spiracles. B. The mercury in the cannulae acts as an air seal and allows a syringe needle to be inserted into the spiracle for withdrawal of an air sample. (Reprinted with permission from R. I. Levy and H. A. Schneiderman, *J. Insect Physiol.* **12,** 83, Copyright 1966, Pergamon Press, Ltd.)

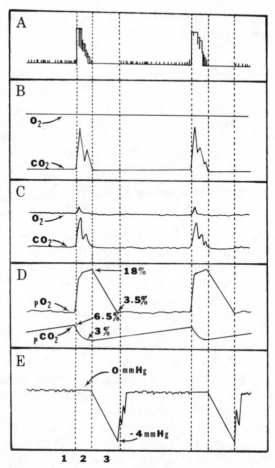

FIGURE 5-17. Correlation of events occurring during the respiratory cycle in a cecropia pupa. A cycle can be divided into major phases corresponding to when the spiracle is fluttering (1), open (2), or tightly closed (3). A. Spiracular movements. B. Gas exchange measured by manometric methods. C. Gas exchange measured by diaferometric techniques. D. Tracheal gas composition. E. Intratracheal pressure. (Reprinted with permission from R. I. Levy and H. A. Schneiderman, *J. Insect Physiol.* **12,** 465, Copyright 1966, Pergamon Press, Ltd.)

about 3.5% until the next burst. A burst is triggered by the accumulation of CO_2 after it reaches a tension somewhat greater than 5%.

Schneiderman (1960) showed that the spiracular muscles receive a continuous barrage of nerve impulses from the central nervous system, and this tends to keep the spiracles closed. The low O_2 tension that develops within the tightly closed system causes the muscles to relax slightly, but as soon as some air has moved in and raised O_2 tension slightly, the nervous control causes the valves to shut again. This dynamic state of O_2 tension results in the flutter of the valves as the actions are repeated frequently. The interac-

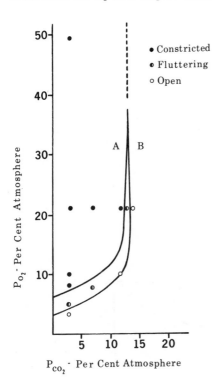

FIGURE 5-18. Interaction of oxygen and carbon dioxide on spiracles of diapausing cecropia pupa. Spiracle valves are constricted for points to the left of curve A, fluttering at points between curves A and B, and open at points to the right of curve B. [From B. N. Burkett and H. A. Schneiderman, *Science* **156,** 1604 (1967). Copyright 1967 by the American Association for the Advancement of Science.]

tion of O_2 and CO_2 on the closer muscles is shown in Fig. 5-18. By using isolated, perfusion ganglia or isolated spiracular closer muscles, it was possible to show that O_2 tension acts centrally at the ganglionic level to inhibit the string of nerve impulses coming out of the ganglion, while CO_2 acts directly on the closer muscle (Burkett and Schneiderman, 1967). CO_2 seems to act on the neuromuscular junctions of the muscles, resulting in relaxation and opening of the spiracles in spite of the incoming nerve impulses. Discontinuous release of CO_2 continues in cecropia pupae at temperatures down to $-5°C$, and the valves respond to CO_2 and O_2 tensions in the expected way. These functions are highly adaptive for pupae that spend the winter in their natural habitat in northern climates. The valves freeze in the closed position at some temperature between $-5°$ and $-10°C$, and any further gas exchange must be by a cutaneous process.

Notwithstanding the importance of discontinuous respiration to water conservation, Kanwisher (1966) found that the spiracular valves of cecropia pupae stayed open slightly longer than necessary for effective CO_2 release. This resulted in a greater loss of water than CO_2 per unit time during the last part of the cycle, as indicated by the combined curve for CO_2 and H_2O in Fig. 5-19 dropping below the 0 horizontal line just before the valves closed. On the other hand, the total water lost is about equal to the metabolic water produced, so the pupa stays in water balance.

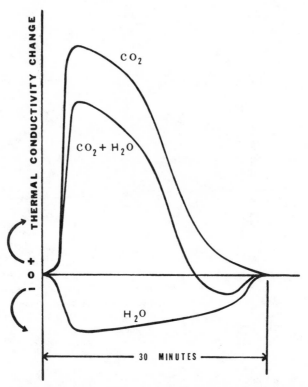

FIGURE 5-19. Loss of carbon dioxide and water from a cecropia pupa during a single burst. A thermal conductivity detector consisting of a heated thermister with a reference gas flowing over it was connected to a recorder. When only the reference gas passed over the thermister, a constant baseline response, the 0 line, was obtained. When carbon dioxide or water vapor from a pupa mixed with the reference gas and then passed over the heated thermister, an upward or downward deflection of the recorder pen occurred, respectively, indicating a change in the temperature and hence in conductivity of the thermister. Since water vapor and carbon dioxide normally come from a pupa at the same time, an appropriate scrubber was inserted in the air stream to selectively remove water or carbon dioxide as desired. [From J. W. Kanwisher, *Biol. Bull.* **130,** 96 (1966). With permission of Marine Biological Laboratory.]

5.4. RESPIRATION IN AQUATIC INSECTS

Aquatic insects utilize a variety of adaptations for respiratory gas exchange. Cutaneous respiration is probably important in many aquatic insects. An elaboration of large surface areas, the tracheal gills, has expanded the surface for cutaneous gas exchange in larvae of Trichoptera, Plecoptera, Odonata, and other orders of insects. Most aquatic insects, however, breathe air in much the same ways as terrestrial insects do, and many simply come frequently to the water surface. Some carry a bubble or film of air beneath the water on their bodies. The bubble may have to be renewed frequently by a return to the surface, but some can hold a nonwetted air space so effectively with a plastron network that it continues to extract O_2 from the water indefinitely. A plastron consists of an extensive physical meshwork or hair pile that can hold a volume of air and can present a large water–air interface. Plastrons, which are normally incompressible, are one of the most common adaptations of insects living in aquatic arrangements and can take many physical forms.

5.4.1. Cutaneous Respiration: Closed Tracheal System

A closed tracheal system without functional spiracles is present in many aquatic insects. The lack of functional spiracles is an adaptation to prevent water from entering the system, and all O_2 is taken into the tracheae by diffusion through a thin cuticle. Often a great increase in body surface area occurs in richly tracheated filamentous or flaplike extensions from the body. These have been called the tracheal gills (Fig. 5-20).

5.4.2. Tracheal Gills

Tracheal gills are highly variable in structure and in location on the body. They occur in several orders of insects. Three caudal gills are characteristic of larvae of the Zygoptera (Fig. 5-21), while tufts of thin gill filaments are located on the head, thorax, abdomen, and coxae of some Plecoptera. Fila-

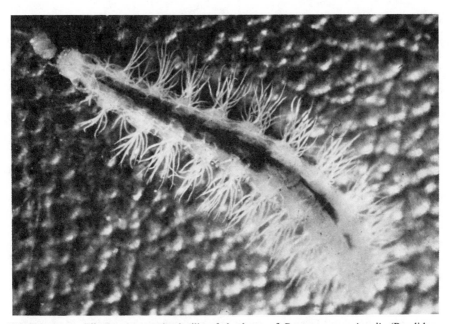

FIGURE 5-20. Filamentous tracheal gills of the larva of *Parapoynx seminealis* (Pyralidae: Lepidoptera) that lives under water. (Photograph courtesy of Dr. Dale Habeck.)

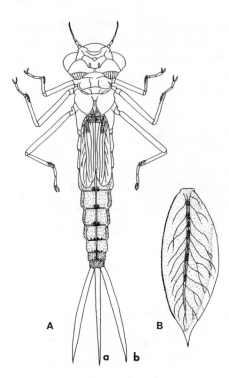

FIGURE 5-21. A. Caudal gills of a damselfly larva; a, medial gill; b, lateral gill. B. Enlarged lateral view of a lateral gill showing tracheal pattern. [Modified with permission from *An Introduction to the Aquatic Insects of North America* by R. W. Merritt and K. W. Cummins, Copyright © 1978, Kendall/Hunt Publishing Co.]

mentous gills also occur on the abdomen of some Trichoptera, Diptera, and Lepidoptera and on the thorax and abdomen of a few Coleoptera. Rectal gills are characteristic of some dragonfly larvae (Anisoptera). Tracheal gills generally are larval structures, but some trichopterous pupae have them, and they persist as atrophied, probably nonfunctional, structures in some adults of Trichoptera and Plecoptera.

Oxygen diffuses into the tracheae of the gills, and into other body regions as well, from the surrounding water. Once inside the tracheae, it probably is distributed by diffusion as in terrestrial insects, and by some body and muscular movements. Some investigators have questioned the respiratory function of tracheal gills since some insects from which the gills were removed showed little, if any, change in behavior or in O_2 consumption. Probably, in the cases of these insects there is a high level of general cutaneous respiration assisted by the gills.

Larvae of Trichoptera, Odonata, and Coleoptera display a highly structured arrangement of tracheoles in the gills as an adaptation for optimal functional efficiency with a minimum of tracheoles (Wichard and Komnick, 1974). The tracheoles run parallel to the gill surface and are just beneath the thin cuticle. In Trichoptera, Odonata, and Coleoptera they are uniform in size and in spacing. These characteristics appear to be highly adaptive for trapping a high percentage of the O_2 that diffuses through the cuticle. By

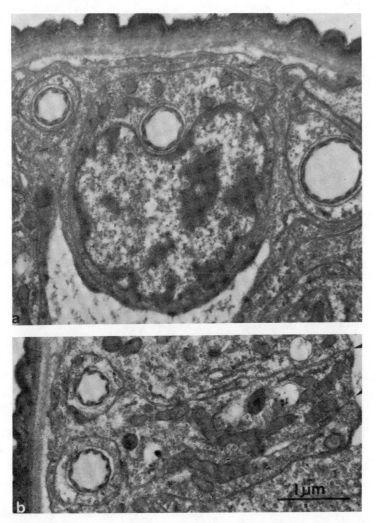

FIGURE 5-22. Electron micrograph of a cross section through the tracheal gill of *Perla marginata* (Perlidae: Plecoptera). A. Tracheoblast with nucleus and a tracheole just under the cuticle. B. Tracheoles just beneath the cuticle and deeply indenting the basal plasmalemma of the epidermal cell (arrow). [From W. Wichard and H. Komnick, *J. Insect Physiol.* **20**, 2397 (1974).]

contrast, the gill tracheoles (Fig. 5-22) of Plecoptera vary in diameter from 0.2 to 1.0 μm. Their spacing and distribution (Fig. 5-23a) are also less uniform. The gills have a tubelike shape (Fig. 5-23b) with the center cavity filled with hemolymph in communication with the hemocoel of the body (Wichard and Komnick, 1974). The cuticle is thin (0.2–1.2 μm thick) and lies over a single layer of epidermal cells. The tracheoles have a diameter less

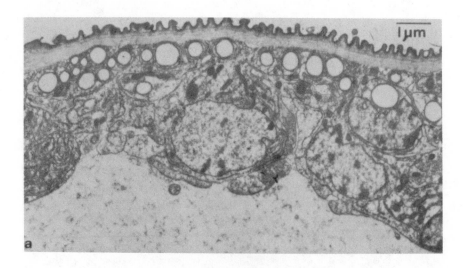

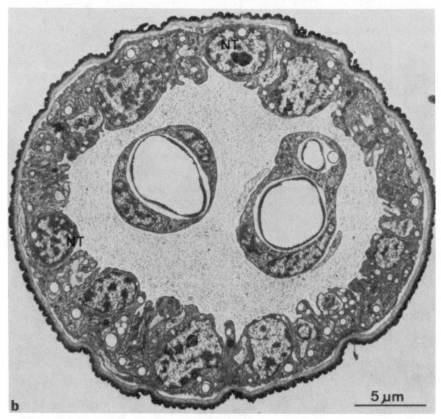

FIGURE 5-23. A. Cross section through a gill filament of *Perla auberti*. Note the size variation in tracheoles. B. Cross section through a distal gill filament of *Perla marginata* showing the hemolymph-filled central cavity and two large tracheae. NT, Nuclei of two tracheoblasts within the epithelium. (Reprinted with permission from W. Wichard and H. Komnick, *J. Insect Physiol.* **20,** 2397, Copyright 1974, Pergamon Press, Ltd.)

than 1 μm, and they indent the epidermal cells so that they come to lie immediately beneath the cuticle.

Rectal gills are found only in dragonfly larvae (Anisoptera). Typically there are six main gill folds in the anterior part of the rectum. Water is drawn into the rectum by elastic expansion of the body as dorsoventral compressor muscles relax. The gills extract O_2 from the water, which is then pumped out by dorsoventral compression of the abdomen. The rate of ventilation varies with several factors including the O_2 content of the water. About 85% of the water in the rectum is renewed during each pumping cycle, and 25–50 cycles/min have been recorded (Chapman, 1969). Larvae will also come to the surface and ventilate the rectum with air when the oxygen content of the water is very low.

Movement of water over the gill and body surface of aquatic insects is important in maintaining a fresh supply of oxygenated water in contact with the body, and most use undulations of the body and/or movements of the gills themselves to create ventilatory currents of water.

5.4.3. Air-breathing Aquatic Insects: Open Tracheal System

The majority of aquatic insects breathe air in much the same manner as terrestrial insects and have an open tracheal system. Many are metapneustic, with only one functional pair of spiracles near the tip of the abdomen. Some have a siphon at the tip of the abdomen that is pushed above the surface of the water for gas exchange. During submergence the spiracles are kept closed. Probably a small vacuum is created within the tracheal system due to O_2 use while submerged, as in discontinuous release of CO_2 in cecropia pupae, and this aids the exchange of gas when the larva or pupa comes to the surface.

An insect must be able to break the surface film of water to get air from the surface. Virtually all aquatic insects have a difficult-to-wet region of the body called a hydrofuge that has a greater affinity for air than for water. Often a spiracle is surrounded by a tuft of long hairs that acts as a hydrofuge. Probably, oily secretions from cuticular glands coat these hairy regions and aid in maintaining their hydrofugic nature.

5.4.4. Use of Aquatic Plants as Air Sources

Some insects with a hydrofuge are able to capture and utilize the gas bubbles released by aquatic plants. Others bite into or penetrate the air spaces of plants with a piercing siphon. The Donaciinae (Coleoptera) live in the mud around the roots of aquatic plants, and larvae penetrate the plant roots with a pointed, posterior siphon. A spiracle at the end of the siphon allows gas entry. Larvae also bite into the root before pupation and construct a pupal case over the lesion. Air probably continues to be released from the lesion into the pupal case to support the pupa. *Hydrocampa* (Lepidoptera) also inserts a respiratory siphon into the air spaces of aquatic plants, while some other lepidopterans bite into the plants. Three families of Diptera and two families of Coleoptera have independently evolved modifications for piercing aquatic plants for air (Wigglesworth, 1972).

5.4.5. Gas Gills

A large number of aquatic insects that breathe air carry a bubble or film of air with them when they submerge. In some the bubble or film must be periodically renewed, while in others a highly effective plastron continues to extract O_2 from the water into an air space so they can stay submerged indefinitely. Such air stores are called gas gills, and they may be compressible or incompressible. Compressible gills slowly collapse as the gas is used up, though additional O_2 is usually extracted from the water into the gill before it must be renewed at the surface. On the other hand, incompressible gills (plastrons) do not collapse and O_2 can continue to be extracted from the water into the gill, allowing the insect to stay under water indefinitely.

5.4.5.1. Compressible Gas Gills.   Air stores in compressible gas gills may be carried in a fine pile of hydrofuge hairs; sometimes there is a fine, dense set of hairs that form a microplastron nearest the body surface from which the gas volume is very slowly used, and a set of longer, larger hairs over these, the macroplastron, from which the gas is rapidly used. Alternately, a gas gill may be carried beneath the elytra, as in dytiscids, or as a gas bubble on the posterior part of the abdomen. Compressible gas gills are present in members of the Coleoptera (Dryopoidea and Hydrophilidae), Hemiptera (*Gerris* and *Velia*), and Lepidoptera (some arctiids and pyralids).

Gas gills serve a hydrostatic function as well as a respiratory function. When the insect comes to the surface to renew the air store, the buoyancy of the remaining bubble allows the part of the body carrying the bubble to come to the top first; thus the air in the bubble is restored with a minimum of exposure of the insect at the surface. Immediately upon surfacing, the air bubble will contain approximately 21% O_2 and 79% N_2 as a result of equilibrating with the atmosphere. As the insect submerges, O_2 will begin to be used by metabolic processes; pO_2 in the air bubble will fall, while pN_2 will rise. In well-aerated water the gas composition is approximately 33% O_2, 64% N_2, and 3% CO_2 (Chapman, 1969). As a result of partial pressure differences between gases in the bubble and those in the water, O_2 from the water will diffuse into the gas bubble and N_2 of the bubble will begin to diffuse out into the water in an attempt to restore equilibrium. Equilibrium cannot be restored, of course, since O_2 is continually being used by the insect, but this dynamic exchange allows the insect to gain up to 13 times the quantity of O_2 originally carried within the bubble. Eventually the insect must surface and renew the bubble. In *Dytiscus* the O_2 content of the bubble dropped from 19.5% at the surface to 2% in 3–5 min (Wigglesworth, 1972).

The N_2 in the bubble plays a crucial role. Its relatively low solubility in water causes it to move out of the bubble more slowly than O_2 enters the bubble. The N_2 remaining in the bubble prevents it from collapsing and allows O_2 from the water to diffuse into the bubble. Beetles allowed to fill the bubble at the surface with pure O_2 cannot stay submerged as long as normal because, without N_2, the bubble shrinks rapidly in size as the O_2 is used by the insects.

5.4.5.2. Incompressible Gas Gills.

An incompressible gas gill, or plastron, consists of a film of air so tightly guarded by a dense network of nonwettable hairs that even though the gas equilibrium may change due to O_2 use, water cannot invade the hair pile and compress the air space. The minimum amount of water–air interface in a meshwork necessary for it to function as a plastron has not been determined, but Hinton (1969) suggested that a water–air interface to weight ratio of 15,000 $\mu m^2/mg$ was sufficient to qualify as a plastron. He based this conclusion on the water–air interface/mg ratio found in the pupa of the fly *Eutanyderus wilsoni*. This fly has the poorest ratio known for an insect obviously adapted for living in water. Most insects with a plastron have a water–air interface of from 10^5 to 10^6 $\mu m^2/mg$.

The thickness of the hair pile will obviously help determine the efficiency of the plastron. *Aphelocheirus aestivalis* (Hemiptera), for example, has from 2 to 2.5 $\times$ 10^8 hairs/cm^2 forming the plastron. Insects that have 10^6–10^8 hairs/cm^2 generally have a very efficient plastron and usually can stay submerged for months (Mill, 1974). Some Coleoptera and Lepidoptera also have very efficient plastrons and often have eight or nine spiracles that open into the plastron air space.

Agents that lower the surface tension of the water (for example, soap or alcohols) will cause wetting of the plastron and failure to retain the air space. High pressure, if applied over long enough periods of time, will cause wetting of the plastron, but these high pressures are not likely to occur in the insect's natural habitat. It should be noted that a plastron can work in reverse and extract O_2 from the insect tissue and pass it into water that is very low in O_2 content, as might occur in cases of severe pollution. Insects utilizing a plastron usually live in the well-aerated water of streams, lake edges, and intertidal zones.

5.4.6. Spiracular Gill Bearing a Plastron

A gill bearing a plastron that is adapted for taking oxygen from water or the air directly has evolved as an extension from the spiracle in some Diptera (Simuliidae) and Coleoptera (Psephenidae). These insects frequently live in aquatic habitats subject to drying. In the water the gills provide a large surface area for extraction of O_2. In the air there is a direct route for O_2 entry through the interstices of the plastron. Water loss in air is reduced

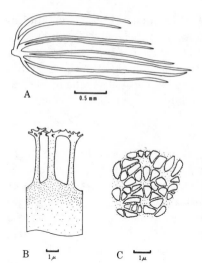

FIGURE 5-24.   A. Spiracular gills of *Simulium ornatum* (Simuliidae: Diptera). A plastron network covers nearly the entire gill surface. B. Diagram of plastron. C. Surface view of plastron network. (Reprinted with permission from H. E. Hinton, *J. Insect Physiol.* **10**, 73, Copyright 1964, Pergamon Press, Ltd.)

because the plastron opens into the atrium of the spiracle and the interface at the spiracle is small.

In the simuliids the spiracular gills are pupal structures formed at the larval–pupal ecdysis. They serve a respiratory function for pupal development and for the pharate adult until pupal ecdysis occurs. The plastron fills with gas shortly before the larval–pupal ecdysis, and after the ecdysis it quickly swells and straightens into its functional appearance (Fig. 5-24). The entire structure of the gill, except a small area at the base, bears a plastron in gills of *Simulium ornatum* (Hinton, 1964). The total surface area of the plastron of both gills is about 4 mm^2, and the effective water–air interface is nearly as large. This large water–air interface readily takes O_2 from the water by diffusion.

5.5.   RESPIRATION IN ENDOPARASITIC INSECTS

Cutaneous respiration is very important in some parasitic Hymenoptera and Diptera that have a fluid-filled tracheal system in the first instar. The tracheae generally fill with air in the second instar, but even then the spiracles become functional only just before the larva is ready to leave the host.

Some parasitic hymenopterans, including many chalcids and tachinid flies, are metapneustic and orient the posterior spiracles at the body surface of the host where they breathe air directly. Chalcid wasps remain in contact with the hollow pedicel of the egg from which they hatched at the host's integumental surface. Some tachinids stimulate the host's integument to become invaginated into a sheath around the parasite, leaving it with an air opening at the host's surface. The larva of the bot fly *Hypoderma* migrates to the skin of the vertebrate host where it bores a tiny opening to the surface through which gas exchange occurs.

5.6. RESPIRATORY PIGMENTS

Only a few insects have respiratory pigments in the hemolymph or cells. *Chironomus* larvae have the smallest known hemoglobin molecule (two units) with a molecular weight of 31,400. It is present in the plasma of the hemolymph and not in the hemocytes. It has a high affinity for O_2 and is 50% saturated at a pO_2 of 0.6 mm Hg at 17°C. Neither CO_2 tension nor temperature changes shift the O_2 loading curve. By contrast, 50% saturation of vertebrate hemoglobin occurs at a pO_2 of about 25 mm Hg at 37°C. The main function of the hemoglobin appears to be to increase the rate of recovery from anaerobic conditions and to keep larvae active at low O_2 tensions (Mill, 1974).

Hemoglobin does occur within certain cells of *Gastrophilus* larvae and in larvae and adults of *Anisops* and *Buenoa* (Notonectidae) (Mill, 1974). The hemoglobin of *Anisops pellucens* has a low affinity for O_2, showing 50% saturation at a pO_2 of 28 mm Hg at 24°C. The CO_2 tension (Bohr effect) causes little change in the curve, but temperature increase does shift the curve sharply to the right. Miller (1966) experimentally found that *A. pellucens* may get about 75% of the O_2 consumed during a normal dive from its hemoglobin.

5.7. RESPIRATION IN EGGS

Insect eggs have the same basic needs as other forms: to obtain sufficient oxygen for development without losing too much water in the process. The majority of aquatic and semiaquatic insects lay eggs with no special respiratory structures incorporated into the shell. Gas exchange occurs by simple diffusion through interstices in the egg shell (Hinton, 1969).

Since the eggs of most aquatic insects have no special respiratory structures, it is somewhat surprising to learn that eggs of a majority of terrestrial insects do contain special structures for respiration. They tend to have an extensive, inner chorionic meshwork that can function as a plastron when the egg is submerged in well-aerated water (Fig. 5-25) (Hinton, 1969). Aeropyles, tubes of very small diameter, connect the meshwork in the inner chorion with the outside atmosphere. Because the openings of the aeropyles present so little surface (for example, 536 μm^2 for a *Rhodnius* egg), water loss from the egg is not increased greatly in an air environment over what might be lost without the plastron network. The plastron is an adaptation to prevent an O_2 deficiency when the eggs are subjected to periodic wetting from dew, rain, or temporary flooding. Furthermore, the eggs of many ter-

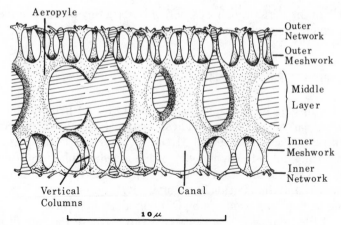

FIGURE 5-25. Diagrammatic representation of a section through the plastron region of the eggshell of a muscid fly. [From H. E. Hinton, *Ann. Rev. Ent.* **14**, 343 (1969).]

restrial insects frequently are laid in wet environments such as decaying organic matter, animal manure, fruits, or other plant tissues.

Ideally a plastron would allow oxygen uptake under water to equal that in air, but generally the actual efficiency is much less than this. Furthermore, an effective plastron must resist wetting of the air-containing structures at the hydrostatic pressures prevailing in the egg's natural habitat. There must also be resistance to wetting in the presence of surface-active agents frequently present in situations where eggs are laid as, for example, in cow manure or in decaying flesh. Resistance of the plastron to wetting is determined by both the pressure exerted and the length of time it is exerted.

Tests have shown that even the least-resistant plastrons will withstand a pressure equivalent to about 31 cm Hg for 30 min. A 4 mm diameter raindrop will exert about the same pressure, but for only a millisecond or so. Thus, the plastron of eggs exposed to raindrops will not become wet. Tests further show that eggs normally laid in environments containing surface-active agents resist wetting better than those eggs laid in sites where surface-active agents are less likely to be encountered. Clearly this greater resistance to wetting in the presence of surface-active agents is an adaptation that has provided a selective advantage to those eggs having it. Nevertheless, it has been shown in laboratory tests that wetting of plastrons of all kinds will occur before there is physical destruction of the plastron or hydrofuge structure. This in itself is an adaptation for survival, provided that the egg or insect can escape and dry out again.

It turns out that a large number of the eggs of terrestrial insects do not have a large enough water–air interface per unit weight to enable effective function of the air space as a plastron when the eggs are submerged under water. Little is known about specific cases in these eggs, but some may be

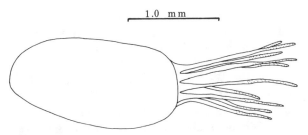

FIGURE 5-26.   Egg of *Nepa cinerea* with 10 respiratory horns. The plastron area is indicated by stippling. (Reprinted with permission from H. E. Hinton, *J. Insect Physiol.* **7**, 224, Copyright 1961, Pergamon Press, Ltd.)

able to tolerate a reduction in metabolic rate from growth to simple maintenance during periods of submersion.

In most eggs the plastron is a part of the shell and thus is very close to the developing tissue. The gradient for O_2 diffusion from air to tissue will usually be great enough over the short distance involved to make the entire plastron surface effective in supplying O_2. In contrast, some eggs have the plastron elevated on a respiratory horn (Hinton, 1961) (Fig. 5-26) that may be up to 10–13 mm long. The plastron may cover most of the surface of the horn. When the egg is submerged completely in water, the length of the longer horns makes the gradient for gas exchange favorable only in the proximal part of the horn. It may be that long horns are useful as a conduit to atmospheric air when the egg is not too deeply submerged in water.

5.8. CONCLUSIONS

Part of the success of insects in effectively exploiting so many different habitats must be attributed to the highly plastic respiratory system. The tracheae are not simply a series of inert tubes ramifying throughout the body of an insect; they are a living tissue with a cellular component, the epithelium. Evidence of this plasticity may be found in the rich array of morphological variations in terrestrial and aquatic insects. For example, there are closed and open tracheal systems, and many variations in structure, number, and position of spiracles and in number and size of air sacs. Although eggs do not have tracheae, they frequently have specialized respiratory surfaces that adapt them for the particular environment, often as not a rapidly changing one, wherein they are laid. The morphology of the respiratory system grows and changes at each molt, especially at metamorphosis. For example, the closed tracheal system of an aquatic damselfly larva becomes the open system of the terrestrial adult.

The tracheal system is an effective way to provide for gas exchange in small animals. Insects not much larger than a drosophilid fly may depend almost entirely on diffusion of gases within the system, but larger insects actively ventilate the body and, especially, the flight musculature. The high efficiency of the tracheae in providing gas exchange is directly related to the extensive network of tracheoles bringing an air path to within a few micrometers of mitochondria. Weis-Fogh conceptualized this relationshp mathematically as the hole fraction that varies from 10^{-1} to 10^{-3}.

Adaptations for water conservation provide additional examples not only of the plasticity within the tracheal system but also within insects as a group. The spiracles are not open all the time in most insects, for that would allow excessive water evaporation. Instead they open at intervals determined in some insects by low oxygen tension and in others by carbon dioxide accumulation. Probably, as in cecropia pupae, there is a respiratory compromise, mediated by the need to renew oxygen and to remove accumulating carbon dioxide, that affects spiracular behavior in many insects. In flying insects, ventilation of the thorax with large volumes of air during flight inevitably causes loss of water. Some insects compensate for this loss by producing substantial amounts of water from the metabolism of fat utilized as the fuel for flight. Behavioral adaptations also aid in water conservation.

Many major areas of respiratory physiology are virtual *terra incognita*! The forces, gradients, and pathways that act on the tracheal epithelium to guide its growth are, for the most part, unknown. The possibility of endocrine influences on the growth and development of the tracheal system needs investigating. Few of the many aquatic insects have been studied in detail with respect to respiration, and very little, indeed, is known about respiration of endoparasitic insects. To be sure, many opportunities still exist for significant contributions to our knowledge of the respiratory system and for documentation of adaptations within it.

REFERENCES

W. E. Beckel and H. A. Schneiderman, *Science* **126**, 352 (1957).

C. Bordereau, *Int. J. Insect Morph. Embryol.* **4**, 431 (1975).

B. N. Burkett and H. A. Schneiderman, *Science* **156**, 1604 (1967).

J. F. Case, *J. Insect Physiol.* **1**, 85 (1957).

R. F. Chapman, *The Insects, Structure and Function,* American Elsevier, New York, 1969.

H. E. Hinton, *J. Insect Physiol.* **7**, 224 (1961).

H. E. Hinton, *J. Insect Physiol.* **10**, 73 (1964).

H. E. Hinton, *Ann. Rev. Ent.* **14**, 343 (1969).

G. Hoyle, *J. Insect Physiol.* **7**, 305 (1961).

J. W. Kanwisher, *Biol. Bull.* **130**, 96 (1966).

M. L. Keister, *J. Morph.* **83**, 373 (1948).

R. I. Levy and H. A. Schneiderman, *J. Insect Physiol.* **12**, 83 (1966a).

R. I. Levy and H. A. Schneiderman, *J. Insect Physiol.* **12**, 465 (1966b).

M. Locke, *Quart. J. Micr. Sci.* **99**, 29 (1958).

P. J. Mill, "Respiration: Aquatic Insects," in M. Rockstein, Ed., *The Physiology of Insecta,* Vol. 6, 2nd ed., Academic Press, New York, 1974.

P. L. Miller, *J. Exp. Biol.* **37,** 237 (1960).

P. L. Miller, *J. Exp. Biol.* **44,** 529 (1966).

P. L. Miller, "Respiration: Aerial Gas Transport," in M. Rockstein, Ed., *The Physiology of Insecta,* Vol. 6, 2nd ed., Academic Press, New York, 1974.

H. Oberlander, "Morphogenesis in Tissue Culture: Control by Ecdysteroids," in M. Locke and D. S. Smith, Eds., *Insect Biology in the Future,* Academic Press, New York, 1980.

W. F. Pickard, *J. Insect Physiol.* **20,** 947 (1974).

O. W. Richards and R. G. Davies, "The Respiratory System," in Imm's *General Textbook of Entomology,* Vol. 1, *Structure and Development,* 10th ed., Chapman and Hall, London, 1977.

H. A. Schneiderman, *Biol. Bull.* **119,** 494 (1960).

H. A. Schneiderman and C. M. Williams, *Biol. Bull.* **109,** 123 (1955).

D. S. Smith, *J. Morph.* **114,** 107 (1964).

R. E. Snodgrass, *Principles of Insect Morphology,* McGraw-Hill, New York, 1935.

W. G. Van Der Kloot, *Comp. Biochem. Physiol.* **9,** 317 (1963).

T. Weis-Fogh, *J. Exp. Biol.* **41,** 207 (1964a).

T. Weis-Fogh, *J. Exp. Biol.* **41,** 229 (1964b).

T. Weis-Fogh, *J. Exp. Biol.* **47,** 561 (1967).

A. T. Whitehead, "Respiration," in V. J. Tipton, Ed., *Syllabus Introductory Entomology,* Brigham Young University Press, Provo, Utah, 1973.

W. Wichard and H. Komnick, *J. Insect Physiol.* **20,** 2397 (1974).

V. B. Wigglesworth, *J. Exp. Biol.* **36,** 632 (1959).

V. B. Wigglesworth, *The Principles of Insect Physiology,* 7th ed., Chapman and Hall, London, 1972.

V. B. Wigglesworth, *J. Cell. Sci.* **26,** 161 (1977).

6

MUSCLE SYSTEMS

T. SMYTH, JR.
Department of Entomology
Pennsylvania State University
University Park, Pennsylvania

CONTENTS

SUMMARY

The muscles of insects (and the muscles of other animals) represent a vast array of variations on a common theme. All are based on interacting action and myosin protein filament systems arrayed in parallel for directional pull. They are powered by ATP and are controlled mainly by mechanisms regulating the free internal calcium ion concentration. Insect muscles differ from typical vertebrate counterparts in many ways, including details of fine structure, the molecular mechanisms of activation–contraction coupling, and their neuromuscular chemical transmitters. Typically, they receive multiterminal or distributed innervation, often by two or more axons having different functions such as fast or slow activation or inhibition. Many insect muscles are greatly modified for specialized kinds of performance such as extreme shortening, rapid activation and relaxation, or graded tension development. The asynchronous fibrillar indirect flight muscles of many of the most effective fliers are capable of especially high rates of metabolism, combined with high efficiency of energy utilization for mechanical work. The extraordinary motor performance of many insects derives partly from the favorable strength–weight relationship that goes with small size and also from mechanisms that store mechanical energy in elastic structures from which it can be recovered to provide power for effective movements.

6.1. INTRODUCTION

A major characteristic of insects is their great mobility. Most insects can run, jump, fly, or swim rapidly and extensively, at least at some stage in their life cycle. This capability has undoubtedly contributed to their vast ecological diversity and evolutionary success.

Effective and efficient movement requires not only a suitable motor apparatus but also coordinated specializations of many other aspects of morphology and physiology—biochemical and respiratory capabilities to supply energy at the necessary rates; regions of integument specialized for strength; elastic storage of energy or almost infinite flexibility; ability to deal with excess metabolic heat by convective, radiative, or evaporative cooling, or to conserve heat and thus increase the metabolic rate; and so forth. Thus, adaptations for rapid locomotion are almost as important as determinants of the physiology of insects as the exoskeleton and the trachael system.

Motility is a general characteristic of eucaryotic cells. Although there are many kinds of movement, some entirely intracellular and others expressed externally, most are effected by one or the other (or both) of two general systems of proteins, the tubulin–dynein microtubular system and the actin–myosin filament system. Both systems are important in insects.

Microtubules are major components of mitotic spindles and are implicated in many aspects of intracellular movement and transport, most dramatically in the rapid movement of materials along nerve axons. They are also central components of flagella and cilia. These are relatively unimportant structures in insects, although functional or rudimentary flagella occur in insect spermatozoa and ciliary elements are found in certain sensory receptors.

Actin and myosin filaments are major components of the contractile system of muscles. However, these same proteins, together with a variety of associated structural and control proteins, are also found in nearly all other kinds of cells. They have been implicated in morphogenetic movements, contractions of microvilli, and several kinds of intracellular movement, and they are reported to be major constituents of the nuclear envelope, the polar regions of spindles, and the protein fraction of chromosomes. Actin, especially, associates with many other proteins that participate in essential cell functions. Consequently, it is not surprising to learn that actin is among the most invariant or evolutionarily conservative of all proteins. Purified actins and myosins can interact even when extracted from organisms in different kingdoms. Consequently, there is an inherent similarity among all muscles, even those that may have evolved independently from nonmuscle cells because their special contractile function has been achieved by marshalling the same general assortment of ubiquitous macromolecules.

No insect muscle has been studied comprehensively; none is as well known as the sartorius of the frog. In order to achieve a general understanding, it is necessary to piece together information from limited studies on many muscles, including muscles of animals in other phyla. There is an inherent unity among all insect muscle systems because they are based on the same general biochemical components and organized into similar subcellular structures. But there is, of course, some danger in generalizing too broadly. Just as the basic contractile system can be modified to subserve many cell functions, organized muscle tissue also is highly variable as a consequence of special functional needs which may emphasize such properties as extreme speed, strength, endurance, range of movement, or precision of control. Indeed, there is an extensive range of variability in nearly all of the functional characteristics of insect muscles. In this chapter the general nature and properties of muscle are summarized and the range of adaptive specializations is indicated. Comparisons with vertebrate muscles are included because these are studied in general biology courses. At the end of the chapter there are brief discussions of the motor mechanisms of flight and jumping, high-performance activities of special importance to insects.

6.2. STRUCTURE AND CHEMISTRY

6.2.1. Structure

Muscles are composed of cellular units called muscle fibers, a name that predates cell theory. Small fibers like the short 1–5 μm diameter fibers associated with insect viscera generally have one nucleus and are branched or cleft at their ends. The cardiac muscle fibers of insects are similar but tend to be larger and multinucleate. Like their vertebrate counterparts, they abut on neighboring fibers through intercalated discs. Skeletal fibers can be much larger, with diameters ranging from a few micrometers to more than 1 mm, and contain many nuclei. These multinucleate fibers may grow by internal mitoses (flight muscles of Orthoptera), fusion of myoblasts (Diptera), or both processes (Hemiptera). Although it is common to classify insect muscles by anatomic position as visceral, cardiac, or skeletal, these are not discrete categories in terms of structure, biochemistry, and control or functional behavior, like their vertebrate counterparts. Rather, there exists a vast assortment of fiber types with structures and properties appropriate to the pattern of use of the particular muscle at the specific stage in development.

Many of the subcellular components of muscle fibers were discovered and named before their homologies with parts of other cells were worked out. These terms usually include the roots sarc- or my-, derived from Greek words for flesh and muscle. Thus, the cell membrane is called the *sarcolemma,* a term often used to include not only the true "unit membrane" permeability barrier but also enveloping fibrous layers. The mitochondria are called sarcosomes; smooth endoplasmic reticulum, sarcoplasmic reticulum; and the fluid cytosol, sarcoplasm. Usually the contractile protein filaments are aggregated into discrete longitudinal threadlike or straplike myofibrils. In visceral fibers the contractile material is generally not separated into fibrils but has a "field" organization similar to that in vertebrate tonus (slow) and cardiac fibers.

Skeletal muscles are attached to the integument, at least at one end. The muscle fibers are attached to epidermal cells through desmosomes where their membranes are wrinkled, increasing the contact area about 10-fold. Traversing the epidermal cells are dense clusters of microtubules, which must have a tensile function. At the outer surface of the epidermal cells hemidesmosomes provide connections of chitinous muscle attachment fibers that extend outward through pore canals to end in the cuticulin layer (outer epicuticle). These attachment fibers are resistant to molting fluid and are shed during molting, breaking just distal to the new cuticulin layer. Often the cuticular attachment extends inward as an apodeme (tendon). There are no pore canals at these attachments.

The cardiac fibers comprising the heart and the fibers surrounding visceral organs are termed intrinsic fibers. They attach to each other. Extrinsic fibers such as the alary muscles of the heart and their counterparts related to viscera run between the intrinsic fibers and the exoskeleton. These are essentially thin skeletal fibers, although they are functionally cardiac or visceral dilators. Visceral fibers in the capsules of some glands insert into connective tissue.

Insect skeletal muscles are usually classified according to the appearance and arrangement of their myofibrils. Flight muscles fall into three general groups, but other skeletal muscles can be of additional types (Fig. 6-1). Radial fibers, also called lamellar and tubular, have straplike fibrils orga-

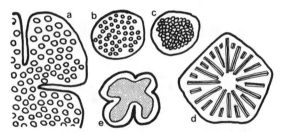

FIGURE 6-1.   Types of muscle fibers, cross-sectional diagrams. A. Fibrillar muscles with 2–4 μm
fibrils. B. Microfibrillar, 1 μm cylindrical fibrils. C. Microfibrillar, mosaic of closely packed
polygonal fibrils. D. Tubular fiber with radial straplike fibrils. E. Field structure—no discrete
fibrils, but contractile protein throughout the central region. Sarcosomes, sarcoplasmic reticulum, transverse tubules, and nuclei occupy the extrafibrillar spaces.

nized about a central core that contains the nuclei. Between the fibrils are
large slablike sarcosomes. In the flight muscles of Odonata and Blattidae the
fibrils are surrounded by a very extensive sarcoplasmic reticulum. Radial
fibers are also found in adults of Diptera and Hymenoptera in various skeletal muscles, but not in the indirect flight muscles.

Microfibrillar, mosaic, or close-packed fibers have threadlike myofibrils
about 1 μm in diameter. These may appear circular or polygonal in cross
section depending on how tightly they are packed. The fibers themselves are
generally 10–100 μm in diameter. In the flight muscles of Orthoptera,
Trichoptera, and Lepidoptera, fibrils alternate with columns of sarcosomes
that give them a light pink hue. Nonflight microfibrillar muscles, on the other
hand, may have few sarcosomes and appear white, with tightly packed
fibrils, as has been reported for some of the muscles of Hemiptera.

Fibrillar flight muscles have larger myofibrils, 2–4 μm in diameter, and
the fibers themselves also are large, ranging to more than 1 mm across. They
are found in Diptera, Hymenoptera, Coleoptera, and most Hemiptera. The
surface areas of the fibers are large because they are deeply cleft and fluted.
Undoubtedly, this facilitates exchange of metabolites with the hemolymph.
Sarcosomes are large and abundant, reflecting capability for intense oxidative metabolism, but the sarcoplasmic reticulum is scanty, paradoxically
indicating slow activation and relaxation.

All insect muscle fibers appear cross-striated under the microscope (Fig.
6-2), although the banding in some visceral muscles is not very distinct. The
repeat pattern of 2–8 μm is a consequence of the segregation of major
protein components along with the fibrils, regions of globular actin and its
associates alternating with predominantly alpha-helical portions of myosin
molecules. The highly parallel molecular order in the latter regions makes
them optically birefringent or anisotropic. Hence, they are called A bands.
Correspondingly, the more isotropic actin-containing regions are called I
bands. The A and I bands are in register in all myofibrils across each fiber,
either like stacks of coins or in a continuous spiral about a central core.

The termination of a fibril at the end of a fiber, or a fibrillar break, occurs

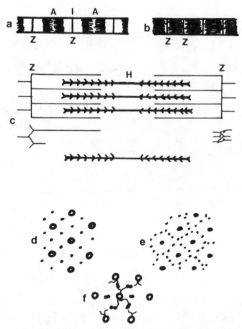

FIGURE 6-2. Diagram of myofibrils and their components. A. Fibril extended, showing alternating A and I bands. B. Fibril shortened. C. Long sections showing thick filaments with cross bridges and thin filaments attached to Z lines. D. Cross section through region of overlapping thick and thin filaments, fast-twitch fiber. E. The same, slow-twitch fiber, indicating a higher ratio of thin to thick filaments. F. Myosin bridge interaction with thin filaments.

at the middle of an I band at a thin, stainable line, the Z line or disc (*Zwischenscheibe*). Thus, the basic repeat unit along a fibril is between two adjacent Z lines. This unit is called a sarcomere. In general, slowly contracting fibers have long sarcomeres. In relaxed, gently stretched muscles, I bands may occupy 30–50% of each sarcomere. When the muscle shortens, the I bands become shorter and may even disappear. In flight muscles, especially the fibrillar type, the I bands are much narrower and the extent of shortening is much less.

At the center of each A band there is usually a lighter staining region, the H zone (*Hellescheibe*). Sometimes an M line (*Mittelmembran*) can be discerned in the middle of the H zone.

6.2.2. The Contractile Protein System

When electron microscopy was effectively used to examine striated muscles in the early 1950s, it was discovered that the A bands are composed of relatively thick double-tapered protein filaments held in register laterally. Each thick filament has crowns of small lateral projections known as cross bridges at intervals of about 14.5 nm. There are no bridges in the H zone. To either side of the H zone, the bridges have opposite polarities.

A second set of filaments, the thin filaments, originates at the Z lines. Alternate thin filaments project into adjacent sarcomeres. In fibrillar flight muscles, the thin filaments leave the Z lines in a regular hexagonal pattern. This is quite different from the square pattern characteristic of vertebrate muscles.

Except when stretched beyond normal limits, thin filaments interdigitate between thick filaments. During shortening, the thin filaments slide further between the thick filaments, bringing the Z lines closer together. Ordinarily, contraction can proceed until the ends of the thick filaments are jammed against the Z lines and the I bands have disappeared. However, some insect muscles, such as the body wall muscles of blowfly larvae and various visceral and cardiac muscles, are capable of further shortening because the thick filaments pass through holes in the Z lines and extend into the adjacent sarcomeres.

In cross sections of fast-acting skeletal muscles and certain slower-flight muscles, the thick and thin filaments are in very precise hexagonal arrays. Where the filaments overlap in a partly contracted muscle, each thick filament is surrounded by six thin filaments and each thin filament lies midway between two thick filaments, giving an overall thin–thick ratio of 3 : 1. (In vertebrates the thin filaments lie equidistant from three thick filaments and the overall ratio is 2 : 1.) Slow muscles have more thin filaments surrounding each thick one, and ratios of 4 : 1, 5 : 1, 6 : 1, and even higher have been observed.

Thick filaments are composed of tadpole-shaped myosin molecules. These have been dissected by enzymes and other biochemical reagents to yield fragments that have been individually studied. The tails are paired alpha-helical peptides that associate with the tails of other myosin molecules to form the shafts of the thick filaments. The heads form the cross bridges and contain paired extensible and compact regions and also four light chain peptides. Functionally, the bridges contain a Mg^{2+}-ATPase that is activated by interaction with actin, a process that is partly regulated by Ca^{2+}-binding and phosphorylation at other sites on the bridges. The packing of myosin molecules in insect thick filaments differs from that in vertebrates, at least in flight muscles, because the filaments are much thicker and have more bridges at each crown. In fibrillar flight muscles the thick filaments apparently have a core of another protein, paramyosin. Cross sections of thick filaments from muscles of several insect groups differ in size and shape, implying a variety of molecular packing patterns.

Thin filaments consist of double chains of globular actin molecules twisted slowly about each other like two strings of beads. The thin filaments also contain two regulatory proteins, tropomyosin and troponin, one of each for every seven actin molecules. Tropomyosin molecules extend along the actin chains where they apparently control the access of myosin bridges to the binding sites of the actin molecules. The position of tropomyosin is determined by troponin which, in turn, responds to the level of free Ca^{2+}.

Activation of myosin ATPase, which supplies the energy for contraction, depends, therefore, on three processes in insects: myosin cross-bridge activation by phosphorylation and calcium binding, and activation of the myosin-binding sites of actin by calcium binding to the troponin–tropomyosin complex.

Although actin, myosin, tropomyosin, and troponin have absorbed most of the attention of students of muscle proteins, other proteins have been isolated from myofibrils. The functional roles of most of these are neither firmly established nor generally accepted. There have been recurring reports of super thin filaments (about 2.5 nm) extending between the Z lines. They undoubtedly contribute to the resting tension of the fiber. Some investigators believe these filaments have a central role in active tension development. Alpha-actinin is reported to be the major protein of the Z lines. Other proteins are reported to lie in the M line region, possibly holding thick filaments in register. Still others surround the fibrils, perhaps keeping adjacent fibrils in register. In certain fibrillar flight muscles the thick filaments appear to be connected to the Z lines by what has been called C protein, and it has been proposed that this accounts for the great stiffness and limited extensibility of these fibers.

It is generally assumed that the mechanism of force generation resides in the myosin cross bridges. According to various models, contact of a bridge with an active site on an actin molecule causes a configurational change in which the end of the bridge rocks or its length changes, pulling the thin filament along the thick about 10 nm. ATP is required to detach the bridge from the thin filament. Under low Ca^{2+} conditions the bridge is unable to reattach and relaxation occurs. With high Ca^{2+} the bridge may react with another actin, more ATP being hydrolyzed and contraction continuing. Such models predict that the ability to develop active tension should be directly proportional to the degree of overlap of thin filaments with correctly oriented bridge-bearing portions of thick filaments. An excellent correlation of this sort has been reported for a frog muscle.

Several observations of length-versus-tension relationships on insect muscles have not been fully reconciled with this kind of interpretation. There have been reports of insect muscles retaining significant ability to develop an active tension increment even when stretched so far that there should be no overlap of thick and thin filaments. In other studies, the relationship between active tension and overlap was found to be nonlinear. There have also been reports of changes in the dimensions of the filaments during contraction of certain insect muscles. An alternative (minority) interpretation holds that the sliding filaments and cross bridges are essentially a mechanism to control enzyme activity, and that force generation must involve additional elements that have not been included in the various versions of the sliding filament model.

6.2.3. Energy

The ability of muscles to do mechanical work by conversion of biochemical energy stores has stimulated a vast amount of inquiry. Insect flight muscles have attracted special interest as they far outperform other tissues in metabolic rate, with peak rates reaching 100 times resting levels. Actually, most of the biochemical energy liberated appears as heat rather than mechanical work. Since rates of heat production and work performed are easier to measure than the rates of intracellular biochemical events, the energy relations of muscle have been studied mainly with the aid of such measurements. For certain vertebrate muscles it has been possible to identify specific energy costs associated with activation, maintenance of contraction, shortening, relaxation, and recovery. The fractional costs related to these various processes differ for different muscles. Recently, as the molecular events have become better defined, there have been efforts to assess the costs of such processes as membrane excitation and the restorative active transport of Na^+, K^+, and Ca^{2+} across the sarcolemma, Ca^{2+} release and active reuptake by the sarcoplasmic reticulum, and myosin cross-bridge attachment cycles, as well as the costs of general energy supply metabolism. Viewed in this way, as will become apparent later, fibrillar flight muscles spend a minimal amount of energy on noncontractile processes.

As metabolic stores are utilized, about two-thirds of their energy is released immediately as heat. The remainder is trapped as phosphate bond energy in ATP or some energetically equivalent compound. This is available for contraction and other energy-requiring processes such as active transport, growth, and repair. Always, some fraction appears as heat.

Flight muscles can expend ATP at rates equivalent to several times their weight per day, but the total amount of ATP present at any moment is very small, enough for about 1 sec of activity. An additional small store of readily mobilized phosphate bond energy is available in the form of arginine phosphate (counterpart of the vertebrate's creatine phosphate). When activity continues beyond the few seconds these phosphate bonds can support, ATP must be regenerated through metabolism of major metabolic reserves. In flies glycogen supplies this energy. Migratory insects may start with carbohydrates and then switch to fats for energy after a few minutes. The tsetse fly burns proline. Honeybees depend on carbohydrates from the crop. (See Chapter 12.)

Just as in vertebrates, the muscles of insects differ widely in their capacity for oxidative metabolism. In general, flight muscles are so well tracheated and so abundantly supplied with mitochondria that they function altogether aerobically. On the other hand, muscles that are used only briefly for intense effort, like the jumping muscles of grasshoppers, have a low metabolic capability. When called on for extended or repetitive activity, they quickly go anaerobic, generating an oxygen debt.

There has recently been a surge of interest in the thermal relations of insects. Although they are basically heterothermic with resting temperatures following the ambient temperature, many can regulate the thoracic temperature to a higher range when undertaking intense activities such as flight. Muscular activity, especially by the flight muscles, generates most of the heat. Heat is conserved by insulating structures, such as superficial air sacs and cuticular scales and hairs, and by mechanisms for limiting cooling due to evaporative water loss. The advantage of this arrangement is to minimize the energy metabolism of the resting insect while allowing energy to be mobilized at high rates on demand.

6.3. CONTROL OF CONTRACTION

6.3.1. Neuromuscular Control

The control of vertebrate skeletal muscle has been extensively studied and can be summarized in a straightforward way. The summary will serve as a basis for discussing the more variable and complex control patterns that have been found in insect neuromuscular systems.

The vertebrate muscle is innervated by a motor nerve containing many motoneuron axons. Each axon branches to innervate a number of muscle fibers that are activated in unison. The neuron and the muscle fibers it controls constitute a motor unit. The strength of a contraction depends on the number and assortment of units recruited. During prolonged moderate contractions, motor units may be alternated as particular units become fatigued. This is controlled by the central nervous system.

Generally, vertebrate twitch-type muscle fibers are innervated at one spot, the motor end plate. On arrival of a nerve impulse, the nerve ending releases about 200 packets of the transmitter acetylcholine (ACh), each consisting of several thousand molecules. After diffusing across a narrow space, the synaptic cleft, some of these ACh molecules combine with receptor molecules in the muscle (postsynaptic) membrane, opening gates to cation-selective channels through the membrane. Since active transport has previously created a sarcoplasm that is high in K^+ but low in Na^+, these ions now tend to diffuse through the open channels down their concentration gradients. Locally, the membrane potential of the fiber is reduced from its resting level of about -70 mV (inside negative) to near zero, a compromise between the Nernst potentials for the Na^+ and K^+ concentration cells. This local depolarization is called an end-plate potential. Current spreads passively from this point, depolarizing adjacent electrically excitable membranes and generating an action potential. This propagates like a nerve impulse, exciting the whole length of the muscle fiber. The fiber responds to a single stimulus with an all-or-nothing twitch and to a high-frequency train of stimuli with a sustained tetanus.

The innervation and control of insect skeletal muscle differs from this typical vertebrate pattern in several ways. The motoneurons generally run along the muscle fibers, synapsing repeatedly at intervals of 10–100 μm, thus providing distributed rather than point innervation. The actual synaptic junctions may be elevated above the muscle surface, lie on the surface, or be indented into the muscle. Surface junctions are usually semi-isolated from the hemolymph by a covering of glial cells. This longitudinally distributed innervation provides for synaptic activation of the muscle fibers locally, all along their length, making propagated muscle action potentials unnecessary. Indeed, many insect muscle fibers show only limited or graded electrical excitability and ordinarily do not produce propagating action potentials.

All fibers in a muscle may be innervated by branches of the same motoneuron, in which case the whole muscle could be regarded as a single motor unit. Other muscles are really composites of several synergistic muscles, each with its own innervation. In these cases, the component units either may be separately controlled or the motoneurons may be tightly coupled in the central nervous system so that the whole complex functions like one unit.

Many insect muscle fibers receive more than one motoneuron (multiple or polyneuronal innervation) (Fig. 6-3). The motor units as defined by the fields of innervation derived from these fibers can be different. For example, one motoneuron may innervate all the muscle fibers in a muscle. Another may innervate a few of these fibers plus some in another muscle. Thus, it is possible for a group of muscle fibers shared by different units to contribute to different actions, depending on which system activates them. Some visceral muscles and some insect hearts have no innervation. Others may receive neurons that release neurosecretory modulators or hormones rather than fast-acting transmitters.

When there is multiple innervation, the different motoneurons generally have different actions (Fig. 6-4). One excitatory neuron may release numerous packets of transmitters, evoking maximal end-plate potentials and a strong, rapid twitch or tetanus. This is called a fast excitor. Slow excitors release few packets of transmitters and evoke small depolarizations. There is little or no contractile response to a single impulse. When the slow fiber fires repetitively, the end-plate potentials may show summation (addition) or facilitation (a more than additive effect). The greater depolarization produces greater muscle activation and tension. Thus, the rate of firing of slow

FIGURE 6-3.   Diagram of muscle fiber with dual innervation. Synaptic contact is made at many points by the two axons which often run together and branch in parallel.

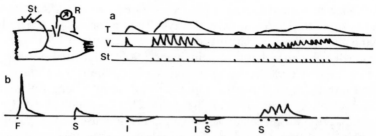

FIGURE 6-4. Electrical (V) and mechanical (T) responses to stimulation. A. Responses to single and repetitive stimuli via fast (F) and slow (S) excitatory motoneurons. Repetitive stimulation of the F axon produces a declining plateau of tension (tetanus) that is greater than peak-twitch tension. Repetitive stimulation of the S axon produces intermediate tensions that vary with the stimulation frequency. B. Electrical responses to F, S, and inhibitory (I) stimulation. End-plate potentials evoked by S excitors during inhibitory postjunctional potentials are reduced in magnitude and duration. Repetitive S end-plate potentials may show facilitation.

motoneurons is responsible for slow movements and the maintenance of tonic intermediate or partial contractions. In addition to fast and slow excitors, intermediate types are commonly found. It is possible for muscle fibers to receive three or even four excitors with differing actions.

Some insect muscles also have distributed inhibitory innervation. Inhibitors reportedly go only to muscle fibers that receive slow excitation. However, some of these receive up to four inhibitory neurons. The effect of the inhibitors is to reduce the amount of depolarization evoked by the slow excitors, reducing the strength of contraction or speeding relaxation at the end of a movement.

It seems probable that all excitors release glutamic acid (GA) as their transmitter, and it is generally agreed that gamma-aminobutyric acid (GABA) is the inhibitory transmitter. The ionic basis for excitatory (depolarizing) end-plate potentials is not well worked out and appears to differ among insect groups, reflecting the diverse and variable composition of hemolymph. In the more primitive orders, hemolymph Na^+ and sarcoplasmic K^+ tend to be high, much as in vertebrates. Excitation makes the postjunctional membrane more permeable to these ions, and probably also to Ca^{2+}, which is relatively high in hemolymph. In the higher orders, and especially in phytophagous insects, the hemolymph composition is quite different, tending to be lower in Na^+, higher in K^+, Ca^{2+}, and Mg^{2+}, and also high in organic ions. In vertebrate muscles the resting potential shows a strong dependence on the K^+ concentration cell, declining as the extracellular K^+ increases relative to the intracellular level. The membrane potential and contraction of many insect muscles shows less of a K^+ dependence. In some cases bathing with a high K^+ saline evokes a sustained depolarization and contraction. In other cases the effects are transient.

The action of GABA is less ambiguous. It appears to open Cl^--selective channels. Since the equilibrium or Nernst potential (E_{Cl}) for the Cl^- concentration cell is near the value of the resting potential, the effect of inhibitory transmitter is to hold the membrane potential more firmly at this level, making it more difficult to depolarize and thus excite the muscle. Depending on the exact value of E_{Cl} relative to the resting potential, GABA may evoke a slight hyperpolarization or depolarization. Avermectins, which are new broad-spectrum insecticides and antihelminthic agents, mimic the action of GABA.

Both GA and GABA are released as multimolecular packets or quanta, like ACh in vertebrates. The rate of release of these from nerve endings similarly is increased by depolarization and osmotic shrinkage and by the action of venoms of several spiders. It appears to be strongly dependent on the level of free Ca^{2+} in the nerve terminals.

In addition to excitatory glutamergic and inhibitory GABAnergic neurons, many skeletal muscles receive branches of dorsal unpaired medial (DUM) neurons. These release a transmitter (possibly dopamine or octopamine) which modulates the effect of the slow excitors, increasing the speed and vigor of contraction and also accelerating relaxation. This transmitter or modulator probably acts by stimulating the enzyme adenyl cyclase in the muscle membrane, increasing the production of cyclic adenosine monophosphate (cAMP), which, in turn, stimulates metabolism, increasing the ATP supply.

Many insect visceral and cardiac muscles are spontaneously active, showing rhythmic myogenic contractions much like those of vertebrate cardiac muscle. Some, including the muscle bands of Malpighian tubules in many insect groups, are uninnervated. Their level of activity is modulated by hemolymph chemistry and mechanical stretch. When innervation is found, the neurons often have characteristics of fine structure that suggest neuroendocrine or modulator rather than transmitter action. The neural pentapeptide proctolin is such a modulator, increasing the activity of hindgut musculature. It has also been reported to have a myotropic action on several other insect skeletal and cardiac muscles. Although no nerve endings have been found in the hearts of a few insects (some larval Diptera and adult *Hyalophora*), most insect hearts are innervated by lateral cardiac nerves of the anterior stomatogastric nervous system or by segmental ventral unpaired medial nerves, or both. Branches of the segmental nerves may also innervate alary muscles. Additional nerve cell bodies are sometimes present in the lateral cardiac nerves. Most of these neurons are filled with colloidal material and appear to be neurosecretory cells. There may also be nonneurosecretory axons which innervate the cardiac muscle. It has been suggested that these are inhibitors.

The resting potentials of skeletal muscle fibers vary widely. As an example, consider a hypothetical fiber with a resting potential of -60 mV (inside negative). When the membrane potential is reduced, the tension-generating mechanism is activated, although there is little effect until the potential is 10–20 mV below the resting level. With further depolarization, contraction increases, reaching a maximum when the membrane potential is somewhere in the vicinity of -10 mV. In skeletal muscles depolarization is due to the distributed excitatory innervation that produces a distributed end-plate potential. This is augmented by any local electrical excitability and modified by the actions of hormones and modulators and by the ionic composition of the hemolymph.

Many visceral and cardiac muscles differ from such behavior in not having constant resting potentials or discrete thresholds for stimulation. They may behave as if continually partially activated to a variable extent or may show cyclical changes in excitation as transmembrane currents of ions such as Ca^{2+} increase and then become self-limiting.

6.3.2. Internal Activation

In all muscle fibers except the smaller visceral fibers, the level of excitation is communicated throughout the interior by a transverse tubule system (TTS) consisting of invaginations of the sarcolemma (Fig. 6-5). The tubules ramify around the myofibrils, providing very local activation. In most insect muscles the TTS has a precise location relative to the sarcomere pattern. It may lie in the plane of the Z line or the M line, or there may be two sets at intermediate levels. In some cases there are two kinds of tubules. Presumably, one set has the usual activating function. The other, lined with base-

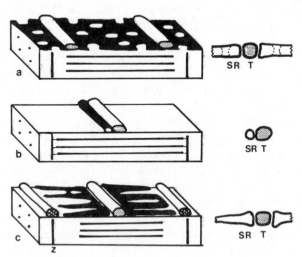

FIGURE 6-5. Internal activating systems. Excitation travels inward along transverse tubules (T) which are invaginations from the sarcolemma. Sarcoplasmic reticulum (SR), which is coupled to these, releases stored calcium ions into the sarcoplasm. A. Extensive SR envelops the fibrils of synchronous flight muscles. B. There is very little SR in asynchronous flight muscles. C. Most other muscles have an intermediate amount of reticulum. Basement membrane–lined Z invaginations resemble T-tubules but are not intimately associated with SR and are probably not involved in excitation–contraction coupling.

ment membrane and lying at the level of the Z lines, appears to have a mechanical role. In different muscles the tubules may enter directly from the outer muscle fiber surface, from tracheal invaginations, or from deep clefts in the sarcolemma. Occasionally, they make longitudinal connections with tubules at other levels.

The TTS is in direct contact with the sarcoplasmic reticulum (SR), which serves as a storage reservoir for Ca^{2+}. When excitation is communicated by the TTS to the SR, it releases stored Ca^{2+} into the sarcoplasm. Diffusing into the myofibrils, Ca^{2+} combines reversibly with troponin and with myosin light chains, permitting the interaction of actin and myosin and activating the myosin ATPase and tension development. The muscle fiber also responds to elevated free intracellular Ca^{2+} by transporting it actively back into the SR and out across the sarcolemma. In relaxed muscles the concentration of free Ca^{2+} in the sarcoplasm is very low, especially in flight muscle. Values of $10^{-7}-10^{-8}$ M have been reported. When fibers are fully activated, the concentration may increase by up to two orders of magnitude.

The amount of sarcoplasmic reticulum reflects the speed of contraction and relaxation of a muscle. Slow or tonic muscles, including asynchronous fibrillar flight muscles, have very little SR, often just simple tubes or cisternae adpressed to the T tubules. In cross section these appear as paired vesicles or dyads. When a T tubule lies between two cisternae, a triad of vesicles results. It may take a train of nerve impulses to cause release of sufficient Ca^{2+} from a sparse SR system (supplemented by inward diffusion across the sarcolemma) to activate a full contraction. After stimulation ends, it takes time for these fibers to sequester or remove enough Ca^{2+} to allow relaxation.

In fast-acting or phasic muscles like the jumping or flight muscles of locusts, the SR is extremely extensive and takes the form of fenestrated membranous compartments completely enveloping the myofibrils. Here Ca^{2+} release and uptake are extensive and rapid, causing quick activation and relaxation. However, the metabolic cost of sequestering or ejecting Ca^{2+} is high.

In visceral, cardiac, alary, and other slowly contracting muscles, the TTS may be much less extensive and regular and the SR is generally sparse.

6.4. MECHANICAL BEHAVIOR

6.4.1. Morphological Aspects

The shapes and locations of muscles correlate nicely with their functions (Fig. 6-6). Where a considerable range of movement is needed, as in the telescoping of body segments, the muscle fibers are long and run in parallel. Where power is needed, as in the mandibular muscles of chewing insects, many short fibers converge on a central tendon or insertion.

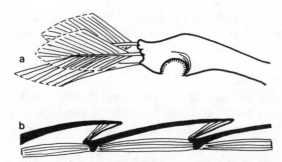

FIGURE 6-6. Muscle fiber arrangements for large force development or for a great range of movement. Honeybee. A. Base of first tarsomere, prothoracic leg. B. Retracted abdominal terga. (From R. E. Snodgrass: *Anatomy of the Honey Bee.* Copyright © 1956 by Cornell University. Used by permission of the publisher, Cornell University Press.)

The ability of a muscle to generate force is a function of the aggregate cross-sectional area of the muscle fibers. Values of about 2 kg/cm² have been reported for both insect and vertebrate muscles. However, insects tend to be much smaller than vertebrates. As overall size is reduced, the cross-sectional areas of muscles, and hence their power, will be reduced in proportion to the square root of linear dimensions, but the mass they must move will diminish as the cube root. Thus, the small insect can produce great power relative to its weight. For very small sizes, viscous drag by surrounding water or air limits the rapidity of movement. This is obviously a greater problem for aquatic than for terrestrial insects.

Limb muscles tend to be located proximal to the segments they move. Thus, the mass of the limb is kept low, reducing the force needed to move it and increasing the speed of movement. Another adaptation for rapid movement commonly found in insects is the limiting of the range and speed of movement of the more powerful and massive muscles. This reduces inertial losses and turnaround times. The necessary speed and range of movement of light skeletal elements like wings and tarsi are then achieved by mechanical amplification through skeletal elements.

Since the actions of muscles are generally exerted through the integument, brief mention of some of its mechanical properties is appropriate. The surface of most insects is covered with relatively stiff, sclerotized cuticle. The basic shape is the tube, strong for its weight and reinforced where necessary to sustain major loads. In contrast, the arthrodial cuticle, connecting sclerotized plates at many joints and forming intersegmental membranes, is extraordinarily flexible and may also be capable of recovery after considerable stretch. A flying insect may move its wings several million times in a single prolonged flight without mechanical failure.

Elastic cuticle containing the protein resilin is important for energy storage, especially in certain joints and tendons. Sclerotized cuticle, although much stiffer, can also contribute to elastic energy storage. Elasticity is important in movement for two reasons. First, elastic cuticle reversibly absorbs energy over a time interval, smoothing peak force transients. This allows certain tendons and skeletal elements to be relatively light in weight and to function routinely near their ultimate breaking strengths. Second, elastic energy storage makes it possible for muscles of ordinary strength and speed to contract over a period of time, storing energy that is released suddenly when a mechanical stop is removed or a point of maximum resistance to movement is passed.

The mechanical system can be described as having a metastable point between two or more stable configurations. Passing this transition releases energy very suddenly and may produce an audible sound as in the snap of a click beetle. Such systems are sometimes called click mechanisms. This energetic principle underlies the prodigious athletic feats of fleas, springtails, and grasshoppers and also delivers peak power at the most effective points in the cycles of wing movement during flight in many of the most effective flyers.

6.4.2. Length, Tension, and Velocity

The essential function of muscle is to contract. When shortening is resisted by a load, tension rises. If the load is greater than the tension the muscle can develop, the contracting muscle will lengthen. In normal use the load on a muscle usually changes through the contraction cycle because (a) leverage factors change as the joint angles are altered; (b) energy is used in stretching viscoelastic elements and may be subsequently partly recovered; and (c) synergistic and antagonistic muscles may be called into action at various times. Therefore, the length and tension relationships vary in a complex way.

To simplify analysis, experimental observations are usually made on isolated muscles contracting against constant loads (isotonic contractions) or at fixed lengths (isometric contractions). Or the muscle may be allowed to go from one preselected length or tension to another during an experiment. Some studies, particularly with flight muscles, have used systematically varying conditions like sinusoidal stretch. From such studies have come descriptions of the mechanical behavior of muscles, but not all of the properties that have been described can be attributed to alterations in specific molecular components.

The maximum speed of shortening has been determined for muscles contracting against zero load. The high value of 13 muscle lengths/second reported for locust flight muscle is very similar to the values reported for various vertebrate skeletal muscles. At one time it was believed that the speed of shortening might be limited by viscous drag. It now appears that time-consuming processes related to excitation and activation are more important. They also introduce a delay between stimulation and the onset of mechanical response.

Muscles show considerable elasticity, both in parallel with the contractile elements and in series with them. Parallel elasticity may be due mainly to the contractile elements themselves, especially in flight muscles, and can also be produced by stress to noncontractile cytoplasmic elements and the fibrous coats of the sarcolemma. Series elasticity can be due to muscle attachments. In insects these tend to be short, except where there are apodemes. Energy is also absorbed by elastic deformation of the integument and antagonistic muscles in many activities. The tibia bends during the jump of a locust, and the wings bend during each cycle of movement. The work flight muscles do in stretching their antagonists is probably largely recovered because the muscles do not relax completely between successive contractions.

When a muscle twitches in response to a single motor impulse, much of the force generated is absorbed in stretching the elastic elements in the muscle and its attachments. Typically, active state (ability to develop additional tension) declines before the elastic elements have been fully stressed. Because of this internal energy storage, the tension observed externally is less than the maximum the muscle can generate. Repeated stimulation prolongs the active state, allowing the elastic elements to be fully stretched. Externally observed tension climbs to a plateau. Such a contraction is called a tetanus. Muscles vary greatly in the relationship between twitch and tetanus tension. Locust flight muscle is quite stiff, with a high twitch/tetanus ratio (0.76). Peak tension is reached quickly in response to a brief volley of nerve impulses. Muscles innervated only by slow excitors and used for slow or tonic contractions have extremely low ratios (e.g., 0.03) or may even show no mechanical response to a single stimulus. The locust jumping muscle, which has fast and slow (dual) innervation and is used for both phasic and tonic activity, has an intermediate ratio (0.15–0.2).

When they are lightly loaded, muscles shorten rapidly; with heavier loads, more slowly. The force–velocity curve for frog sartorius in an ice bath takes the form of a rectangular hyperbola described by the famous Hill equation. Other equations with different theoretical implications provide better fits to data collected from other muscles or under other conditions. The form of the force–velocity curve for locust flight muscle is strongly influenced by the starting length and is not satisfactorily described by Hill's equation.

When a muscle shortens and then is allowed to relax, it tends to remain at a short length unless a small force is applied to extend it. (*Note:* After extreme shortening, there may be partial elastic return.) As the muscle is pulled out, the force required is essentially constant until the muscle reaches its maximum relaxed length. Beyond this length the force required (passive tension) increases until the hyperextended muscle tears. Vertebrate skeletal muscles and many insect muscles can be extended considerably before the passive tension climbs excessively. By contrast, flight muscles, especially

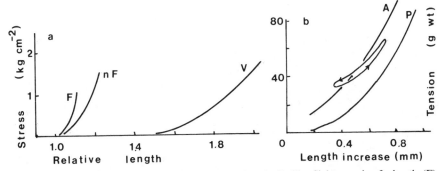

FIGURE 6-7. Length–tension curves. A. Passive stretch: fibrillar flight muscle of a beetle (F), nonfibrillar flight muscle of a locust (nF), and a vertebrate muscle, frog semitendinosus (V), which is much less stiff. B. Passive (P) unstimulated, active (A) stimulated non-oscillatory and self-oscillatory length–tension diagrams from asynchronous flight muscle of a rhinoceros beetle (Machin and Pringle, 1959). (From J. W. S. Pringle, in G. H. Bourne, Ed., *The Structure and Function of Muscle*, Academic Press, New York, 1972.)

the fibrillar type, are quite stiff and inextensible. Their passive tension climbs quite steeply beyond rest length (Fig. 6-7a).

When a muscle is stimulated, additional force is needed to extend it. The difference between the force required to hold a muscle at a particular length when activated and when relaxed can be called the active increment. (It is often called simply the active tension, but confusion can result because certain authors use that term for the total tension of an activated muscle.) The active increment is greatest at the normal maximum relaxed length. In frog sartorius this is the length at which thin filaments just reach to the central bridge-free region of the thick filaments, affording maximum opportunity for effective myosin cross-bridge interaction with actin. Beyond this length, bridge overlap and the active increment decrease linearly with distance until the thick and thin filaments are pulled out of register and the active increment is zero. As mentioned previously, several length–tension studies of insect muscles have disclosed more complex relationships.

6.4.3. Stretch Activation

It has long been known that certain types of vertebrate muscle are responsive to stretch. The heart contracts move vigorously when distended by a large volume of returning blood, and various smooth muscles contract when stretched, sometimes developing rhythmic pulsations. Skeletal muscle also shows some stretch sensitivity. Among insect muscles this property is most highly developed in fibrillar flight muscles where a delayed contractile response to stretch is essential in generating rhythmic wing movements. Indeed, these muscles can be fully activated only when a high internal free Ca^{2+} level is combined with stretch.

6.5. FUNCTIONAL MOTOR SYSTEMS

This section summarizes special characteristics of certain insect muscle systems that have been extensively studied or that show unique specializations. Some of the information has been introduced earlier in this chapter and will be restated here in a functional context. Comprehensive discussions will not be possible in this limited space and without all the relevant morphological details.

6.5.1. Flight

Mechanically, a thoracic segment can be regarded as a box with dorsal tergum, ventral sternum, and lateral pleura. The wings are movable lateral projections at the junctions of the tergum and pleura, articulating with each. Although the basic wing movement is up and down from this horizontal hinge, the hinge itself rocks forward and back on a central post at the top of the pleural suture. Therefore, the angle of attack of the wing relative to the oncoming airstream changes during the wing stroke. In extreme cases, the wings twist so far that the lower surface becomes momentarily uppermost during the upstroke, and it is possible for both up and down movements to contribute to both lift and thrust.

In the most primitive orders two sets of direct flight muscles contract to depress the wings. The basalar muscles, which are attached anterior to the pivot point, contract earlier than the subalars, which are posterior, twisting the wings about their long axes. In the most advanced flyers the basic twisting motion is caused by the complex wing articulation itself.

In all insects the upstroke is produced by contraction of tergosternal (vertical) muscles, which depress the tergal wing articulation relative to the more lateral pleural support (Fig. 6-8). Since these muscles insert on the tergum rather than on the wings, they are called indirect flight muscles. In Orthoptera, Odonata, and other more primitive groups, the downstroke is produced by the previously mentioned direct flight muscles, which insert on the wings lateral to their tergal articulations. In Diptera and Hymenoptera the downstroke is produced by another set of indirect flight muscles, the dorsal longitudinal muscles (horizontal), which shorten the tergum in an anterior–posterior direction, bowing up the tergal wing articulations relative to the pleura.

If the wings are regarded as levers, it can be seen that the driving force is applied very close to the fulcrum and must therefore be a large force acting through a very short distance. This is transformed by the wings into a large movement, accelerating a large volume of air, but air has low mass and viscosity and offers little resistance to the movement. A widely publicized study of the aerodynamics of bumblebee flight once concluded that it should be physically impossible for the bumblebee to fly. This analysis considered the bee as a fixed-wing aircraft, an inappropriate model. Insects with flap-

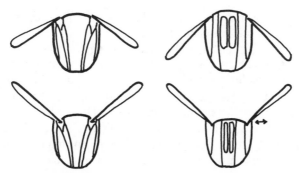

FIGURE 6-8. Diagram of alternative mechanisms for supplying power for wing movement. Left, in the Orthoptera and Odonata direct flight muscles rock the lateral wing base across a central pivot. Right, in the Diptera indirect flight muscles warp the tergal wing articulations up and down relative to the pleural articulations. The pleura must be forced laterally (double arrow) to allow movement.

ping flight more closely resemble helicopters with counterrotating blades. At the top and bottom of each stroke, the wings pivot to complete the circuits begun by the opposite wings.

When there are two sets of independently driven wings, as in the dragonflies, the hind wings avoid most of the turbulence created by the fore wings by starting their movements first. The coordination of wing movement is, therefore, from posterior to anterior, just as it is in most insect walking and running patterns.

The flight muscles, direct and indirect, are specialized to deliver considerable force over a short range of movement. This is reflected in their fine structure. Thick filaments extend nearly the whole length of the sarcomeres, providing a maximal number of cross bridges to interact with thin filaments. However, the contractions are nearly isometric. The muscles usually shorten by only 1–2% of their extended length or about one bridge movement per sarcomere. Since the range of movement is small, the velocity of shortening can be quite ordinary even though the wingbeat frequency is high. Therefore, there is little energy loss to internal viscous drag or delay due to inertia.

There are two very different patterns of neural control of flight muscles, with concomitant differences in the activation and contractile properties of the muscle fibers. In the more primitive pterygote orders, groups with relatively large wings and low wingbeat frequencies, each flight muscle contraction is driven by a short volley of nerve impulses. The muscle fibers are well endowed with T tubules and have a very extensive SR. After being activated, they contract quickly and strongly and then relax, at least partially, in each cycle of wing movement. These muscles are classified as fast, and their contractions are synchronized with the pattern of neural activation. Undoubtedly, a large fraction of the total energy spent for flight is required for active transport of ions to restore membrane excitability and sequester Ca^{2+} in the SR in each cycle movement.

By contrast, the insects with the highest wingbeat frequencies have slow muscles whose movements are not phase-related to motor impulses. Asynchronous flight muscles have been extensively studied in various Diptera, Coleoptera (Fig. 6-7), Heteroptera, and Hymenoptera (Apocrita) and have also been reported in Thysanoptera and some Homoptera and Psocoptera. These muscles, although endowed with a fairly extensive TTS, have very little SR. At the beginning of flight, volleys of motor impulses are sent to all flight muscles. These gradually become activated and isometric tension mounts. In this dynamic system the two lowest-energy positions are where the wings are at the upper and lower limits of their range of movement. Here, their pleural articulations are closest to the medial sagittal plane. To move the wings from one extreme position to the other, the pleura must be forced laterally, demanding extra energy from the appropriate muscles. Much of this energy is recovered elastically when the wing hinge passes the point of highest resistance and snaps toward the other low-energy position. This sudden movement not only aids the wings in accelerating air at the most effective point in the cycle of wing movement but also delivers a sudden tug to the already activated antagonistic muscles. These respond, after a delay, with additional tension, and the cycle is repeated. Occasional nerve impulses keep the Ca^{2+} activation system effective and the muscles responsive to alternating stretch activation. Wingbeat frequency is determined by the delay in the stretch-activation system and the mechanical characteristics of the total system, including wing loading. Adding weights to the wings reduces the frequency, and shortening the wings increases it. The maximum reported frequency of more than 2200 Hz was obtained with a midge, its wings shortened to stumps, heated to near the thermal death point in an oven.

Sometimes this oscillatory system can operate with the wings over the abdomen in the position of rest. Honeybees, for example, run the flight motor with the wings disengaged, generating metabolic heat and raising the thoracic temperature before flying in cold weather. At low temperatures the rate of supply of biochemical energy is too low to sustain flight. Such movements also produce audible sounds or vibrations that can be used for communication.

At the end of flight, the wings are disengaged and impulses to the flight muscles stop, but there may be several more cycles of thoracic oscillation as the muscles slowly relax.

Maneuvering during flight is effected by muscles that modify the geometry of the articular hinge, altering the path and excursion of wing movement and the angle of attack of the wing relative to the oncoming air. In flies the fields of chordotonal organs near the base of the halteres respond to deviations from stable flight and provide sensory input for dynamic reflex adjustments of wing movements. In dragonflies the head serves as a static organ, deflecting fields of sensory hairs when its position changes relative to the thorax. In many insects, the legs serve this function.

Power output during flight is controlled in flies by sternopleural muscles, which set the tension on the pleural walls. The greater the lateral tension, the more force the flight muscles must generate to overcome it and, therefore, the more energy there is available to the wings when they move through the click point. Maximum force is required at take-off. Typically, the energy output is reduced as the insect gets up to cruising speed.

In flies deflection of the antennae by the moving air reflexly reduces the motor output to the sternopleural tensors, and detection of forward movement by the compound eyes reduces the impulse frequency to the indirect flight muscles. Also, the amplitude of wing movement is reduced and the path of wing movement and angle of attack are modified. In contrast to flies, hairs on the head are the wind speed detectors of locusts.

6.5.2. Jumping Mechanisms

A flea may jump 100 times its body length. The athletic feats of locusts, springtails, and click beetles are also spectacular, especially considering the fact that their muscles are quite ordinary in terms of ability to generate force. The jumping muscles are strong muscles with large cross-sectional areas because they are composed of large numbers of short fibers that converge on strong central apodemes. When the muscles start to contract, the skeletal joints across which they act are locked into positions that prevent movement. Initially, the contraction is isometric and energy is stored over time in elastic elements. Then a mechanical constraint or catch is removed and the full force of the muscle plus the elastically stored energy is available to accelerate the mass of the insect.

In the case of the locust, the extensors contract initially against the weaker flexors, which are able to prevent extension because they have a mechanical advantage when the leg is flexed (Fig. 6-9). Energy stored in the isometrically contracting extensor system is released when the flexor sud-

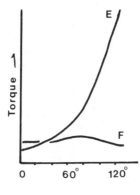

FIGURE 6-9. Effective torque exerted by the flexor (F) and extensor (E) muscles in the jumping leg of the locust as a function of degree of extension. In the fully flexed leg, the small, weak flexor can prevent movement, allowing the extensor to develop full isometric tension before the jump begins. (Reprinted by permission from R. H. J. Brown, *Nature* **214**, 939. Copyright © 1967 MacMillan Journals Limited.)

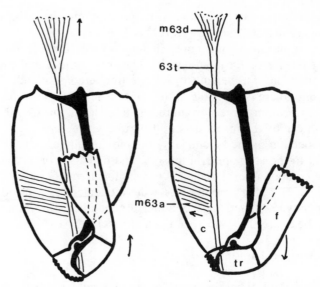

FIGURE 6-10. Jumping mechanism of a pulicoid flea. When the femur is fully raised, force developed by muscle 63d is applied by tendon 63t beyond the coxa-trochanter pivot, locking the leg in the flexed position. Energy is stored in a resilin pad proximal to the coxa (not shown). Contraction of muscle 63a pulls the tendon laterally relative to the pivot, allowing femoral extension and release of the energy stored in muscle 63d, tendon 63t, and most importantly, the energy stored in the resilin pad. [After H. C. Bennet-Clark and E. C. A. Lucey, *J. Exp. Biol.* **47,** 59 (1967).]

denly relaxes. Since the legs are long, their extension provides acceleration over a long time and short distance. It has been calculated that an extensor muscle in a 3-g locust may exert a tension of 900 g on its apodeme, bending the tibia during jumping. This force is near the breaking strength of cuticle. Indeed, if a locust tries to jump while restrained, it may snap its tendons.

In the case of a flea (Fig. 6-10), the coxa–trochanter joint is flexed so far before the jump that the force of the jumping muscle is applied slightly on the wrong side of the pivot. Energy generated by the muscle is stored in a resilin pad. Subsequent contraction of another part of the muscle pulls its tendon sideways so the force is now applied on the other side of the pivot, allowing sudden release of the stored energy. This movement and the maximum velocity of the jump are achieved within 1 msec. Although the muscles are capable of generating the energy necessary for the jump, it is very unlikely that they can do so within 1 msec. However, the resilin pads are capable of storing sufficient energy and can release it almost instantaneously.

The jumping mechanisms of springtails and click beetles work on the same principle: powerful muscles contract isometrically against a mechanical stop that is suddenly removed, releasing elastically stored energy.

6.6. FURTHER READING

For the student who wishes to look farther into the subject, several reviews are suggested (see the references). Huddart's book is a recommended general introduction to muscle because it contains a generous amount of information on insect muscles. Pringle's review of arthropod muscles assumes a little more background but considers mainly insect muscles. There are excellent reviews with numerous references in the books edited by Rockstein and Usherwood. In the former, locomotor systems are discussed by Hughes and Mill and Nachtigall and Pringle and neural control by Hoyle. In the latter there are reviews on many aspects of insect muscle. The book edited by Tregear deals entirely with the fibrillar flight muscles of belostomatid water bugs of the genus *Lethocerus*. Pringle's little booklet on insect flight provides a highly readable and well-illustrated introduction to this subject.

6.7. INSECT MUSCLES IN RETROSPECT

The study of insect muscle systems has yielded both important basic information about muscles and their functional behavior and also interesting details about the motor capabilities of specific insects. Flight muscles, especially the asynchronously excited types, are the most regular of all muscles in their molecular architecture and in the synchrony of their cross-bridge movements. Therefore, they have been favored objects for investigation by X-ray and optical diffraction and by electron microscopy. Wing-movement systems have been especially useful in the study of the stretch sensitivity of muscles. The notably high performances of many insect flight and jumping mechanisms led to the description of the elastic storage and recovery of large amounts of energy in skeletal elements and antagonistic muscles.

Many characteristics combine to make insect muscles especially favorable as materials for study. Some muscles are easy to expose or dissect out and attach to mechanical recording systems. There is almost no connective tissue. Superfusion with blood substitutes presents no problems. As long as tracheal trunks are open to the atmosphere, adequate gas exchange is assured. The motor innervation is simple in that there are few neural elements. Often the motor neurons can be identified and their activity monitored. Or they may be controlled, either by direct stimulation or through reflex pathways. When a muscle fiber is short, a microelectrode placed anywhere in it can be used to record its total electrical activity. Some insect muscle preparations are quite rugged and can continue to provide useful information for many hours without elaborate maintenance procedures.

In the future insect muscle systems will surely give up many more of their secrets. Undoubtedly there will be considerable progress in the areas where there is extensive study. Hence, we can expect to gain a far more extensive and sophisticated understanding of the interactions of transmitters and modulator substances in muscle excitation. The catalog of the diversity of muscle types with unique functional adaptations will continue to expand. More will be learned about the development of muscle and about functional changes associated with metamorphosis. And the control of muscle metabolism will be elucidated for many additional species. One can seldom predict the big breakthroughs in science, but progress is most likely when important questions are investigated with the aid of favorable experimental material. Therefore, we can expect significant new basic knowledge to accrue from the continued study of insect muscle systems.

REFERENCES

H. C. Bennet-Clark and E. C. A. Lucey, *J. Exp. Biol.* **47,** 59 (1967).

R. H. J. Brown, *Nature* **214,** 939 (1967).

G. Hoyle, *Muscles and their Neural Control,* Wiley-Interscience, New York, 1983.

H. Huddart, *The Comparative Structure and Function of Muscle,* Pergamon, Oxford, 1975.

K. E. Machin and J. W. S. Pringle, *Proc. Roy. Soc. Lond.* **151B,** 204 (1959).

J. W. S. Pringle, "Arthropod muscle," in G. H. Bourne, Ed., *The Structure and Function of Muscle,* 2nd ed., Vol. 1, Part 1, Academic Press, New York, 1972, pp. 491–541.

J. W. S. Pringle, *Insect Flight,* Oxford, London, 1975.

M. Rockstein, Ed., *The Physiology of Insecta,* 2nd ed., Vols. 3 and 4, Academic Press, New York, 1974.

R. E. Snodgrass, *Anatomy of the Honeybee,* Comstock, Ithaca, 1956.

R. T. Tregear, Ed., *Insect Flight Muscle,* North-Holland, Amsterdam, 1977.

P. N. R. Usherwood, Ed., *Insect Muscle,* Academic Press, London, 1975.

7

NERVOUS SYSTEM: ELECTRICAL EVENTS

D. L. SHANKLAND
Department of Entomology and Nematology,
University of Florida
Gainesville, Florida

J. L. FRAZIER
Agricultural Chemicals Department,
E. I. DuPont de Nemours and Co.,
Wilmington, Delaware

CONTENTS

SUMMARY

The processes that produce electrical potentials across membranes require active transport, with the inside being negative with respect to the outside. A variety of biochemical mechanisms have been evolved to maintain a bioelectric potential across the cell membrane. Cells of the nervous system have evolved mechanisms of active response to various stimuli, endowing them with the property of irritability, by which active changes in membrane potentials are used as means to transfer and process information within the nervous system and in communication with other systems. The roles played by these active events in reception, transmission, signal processing, and effecting are described to provide the reader with an appreciation for the integrative role of the nervous system and a basis for understanding other related material in this volume. Since the insect nervous system offers a tempting target for the development of highly specific insecticides, investigations on neurophysiology should continue to be exploited in order to expose selected sites that are susceptible to insecticide-induced biochemical lesions.

7.1.　NERVOUS INTEGRITY AND TRANSMISSION

The survival of even very simple animals depends on, among other things, their receiving and responding to information about their environment. The quantity and quality of information required become greater and more diverse with more complex behavior among higher animals. Behavioral output varies with temporary conditions and frequently involves tightly controlled and sequenced activities of great variety and complexity. In addition, there is need to monitor and coordinate internal events having to do with growth and development, homeostasis, and various vital processes.

These integrative functions are served by the nervous system, which possesses electrical and chemical mechanisms for information reception, transmission, and processing, which have evolved to effect desired output and control. The quintessence of nervous systems is prescribed in the neuron doctrine, mainly attributed to Ramon y Cajal, and briefly stated by Bullock and Horridge (1965): "All nervous systems consist in essence (whatever other non-nervous elements may be present) of distinctive cells, called neurons, which are specialized for nervous functions and which produce prolongations and branches."

The prolongations and branches are processes that are especially adapted for neural transmission. Neurons in central nervous systems (CNS) of insects are characteristically monopolar, that is, the cell body, or soma, has a single process that produces an axon with dendrites at one end and a terminal arborization at the other, as illustrated in Fig. 7-1. Commonly, a neural transmission proceeds from some input source (e.g., a sensory structure) over an *afferent* axon (fiber) to an *internuncial* neuron in the CNS and thence via an *efferent* neuron fiber to an effector (e.g., neurosecretory or muscle cells). Transmission between cells is effected by way of synapses, or points of intimate contact between the presynaptic terminal arborization and the postsynaptic dendrites. Transmission across synapses occurs in only one direction and is chemically mediated. Electrical signals arriving presynaptically trigger the release of a chemical transmitter from specialized sites on the presynaptic membrane. The compound diffuses across an intercellular gap (synaptic cleft) of 200–300 Å and activates specialized receptor sites on the dendrites of the postsynaptic fiber. The dendrites respond by producing graded electrical signals which, in turn, trigger the production of nerve impulses in the axon. The impulses are transmitted to the terminal arborization where, again, a chemical transmitter mediates transmission to an efferent fiber.

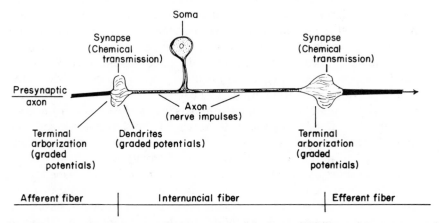

FIGURE 7-1. Monopolar nerve cell characteristic of the insect CNS illustrating structures and processes involved in nervous transmission from afferent, through internuncial to efferent fibers.

The nervous system of insects appears to be relatively simple compared to those of higher animals. Nevertheless, they contain tens of thousands of cells in extremely complicated networks that are exquisitely integrated to produce functional integrity in virtually every activity. Besides being able to receive, process, and effectively use enormous amounts of information, the CNS contains genetically defined programs of behavioral control. These programs, called out by certain circumstances, may orchestrate series of complex activities in precisely arranged sequences, for example, courtship and mating or nest building and provisioning. In these functions, communication in the nervous system depends on certain electrical and chemical events, which form the subject of this chapter.

7.2. BIOELECTRICAL EVENTS

7.2.1. Origin of Bioelectrical Potentials

Characteristically, living cells show a voltage called the membrane potential across the cytoplasmic membrane, the inside being negative with respect to the outside. The basis of this membrane potential is the active transport of ions across the cell membrane.

Many intracellular processes require a stable ionic environment different from the outside medium. The outside medium may be fresh or saline water for lower aquatic animals or body fluids in higher aquatic or terrestrial forms. The degree to which animals regulate the composition of body fluids varies widely, but the cells must have mechanisms to regulate their intracellular composition. There appears to be a common need in animal cells for relatively high internal concentrations of potassium and low concentrations of sodium. Furthermore, cells must regulate the osmolarity of their cytoplasm, partly because large organic anions cannot diffuse through the cell membrane. If the cell membrane were freely permeable only to small ions and water, the cell would swell due to osmosis, whatever the osmolarity of an external medium, because of these internally fixed organic ions. The swelling could be prevented by (1) a rigid cell wall, as in plants; (2) the active extrusion of water at exorbitant energy cost; or (3) the active extrusion of ions at more reasonable energy cost. Animals use the last mechanism, combined with active uptake of potassium, to regulate both osmolarity and ionic content of the cytoplasm. Generally, while potassium is taken up, sodium, calcium, and magnesium are actively extruded from the cells.

Active transport by membranes of cells and intracellular organelles is a well-known phenomenon. A variety of substances may be transported by different membranes to meet the specific needs of different cells, tissues, or organs (Semenza and Carafolr, 1977). In this discussion of electrical events in nervous tissue, we are concerned primarily with ion transport. Ion transport mechanisms, referred to as ion pumps, draw upon ATP for energy to transport ions across membranes against concentration gradients. The energy from ATP is released through the action of specific ATPases. For example, nerve cell function depends on active outward transport of Na⁺ coupled with inward transport of K⁺, and this process depends on a specific enzyme (ATPase), which requires Mg^{2+}, Na^+, and K^+ for optimal activity (Skou, 1965). In addition to active transport, membranes show selective permeability to different ion species. These membrane processes lead to the production of the membrane potential, which in irritable cells (e.g., nerve and muscle) is called a resting potential.

7.2.2. Resting Potential

The processes that produce the resting potential are illustrated in Fig. 7-2. Present understanding of nerve membrane function is due in large part to the

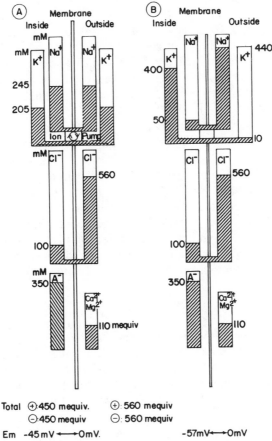

FIGURE 7-2. Model illustrating the generation of a resting membrane potential through the active transport of Na^+ and K^+ by the ion pump.

pioneering work on the squid giant axon by Hodgkin and Huxley (1952). The ion concentrations shown in Fig. 7-2B are based on those reported for that axon (Hodgkin, 1964). However, for the purpose of this discussion assume an initial state (Fig. 7-2A) in which both cytoplasm and extracellular fluid contain 205 mM K$^+$ and 245 nM Na$^+$. The cytoplasm (inside) contains 450 meq of cation (K$^+$ and Na$^+$) which must be balanced by an equal concentration of anion; that is, the solution must be electrically neutral. This balance is provided by 350 meq of organic anion, A$^-$, and 100 meq of Cl$^-$. The extracellular fluid contains 450 meq of Na$^+$ and K$^+$ combined and 110 meq of Ca^{2+} and Mg^{2+}, all of which are balanced by 560 meq of Cl$^-$. The membrane is impermeable to A$^-$, Ca^{2+}, and Mg^{2+}. Under these assumptions, and with the inviolable requirement that the bulk solutions on either side of the membrane be electrically neutral, it can be seen that a concentration gradient of Cl$^-$ exists across the membrane (Fig. 7-2A). The gradient creates a diffusion pressure that drives Cl$^-$ inward, carrying a negative charge with each ion that crosses the membrane. Assume for the moment that K$^+$ and Na$^+$ cannot diffuse through the membrane, so the only transfer of charge across the membrane is due to Cl$^-$. As Cl$^-$ diffuses inward, an excess of negative charge accumulates at the inner surface of the membrane and opposes further diffusion. These two opposing forces—diffusion pressure and electrical potential—come to equilibrium when the electrical potential is just sufficient to balance the diffusion pressure. This so-called "equilibrium potential" is defined by the Nernst equation.*

$$E_{Cl} = \frac{RT}{nf} \ln \frac{[Cl^-]_o}{[Cl^-]_i}. \tag{1}$$

where E_{Cl} is the equilibrium potential of chloride in volts; R is the international gas constant (8.314 J/degree/mol); T is temperature in degrees Kelvin; n is the number of unit electrical charges per ion; F is the Faraday constant (96,500 C); and $[Cl^-]_o/[Cl^-]_i$ is the concentration gradient of Cl$^-$ across the membrane from outside to inside; and ln is the natural logarithm.

Assuming a temperature of 30°C, converting to common logarithms, and expressing E_{Cl} in millivolts, equation (1) reduces to

$$E_{Cl} = 60 \log \frac{[Cl^-]_o}{[Cl^-]_i}.$$

Under the conditions shown in Fig. 7-2A, $E_{Cl} = -45$ mV. It is conventional to express membrane potentials as inside voltage relative to an outside voltage of zero. It requires a transfer of only 0.45 peq of Cl$^-$ (i.e., 0.45×10^{-12} eq) across 1 cm^2 of membrane to produce this potential (Lakshminarayanaia, 1969). The small excess of internal Cl$^-$ is confined to the cytoplasm–mem-

---

* Its application to physical and biological systems is discussed by many authors (Adamson, 1969; Cole, 1968; Davson, 1964; Lakshminarayanaia, 1969; Plonsey, 1969; Rodahl and Issekutz, 1966; Sheehan, 1961; Wilson, 1972).

brane interface and does not alter the concentrations in the bulk solutions on either side of the membrane.

Presume that from this initial stage (Fig. 7-2A) the ion pump resident in the membrane begins to pump Na^+ outward and K^+ inward in a one-for-one exchange, eventually producing the concentration gradients shown in Fig. 7-2B. These ions, now also subject to diffusion pressure, will diffuse down their respective concentration gradients to the extent that the membrane is permeable to them. Staverman (1952, 1954) has rigorously defined the factors controlling the diffusion of ions through membranes, and in this discussion only three need be considered: (1) chemical potential, referred to here as diffusion pressure and equated to concentration gradient; (2) electrical potential gradients; and (3) membrane permeability. Given the concentration gradients of the three diffusable ion species (Na^+, K^+, Cl^-) in Fig. 7-2B, each species will contribute to the membrane potential through the interactions of the three factors listed above. The way in which the factors interact is defined by the Goldman constant field equation:

$$E_m = \frac{RT}{nf} \ln \frac{P_K[K^+]_i + P_{Na}[Na^+]_i + P_{Cl}[Cl^-]_o}{P_K[K^+]_o + P_{Na}[Na^+]_o + P_{Cl}[Cl^-]_i},\tag{2}$$

where E_m is the membrane potential; R, T, n, F, ln, and the ionic concentration gradients are as in equation (1); and P is the relative permeability for each respective ion. Solving equation (2) for the ionic gradients in Fig. 7-2B at 30°C and using the values of P for the squid giant axon from Hodgkin and Katz (1949), ($P_K : P_{Na} : P_{Cl} = 1 : 0.04 : 0.45$) gives a value for E_m of -57 mV, that is, inside negative. This is the so-called "resting potential" which with suitable instrumentation is measurable in the living cell. It bears emphasis that the permeabilities used in equation (2) are relative and not in absolute units. A relative permeability of 1.0 for K^+ does not mean that the membrane in the resting state is maximally permeable to that ion. This has important implications in the transient processes to be discussed later.

The amounts of the three ion species required to produce the membrane potential are too small to affect the concentrations in the bulk solutions on either side of the membrane. However, if after the conditions shown in Fig. 7-2B are established, no more energy is put into the system (i.e., the ion pump is inactive), the ions would leak through the membrane. Eventually a Donnan equilibrium would be attained, and the distribution of the ions would differ from those found experimentally, as shown in Table 7-1. However, the ion pump operates continuously to counteract ion leakage, so the resting condition shown in Fig. 7-2B is actually a dynamic equilibrium, and the maintenance of the characteristic ion distribution and membrane potential requires continual energy input.

Because of the abundant data available, this discussion has been based on the squid giant axon. However, there is evidence that excitation and conduction in giant axons of the American cockroach and nerves of other insects

TABLE 7-1. Comparison of Ion Distribution Found Experimentally in Squid Axon (Hodgkin, 1964) with That Predicted by a Donnan Equilibrium for the Model in Fig. 7-2.

	Ion Concentrations (mM)			
	Inside		Outside	
Ion	Experimental	Donnan	Experimental	Donnan
Na^+	50	315	440	321
K^+	400	264	10	146
Cl^-	100	229	560	431

(Pichon, 1974; Treherne, 1974) are mediated by ionic mechanisms essentially the same as those in squid giant axon.

7.2.3. Transient Electrical Events

7.2.3.1. Graded Potentials. Nerve membranes can produce active transient changes in membrane potential in response to suitable stimuli. Graded responses, in which the amplitude of the response is a function of the strength of the stimulus, are basic to the ability of nervous systems to receive, transmit, and process information. The nongraded, all-or-none response of nerve axon membrane is a specialized adaptation for transmission of information over distance (Bishop, 1956). Stimuli may be in a sensory mode (taste, odor, vision, etc.) or in the form of electrical or chemical events involved in transmission within the nervous system. Whatever the nature of the stimulus, a graded response can be only in the form of reduction or increase in the membrane potential, that is, depolarization or hyperpolarization. The time course of the response may be the same or different from that of the stimulus. These two variables—amplitude and time course of response—occur in various ways in many elements of a nervous system, endowing even simple nervous systems with enormous capacity to extract and process information from inputs of various kinds.

Equation (2) defines how ionic gradients and membrane permeabilities to the respective species of permeant ions determine the membrane potential in a system such as that depicted in Fig. 7-2B. The membrane potential can be viewed as a compromise among the concentration gradients of the respective ion species and the relative permeabilities of the membrane to each of them. An increase or decrease in permeability to one of the ions would allow that ion to make a greater or lesser contribution to the membrane potential. As a result of changed permeability, the membrane potential would arrive at a new compromise value by way of ionic fluxes through the membrane. If the permeability change were transitory, the change in membrane potential also would be transitory.

Conventionally, ionic fluxes are described as current, symbolized by I and measured in amperes. Permeability is described as a conductance, symbolized by g and expressed in reciprocal ohms or mhos. Conductance changes are caused by appropriate stimuli that vary with different parts of the same neuron. For example, axonal membrane is responsive to electrical stimulation, while subsynaptic membranes of chemical synapses are responsive to certain compounds and not to electrical stimulation. Furthermore, stimuli may produce changes in conductance to only one ion species. Altered conductance caused by a stimulus is called transducer action, and the potential change is called electrogenesis (Grundfest, 1961).

The effects of selective changes in conductance are shown in Fig. 7-3, which is based on the conditions illustrated in Fig. 7-2B. The Nernst equation (1) was used to calculate the equilibrium potentials for the respective concentration gradients of Na^+, K^+, and Cl^-. Those voltages, and the resting potential E_m (R), are shown on the ordinate in Fig. 7-3. Under these conditions, a selective increase in K conductance (g_K), called K activation, would cause a shift in E_m toward E_K. This would be a hyperpolarization due to K activation. Conversely, a decrease in g_K would cause a shift in E_m toward E_{Cl} or a depolarization due to K inactivation. In like manner, Cl activation or inactivation would cause E_m to shift respectively toward or away from E_{Cl}.

In the resting nerve membrane g_{Na} is relatively very low [cf. equation (2)], so Na inactivation would be expected to produce only a minor change in E_m. However, E_{Na} is 113 mV more positive than E_m (R) (Fig. 7-3), so Na activation can produce very large depolarizations of E_m. In the special case of nerve action potentials in axons, discussed below, Na activation is not graded but is all-or-none and may cause the polarity of E_m to reverse, the

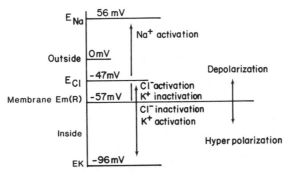

FIGURE 7-3. Depolarizing and hyperpolarizing deviations from the resting membrane potential, $E_m(R)$ are produced by increases (activation) and decreases (inactivation) in membrane conductance to specific ions.

inside becoming positive with respect to the outside. In other electrogenic processes at other sites in neurons, Na activation is graded and often occurs in conjunction with K or Cl activation. When that happens, the resulting change in membrane potential may be hyperpolarizing or depolarizing, the direction depending on the relative changes in the respective conductances.

The graded potentials occurring in sensory structures or synapses cause the production of nerve impulses in axons of the same cell (Fig. 7-1). The ionic basis for these action potentials was first determined in studies on the squid giant axon (Cole, 1968) and subsequently shown to apply also to insects.

7.2.3.2. Nerve Action Potentials. The Hodgkin–Huxley squid axon model (1952) will be used as a basis for the explanation of the nerve action potential.

The nerve action potential is the propagated all-or-none nerve impulse used in the transmission of pulse-coded information throughout the nervous system. Figure 7-4 shows how transient changes in sodium conductance (g_{Na}) and potassium conductance (g_K) lead to inward and outward currents

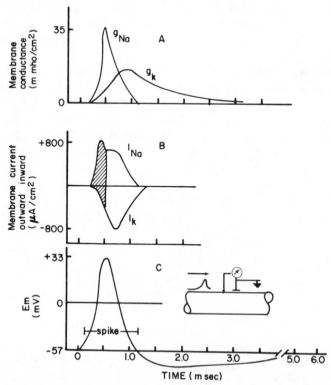

FIGURE 7-4. Conductance, current, and voltage changes occurring during the production of a nerve action potential according to the Hodgkin–Huxley nerve membrane model.

of Na^+ and K^+, respectively, producing a transient change in the membrane potential. The change is characterized by a prominent initial voltage wave form called a *spike* (Fig. 7-4C). Chloride is not directly involved in action potential production, and chloride conductance does not change during the process. (The inset in Fig. 7-4C shows the changes that would occur at a recording site as a propagated action potential passes a recording electrode.)

Spike production is initiated by Na activation, which is triggered by a depolarization of the membrane potential. The depolarization must reach a critical threshold, but once that threshold has been reached, a nongraded Na activation occurs, followed immediately by Na inactivation (these are indicated by rise and fall of the g_{Na} curve in Fig. 7-4A). Once triggered, this activation–inactivation cycle proceeds to completion, Na conductance being turned on and off in a precisely timed sequence. During the time of increased conductance, the concentration gradient of Na from outside to inside (Fig. 7-2B) leads to an inward sodium current, I_{Na}, which displaces the membrane potential, E_m, from its resting value (-57 mV) toward the sodium equilibrium potential, E_{Na} ($+33$ mV). I_{Na} reaches a peak in slightly less than 0.5 msec after the spike is initiated.

The depolarization caused by I_{Na} initiates a process of K activation, producing an increase in g_K that lags behind g_{Na} (Fig. 7-4A). By this time, the membrane potential is no longer at or near the K^+ equilibrium potential, and an outward diffusion pressure of K^+ is no longer balanced by the membrane potential. The increased g_K allows an outward current of K^+ (Fig. 7-4B), which tends to repolarize the membrane (descending portion of the voltage curve in Fig. 7-4C). Note that turning off I_{Na} (Na inactivation) is not sufficient to account for the rapid repolarization of the membrane; the latter requires the outward transfer of positive charge carried by I_K. The total positive charge transferred inward by Na current is a function of the area under curve I_{Na} (Fig. 7-4B). Conversely, the total positive charge transferred outward is a function of the area under the curve I_K (Fig. 7-4B). The algebraic sum of the voltage changes produced by the two currents yields a maximum voltage of $+33$ mV, occurring at about 0.5 msec (Fig. 7-4C). The total excursion of E_m from the resting potential (-57 mV) to the peak of the spike (33 mV) is 90 mV. This voltage change is used to express the amplitude of the action potential.

As I_K continues to increase with increasing g_K until about 0.7 msec, it begins to repolarize the membrane. However, g_K is inversely related to the membrane potential. That is, as the membrane is repolarized, the repolarization itself causes a reduction in g_K. The process can be likened to a float valve in which the rising level of a liquid closes the valve, cutting off further flow. Note, however, that at the time I_K reaches its peak at ~0.7 msec (Fig. 7-4B), there is still a significant g_{Na} (Fig. 7-4A), and the membrane potential is well below the Na equilibrium potential. These conditions produce a slight delay in the reduction of I_{Na}, shown by the inflection in the descending portion of the curve I_{Na} (Fig. 7-4B). The continued decrease in g_{Na}, however, ultimately turns I_{Na} off, leaving only I_K during the final phase of repolarization. At about 1.2 msec the membrane potential has returned to the resting level (Fig. 7-4C), but g_K at that time is still higher than normal (Fig. 7-4A). The higher-than-normal g_K produces a slight hyperpolarization due to K activation, as described, that persists until about 6 msec.

A close inspection of Fig. 7-4C will show that the membrane potential begins to change about 0.25 msec before the Na current begins to flow. This small voltage change is due to a field effect of the approaching spike (inset, Fig. 7-4C). This early potential, referred to as the foot of the action potential, causes the depolarization of the membrane to the threshold required to trigger Na activation. The foot and the distance ahead of the spike that it is detectable are functions of the cable properties of nerve (see Section 7.2.3.3). Before doing this, however, it is worth considering the magnitudes of the ionic fluxes that produce the action potential and the role of the ion pump.

As was pointed out in reference to Fig. 7-2, it requires the net transfer of only 0.45 peq of an ion to produce a change of 45 mV across 1 cm^2 of axonal membrane. The change in an action potential is about 90 mV. However, during a significant part of the action potential, there are both inward and outward currents. Although there may be net transfers of only 0.9 peq inward during depolarization and outward during repolarization, the total inward and outward fluxes are greater than that. It has been estimated that there is an entry of about 4 peq of Na$^+$ and a loss of an equal amount of K$^+$ per cm^2 per impulse in squid axon (Lakshminarayanaia, 1969). These amounts are insignificant compared to the total amounts of the two ions inside the axon. Therefore, even if the ion pump were inactive, the concentration gradients of Na$^+$ and K$^+$ that normally occur across the resting membrane could support the production of thousands of impulses before the gradients were seriously reduced.

The ion pump is active, however, even in the resting membrane, counteracting the slow leakage current of sodium and potassium. The rate of Na$^+$–K$^+$ exchange by the pump is controlled by the internal sodium concentration (Hodgkin and Keynes, 1956). Increased internal sodium resulting from periods of firing in the axon accelerate the pump within seconds to hasten the restoration of normal ionic gradients. Thus, the ion pump is only indirectly related to the spike-generation process.

7.2.3.3. Cable Properties of Nerve. Passive impedance, which has been discussed elsewhere in terms of cable theory, is a quantitative expression of current-voltage relations, and defines how voltage changes spread passively, or electrotonically, along the nerve membrane (Cole, 1968; Plonsey, 1969; Taylor, 1963).

The three parameters of impedance are resistance, capacitance, and inductance. A complete accounting of membrane impedance requires an involved analysis, but for the present purpose a relatively simple consideration of only resistance and capacitance will suffice.

If a stimulus is imposed on axonal membrane, as in Fig. 7-5, triggering sodium activation, the resulting inward Na current can turn either way along the axon but must flow outward at a site some distance from the inward current. If the membrane's resistivity, r_m, to this outward current is relatively low compared to the axoplasm's resistance, r_i (i.e., if the ratio r_m/r_i is low) (Fig. 7-5), positive charge will accumulate on the inside of the membrane close to the stimulus site. This accumulated charge will find ready egress from the axon because of the low r_m. The initial current flowing into the region of accumulated charge is limited by the ability of the membrane surface to accommodate changes in ion density and is called capacitive current. (In electrical terminology, this ability to accommodate charge accumulation is called capacitance.)

If the ratio r_m/r_i is high, the ionic current will flow relatively farther along the axon before flowing outward through the membrane, as shown on the right in Fig. 7-5. Obviously, the distance along the axon the current will flow and the spread of the potential change caused by the current are functions of the relative resistance of the membrane and axoplasm. The relationship given by cable theory is shown in equation (3):

$$\lambda^2 = \frac{r_m}{r_i}, \tag{3}$$

where λ is the membrane space constant (in cm), r_m is membrane surface resistivity (in ohm cm^2), and r_i is axoplasm resistance (in ohms/cm). The space constant is the distance along the membrane from the current source at which the amplitude of the voltage change across the membrane is $1/e$th of the amplitude at the site of stimulation. Equation (4) is the mathematical statement of that relation:

$$V_d = V_0 e^{-d/\lambda}, \tag{4}$$

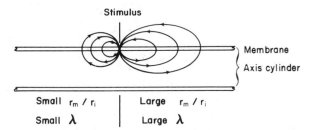

FIGURE 7-5. Cable properties of nerve and the relevance of membrane resistivity (r_m) and axoplasm resistance (r_i) to the space constant, λ, of the nerve membrane. A small space constant is associated with a low ratio of r_m/r_i (on the left), and a large space constant is associated with a high ratio of r_m/r_i.

where V_d is the amplitude of voltage change at any distance d from the current source and V_0 is the amplitude at the current source. The space constant varies directly as the ratio r_m/r_i (Fig. 7-5). The inward Na current at a site in an axonal membrane will affect the membrane voltage at some distance from that site, depending on the space constant. Due to the capacitive current, the initial effect at the remote site will be a depolarization of the membrane, causing the foot of the action potential discussed in reference to Fig. 7-4C. The capacitive current signals the approach of an action potential and triggers the electrogenic process in the membrane ahead of the action potential.

If the space constant is very large, the triggering depolarization occurs far ahead of the action potential, thus increasing the speed of nerve impulse transmission. Large space constants can be produced by greatly reducing axoplasm resistance or greatly increasing membrane resistance, both cases resulting in increased values of the ratio r_m/r_i. Where speed of conduction is essential (e.g., in nerve networks controlling escape mechanisms), invertebrates have giant nerve fibers with low axoplasm resistance, which is inversely proportional to the cross-sectional area of the axon. Higher animals have met the need through nerve myelination. Myelinated nerves are surrounded by layers of membrane materials wrapped closely around the axon, which form a highly resistive sheath with interruptions (nodes of Ranvier) along the length of the axon. The nodes are sites of low resistance at which action potentials are triggered progressively down the length of the axon. This is called saltatory conduction. Because it proceeds from node to node, it allows high-speed transmission.

7.3. CHEMICALLY MEDIATED TRANSMISSION

Transmission at synapses between cells within the nervous system is mediated by chemical transmitters. In addition, many short- and long-term processes in nonneuronal tissues are controlled by neurohormones released from neurosecretory cells into the blood. The latter include such diverse functions as growth and differentiation, ovarian development, cuticular tanning, diuresis, gut motility, and digestive enzyme secretion (Maddrell, 1974).

7.3.1. Synaptic Transmission

Transmission in chemically mediated synapses has been studied in a wide range of animals, but in few insects (Eccles, 1964; McClennan, 1970). Most studies on synaptic function in the insect CNS have used the American cockroach (Callec, 1974), while those on nerve muscle synapses have used mainly locusts, cockroaches, and grasshoppers (Usherwood, 1974).

Although there is no doubt of the occurrence of chemical transmission in the insect CNS, there is no rigorous proof of the identity of any chemical transmitter. However, a great deal of evidence supports the role of acetylcholine (ACh) as an excitatory transmitter between the cercal nerve afferent fibers and the giant internuncial neurons in the sixth abdominal ganglion of the American cockroach (Callec, 1974). [A lesser body of evidence favors the role of gamma-aminobutyric acid (GABA) as an inhibitory transmitter in that same ganglion.] Cholinergic transmission is widespread among animals with CNSs (i.e., flatworms to mammals) and appears to be characterized by similar biochemical elements and pharmacologic properties through the animal kingdom (Florey, 1961; Gerschenfeld, 1973; Sakharov, 1970; Tauc, 1967). The cholinergic synapse will serve as a model for transmission. Although synaptic architecture may involve elaborately branched pre- and postsynaptic structures, a simplified form (Fig. 7-6) will illustrate the events in transmission.

Important structural features of the synapse include the presynaptic terminal with its specialized presynaptic membrane through which ACh is released from vesicles into the synaptic cleft. The vesicles, which range in size from 400 to 600 Å in diameter (Bullock and Horridge, 1965), have been well established as repositories of releasable stores of ACh (Pappas and Purpura, 1972).

A synaptic cleft 200–500 Å wide separates the pre- and subsynaptic membranes, which are approximately coextensive in area. There is cytochemical evidence that both pre- and subsynaptic membranes in some synapses contain membrane-bound AChE which, through the action of detergents, is reversibly dissociable from the membrane. The subsynaptic membrane also contains specialized macromolecules known as acetylcholine receptors

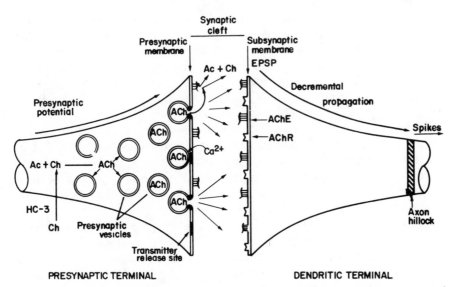

FIGURE 7-6. Schematized excitatory synapse showing pre- and postsynaptic structures and process.

(AChR in Fig. 7-6), which are intimately associated with the membrane and not dissociable from it as is AChE. The subsynaptic membrane is surrounded by a nonspecialized dendritic membrane that extends to the axon hillock where it joins the axonal membrane.

Chemical transmission is initiated by the invasion of the presynaptic terminal by a depolarization (presynaptic potential, Fig. 7-6) resulting from a spike arriving over the afferent fiber. Current theory holds that the depolarization of the presynaptic membrane causes an increased permeability to Ca^{2+}, which diffuses inward, facilitating the attachment of presynaptic vesicles to release sites, as shown in Fig. 7-6. The vesicular and presynaptic membranes fuse, and through a process of exocytosis the ACh is released into the synaptic cleft. During this process, the vesicular membrane becomes incorporated into that of the presynaptic terminal, and the latter expands in area as a consequence. Subsequently, new vesicles are formed from cisternae that develop from the terminal membrane.

The vesicles contain quantal units of ACh ranging from a thousand or so to a few tens of thousands of molecules per vesicle, depending on the particular junction. Each presynaptic potential will cause the release of 100 or so quanta. The details of the release process have been elucidated through studies on higher animals. However, chemical synapses in insects appear to function according to the same scheme.

After release into the synaptic cleft, ACh molecules appear to diffuse randomly as in free solution. The distribution of AChE and AChR on the subsynaptic membrane is such that it is about equally probable that molecules of ACh will collide with an enzyme or a receptor molecule. Those colliding with the enzyme are hydrolyzed to acetate and choline, which diffuse out of the synaptic cleft. Those colliding with AChR combine reversibly with it, inducing during the association a configurational change in the molecule which in turn alters the permeability (i.e., conductance) of the subsynaptic membrane to specific ions. In excitatory synapses g_{Na} and g_K are increased, and in inhibitory junctions g_{Cl} is generally increased. Depending on the relative changes in g_{Na} and g_K, the potential produced will fall somewhat between the equilibrium potentials for the two ions (cf. Fig. 7-3). In inhibitory junctions increased chloride conductance may lead to a depolarization toward the chloride equilibrium potential (Figs. 7-2 and 7-3). Under that condition, if subsequent increases occur in g_K or g_{Na}, a chloride current can flow to compensate for currents of either or both of those ions. In this way the membrane potential is essentially clamped at the chloride equilibrium potential, thus inhibiting transmission.

The interaction between ACh and AChR is reversible; and ACh released after activating the receptor may collide with another AChR or with AChE. The enzyme prevents indefinite prolongation of this process.

Choline released by the hydrolysis of ACh is resorbed by the presynaptic terminal, where it is used to synthesize ACh to replenish releasable stores (Birks and MacIntosh, 1957, 1961). Choline resorption is mediated by an active choline transport system that is competitively inhibited by the drug hemicholinium-3 (MacIntosh, 1959) (HC-3, Fig. 7-6). This reuse of choline, called transmitter turnover, is essential to prolonged transmission by synapses.

In excitatory synapses, activation of the postsynaptic AChR causes a depolarization called an *excitatory postsynaptic potential* (EPSP) (Fig. 7-6), the amplitude of which is directly related to the degree of receptor activation. The EPSP is transmitted decrementally along the postsynaptic membrane to the axon hillock at the juncture of the dendritic and axonal membranes. The amount of decriment is a function of the space constant of the membrane, but in a given synapse the depolarization at the axon hillock is a constant proportion of the EPSP. Nondecrimentally propagated action potentials arise at the axon hillock at a repetition rate determined by the amplitude of the EPSP.

The entire process between arrival of a depolarization at a presynaptic terminal and the production of a postsynaptic action potential may take from 0.5 to 2.6 msec. Most of this time, called *synaptic delay,* is due to the transmitter release process (Katz and Miledi, 1965).

Transmission at inhibitory junctions appears to follow essentially the same process described for cholinergic excitatory synapses, except that in many such junctions in invertebrates and vertebrates the transmitter appears to be GABA (Gerschenfeld, 1973; Obata, 1972).

7.4. SIGNIFICANCE OF ELECTRICAL EVENTS IN NEURAL COMMUNICATION

In the preceding sections, we have seen that the movements of a few selected ions regulated by the properties of neuronal membranes give rise to the membrane potentials that are the basis for transferring information among neurons. In the not too distant past, researchers were considering how the observed varieties of animal functions could be coordinated and controlled by the arrangement of such neurons and their junctions. Recently, however, sophisticated techniques have been developed that allow the simultaneous recording of membrane potentials from both pre- and postsynaptic cells and viewing of dyed cells by light and electron microscopy so that their three-dimensional structure can be determined. This allows the functional capabilities of the cells to be correlated with their structural features. Since researchers consistently find the same neurons in different animals of the same species, they can fully document their functions and connections.

It is now possible to begin to construct neural circuits underlying specific behaviors. Although such neuroethological studies are far from complete, they have increased understanding of the myriad properties and complexities of neurons and their connections and have given new strength to the neuron doctrine.

The transfer of information in the nervous system is accomplished by the all-or-none spikes and the continuously variable intervals between them. Transfer of information requires an information source. For insects that survive by the adaptive complexity of their behavior, we may consider environmental stimuli as information sources. The sensory receptors are the gates through which information about the outside world must pass to the nervous system, and it is the properties of these receptors that first modify the stimuli from the real world.

Sensory receptors are characterized by having a limited spectrum of response, with a lowered threshold in the preferred spectrum. As such, a given sense cell only responds to certain features of a stimulus, such as intensity, wavelength, direction of movement, and so on. The sense cell thus encodes limited information about some features of the impinging stimulus as a pulse code of spikes, which in insects are carried by sensory axons directly to the CNS. Internuncial fibers of the CNS constitute the transmission channels for the sensory inputs that arrive. The many properties of these interneurons and their connections modulate the sensory input, call forth separate central programs of neural activity, and recombine and redistribute (integrate) this information in many yet undefined ways. As we shall see, integration actually occurs in many neurons as well as internuncials. Motor neurons carry modulated information from the CNS to the muscles, where their varied connections and functions combine with those of muscle cells to give the proper movements. These cells may be considered the receivers and thus the decoders of the pulsed information sent about external stimuli. In the case of behavioral responses, successful information encoding, transmission, and decoding of external stimuli by the elements of the neuromuscular systems result in appropriate action that aids survival.

Current understanding of information transfer in the nervous system indicates that a combination of the cellular form of neurons plus their physiological properties produce neuronal coupling functions (Bullock et al., 1977). These coupling functions determine at successive levels of neuronal organization how information is transferred. Within a neuron, there is a wide variation in the cellular architecture. Even though insect internuncials are monopolar, the variations in their dendritic and axonal branchings are enormous (Bullock and Horridge, 1965; Bullock et al., 1977). Differences in such physiological properties as membrane space constants, thresholds, subsynaptic receptors, and many others, in combination with the distinctive cellular architecture, provide numerous functions at dendrites, somas, and initial and terminal branches and branch points of axons. At the level of neuronal junctions, two cells may join monosynaptically or three or more cells may join polysynaptically in convergent or divergent arrangements. Again, new coupling functions and their modifications of information transfer arise from the new combinations that are possible at this level. At more complex connections of neurons in elementary aggregates and neural networks, still further unique coupling functions arise. Although a full discussion of the multiple properties of neurons and their connections is beyond the scope of this chapter, several excellent reviews are available (Treherne, 1974; Bullock et al., 1977; Hoyle, 1976; Kandel, 1976). Some major variations in neuronal membrane properties and neuronal connections and their effects on information transfer will be presented, citing specific examples from insects when possible.

7.4.1. Reception

Sense cells are specialized neurons. In higher animals, sensory receptors may consist of a specialized cell of nonneural origin, called a secondary sense cell, that is linked synaptically to a nerve cell. In insects all sense cells are primary sense cells, which are derived from epidermal cells. These are true bipolar neurons, each with a specialized dendrite (the receptive area) and the axon (the conductile area) that extends directly to the CNS. The dendrites are specialized in various ways to receive different kinds of stimuli, and these specializations together with the properties of various types of accessory cuticular structures combine to form the functional capabilities of the receptive unit, called a sensillum (for details, see Chapter 8). The external energy that impinges on the insect may take various forms—mechanical, electromagnetic (photo), chemical, or thermal—and sense cells that respond preferentially to these forms are called mechanoreceptors, photoreceptors, chemoreceptors, and thermoreceptors, respectively. Regardless of its form, insect sense cells, like those of all animals, convert stimulus energy into a graded membrane potential, the receptor potential (RP), and are thus transducers. Secondary sense cells in higher animals produce only receptor potentials, while primary sense cells of insects produce both receptor potentials and a corresponding pattern of action potentials.

7.4.2. Sensory Potentials and Their Properties

Impulses were first recorded from pressure and stretch receptors of cats by Adrian (1928) and from photoreceptors of the crab *Limulus* by Hartline and co-workers (1952), who concluded that the frequency of impulses increased with the intensity of stimulation and the time over which it acts. The production of these impulses was explained by regarding the sense cell as a structure having the same properties as a nerve fiber but differing in its rate of recovery and adaptation. It was presumed that stimulation evoked depolarization of the sensory dendrite, but it was not until Katz (1950) recorded potentials from the stretch receptors of the frog that the idea was supported. Although intracellular recordings have been made only with visual cells of insects (Naka and Eguchi, 1962), a variety of extracellular recording on chemo-, mechano-, and thermoreceptive cells of insects support the idea that the same basic mechanisms are operating (Naka and Eguchi, 1962; Schwartzkopff, 1977; Davis and Sokolove, 1975).

The effects of a step stimulus displacement of a cuticular hair on the single mechanoreceptive sense cell are depicted in Fig. 7-7. Deflecting the hair in one direction causes movement at the base of the flexible membrane and imparts a stress on the specialized dendrite inserted at the hair base. In insect mechanoreceptor cells a characteristic tubular body in the distal dendrite and other ultrastructural features of the cuticular processes determine some of the functional response properties of the sense cell (McIver, 1975). The stress on the distal dendrite produces a depolarizing receptor potential for as long as the stress is maintained. Presumably, several events (termed transduction processes) must occur in the dendrite membrane between the arrival of a stimulus and the production of the RP, but these events are not well understood for all sense cells. The RP then propagates decrementally to the soma and beyond to the spike-generating zone where sufficiently maintained depolarization results in the production of action potentials. Although the events of transduction probably differ for sense cells of different modalities, the coupling of the RP and action potential production are more likely the same. It has been suggested for both vertebrates and insects, however, that RP generation may be the result of nonspecific permeability to several ions including Na^+ (Broyles et al., 1976; Corey and Hudspeth, 1979).

It is apparent (Fig. 7-7) that a maintained constant stimulation does not result in a constant amplitude RP but declines with time in a characteristic

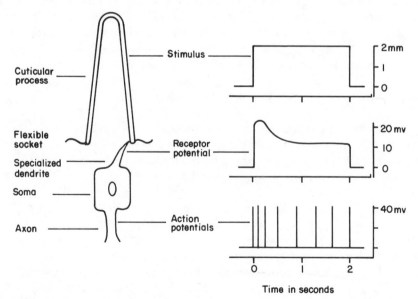

FIGURE 7-7. Generation of responses in a sensory receptor cell as a result of bending a cuticular hair.

manner. This decrease in dendrite membrane potential with a sustained stimulus is called adaptation. As the RP adapts, the frequency of action potentials decreases.

Sense cells that adapt rapidly to stimulation are called phasic receptors; those that do not are called tonic receptors. Phasic receptors are often those involved in high-frequency, short-duration stimuli where they may produce a burst of action potentials at only the onset and offset of stimulation. Acoustical sense cells of insect tympana and auditory hairs are examples of phasic and on–off response types. Tonic receptor cells are often those involved in low-frequency, long-duration stimuli of minutes or hours. Mechanoreceptor cells of insect muscles and joints are commonly of this type, producing continuous low-frequency impulses for long periods. Insect receptor cells having combined adaptation properties are called phasic–tonic. Their initial RP amplitude decreases to a steady-state level maintained throughout stimulation; their action potential frequency varies correspondingly (Fig. 7-7). Many insect chemosensitive cells have phasic–tonic responses, although sense cells of other modalities show similar properties (Schwartzkopff, 1977; Davis and Sokolove, 1975).

The relationships of the membrane potentials of insect sense cells to the stimulus strength is shown diagrammatically in Fig. 7-8 (left). Using the same model sensillum as before, deflections of the cuticular hair of differing amounts in the same direction result in increasing amplitudes of the RP. For RPs of sufficient amplitude to result in depolarization, the frequency of action potentials is linearly related to the amplitude of the RP (Fuortes, 1971) (Fig. 7-7, left). Extracellular records from a wide variety of insect sense cells are consistent with these concepts (Naka and Eguchi, 1962; Schwartzkopff, 1977; Davis and Sokolove, 1975).

The effects of stimulus strength on sensory receptors have been measured on a wide variety of animals, resulting in two fundamental methods for expressing these relationships. One method, derived from the Weber–Fechner relationships, states that the magnitude of the stimulus effect on the frequency of action potentials of the sense cell is related to the logarithm of the stimulus strength (Stevens, 1975). For stimulus strengths above threshold (S_0), any increase results in a proportional increase in the frequency of firing and is often linearly related over the midrange of intensities (Fig. 7-8, upper right). While some mechanoreceptor and photoreceptor cells of insects have very limited ranges of response, so that this linear relationship holds well, many chemoreceptor cells have wide intensity ranges and deviate significantly from linearity at both high and low concentrations (Naka and Eguchi, 1962). For many other receptor cells the relationship of stimulus strength to frequency of firing is related by the Stevens power function (Stevens, 1971). Here the power $n = 1$ results in a linear relationship, while $n < 1$ results in the hyperbolic curve characteristic of many types of sense cells, including those of insects (Fig. 7-8, lower right). While these relationships are neither absolutely correct nor universally applicable, they are use-

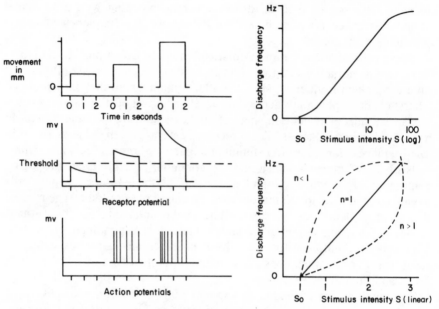

FIGURE 7-8. Relation of stimulus intensity (strength) to receptor cell response and illustration of nerve impulse encoding of receptor potentials.

ful in quantifying the effects of stimulus strength on the response of sense cells.

7.4.3. Transmission

7.4.3.1. Neurons with Action Potentials.
The use of action potentials as the coded form of messages optimizes the rapid flow of information throughout the nervous system. Such potentials allow for a maximum signal amplitude–noise amplitude ratio. Thus, random fluctuations in membrane potentials that are noise are small with respect to the larger action potentials. In this way the integrity of the coded message is maintained through a favorable signal–noise ratio, a key feature of any information-carrying system. Additionally, the constant amplitude of action potentials and their propagation along axons without decrement further ensures the integrity of the coded message. Unlike passive responses that undergo considerable summation of positive and negative polarizations and delete information, action potentials are neither additive nor are they deletive. They are also speedy. Invertebrates, particularly insects, have developed a number of specializations that aid in the rapid conduction of action potentials. Transmission speed, together with the integrity of the coded message, make the insect nervous system highly successful in transferring information.

Invertebrate nervous systems fit into an evolutionary developmental scheme for increased speed of conduction of action potentials with resulting efficiency of information transfer (Bullock and Horridge, 1965). This has been accomplished by two principal developments. One is to reduce the number of neurons in a functional circuit, thereby reducing the number of snyapses. The conduction velocity of action potentials along axons is much greater than the time required for transfer across a synapse, so synaptic delay limits the overall speed of information transfer in a neural circuit. In animals like the coelenterates with nervous systems composed of nerve nets with numerous synapses, the effective conduction velocity of a circuit may be from 0.04 to 0.5 m/sec at 20°C, with response times to external stimuli ranging in tens of seconds (Bullock and Horridge, 1965). In contrast, the evasive response of the cockroach and the tail-flip response of crayfish have axonal conduction velocities of 0.7–44 m/sec at 20°C. The extremely rapid responses result not only from the reduced numbers of synapses but also from the second development for increased speed of transmission, the giant nerve cells. By increasing the axon's cross-sectional area, the cable properties allow for an increased conduction velocity. The conduction velocity of the giant fibers of the American cockroach is 9–12 m/sec at 20°C, representing a 17-fold increase over that of smaller axons (Bullock and Horridge, 1965).

7.4.3.2. Neurons Without Action Potentials.

Interneurons (INs) contained entirely within abdominal or thoracic ganglia of insects have been known anatomically for some time (Zwarzin 1942; Bullock and Horridge, 1965). Only recently, however, have their functional properties been described, using the metathoracic ganglion of the American cockroach (Pearson and Fourtner, 1975). How frequently INs occur in insect ganglia is unknown, but studies of the cockroach mesothoracic ganglion have produced some surprising results. This ganglion contains a little over 300 motoneurons (MNs), with as many as 200 cells sending processes into the anterior and posterior connectives, for a total of about 500 cells with processes leaving the ganglion (Gregory, 1974; Pearson, 1977). The total number of somata in the ganglion is estimated to be about 1500, based on counts of stained whole mounts (Gregory, 1984). By subtracting those cells with axons leaving the ganglion (500), one can estimate 1000 or more INs within the ganglion. Although these neurons represent the major route for information transfer within the ganglion, their functional properties are only beginning to be known.

All intraganglionic INs so far studied have three characteristic features: (1) a low resting potential of −30 to −50 mV, (2) a high level of synaptic activity, and (3) when depolarized, they do not generate action potentials but respond with slow depolarizations of 1–20 Hz (Pearson and Iles, 1970). Since these cells possess short axons and do not transmit information over

long distances, their features are indicative of an integrative function. Those
that have been studied the most connect with MNs (Pearson, 1972). Figure
7-9 shows diagrammatically how the properties of these nonspiking INs are
determined experimentally.

The nonspiking behavior of INs has been confirmed in the locust *Schistocerca* as well as the cockroach, where they are involved in excitation of the
MNs of the flexor muscles of the leg (Pearson and Fourtner, 1975; Pearson,
1977). Nonspiking INs may be both numerous and important in information
transfer in the insect CNS, but at this time their functions are known from
only these few studies.

7.4.4. Processing

7.4.4.1. Within a Neuron.   Individual neurons and even separate branches
of the same neuron can possess distinct membrane properties with a variety
of integrative capabilities. For example, the thresholds for individual neurons can be quite different if we define the membrane threshold as the level
of depolarization necessary to initiate an action potential in 50% of the trials.
In addition, besides the space constant mentioned earlier, another property
of the membrane is the time constant *t*, a measure of the time of decay of a
passive potential to 37% of its original amplitude (Bullock et al., 1977). Both
the time and space constants of individual branches of a neuron can vary, so
that the propagation of both active and passive membrane potentials can be
modulated within the cell.

If the rate of depolarization is slow or subthreshold depolarization is
maintained, the threshold for spike initiation in a neuron rises. This property
of neurons is called accommodation and is another source of modulation

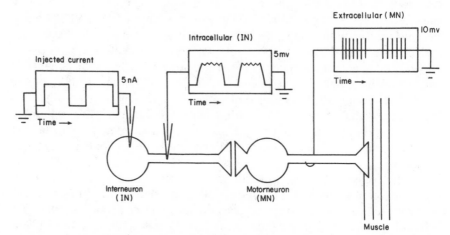

FIGURE 7-9. Determination of nonspiking internuncial neurons within the mesothoracic ganglion of the cockroach.

(Bullock et al., 1977). Neurons may also accommodate maintained hyperpolarization, which sometimes gradually decreases the threshold depolarization.

If the rate of depolarization is held constant but the particular stimulus is presented repeatedly, there may be a decrease in the amplitude or probability of a given response. This is called habituation (Bullock et al., 1977). Individual neurons as well as entire circuits and even entire behaviors can habituate to repeated sitmuli. Withholding the stimulus for characteristic times restores the response through dishabituation.

Neurons also exhibit afterpotentials that follow the production of action potentials but vary greatly in amplitude, duration, and sequence (Bullock et al., 1977). These afterpotentials influence integration within a neuron by being the background on which successive inputs arrive. Thus, while individual neurons clearly possess diverse capabilities for integration, how these function in intact neural circuits remains largely to be determined, especially for insects.

7.4.4.2. Mononeuronal Junctions.   Synaptic junctions occurring between one neuron and a second neuron, muscle cell, or nonnervous sense cell are called mononeuronal junctions. These junctions commonly involve a chemical transmitter (described earlier) or, more rarely, the membranes are closely appressed without a cleft, forming a direct electrical connection called an ephaptic junction. The action of the presynaptic cell on the postsynaptic or follower cell can be only excitatory or inhibitory. If the effect is excitatory, then other features of the junction can modify this action so that multiple effects can result, thus producing integration. If the EPSP lasts longer than the interval between impulses on the presynaptic cell so that a second EPSP rises upon a residue of the first, then temporal summation occurs. Temporal summation, therefore, requires a certain level of presynaptic activity before information passes on via the follower cell. Neurons with differing time constants of their dendritic branches would have unique temporal summation curves. If a second EPSP gives more than an additive summation to the preceding residue, then facilitation occurs. The duration of facilitation in neurons may differ from tens of milliseconds to tens of seconds. The rate of growth and rate of saturation may also differ, so that some junctions can be sensitive to temporal patterns of impulses with the same mean interval (Bullock et al., 1977). Even at the level of single junctions of neurons, integration clearly occurs.

The inhibitory action of the presynaptic cell is defined by the properties of the subsynaptic cell membrane. Evidence suggests that the presynaptic cell produces only one type of transmitter substance. This transmitter may be inhibitory for some follower cells and excitatory for others. In one case, depending on the frequency of presynaptic activity, dual action of excitation and inhibition can be mediated by the same transmitter, acetylcholine, on the same follower cell (Gardner and Kandel, 1972). More commonly, however, a given mononeuronal junction is classified as either excitatory or inhibitory with a constant action exhibited by the follower cell in response to presynaptic stimulation.

7.4.4.3. Polyneuronal Junctions. Synaptic connections of a neuron with two or more neurons, a muscle cell with two or more motoneurons, or one or more sense cells with a single follower neuron are all types of polyneuronal junctions. Given the variable properties of individual neurons, it is easily seen that multiple integrative possibilities exist at such connections. One of the most commonly documented types of connections in the insect nervous system involves multiple sense cells converging as inputs onto a common follower cell as depicted in Fig. 7-10 (upper left). These sense cells connect to the next level of neurons, commonly called second-order neurons, which in turn connect to third-order neurons. Convergence of a few to as many as several hundred sense cells onto a single second-order neuron is documented for mechanoreceptors of the cercus onto giant fibers in the sixth abdominal ganglion of the American cockroach, for olfactory receptor cells of the antenna onto deutocerebral neurons of the American cockroach, and for visual cell inputs onto neurons of the laminar layer of the optic neuropile of flies (Boeckh and Boeckh, 1979; Miller, 1974; Strausfeld, 1976). Depending on the individual system, such high degrees of convergence may result in multimodal responses in the follower cell and extremely high sensitivity or

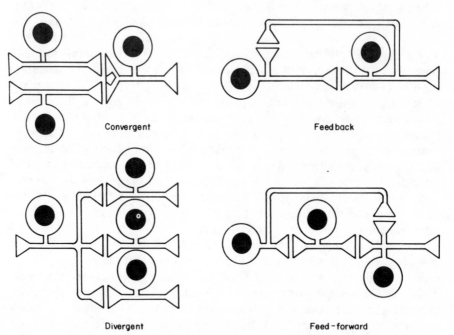

Convergent   Feedback

Divergent   Feed-forward

FIGURE 7-10. Some characteristic neuronal junctions in insects.

require multiple inputs before responding. The sensory inputs usually all have the same effect on the follower cell, so that there is additive excitation or inhibition. Within the CNS, however, such convergent connections may involve both an excitatory and an inhibitory input, so that their opposite effects modulate the output of the follower cell in different fashions depending on the relative frequencies and times of occurrence of their inputs. Such a connection can result in total excitation, total inhibition, or one of several partial effects on the follower cell, thus forming a highly integrative junction. Such junctions are well known in the insect CNS (Hoyle, 1974; Miller, 1974; Wiersma and Roach, 1977).

A divergent polyneuronal connection involves one neuron with its outputs to two or more follower cells (Fig. 7-11, lower left). Here the information from one cell is spread simultaneously to several cells. Such lateral spread of information can have multiple effects on the follower cells, depending on the properties of the synaptic junctions. The large interneuron L10 in the abdominal ganglion of the marine mollusk *Aplysia* mediates diphasic electrical coupling, chemical excitation, two-component excitation–inhibition chemical action, two-component inhibition–inhibition chemical action, and chemical inhibition to five follower cells with the single transmitter substance acetylcholine (Kandel, 1976). These multiple effects are due to the different properties of the subsynaptic membranes of the follower cells. Multiple connections of INs in the laminar layers of the optic neuropiles of

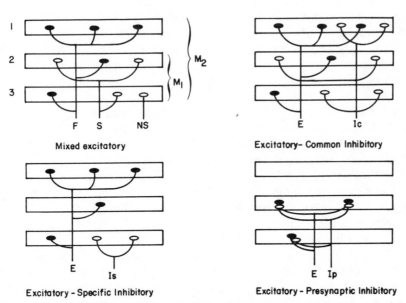

FIGURE 7-11. Types of motoneuron innervation of insect skeletal muscle. Fast motoneuron (F); slow motoneuron (S); neurosecretory neuron (NS); M_1, M_2, motor units; common inhibitory neuron (I_c); specific inhibitory neuron (I_s); presynaptic inhibitory neuron (I_p).

flies and bees are known, but their exact roles in information processing are yet to be determined (Strausfeld, 1976).

In addition to the direct cell-to-cell connections of convergence and divergence, cells may also have branches that extend back to the previous cell or forward to another cell, thus forming secondary loops for information flow. Such feedback loops (depicted in Fig. 7-10, upper right) are common in insect sensory motor systems, where the sense cells monitor movements and stresses, giving feedback to a previous motoneuron or a motor center cell to modulate the output of the motoneuron (Pearson, 1976). Feedback of information flow is also achieved by feed-forward loops (Fig. 7-10, lower right). These loops allow information to be fed forward and added to the effects of an intermediate cell. If the intermediate cell is inhibited, the information fed forward can substitute for it. Feed-forward inhibition has been identified in the visual interneurons of the locust, where it results in the processing of sensory input so that small-area stimuli is preferred over whole-field. This is thought to help protect labile synapses of the processing INs from overstimulation resulting from self-generating head movements (Rowell et al., 1977).

7.4.5. Effecting

7.4.5.1. Insect Neuromuscular Junctions. As the final stage of information decoding in the output of neural messages, insect muscles possess numerous integrative capabilities. This is because of their combination of anatomical diversity and mixed patterns of innervation by motoneurons (MNs). Insect skeletal muscle is composed of one to several muscle fibers that are usually innervated by two or more MNs. A motor unit consists of MNs and the fibers it innervates and, unlike vertebrate muscle, insect muscles are composed of mixed motor units. Figure 7-11 (upper left) gives a diagrammatic representation of a muscle with two motor units, M_1 and M_2, with muscle fibers 2 and 3 in common. Thus, stimulation via the fast (F) MN would cause fibers 1, 2, and 3 to contract, whereas stimulation via the slow (S) MN would cause fibers 2 and 3 to contract. A third neurosecretory neuron (NS) could release secretions onto the muscle that could alter its responses to stimulation of the fast and slow MNs for prolonged periods of time (Maddrell, 1974).

Stimulation arriving via the fast MN presumably causes the quantal release of an excitatory transmitter that reacts with the receptors of the subsynaptic muscle membrane to produce depolarizing EPSPs (Usherwood, 1975). There is a growing body of evidence that the excitatory transmitter in some insect muscles is the amino acid L-glutamate (Beranek and Miller, 1968; Usherwood, 1976). The EPSP is propagated decrementally along the T-tubule system of the muscle, where by an unknown mechanism referred to as excitation–contraction coupling the level of membrane depolarization is transferred into contraction within the muscle cell. As the frequency of stimulation of the fast MN increases, the large EPSPs of up to 5 mV that travel along the muscle membrane recruit more fibers so that larger contractions of the muscle result. While different frequencies of fast stimulation can result in different amounts of muscle contraction, they all tend to be of short duration so that the observed contractions are phasic. Stimulation arriving via the slow MN presumably causes similar events, although the measured EPSPs are smaller, rarely larger than 2 mV (Usherwood, 1976). They result in more prolonged depolarization of the muscle membrane, so that both temporal and spatial summation is important. Muscle contraction resulting from slow stimulation is usually more longlasting or tonic.

Often in conjunction with slow MN innervation of insect skeletal muscle, various types of inhibitory MNs occur, as shown in Fig. 7-11. Depending on the distribution of their terminals, inhibitory MNs can be common to the entire muscle (Ic) specific to a few fiber (Is) or can affect the excitatory MN presynaptically (Ip) (Usherwood, 1975). Stimulation of an inhibitory MN results in the release of an inhibitory transmitter substance that for many insects appears to be GABA (Usherwood, 1975). GABA combines with postsynaptic receptors on the muscle membrane and produces a hyperpolarizing inhibitory postsynaptic potential (IPSP) due to increased Cl^- conductance (Usherwood, 1975). The effects of the IPSP are antagonistic to those of the EPSP, so both excitatory and inhibitory effects can occur on the muscle membrane following MN activation. The location and number of MN as well as the level of presynaptic stimulation determine the final effects of the muscle.

7.4.5.2. Neural Control of Locomotion. Having examined some of the variations in structure and functioning of the neural elements for receiving, transmitting, processing, and effecting information flow in the insect neuromuscular systems, it seems appropriate to consider one example of their coordinated action in regulating behavior. Fortunately, considerable progress has been made in understanding the neural mechanisms controlling locomotion in invertebrates, including insects (Hermann et al., 1976; Hoyle, 1976; Stein, 1976). Both the flight motor system of the locust and the walking motor system of the American cockroach have been studied in sufficient detail to permit the construction of a basic assembly of connected neurons, or elementary neuronal network, that control movements of the appendages (Pearson, 1977; Wilson, 1968).

A circuit for the neurons regulating the movements of the metathoracic leg of the American cockroach (Fig. 7-12) is based largely on the work of Pearson and his associates (Pearson, 1972; Pearson and Fourtner, 1975; Fourtner and Pearson, 1976). This diagram incorporates two central components that control MN activity and two peripheral sensory components that modulate the activity of both the central and MN components.

One central component is a command neuron. A command neuron is an IN that when stimulated will produce a complex, yet stereotyped behavior that is an excellent or identical replica of the natural behavior (Wiersma and

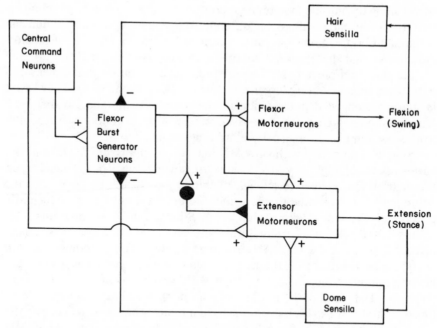

FIGURE 7-12. Elementary neuronal network regulating movements of the metathoracic leg of the American cockroach, (+) denotes excitation, (−) denotes inhibition. (Modified from Pearson, 1977.)

Ikeda, 1964; Kennedy, 1969). Command neurons have been described for a number of stereotyped behaviors in several invertebrates; those controlling the positions of appendages are called locomotory command neurons (Wiersma and Ideka, 1964; Larimer and Gordon, 1976). In the cockroach these central locomotory command neurons, found within the segmental ganglia, connect to the extensor MNs and to a second set of INs called rhythmic pattern, or flexor burst, generators. When excited, this group of neurons provide rhythmic outputs to the MNs. Located in each ganglion in the cockroach, these command neurons act as a local control center by connecting to the MNs for a single leg. The neurons of this center can furnish rhythmic output to the MNs in the absence of any sensory or other CNS input (Pearson, 1976; Pearson and Iles, 1970, 1973). In Fig. 7-12 it can be seen that the flexor burst generator neurons directly excite the flexor MNs and indirectly inhibit the extensor MNs. There are thus six such local control centers in the cockroach, one for each leg, with coordination among them resulting from mutual inhibition (Pearson and Iles, 1970, 1973).

One of the unexpected coupling functions of this system is the asymmetrical excitation of the antagonistic MNs. The command cells excite the extensor MNs and the flexor burst generator neurons. These, in turn, excite the flexor MNs but simultaneously inhibit the extensor MNs via a second IN.

Thus, the flexor MNs receive rhythmic excitation, while the extensor MNs receive constant excitation and rhythmic inhibition. The excitation of flexor MNs activates the muscles that produce flexion, swinging the leg forward, while the extensor MNs are being inhibited.

The completion of the stepping cycle requires additional inputs from peripheral sensory cells. In the American cockroach, there are fields of mechanoreceptive hairs at the coxal bases and fields of dome sensilla near the femoral–tibial joint (see Fig. 8-4, lower right). As the leg swings forward to its full extent, the coxal hair sensilla are activated. Their inputs inhibit the flexor burst generators, thereby reducing excitation to flexor MNs and exciting the extensor MNs (Fig. 7-12). Thus, the hair sensory inputs are a negative-feedback loop to flexor MNs and a positive-feedback loop to extensor MNs. Such reciprocal inhibition is a common feature of sensory feedback in motor systems (Hoyle, 1974; Stein, 1976; Hermann et al., 1976). The input of the sensory hairs ensures that the swing phase of leg movement will be ended at the same point each time regardless of where it was begun.

As the leg is placed on the substrate, excitation of the extensor MNs causes the muscles to propel the leg backward and move the insect forward. As the weight of the insect increases on the leg, the dome sensilla detect the increasing stress on the cuticle and their inputs give positive feedback to the extensor MNs and negative feedback to the flexor burst generator. Their role is exactly antagonistic to the hair inputs. As other legs begin to take the weight of the insect, the dome sensilla inputs decrease to the point that the flexor MNs become activated and the cycle repeats. If there are unexpected variations in the load in the leg during extension, the dome sensilla inputs change and provide for compensating MN activity.

The basic rhythm of appendage movements is thus determined by central pattern generator neurons and command neurons, but the amplitude and phase of leg movements depends on sensory feedback. Coordination of the local control centers is achieved by mechanisms less well understood, but they result in only one leg moving at a time. This represents one of the best-documented examples of centrally generated patterns of motor activity usually referred to as motor programs. Such motor programs have been described for insect respiration and flight, as well as locomotion, and represent a major achievement in our continual efforts to understand the neural basis for behavior in insects as well as ourselves.

7.5. CONCLUSIONS

This chapter has focused on the ionic basis of membrane potentials in excitable cells and some principles of the reception, transmission, and processing of information by insect neurons. Many details are yet to be deciphered and will undoubtedly form the focus of research in the immediate future. Biochemical characterization of membrane receptor and ionic channel proteins from insect neurons is underway and will likely expose some properties unique to insects. Chemical transmitters are far from fully described in insect nervous systems, and current results indicate that small neuropeptides may be extremely widespread and important in modulating information flow. New types of neuronal coupling functions will undoubtedly be described as higher levels of processing are studied in networks of fully characterized neurons.

As the principal site of action of the vast majority of insecticides, the insect nervous system will remain of central importance for the development of new compounds for insect control. Descriptions of neuronal membrane proteins and new transmitters and their synthetic, storage, and degradation routes will provide many new target sites. More subtle alterations in neuronal functions may result in overt loss of coordination, or decision-making processes, so that the insect can no longer successfully meet the challenges of survival. In this way, perhaps control can even be achieved at a species-specific level, allowing integration into more tailored pest management strategies.

REFERENCES

A. W. Adamson, *Understanding Physical Chemistry*, 2nd ed., W. A. Benjamin, New York, 1969.

E. D. Adrian, *The Basis of Sensation: The Action of Sense Organs*, Christophers, London, 1928.

R. Beranek and P. L. Miller, *J. Exp. Biol.* **49**, 83 (1968).

R. Birks and F. C. MacIntosh, *Brit. Med. Bull.* **13**, 157 (1957)

R. Birks and F. C. MacIntosh, *Can. J. Biochem.* **39**, 787 (1961).

G. H. Bishop, *Physiol. Rev. Camb. Phil. Soc.* **36**, 376 (1956).

J. Boeckh and V. Boeckh, *J. Comp. Physiol.* **132**, 235 (1979).

J. L. Broyles, F. E. Hanson, and A. M. Shapiro, *J. Insect Physiol.* **32**, 1581 (1976).

T. H. Bullock and G. A. Horridge, *Structure and Function of the Nervous Systems of Invertebrates*, Vols. 1 and 2, W. H. Freeman, San Francisco, 1965.

T. H. Bullock, R. Orkand, and A. Grinnell, *Introduction to Nervous Systems*, W. H. Freeman, San Francisco, 1977.

J. Callec, "Synaptic transmission in the central nervous system," in J. E. Treherne, Ed., *Insect Neurobiology (North-Holland Research Monographs, Frontiers of Biology*, Vol. 35), North-Holland Publishing Co., Amsterdam and Oxford, 1974, p. 119.

K. S. Cole, *Membranes, Ions and Impulses*, University of California Press, Berkeley, 1968.

D. P. Corey and A. J. Hudspeth, *Nature* **281**, 675 (1979).

E. E. Davis and P. G. Sokolove, *J. Comp. Physiol.* **96**, 223 (1975).

H. Davson, *A Textbook of General Physiology*, 3rd ed., Little Brown and Co., Boston, 1964.

J. C. Eccles, *The Physiology of Synapses*, Springer-Verlag, New York, 1964.

E. Florey, *Ann. Rev. Physiol.* **23**, 501 (1961).

C. R. Fourtner and K. G. Pearson, "Morphological and physiological properties of motor neurons innervating insect leg muscles," in G. Hoyle, Ed., *Identified Neurons and Behavior of Arthropods*, Plenum, New York, 1976, p. 87.

M. G. Fuortes, "Generation of responses in receptors," in W. R. Loewenstein, Ed., *Principles of Receptor Physiology*, Vol. 1, Springer-Verlag, Berlin, 1971, p. 243.

D. Gardner and E. R. Kandel, *Science* **176**, 675 (1972).

H. M. Gerschenfeld, *Physiol. Rev.* **53**, 1 (1973).

G. E. Gregory, *Phil. Trans. Roy. Soc. Lond. Ser. B.* **267**, 421 (1974).

G. E. Gregory, *Phil. Trans. Roy. Soc. Lond. Ser. B.* **306**, 191 (1984).

H. Grundfest, *Ann. N.Y. Acad. Sci.* **94**, 405 (1961).

H. K. Hartline, H. G. Wagner, and E. F. MacNichol, *Cold Spr. Harb. Symp. Quant. Biol.* **17**, 125 (1952).

R. H. Hermann, S. Grillner, P. S. G. Stein, and D. G. Stuart, Eds., *Neural Control of Locomotion*, Plenum, New York, 1976.

A. L. Hodgkin, *The Conduction of the Nervous Impulse*, Charles C. Thomas, Springfiled, Ill., 1964.

A. L. Hodgkin and A. F. Huxley, *J. Physiol.* **117**, 500 (1952).

A. L. Hodgkin and B. Katz, *J. Physiol.* **108**, 37 (1949).

A. L. Hodgkin and R. D. Keynes, *J. Physiol.* **131**, 592 (1956).

G. Hoyle, "Natural control of skeletal muscle," in M. R. Rockstein, Ed., *Physiology of Insecta*, Academic Press, New York, 1974.

G. Hoyle, *Identified Neurons and Behavior of Arthropods*, Plenum, New York, 1976.

E. R. Kandel, *Cellular Bases of Behavior*, W. H. Freeman, San Francisco, 1976.

B. Katz, *J. Physiol.* **111**, 261 (1950).

B. Katz and R. Miledi, *Proc. Roy. Soc. Lond. Ser. B.* **161**, 483 (1965).

D. Kennedy, "The control of output by central neurons," in M. A. Brazier, Ed., *The Interneuron*, University of California Press, Los Angeles, 1969, p. 21.

N. Lakshminarayanaia, *Transport Phenomena in Membranes*, Academic Press, New York, 1969.

J. L. Larimer and W. H. Gordon, "Circumesophageal interneurons and behavior in crayfish," in G. Hoyle, Ed., *Identified Neurons and Behavior of Arthropods*, Plenum, New York, 1976, p. 243.

W. R. Loewenstein, C. A. Terzuolo, and Y. Wshizu, *Science* **142**, 1180 (1963).

H. McClennan, *Snyaptic Transmission*, 2nd ed., Saunders, Philadelphia, 1970.

F. C. MacIntosh, *Can. J. Biochem. Physiol.* **37**, 343 (1959).

S. H. P. Maddrell, "Neurosection," in J. E. Treherne, Ed., *Insect Neurobiology (North-Holland Research Monographs, Frontiers of Biology*, Vol. 35), North-Holland Publishing Co., Amsterdam and Oxford, 1974, p. 307.

P. L. Miller, "The neural basis of behavior," in J. E. Treherne, Ed., *Insect Neurobiology (North-Holland Research Monographs, Frontiers of Biology*, Vol. 35), North-Holland Publishing Co., Amsterdam and Oxford, 1974, p. 359.

K. Naka and E. Eguchi, *J. Gen. Physiol.* **45**, 633 (1962).

K. Obata, *Int. Rev. Neurobiol.* **15**, 167 (1972).

G. D. Pappas and D. P. Purpura, Ed., *Structure and Function of Synapses*, Raven Press, New York, 1972.

K. G. Pearson, *J. Exp. Biol.* **56**, 173 (1972).

K. G. Pearson, *Sci. Amer.* **235**, 72 (1976).

K. G. Pearson, "Interneurons in the ventral nerve cord of insects," in G. Hoyle, Ed., *Identified Neurons and Behavior of Arthropods*, Plenum, New York, 1977, p. 329.

K. G. Pearson and C. R. Fourtner, *J. Neurophysiol.* **38**, 33 (1975).

K. G. Pearson and J. F. Iles, *J. Exp. Biol.* **52**, 134 (1970).

K. G. Pearson and J. F. Iles, *J. Exp. Biol.* **58**, 725 (1973).

Y. Pichou, "Axonal conduction in insects," in J. E. Treherne, Ed., *Insect Neurobiology (North-Holland Research Monographs, Frontiers of Biology,* Vol. 35), North-Holland Publishing Co., Amsterdam and Oxford, 1974, pp. 73–117.

R. Plonsey, *Bioelectrical Phenomena,* McGraw-Hill, New York, 1969.

K. Rodahl and B. Issekutz, Jr., *Nerve as a Tissue,* Harper & Row, New York, 1966.

C. H. F. Rowell, M. O'Shea, and J. L. D. Williams, *J. Exp. Biol.* **68**, 157 (1977).

D. A. Sakharov, *Ann. Rev. Pharmacol.* **10**, 431 (1970).

J. Schwartzkopff, *J. Comp. Physiol.* **120**, 11 (1977).

G. Semenza and E. Carafolr, Eds., *Biochemistry of Membrane Transport,* Springer-Verlag, New York, 1977.

W. F. Sheehan, *Physical Chemistry,* Allyn and Bacon, Boston, 1961.

J. C. Skou, *Physiol. Rev.* **45**, 596 (1965).

A. J. Staverman, *Trans. Faraday Soc.* **48**, 176 (1952).

A. J. Staverman, *Acta Physiol. Pharmacol. Neer.* **3**, 522 (1954).

D. G. Stein, "A comparative approach to the neural control of locomotion," in G. Hoyle, Ed., *Identified Neurons and Behaviors of Arthropods,* Plenum, New York, 1976, p. 227.

S. S. Stevens, "Sensory power functions and neural events," in W. R. Loewenstein, Ed., *Handbook of Sensory Physiology,* Springer-Verlag, New York and Berlin, 1971.

S. S. Stevens, *Psychophysics,* Wiley, New York, 1975.

N. V. Strausfeld, "Mosaic organizations, layers, and visual pathways in the insect brain," in F. Zittler and R. Weiler, Eds., *Neural Principles in Vision,* Springer-Verlag, New York, 1976, p. 260.

L. Tauc, *Physiol. Rev.* **47**, 521 (1967).

R. E. Taylor, "Cable theory," in W. L. Nastuk, Ed., *Physical Techniques in Biological Research,* Vol. 6, Academic Press, New York, 1963, p. 219.

J. E. Treherne, "The environment and function of insect nerve cells," in J. E. Treherne, Ed., *Insect Neurobiology (North-Holland Research Monographs, Frontiers of Biology,* Vol. 35), North-Holland Publishing Co., Amsterdam and Oxford, 1974, p. 187.

P. N. R. Usherwood, "Nerve-muscle transmission," in J. E. Treherne, Ed., *Insect Neurobiology (North-Holland Research Monographs, Frontiers of Biology,* Vol. 35), North-Holland Publishing Co., Amsterdam and Oxford, 1974, p. 245.

P. N. R. Usherwood, *Insect Muscle,* Academic Press, New York and London, 1975.

P. N. R. Usherwood, "Neuromuscular transmission in insects," in G. Hoyle, Ed., *Identified Neurons and Behavior of Arthropods,* Plenum, New York, 1976, p. 31.

C. A. G. Wiersma and K. Ikeda, *Comp. Biochem. Physiol.* **12**, 509 (1964).

C. A. G. Wiersma and J. L. M. Roach, "Principles in the organization of invertebrate nervous systems," in E. R. Kandel, Ed., *Handbook of Physiology (The Nervous System,* Vol. 1), Part 2, American Physiological Society, Bethesda, 1977, p. 1089.

D. M. Wilson, "The nervous control of insect flight and related behavior," in J. W. L. Beament, J. E. Treherne, and V. B. Wigglesworth, Eds., *Advances in Insect Physiology,* Vol. 5, Academic Press, New York, 1968, p. 289.

J. A. Wilson, *Principles of Animal Physiology,* MacMillan, New York, 1972.

Z. Zwarzin, *Z. Wiss. Zool. Abt.* **12**, 323 (1942).

8

NERVOUS SYSTEM:
SENSORY SYSTEM

J. L. FRAZIER
Agricultural Chemicals Department
E. I. DuPont de Nemours and Co.
Wilmington, Delaware

CONTENTS

SUMMARY

Three basic types of insect sensilla are recognized as elements in different receptor systems. These sensilla—without pores, uniporous, and multi-porous—are components of mechanoreceptor, photoreceptor, and chemo-receptor systems. Feature extractions, which are characteristic of each of these sensory modalities, are characteristic of the receptor systems, providing them with optimal characteristics vis-à-vis their evolved functions. The adaptive radiation that has evolved in the major sensory systems has been a major factor in the success of Insecta, and selected examples illustrate this vividly. The existence of other sensory systems (e.g., thermoreceptor, hygroreceptor, geomagnetic) further emphasizes the evolutionary importance of these nervous elements.

8.1. INTRODUCTION

The sensory systems of insects, like those of all highly mobile animals, have evolved with numerous specializations that permit them to detect important features of the external environment, to monitor constantly the internal states of the organism, and to provide information on the positions of the appendages. Considering the vast numbers of insect species and the variety of habitats they successfully occupy, it is not surprising that they possess equally diverse specializations of their sensory systems. It is in these, perhaps more than in any other physiological system of the insect, that we find specialization of structure, with its concomitant increased efficiency of function in response to selective pressures for survival. These evolutionary developments have resulted in such an adaptive radiation of insect sensory systems that the diversity and subtleties of their structure and function are only beginning to be unraveled.

Although insects are small in size, the external rigid construction of their sensory receptors and the segmental organization of their nervous system have made them favored research subjects for scientists from a variety of disciplines. This interest has resulted in a wealth of information on the sensory systems of insects, especially within the last five years. As a result, we know a great many details about a few model systems and some scattered details of a few other species. About the vast majority of insects, we still know relatively little. We do know that the differences in the filtering capacities of insect senses makes their perceived world not only vastly different from our own perceptions but also uniquely different for each species and often for different life stages.

This chapter will focus on the common properties of sensory receptors, using selected examples to illustrate the structural and functional adaptive radiation that has rendered insects so successful.

8.2. THE INSECT SENSILLUM

8.2.1. Origin and Basic Structure

The sensillum is the structural unit from which the majority of insect sensory organs are derived. Ectodermal in origin, it develops by differentiation from a mother epidermal cell. It consists of cuticular parts, one or more sense cells, and two or more sheath cells (Fig. 8-1) (Altner and Prillinger, 1980; Zacharuk, 1980). The cuticular parts are either above the surface of the body or sunken in pits or depressions. Of numerous shapes and sizes, they may be attached either rigidly or through flexible cuticle in a socket.

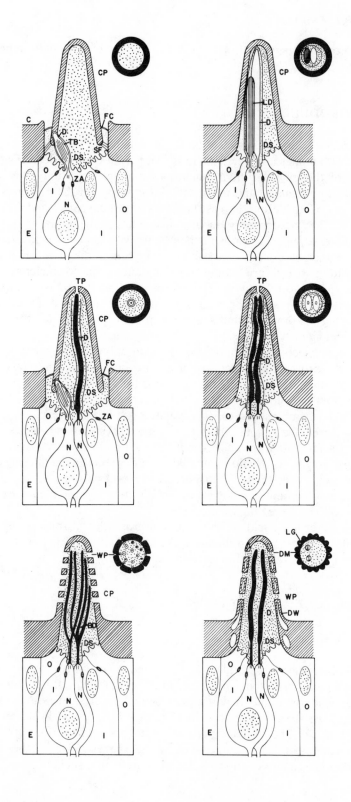

The sense cell's vary in number from 1 to 40 or more and have large nuclei located in or immediately below the epidermis. These bipolar sense cells send their dendrites to the cuticular parts where their form, ultrastructural features, and methods of attachment are characteristic for cells of different modalities. Their axons extend into the sensory nerve parallel with other sensory axons, often extending directly to the central nervous system (CNS) before making synaptic connections to second-order neurons. Thus, these are primary sense cells that contain both a sensory receptor area on their dendrites and an impulse-conducting area on their axons.

The sheath cells vary in number, but usually an inner sheath cell (formerly called a trichogen cell) and an outer sheath cell, also known as a tormogen cell, are present. The inner sheath cell envelops the sense cell directly, while the outer sheath cell encloses both of these. These sheath cells are thought to furnish nutrients and ions that aid in the functioning of the sense cells. Often both an intermediate sheath cell and a basal sheath cell, also called a glial cell, are present, with the latter wrapping around the sensory axons and insulating them from each other and from the hemolymph. Each of these structural units comprising an insect sensillum is subject to a great deal of variation in form. Possible combinations are extremely numerous, providing for adequate solutions to a variety of adaptive requirements.

8.2.2. Types of Sensilla

The cuticular projections of insect sensilla are the most visible portion, and their size, shape, and position have been the basis for classifying them. With wide use of the transmission electron microscope (TEM) and the scanning electron microscope (SEM), coupled with the techniques for recording the nerve impulses from the sense cells associated with a single sensillum, many new features of structure and function have been catalogued. A new simpler terminology that has good correspondence of sensillum structure and function has been proposed recently and is currently being used by increasing numbers of investigators (Table 8-1).

FIGURE 8-1. Diagrammatic representations of six basic types of insect sensilla. Upper left, sensillum in flexible socket with single sense cell containing a tubular body (TB). Upper right, sensillum without flexible socket containing a sense cell with lamellated dendrite (LD). Middle left, uniporous sensillum in flexible socket containing one cell with tubular body and one or more cells with dendrites extending to the terminal pore (TP). Middle right, uniporous sensillum without a flexible socket containing two or more cells with unbranched dendrites (D). Lower left, multiporous sensillum with single wall and multiple cells with branched dendrites (BD). Lower right, multiporous sensillum with double wall and multiple cells with unbranched dendrites. CP, cuticular processes; C, cuticle; FC, flexible cuticle; DS, dendritic sheath; SF, suspensory filaments; O, outer sheath cell; I, inner sheath cell; N, nerve cell; E, epidermal cell; ZA, *zonnula adherens* junction; LG; longitudinal groove; DM; dense material; WP; wall pore; DW, double wall.

TABLE 8-1. Types of Externally Located Insect Sensilla with Comparison of Terminology

	Morphological Properties	Sensitivity	Old Terminology	New Terminology[a]
Pores absent	Socket flexible; 1 sensory cell: dendrite with tubular body	Mechano	S. campaniformia	Dome in flexible socket
			S. chaetica	Hair in flexible socket
	Socket inflexible; 3–4 sensory cells (1 dendrite folded or lamellated)	Thermo and hygro	S. basiconica	Peg
			S. coeloconica	Peg in a pit
Single-terminal pore (thick walled)	Socket flexible; 2–20 sensory cells: 1 dendrite with tubular body, dendrites not branched	Gustatory and mechano (contact chemoreceptor)	S. chaetica	Uniporous hiar in flexible socket
			S. trichodea	Uniporous hair in flexible socket
			S. basiconica	Uniporous peg in flexible socket
			S. styloconica	Uniporous dome in flexible socket
	Socket inflexible; 2–9 sensory cells, dendrites not branched	Gustatory	S. basiconica	Uniporous peg
		Gustatory	S. styloconica	Uniporous cupola
Multiple pores in wall (thin walled, socket inflexible)	Single-walled pore tubules (or similar structure); 1–40 sensory cells, dendrites branched or not branched	Olfactory[b]	S. basiconica	Multiporous peg
			S. coeloconica	Multiporous peg in a pit
			S. trichodea	Multiporous hair
		Olfactory	S. placodea	Multiporous plate
		Olfactory and thermo	S. basiconica	Multiporous grooved peg
	Double-walled, spoke canals with secretion, 2–4 sensory cells, dendrites not branched	Olfactory	S. coeloconica	Multiporous grooved peg in a pit
		Thermo and hygro		

[a] Pore type + shape + socket type. Modified from Altner and Prillinger (1980) and Zacharuk (1980).

[b] In Apterygota multiporous sensilla sometimes with flexible socket and one mechanosensitive cell (tubular body) (Altner and Prillinger, 1980).

Insect sensilla on the outside of the body consist of the six major types shown in Fig. 8-1 and designated in Table 8-1. These sensilla are designated by a term encompassing the shape of the cuticular part, the presence or absence of pores, and the type of attachment to the cuticle (Altner and Prillinger, 1980; Zacharuk, 1980). Thus, a hair without a pore in a flexible socket represents the first type in Table 8-1 and Fig. 8-1 (upper left). The usefulness of this new approach becomes apparent when we see that a uniporous hair in a flexible socket is the third type in Table 8-1 and denotes a sensillum that characteristically has 2–20 sensory cells and a dual modality. One of the sense cells is mechanosensory and one or more are chemosensory. This type was also called sensilla chaetica in the older terminology (Table 8-1), so that this former designation included two functionally different types of sensilla. Much earlier confusion arose from attempting to assign one function to a sensillum. The sensillum is the structural unit, but the sense cells are the functional inputs, thus sensilla with several sense cells may have more than one function.

8.2.2.1. Sensilla Without Pores. These sensilla may be attached in flexible sockets (Fig. 8-1, upper left) or without sockets (Fig. 8-1, upper right). The sensilla with flexible sockets are usually hairs of various lengths, curvatures, and tip configurations, which at their extreme variation may be only domes (Altner and Prillinger, 1980). A cross section of the hair at its midpoint reveals a hollow shaft. Its flexible base is enclosed in a socket of various dimensions that may limit the degree of bending of the hair. These hairs are innervated by a single sense cell that has a characteristic tubular body in the distal dendrite. The tubular body is composed of 50–1000 closely packed microtubules that lend rigidity to the dendrite area and are thought to be involved in the transduction process. The dendrite may be inserted at various points at the hair base. Suspensory filaments connect the base of the hair with the cuticle and together with the tubular body attachment and the socket dimensions build in a directional sensitivity for the sense cells response. Such mechanosensitive hairs are extremely numerous on the surface of the insect's body where they detect movements of the air, substrate, and body parts.

Comparatively little is known about the few nonporous sensilla that are rigidly attached (Fig. 8-1, upper right). They are usually pegs located on the surface of the cuticle or enclosed in pits and contain three to four sense cells that have folded or lamellated dendrites within the lumen. These cells have been shown electrophysiologically to respond to temperature and humidity, resulting in a bimodal sensillum (Altner and Prillinger, 1980).

8.2.2.2. Uniporous Sensilla. Uniporous sensilla in flexible sockets occur in a wide variety of sizes as hairs, pegs, or papillae. They are numerous and widely distributed on the insect body (Fig. 8-1, middle left). Thick-walled, they contain one dendrite with a tubular body that attaches at the base of the cuticular part and one or more sense cells with unbranched dendrites that extend into the hair lumen to the terminal pore (Zacharuk, 1980). The dendrites are divided into the proximal dendrite, containing a ciliary basal body, and the distal dendrite, extending into the cuticular part; the juncture of these two is marked by a constricted ciliary region. The distal dendrite is enclosed in a cuticular sheath, and both of these are shed at molting. There are usually the full compliment of sheath cells, with the inner cell enclosing the proximal dendrites and the intermediate and outer sheath cells terminating in microvillar projections into the sensillar sinus. The sheath cells form *zonula adherens* junctions with each other and with the sense cells so that cellular communication is possible. The other sheath cells pass materials, perhaps by active transport, into the sensillar sinus, where they furnish ions and perhaps other materials important for transduction (Zacharuk, 1980).

There are fewer uniporous sensilla with inflexible sockets, but cupola or dome-shaped sensilla occur often in the mouth region, where they serve to monitor the food being eaten (Fig. 8-1, middle right). They typically contain two to nine sensory cells, have thick walls, and are otherwise structurally similar to other uniporous sensilla.

8.2.2.3. Multiporous Sensilla. Multiporous pegs, plates, and hairs of wide variation in size may be found in extremely large numbers on the body surface, particularly on the antennae and to a lesser extent on the maxillary and labial palpi where they serve an olfactory function (Fig. 8-1, lower left). The cuticular wall is thinner than that of uniporous sensilla, and these sensilla contain up to 40 sense cells, with large numbers not uncommon (Altner and Prillinger, 1980). The distal dendrites are typically branched within the sensillar sinus, where they contain a variety of connections to the numerous cuticular pores. These connections are thought to represent channels for conducting stimulus molecules to the dendrite membranes of the sense cells where transduction occurs. The cuticular sheath is usually much shorter than in uniporous sensilla, while other structural features are the same.

Multiporous pegs with longitudinal grooves and a double-wall construction, less numerous than other multiporous sensilla, are found on the antennae, where they may have the dual function of olfactory and temperature reception (Fig. 8-1, lower right). They typically contain two to four sense cells, with distal dendrites unbranched, and show secretory products in the channels connecting the cuticular pores and the dendrites. Although the cuticular parts may be variously shaped, other structural details, with the above exceptions, are similar to those of other multiporous sensilla (Altner and Prillinger, 1980).

8.3. MECHANORECEPTOR SYSTEM

Mechanoreception in insects serves to detect mechanical distortions resulting from contacting an object; vibrations of the air, water, or substrate;

distortions of internal structures resulting from muscular activity; or forces of gravity (Schwartzkopff, 1974). Thus, the external senses of touch and hearing are accomplished by various modifications of sensilla without pores and uniporous sensilla in flexible sockets. The internal sense of stretch is accomplished by a separate type of sensory neuron that is multipolar and has its dendrites attached in muscles, suspensory ligaments, or other internal structures. The sense of vibration includes still another type of internal sensillum, a chordotonal sensillum. These occur singly or in groups with attachments to trachea and other internal structures (Dethier, 1964). The sense of gravity, important for the primary orientation of the body, is the result of CNS integration of various mechanoreceptor inputs from both external and internal locations (Schwartzkopff, 1974; Dethier, 1964).

8.3.1. Structure and Distribution

8.3.1.1. Tactile Sensilla. Nonporous and uniporous hairs in flexible sockets are the most common types of tactile, or touch, sensilla on the insect's surface. Their single sense cell with the tubular body has the distal dendrite inserted at various points of the hair base (Fig. 8-2). The cuticular sheath that covers the dendrite at its insertion is associated with a molting pore in the side of the hair. As the insect passes through successive stages of development, the mechanosensitive hair remains functional until a few hours before ecdysis when this sheath is shed together with the previous cuticular structure.

Tactile hairs are widely distributed over the surface of the body and are particularly numerous on the head, at points of articulation of the body, and of the appendages (Fig. 8-3) where they occur singly or in dense hair beds (Fig. 8-4, upper left). The antennae of many insects are richly supplied with tactile hairs, especially those species for which the tactile sense is important for orientation during locomotion [e.g., species that are nocturnally active, like

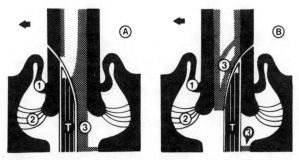

FIGURE 8-2. Simplified models of mechanosensitive sensilla showing structural basis for directional sensitivity. A, molting channel points in direction of maximal sensitivity (*Locusta*). B, molting channel points away from direction of maximal sensitivity (*Dysdercus*), see arrows. 1, joint membrane; 2, suspension filaments; 3, socket septum; T, tubular body. [From Gaffal et al., *Zoomorphologie* **82,** 79 (1975).]

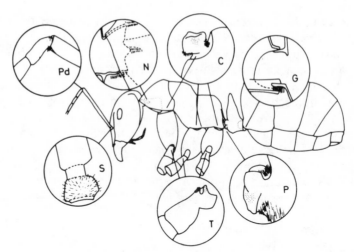

FIGURE 8-3. Worker of the ant *Formica polyctena*. The circles give enlarged view of the different joint-bearing hair plates; the lines indicate where the joints are situated. Coxal joints (C), gaster (G), neck (N), petiole (P), joint between first and second antennal segments (Pd), joint between antenna and head (S), joint between coxa and trochanter (second segment of the leg) (T). (From Markl, 1971.)

the American cockroach (Fig. 8-4, upper right), or that live in caves, covered nests, or underground]. The cerci of cockroaches and other Orthoptera (Fig. 8-4, middle right) are covered with long filiform hairs and more robust hairs in sockets, which function as vibration receptors when held against the substrate or as sound detectors when held aloft (Schwartzkopff, 1974).

8.3.1.2. Vibration Sensilla. On the other hand, nonporous sensilla in flexible sockets may be modified to the extent that the cuticular part is only a dome of thin flexible cuticle surrounded by a ring of raised cuticle (Fig. 8-4, lower left). They often occur in groups, such as on the femur of the American cockroach (Fig. 8-4, lower right) where their axes are aligned within a field. The proximal dendrite bearing a tubular body is attached either in the center or near one end.

These sensilla are dome-, bell-, or cupola-shaped and either round or elliptical at the base. The sense cell has a preferential response along this axis, and neighboring groups are oriented differently so that each group can best determine differential compressions of the cuticle. Some sensilla commonly occur near points of stress such as at appendage joints where they monitor cuticular deformations. Dome sensilla are also found at the base of tactile hairs, on the tibia of the American cockroach (Fig. 8-4, middle left), and at the base of filiform hairs on the cercus of crickets where they monitor some hair movements (Palka et al., 1977; Dumpert and Gnatzy, 1977).

Internal (chordotonal) sensilla for vibration reception occur widely in insects (Schwartzkopff, 1974; Dethier, 1964). Their structure is shown dia-

FIGURE 8-4. Insect mechanosensitive sensilla. Upper left, field of hairs at the base of the hind wing of the American cockroach (230×). Upper right, numerous long sensilla on an antennal segment of the American cockroach (100×). Middle left, tactile spines in flexible sockets on the tibia of the American cockroach (150×). Insert, directional limitation imposed by the shape of the cuticular base (300×). Middle right, ventral surface of the cercus of the American cockroach showing long vibrational sensilla (115×). Insert, detail of cuticular base imposing no directional limitation to movement (345×). Lower left, dome sensilla for detecting cuticular stress on the antenna of the tarnished plant bug (15,000×). Lower right, field of dome sensilla near femoral–tibial joint of a leg of the American cockroach (805×). (Scanning electron micrographs courtesy of D. Ave and L. D. Hatfield.)

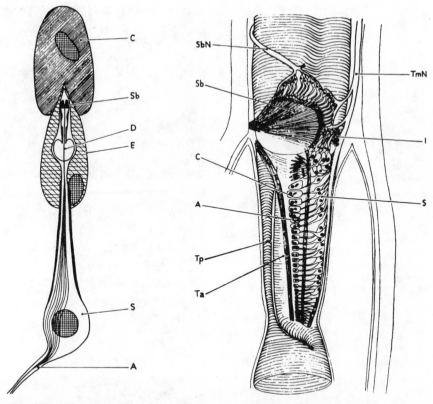

FIGURE 8-5. Diagrammatic structure of a chordotonal sensillum (left) with tympanic organ of the prothoracic leg of Orthoptera (right). C, cap cells, S, scolopoid sense cell; Sb, scolopoid body; D, dendrite; E, enveloping cell; A, axon; (right) Ta, anterior trachea; Tp, posterior trachea; Sb, subgenual organ; I, intermediate organ; A, acoustic organ; SbN, subgenual nerve; TmN, tympanic nerve. (From V. G. Dethier, *The Physiology of Insects*, Methuen, Inc., 1964.)

grammatically in Fig. 8-5 (left). They consist of a single sense cell with a distal dendrite containing a scolopoid body. This is enclosed in an enveloping cell that connects with the cap cell. The cap cell connects the sensilla, often through accessory cells, to the body wall, muscles, tracheae, or other internal structures. These sensilla serve to detect movements of the cuticle transmitted to other internal structures or movement resulting from muscle activity, such as respiratory movements, so that they form one of the main types of internal receptors (proprioceptors). The best-studied functions of chordotonal sensilla are associated with a special organ located in the tibia of the prothoracic leg of many species called a subgenual organ (Fig. 8-5, right). This organ, which consists of 20–40 chordotonal sensilla attached to the cuticle of the leg, detects substrate vibrations of 10–50 Hz transmitted through the cuticle. Termites, ants, and honeybees communicate through substrate vibrations believed to be received by this organ (Schwartzkopff, 1974; Dethier, 1964). Other chordotonal sensilla in the leg have an acoustical function as discussed in the next section.

8.3.1.3. Acoustical Sensilla. Insect sensilla that respond to long-range vibrations of the air or water at frequencies greater than 50 Hz are considered to be acoustical receptors. Acoustical receptors may be of two fundamental types, movement or pressure receivers, depending on how the movement is received. A movement receiver responds to sound through viscous interactions with vibrating air particles. An example is a long filiform hair on the cercus of the American cockroach that can vibrate in response to air movement. Such sound receivers are good detectors for the direction of the sound source but are not very sensitive, with high thresholds around 90–100 dB of sound pressure level (SPL).

Sound receivers that have a vibrating membrane backed by a closed air space are pressure receivers. The ear of noctuid moths is such a receiver (Fig. 8-6). If the vibrating membrane is backed by an air cavity open to receive air vibrations out of phase with the sound in front, then it is a pressure difference receiver. Such is the case with the tympanum of many Orthoptera. These types of receivers have all evolved in insects, with many details of their functioning yet to be unraveled.

Acoustical sensilla of insects, consisting of hairs, chordotonal organs, and tympanal organs (Table 8-2), may respond to vibrations of the air at frequencies above 50 Hz and are acoustical receptors. The long hairs of some caterpillars and the filiform hairs on the cerci of Orthoptera are especially well known for their acoustical functioning (Michelsen, 1974).

Chordotonal organs consist of groups of a few to 1500 chordotonal sensilla, with various modifications of their structures at the point of attachment. Otherwise, they are similar in structure to the proprioceptive chordotonal sensilla, which occur singly or in several groups and are widely distributed throughout the body. Many of these may serve a dual function of proprioception and acoustical reception, but only a few examples have been studied in any detail (Weise, 1972; Michelsen, 1974).

The chordotonal organ at the junction of the antennal flagellum and the pedicel is called Johnston's organ and is found in adults of all orders except Collembola and Diplura (Schwartzkopff, 1974; Dethier, 1964). It is particularly well developed in male mosquitoes and midges, where it detects species-characteristic vibrations of the flagellum in response to the sound of female wings. Johnston's organ can serve a variety of functions in other species and perhaps more than one function within the same animal. For example, female *Drosophila* use the antennae to respond to the wing buzzing of conspecific males during courtship (Ewing, 1978), while in many species (e.g., *Locusta*) it offers important feedback information during flight (Gewecke and Philippen, 1978). In the aquatic gyrinid beetles the flagellum is

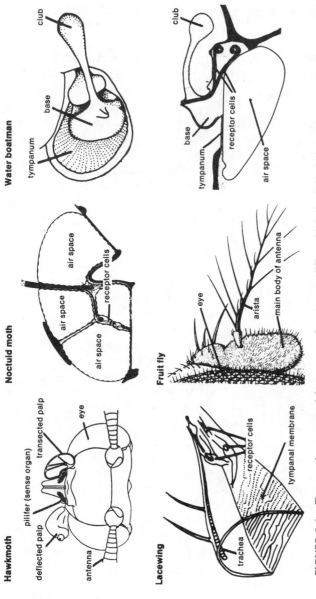

Hawkmoth

pilifer (sense organ)
transected palp
deflected palp
eye
antenna

Noctuid moth

air space
air space
air space
air space
receptor cells

Water boatman

club
base
tympanum

club
base
tympanum
receptor cells
air space

Lacewing

receptor cells
trachea
tympanal membrane

Fruit fly

eye
arista
main body of antenna

FIGURE 8-6. Tympanal organs and insect ears. Insect ears differ widely both in their anatomy and in the physical mechanisms they exploit. The hawkmoth uses balloon-shaped palps together with pilifer organs to detect the ultrasonic cries of hunting bats. The palps, filled with air sacs, are pressure difference receivers that act in concert with the pilifer sense organs. The lacewing ear, also a bat detector, is situated in a swelling of a wing vein and is probably a pressure receiver. Most of the ear, except for a tracheal tube, is filled with hemolymph. Only 3 of about 25 receptor cells are shown here. The ear of the noctuid moth is a pressure receiver with a tympanal membrane backed by a tracheal air space. Two receptors cells attach to the tympanum. The feathery arista on the third antennal segment of the fruit fly points sideways from an apical joint. A movement receiver, responding to sound through viscous interaction with vibrating air particles, it is used to detect the species' calling song at some hundred hertz. In the curious ear of the water boatman, seen both from the outside (upper right) and transected (lower right), most of the tympanum is covered by the base of a club-shaped cuticular body that protrudes into the air outside the animal. The club performs rocking movements which allow some frequency analysis of the songs of other water boatmen (Michelsen, 1979).

300

TABLE 8-2. Selected Examples of Frequency Range and Sensitivity of Hearing in Insects

Organ and Insect	Order	Family	Frequency Range (kHz)	Threshold (db SPL)	Reference
1. Hair sensilla					
Barathra brassicae (caterpillar)	Lepidoptera	Noctuidae	0.15–1.0	91.1	Tautz (1978)
Locusta migratoria	Orthoptera	Acrididae	0.4–10	80	Pumphrey and Rawdon-Smith (1936)
Gryllotalpa hexadactyla	Orthoptera	Gryllotalpidae	0.1–6	77	Suga (1968)
2. Chordotonal organ					
Anopheles subpictus	Diptera	Culicidae	0.15–0.55	0	Tischner (1953)
Calliphora erythro-cephala	Diptera	Calliphoridae	0.03–0.5	0	Burkhardt and Schneider (1957)
Notonecta glauca	Heteroptera	Notonectidae	0.005–10		Wise (1969)
3. Tympanal organ					
Locusta migratoria	Orthoptera	Acrididae	0.6–45	54	Katsuki and Suga (1960)
Conocephalus saltator	Orthoptera	Tettigoniidae	6–100	30	Suga (1966)
Gryllus domesticus	Orthoptera	Gryllidae	2–18	40.50	Popov (1969)
Cicada orni	Homoptera	Cicadidae	2–15	24	Popov (1969)
Heliothis zea	Lepidoptera	Noctuidae	2–200	45	Roeder and Treat (1961)
Chrysopa carnea	Neuroptera	Chrysopidae	13–120	60	Miller (1971)

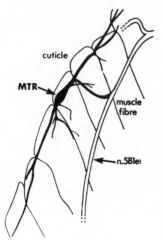

FIGURE 8-7. A multiterminal stretch receptor neuron on the tibial flexor muscle of the locust. [From G. Theophilidis and M. D. Burns, *J. Comp. Physiol.* **131,** 247 (1979).]

held above the water and used to detect Doppler shifts in reflected waves as a cue of approaching objects, while in *Notonecta* the shape of an air bubble carried under the antennae below water aids in orientation during swimming.

8.3.1.4. Stretch Receptors. Stretch receptors differ from other mechanosensitive sensilla in consisting of only a multipolar neuron with free dendrites. These receptors are widely distributed throughout the body and are associated with connective tissue or muscles (Schwartzkopff, 1974; Dethier, 1964), being particularly numerous on the surface of the latter (Fig. 8-7). Although their structures vary, they all seem to function similarly to proprioceptors. Although some stretch receptors are not spontaneously active, others are (Theophilidis and Burns, 1979; Anderson and Finlayson, 1978).

8.3.2. Feature Extraction

Like all sensory receptors, insect mechanoreceptors act as filters by responding to a relatively narrow range of adequate characteristic features of the stimulus, such as intensity, frequency, direction, or time course. Since individual sensory units may have distinct but overlapping ranges of responsiveness, collectively they provide a sensory system with a broad range of capabilities. For any given stimulus, only a portion of the receptor population need respond.

The functioning of insect mechanoreceptive sensilla depends on a combination of their position on the body, the structural features of the sensillum and associated structures, and the intrinsic response properties of the sense cells. The position of the sensillum may determine either part or all of the stimuli that impinge on it. Thus, the hair sensilla in the compound eye or on the head may be stimulated by air movements during locomotion, while those at the base of the wings (Fig. 8-4, upper left) detect only movements of the appendage. The cercal hairs of the American cockroach (Fig. 8-4, middle right) may detect either airborne sound or substrate vibrations depending on the position of the cercus. Dome sensilla of the legs (Fig. 8-4, lower right) may detect stress at the nearest joint when the animal is moving or vibrations of the substrate when at rest. In addition, the distribution of various mechanoreceptive sensilla allows for spatial and directional information to be provided to the CNS, including those from the forces of gravity.

The structural features of the cuticular apparatus together with those of associated structures like sockets, suspensory filaments, tracheal horns, or air sacs combine with the positional limitations to restrict or to enhance the response properties of the sensillum (Schwartzkopff, 1974; Dumpert and Gnatzy, 1977; Tautz, 1978). Thus, the filiform hairs of the cockroach cercus appear structurally omnidirectional, while the tibial spines have a highly directional limitation imposed by the socket (Fig. 8-4, middle left insert). The curvature of hairs may impose a directional limitation for response, while the cuticular thickenings of dome sensilla bias compressions along specific axes (Fig. 8-4, lower). Such ultrastructural variations in insect mechanoreceptive sensilla are only beginning to be understood in functional terms.

Intrinsic response properties of the sense cell are discussed in the sections that follow.

8.3.2.1. Intensity. Whether the adequate stimulus is movement of the air, water, substrate, or body part, the absolute amount or rate of this movement is coded as the intensity of stimulation by an increase in frequency of spikes. Where the intensity increase may be long-lasting, as with stretch receptors or sensilla at joints of the appendages, the sense cells respond tonically. Where intensity changes are relatively rapid, as with acoustical or wing receptors, the sense cells respond phasically or phasic-tonically. Thresholds vary widely among insect mechanoreceptors, with chordotonal sensilla reported to be the most sensitive, responding to movements of 10^{-9} cm, which is near the calculated theoretical limits for a mechanoreceptor (Schwartzkopff, 1977a). While acoustical receptors, movement receivers like hair in sockets or antenna that vibrate due to the small forces of air particles, have high thresholds (Section 8.3.1.3), those of pressure gradient receivers (e.g., tympanic organs) are much lower (Table 8-2).

Cercal hairs of the cricket *Gryllus bimaculatus,* like those of many Orthoptera, have a single sense cell that responds to movement of the hair by an increased frequency of spikes (Fig. 8-8, upper). For wind speeds up to 1.6 m/sec, the cell responds throughout the duration of stimulus, but at 1.9 m/sec the response becomes shortened, indicating that the upper response range of the cell has been saturated. Unlike other cercal hairs, this class has a thickened area of hair shaft that coincides with the rim of a raised socket, which is surrounded by flexible cuticle. A dome sensillum is set in this

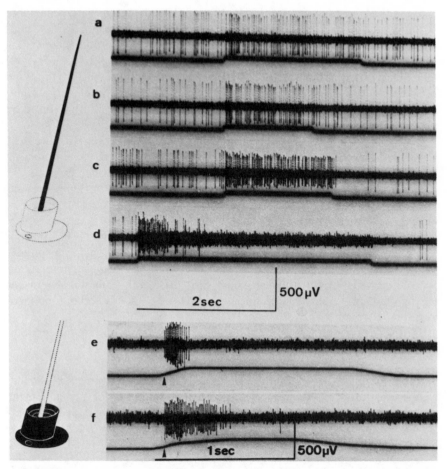

FIGURE 8-8. Mechanoreceptive properties of one filiform hair and two campaniform sensilla of the cercus of the cricket, *Gryllus bimaculatus*. a–d, recordings from one filiform hair with successive increase of the stimulation intensities by air currents. Stimulation duration: see stimulus registration. a, 0.5 m/sec: b, 0.9 m/sec: c, 1.6 m/sec: d, 1.9 m/sec, e and f, recordings from two different campaniform sensilla. Stimulation was applied by deflecting the filiform hairs and their sockets with a fine wire hook connected to a manipulator. The beginning of socket deflection is indicated by an arrow. [From K. Dumpert and W. Gnatzy, *J. Comp. Physiol.* **122**, 9 (1977).]

flexible cuticle (Fig. 8-8, lower). This cell becomes a phasic detector at wind speeds of 1.9 m/sec, as the hair's greater deflection brings it into contact with the socket rim and causes flexing of the surrounding thin cuticle. Dome cell responses increase to wind currents parallel to the cercal axis up to 2.3 m/sec or up to 3.6 m/sec for stimulation in the perpendicular direction (Dumpert and Gnatzy, 1977). Thus, two sense cells have distinct intensity response ranges that permit them to act in concert to increase the dynamic response range of the single hair sensillum. Stimulation in the upper range only causes an increase in the kicking response of the hind leg, which would result in escape behavior (Dumpert and Gnatzy, 1977).

Acoustical, vibration, and stretch receptors also show varying intensity response curves for individual cells. For example, recordings from the tibial neurons of the bush cricket show curvilinear responses of individual receptors, with distinct but widely overlapping ranges representative of the type found in many species of insects (Klamring et al., 1978). On the other hand, stretch receptors usually show a more limited range of responsiveness.

A clear example of the varying intensity response of cells is shown by the tibia of the locust with response properties of chordotonal organs and tibial subgenual organs, compared to the tonic increase in firing of the multipolar stretch receptors of the muscle and joint receptors. The muscle receptor firing increases linearly with an increase in muscle tension but has a lag of about 40 m/sec. The receptor responds to a decrease in muscle tension by stimulating inhibitory fibers in the flexor motor nerve (Theophilidis and Burns, 1979).

8.3.2.2. Directionality. Many insect mechanoreceptors show a preferential response to stimulation from a specific direction. This results from a combination of the structural variation in the cuticular projection, the ultrastructural features of the base, tubular body, and sense cell attachment (Gaffal et al., 1975), as well as the position on the body. By holding the frequency and intensity of stimulation constant, one can measure the change in response as a function of the stimulation direction and plot this in a polar coordinate diagram. Figure 8-9 is a polar diagram of the directional response of the cercal hairs of the cricket. Although each hair was a preferred plane of responsiveness, these planes are arranged in various directions on the cercus (Palka et al., 1977). Thus, the system of cercal hairs collectively possesses an omnidirectional capability, even though each unit shows optimal responsiveness in only one direction. Presumably, the input from a few cells with the same orientation could serve to signal the direction of stimulus, but details of the CNS processing are only beginning to emerge (Gaffal et al., 1975). Such directional polarity is common among tactile and vibrational sensilla.

Detecting a sound source direction may be more difficult for insect auditory receptors than the direction of stimulation for tactile receptors. In males of the mosquito *Aedes aegypti* complex summed receptor potentials can be recorded from Johnston's organ, indicating that individual receptor cells have directional sensitivities with constant latencies and that the mechanical elasticity of the cuticle provides for rebound of the deflected flagellum. A given receptor cell thus responds to movement of the flagellum in one direction (push), but not the opposite (pull). The extremely large number of 30,000 receptor cells/organ indicates that a very complex information-processing system is probably working. This system permits determining the

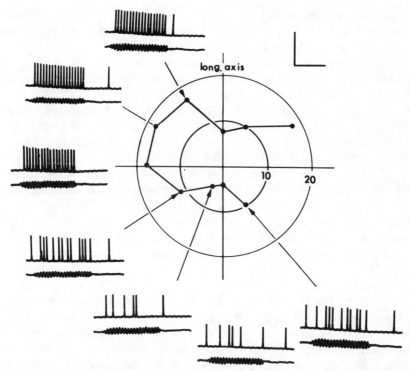

FIGURE 8-9. Polar diagram of the directional response of a cercal hair from the cricket, *Acheta domesticus*. The number of spikes in response to a tone pulse is plotted against speaker position in relation to the longitudinal axis of the cercus. Insets show the receptor's response to a constant intensity and duration tone pulse from seven speaker positions 30° apart. Note that this receptor shows a preference for transversely oriented stimuli. Long axis, longitudinal axis of cercus; tone frequency, 100 Hertz; intensity, 71 dB; scale, 100 msec and 20 mV; stimulus marker, 200 msec. [From J. Palka et al., *J. Comp. Physiol.* **119,** 267 (1977).]

direction of sound while in flight by detecting very small phase differences between pulses arriving at the left and right antennae (Belton, 1974).

8.3.2.3. Frequency. Insects produce and detect vibrations and sounds conducted at either relatively short distances through various substrates like ground, bark, or water, or relatively long distances through the air (Michelsen, 1974; Weise, 1972). Separating these stimuli on the basis of threshold, frequency, or distance is highly arbitrary, as there are many overlaps in the functional ranges of receptors of insects from diverse habitats. For the reception of a pure tone of relatively low frequency, hairs in sockets seem to work well, with hairs at different positions on the body serving to code for different phase parameters of the signal (Schwartzkopff, 1977a). This method is utilized by insects using auditory signals for communication, where the sensitivity maximum of the frequency is around 300 Hz (Table 8-2, group 1). Where the frequency of sound is composed of mixed frequencies, some of which may be very high, reception and discrimination are much more complex.

Recent electrophysiological studies of individual receptor neurons in the tibia of the bush cricket revealed distinct response characteristics of the cells (Klamring et al., 1978). These cells—characterized as pure sound, mixed sound and vibration, and pure vibration—showed different, but widely overlapping frequency responses (Fig. 8-10). Threshold curves show that for airborne sound the bush cricket can cover the range of 2–40 kHz with just

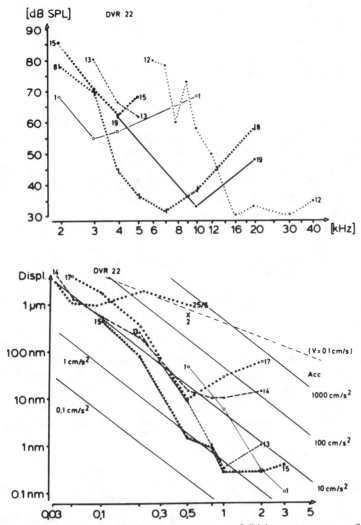

FIGURE 8-10. Threshold curves for frequency responses of tibial receptor neurons of the bush cricket, *Decticus verrucivorus* L. (Tettigoniidae). Upper, airborne sound stimuli. Lower, vibration stimuli. Unit 25/6 is a campaniform sensillum. [From K. Klamring et al., *J. Comp. Physiol.* **127,** 109 (1978).]

two cells (8 and 12), while for vibration cell 15 covers the range of 0.1–3 kHz. Such economy of receptor cells with a broad range of frequency responsiveness seems to be characteristic of insect auditory receptor systems where high frequencies are involved (Klamring et al., 1978; Rheinlaender and Morchen, 1979). For the high range of frequencies, the receptors of the bush cricket also show an afterdischarge of the cell beyond the time of the sound pulse, suggesting that the receptor cells, and possibly parts of the CNS as well, may produce neuronal enhancements of the signal parameters permitting more reliable processing (Klamring et al., 1978). Detailed studies of auditory interneurons of crickets, moths, and grasshoppers indicate that signal enhancement of sensory input is widespread, but many of the integrative roles have yet to be determined (Wiersma and Roach, 1977).

8.3.3. Adaptive Radiation of Mechanoreceptor Systems

Adaptations to optimize survival and reproduction have resulted in some remarkable developments in insect auditory systems. Several groups of insects have apparently evolved the capability of detecting the ultrasonic pulses of hunting bats before the reflected signals inform the bat of their presence. This adaptation allows them to avoid the faster-flying adversary by evasive maneuvers. This is truly remarkable from two standpoints: first, that insects should evolve hearing organs for predator avoidance—exclusive of communication with conspecifics—and, second, that this can be accomplished with as few as four to six receptor cells. That such selection pressure exists is further documented by recent studies indicating that bats in North America utilize high-intensity, frequency-modulated pulses in the 20–40 kHz range which can be detected by moths 40 m away, while in Africa bats use low-intensity calls with high-frequency components which are not detectable by moths at over 2 m. The hearing ability of moths may thus significantly influence the feeding efficiency of bats, resulting in counteradaptation (Fenton and Fullard, 1979). Furthermore, arctiid moths can produce ultrasonic clicks that may jam the bats' signals (Fullard et al., 1979). Insects that utilize ultrasonic communication signals for attracting mates and for courtship have numerous advantages. While on the ground, they are communicating at a frequency range beyond that of most stalking predators, and in the air they are able also to perceive bat signals and to avoid them (Moiseff et al., 1978).

Recent examples of special antipredator roles evolved by insect mechanoreceptor systems other than acoustical ones indicate that many other such examples are likely to be discovered. Water striders can generate and detect sex-specific vibrations on the water surface (Wilcox, 1979), while *Notonecta*, water boatman predators, can perceive prey by detecting surface waves with specialized tarsal scolopidial organs (Weise, 1972). The roles of the subgenual and chordotonal organs in most species of insects have yet to be determined (Schwartzkopff, 1974, 1977a).

8.3.3.1. Detection of Predators and Parasites. Night-flying moths, principally of the Noctuidae, Geometridae, Arctiidae, and Cerocampinae, possess tympanal or palp-pilifer organs that have low thresholds of response to the frequencies of ultrasonic pulses used by bats to locate food. The bat can detect flying moths at a distance of about 3–4 m, but the moth responds to bat signals at 40 m by turning and flying away. If the intensity of the bat signal is high, the moth folds its wings and drops to the ground or it makes erratic movements, for which the bat makes compensating movements, but with reduced capture success. The bat pulses are detected by the noctuid moth tympanum with two receptor cells (Roeder, 1966). The A_1 cell is slower adapting, has a lower threshold, and responds over a 20-dB range with a marked intensity–response curve within the frequency range of 10–20 kHz. The A_2 is fast adapting, with a 10-fold higher threshold to the same frequency range as the A_1 but with a reduced response area (Roeder, 1966). Thus, both cells respond not to the frequency, but principally to the change in intensity of the signal over a 40-dB range.

As few as three to four impulses in the A_1 cell can trigger escape behavior in the moth (Roeder, 1966). The intensity-coded input of the A cells is fed to three classes of thoracic acoustical interneurons that show further specialization of response. A "repeater" neuron shows repetitive discharge during each pulse, a "train marker" neuron fires at an intrinsic frequency of 20/sec only as long as the pulses continue to arrive, and a "pulse marker" neuron fires a single spike for each pulse up to rates of 40/sec with an intensity-dependent latency (Roeder, 1966). Additional interneurons with connections to both tympana, as well as brain interneurons, have been studied. The exact roles of these interneurons in information processing as well as the importance of a third cell (the B cell) in noctuids and the four cells of geometrids remain to be determined (Roeder, 1975).

More recently it has been shown that the green lacewing, *Chrysopa carnea,* also detects the pulses of hunting bats with a tympanal organ located in the radial veins near the base of each forewing (Fig. 8-6) (Miller and Olesen, 1979; Olesen and Miller, 1979). These tympana consist of two chordotonal organs with a total of 25 cells. They respond in the 13–120 kHz range with 20 dB less sensitivity than the A cells of noctuids. While hunting, a bat produces search pulses at about 7–10/sec, increases this frequency during the approach phase after the prey is detected, and closes the last 1–2 m with the terminal buzz of even higher frequency. Lacewings fold their wings and nose dive in response to search pulses; thus, this early-warning system works similar to that of the moth but at a reduced distance (Fig. 8-11). Such variability in avoidance behavior by the lacewing may represent an adaptation to overcome learning by the bat (Olesen and Miller, 1979).

It is well known that caterpillars of several species possess filiform hairs on their surface and that they respond behaviorally to air vibrations of 30–1000 Hz, but only recently has the sensory basis for this response been electrophysiologically characterized (Markl and Tautz, 1975; Tautz, 1977,

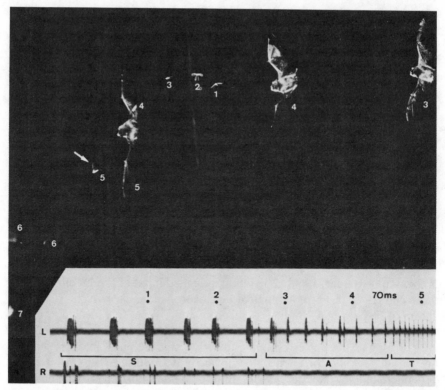

FIGURE 8-11. Last-chance response from a falling green lacewing. Bat and insect detect each other at about the same time (between flashes 2 and 3). Note that the insect's wings are folded at the start of the fall (flash 3). The bat begins its terminal buzz about halfway between flashes 4 and 5, indicating an attempted catch. The insect's wings are partially extended in a last-chance response (arrow) about midway through the buzz (5 in inset). The bat misses the insect and flies out of the camera field (flash 6) while the insect folds its wings again and continues to fall (flashes 6 and 7). L of the inset shows the three phases of bat's orientation sounds: S, searching phase; A, approach phase; T, terminal phase or buzz. Pause after buzz and return to searching cries not shown in L. [From L. A. Miller and J. Olesen, *J. Comp. Physiol.* **131,** 113 (1979).]

1978). Studies on the cabbage worm, *Barathra brassicae,* demonstrated that it possesses eight filiform hairs on its thoracic segments, with four curved hairs on the first segment and two straight ones on each of the other two segments (Markl and Tautz, 1975). The hairs in flexible sockets have one sense cell each, which show a threshold sensitivity of 0.1° of hair movement within the frequency range of 40–200 Hz (Tautz, 1977, 1978). Although the hairs can bend through a 15° angle, the mechanical properties of the hair make it stiffer at angles greater than 10°, a mechanical limitation that results in an upper frequency limit (Morchen, 1978). The sense cells respond phasically, with their maximal responses just above threshold intensity, and give a flat response over the 40–200-Hz range. They are uniquely different from the

A sense cell of the noctuid moths in that they are threshold detectors only, not intensity detectors. The frequency response of these cells matches that produced by the wing buzzing of such parasites as ichneumonid wasps, tachinid flies, and predatory sphecid and vespid wasps. Figure 8-12 shows the simultaneous recording of the vibration of a wasp and the response of the filiform hair sense cell at a distance of 20 cm. Thus, the caterpillar's eight sense cells, whose thresholds are the lowest for air vibrations among known insect hair sensilla, serve to detect the presence of an approaching parasite or predator (Tautz, 1978). The caterpillar may cease moving and contract or squirm and fall off a stem in response to such stimuli.

8.3.3.2. Acoustical Communication. Insects using acoustical communication utilize both frequency- and time-modulated signals (Schwartzkopff, 1977b; Michelsen, 1979). The perception and integration of such signals present some unique problems. The locust, *Locusta migratoria,* provides a

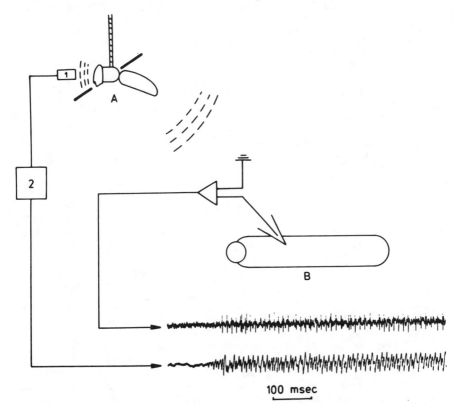

100 msec

FIGURE 8-12. Synchronous registration of the flight sound of a fixed flying wasp (A) and response of the sensory cell of a filiform hair of the caterpillar of *Barathra brassicae* (B). Distance from wasp to caterpillar 20 cm. Impulses from mechanoreceptor retouched. [From J. Tautz, *J. Comp. Physiol.* **125,** 67 (1978).]

good example of how this is accomplished with two tympanal organs each containing 60–80 cells (Michelsen, 1979). The locust tympanum is located laterally on the first abdominal segment, with a spiracle on the anterior edge and a large trachea appressed behind the tympanal membrane. The tympanal nerve connects the metathoracic ganglion to a modified body containing the sense cells having dendrites attached at various orientations and points on the membrane (Fig. 8-13, upper left). The four groups of sense cells are broadly tuned, with greatly overlapping frequency response curves (Fig. 8-13, lower left).

Group a, with a lower threshold than group b, has a second peak of responsiveness at 8 kHz, while group d, with a higher response, was unknown in earlier studies (Wiersma and Roach, 1977; Michelsen, 1979). At frequencies below 10 kHz the sound waves travel from one tympanum through the tracheal air sacs to the back of the other, so that each ear works as a pressure-difference receiver (Michelsen, 1979). At frequencies above 10 kHz little sound reaches the back of the tympanum, and it functions as a pressure receiver. At the lower frequencies of 2–8 kHz the tympanum acts

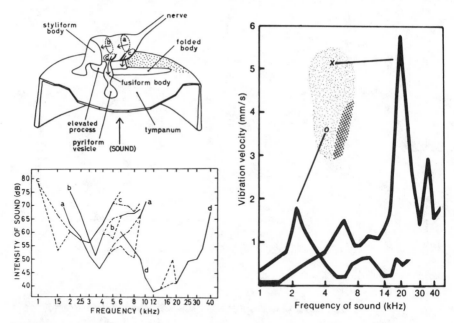

FIGURE 8-13. Structure and response properties of the tympanal organ of the locust. Upper left, left tympanal organ showing four groups of receptor cells (a–d). Arrows indicate direction of dendrites and stippled area is thickened portion of membrane. Lower left, threshold curves of the four groups of receptor cells in an isolated locust ear. Dashed lines show variations measured within each group. Right, like the locust, the mole cricket uses resonances in the tympanal membrane for frequency analysis. Shaded area represents the cricket tympanum with the vibrational velocities at the two points indicated. The hatched area is the point of receptor cell attachment (upper left, modified from Michelsen, 1971; lower left, Michelsen, 1971; remainder, Michelsen, 1979).

as a frequency analyzer. The groups of sense cells attach at points of different thickness of the membrane (Fig. 8-13, upper left). These areas vibrate independently of each other, so that as frequency changes, sense cells can register different modes of vibration (Fig. 8-13, right). With this sophisticated, yet structurally simple, frequency analysis system the locust can detect its species-specific calls with a greatly improved signal-to-noise ratio (Michelsen, 1979).

Determining the direction of the sound source presents a quite different problem. The distance between the two tympana is about 4 mm in *Locusta,* and although the two tympanal nerves show different responses to changing intensity, the physical time difference of about 10 μsec for sound arriving at first one ear and then the other was considered too small to be successfully used by the nervous system (Schwartzkopff, 1977b). Recently, however, it has been determined that there is an intensity-dependent latency of response that increases the time of differential responses of the two organs by as much as 6 msec (Morchen, 1978; Rheinlaender and Morchen, 1979). This neural time cue exceeds the physical time cue by 100- to 1000-fold. There are thus two neural codes available to specify the direction of sound, the latency and frequency of spikes. Both parameters are utilized by the auditory interneurons for further processing (Morchen, 1978). Such enhancement of receptor input by auditory interneurons appears to be quite common in other insect systems that have been studied in detail (Wiersma and Roach, 1977). Thus, adaptive specialization of auditory reception includes not only the receptors but the integrating neurons as well in a tightly coupled system. This system, which is under genetic control, may include both the sound-producing and the sound-receiving systems (Bentley, 1971).

8.4. PHOTORECEPTOR SYSTEM

Like other insect sensory systems, the photoreceptor system is composed of primary sense cells, sheath cells, and associated cuticular structures although extraocular reception is known (Arikawa and Aoki, 1982). Since the receptor cells are covered with cuticle, their function is intimately associated with the optical properties of clear cuticular lenses and directly underlying cones of various shapes, and with the filtering effects of various pigment inclusions, some of which can migrate within the cells. Additional structural features are correlated with the reflection and refraction of light by tracheal tapeta and screening pigments. The functional capabilities of photoreceptors vary widely among insects of diverse habits. For example, flying insects, often with very rapid movement, that are active during daylight periods have distinctly different visual structures and capabilities from those species that walk or are active at night. After a consideration of the basic structure of photoreceptors, the adaptive strategies of flies, active in bright light, and of praying mantids, active in daylight and at night, will be contrasted in terms of their specialized photoreceptor systems and different visually mediated behaviors.

8.4.1. Structure and Distribution

8.4.1.1. Ocelli. Ocelli are simple eyes that occur in adults of all orders and in endopterygote larvae (Dethier, 1964; Goldsmith and Bernard, 1974). Typically, there are three arranged in a triangular pattern on the front of the head (Fig. 8-14, upper left), although paired and single ocelli are well known in certain species. Ocelli are often absent in wingless forms and may be reduced to just clear areas in the cuticle, as in the American cockroach (Fig. 8-8, upper right). The ocelli consist of a transparent cuticular cornea with, in some species, a corneagen cell layer beneath. Groups of a few to several hundred photoreceptor (retinula) cells are located above the focus point of the corneal lens. These cells contain visual pigments (rhodopsins) that absorb light and produce graded receptor potentials. The axons of the retinula cells form convergent synapses with second-order neurons, so that the ocellar nerve that connects to the brain contains far fewer cells than the number of receptors. Often a few of these neurons of the ocellar nerve are much larger than others, but relatively little is known about the details of information transfer in these channels (Patterson and Chappell, 1980).

The three ocelli are separated by angles of 3–10°, and this feature, together with the fact that the light entering each corneal lens is focused at a point below the sense cells, indicates that the ocelli are poorly designed for detecting images. They are, however, well designed for gathering light. Recordings from the retinular cells indicate that the receptor cells respond to a wide variation in light intensity (Hu et al., 1978; Patterson and Chappell, 1980). For example, navigation at dusk by bumblebees is made possible by the sensitivity of ocelli to low light intensity (Goodman, 1975).

8.4.1.2. Stemmata. Stemmata (lateral ocelli) are simple eyes that occur in larval stages of insects with endopterygote development (Dethier, 1964; Goldsmith and Bernard, 1974). They vary structurally from simple pigment spots, like those in fly larvae, to complex structures resembling the dorsal ocelli of adults. In stemmata of caterpillars the one to six units appear as a scattered group on the lateral–ventral region of the head. The corneal lens of one stemmata is secreted by three epidermal cells, resulting in a tripartate appearance (e.g., *Heliothis zea*, Fig. 8-14, middle left). The stemmata are capable of limited color, UV, and form vision (Dethier, 1964; Goldsmith and Bernard, 1974; Ichikawa and Tateda, 1982).

8.4.1.3. Compound Eyes. The major photoreceptors of insects are the often large and prominent pair of compound eyes located on the lateral margins of the head. They each consist of a few to several thousand units called ommatidia (Fig. 8-8, middle right). Compound eyes are absent in the primi-

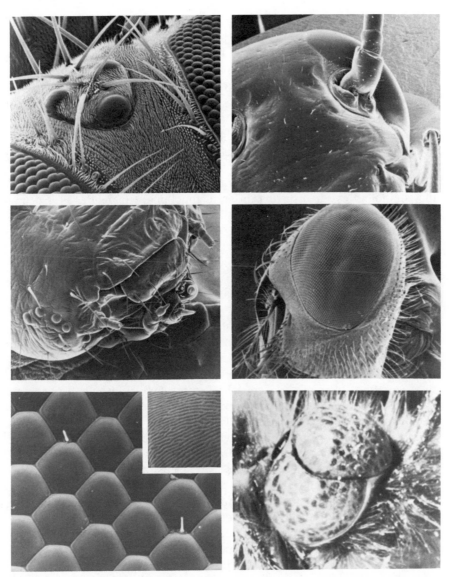

FIGURE 8-14. Insect photoreceptors. Upper left, dorsal ocelli of the house fly, *Musca domestica* (230×). Upper right, compound eye of the American cockroach, *P. americana*, behind the antenna. Ocelli are the indentations at the inner margins of antennal bases (23×). Middle left, dorsal–lateral view of the head of the caterpillar of the cotton bollworm, *Heliothis zea,* showing the six stemmata on the lateral side (lower left) (35×). Middle right, compound eye of the fly *Musca domestica* (46×). Lower left, individual hexagonal ommatidia with mechanosensitive hairs of the compound eye of *M. domestica* (1092×). Insert shows corneal ridges that reduce reflectance (9080×). Lower right, dual compound eye of the owlfly *Ascalaphus.* Frontal region is UV sensitive, while ventral is not. Note presence of pigment spots that may act as interference filters augmenting color vision (10×). (Scanning electron micrographs courtesy of L. D. Hatfield and D. Ave.)

tive orders Protura and Diplura and occur as groups of two to eight single ommatidia in Collembola and some Thysanura (Goldsmith and Bernard, 1974). However, in the vast majority of species the compound eyes occur as structures of various sizes and shapes, either positioned so that the total visual field may be somewhat limited (e.g., American cockroach, Fig. 8-14, upper right) or to permit vision in nearly all directions as in many flies and dragonflies (Fig. 8-14, middle right). Often there is sexual dimorphism with males having larger eyes and very different interneuronal organization, reflecting perhaps the different visually mediated behaviors of the sexes (Bernard and Stavenga, 1979).

The significance of ommatidia having different sizes in different regions of the compound eye and having different functional capabilities is just now being understood (Burkhardt, 1977). For example, the eyes of aquatic beetles (Gyrinidae) are actually separated into a dorsal and a ventral eye, which permits viewing above and below the surface of the water (Goldsmith and Bernard, 1974). Similar specialization is evident in the owl fly, *Ascalaphus,* in which each compound eye has two distinct morphological areas consisting of a frontal eye that is UV sensitive and a ventral region that is not (Gogala, 1978) (Fig. 8-8, lower right).

The structural unit of the compound eye is the ommatidium. Each ommatidium consists of a transparent corneal lens, a crystalline cone, a group of retinula cells, and one or more groups of screening pigment-containing cells (Goldsmith and Bernard, 1974). Figure 8-15 shows some of the structural variations of this basic scheme in representative insects. The corneal lens is usually convex above, assuming a hexagonal shape as seen in the compound eyes of the house fly (Fig. 8-14, lower left). However, a wide variety of polygonal shapes are known in various species. The outer surface of the cornea in many species is sculptured with many tiny ridges or nipples that reduce the reflection of light and increase its transmission to the receptor below (Carlson and Chi, 1979) (Fig. 8-14, lower left insert). Corneal lenses pass light in the near UV region, but do not pass the mid to far UV, thus protecting the receptor cells from damage (Carlson and Chi, 1979).

In most species, the crystalline cone is located immediately below the lens, but in some species a protuberant proximal extension of the lens acts as a cone (exocone) (Carlson and Chi, 1979). The corneal lens and crystalline cone form the dioptric apparatus, gather light (within the angle of acceptance determined by the shape of the ommatidium), and focus it onto the central or optical axis of the ommatidium. The crystalline cone is highly variable in structure and shape: it may be absent (acone) as in the bug *Lethocerus,* rounded (econe) as in the firebrat *Lepisma,* or extremely elongated (exocone) as in the firefly *Photuris* (Fig. 8-15). In many flies the cornea acts as an interference or quarter-wavelength filter, which results in the production of iridescent colors that can be seen on the surface (Bernard and Stavenga, 1979).

The retinula cells are specialized primary receptor cells that are elongate and occur in groups of 7–12 arranged radially about the central optical axis

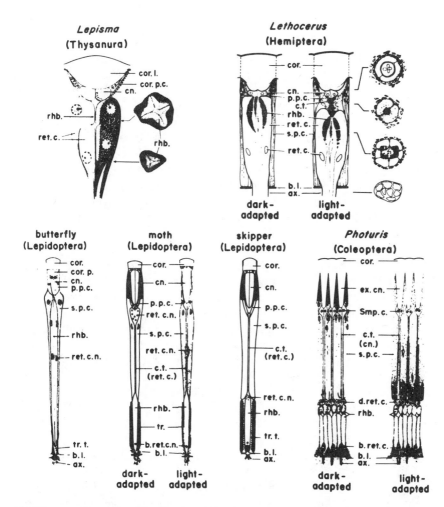

FIGURE 8-15. Examples of insect ommatidia, shown diagrammatically in longitudinal section. Above, *Lepisma* has an eucone eye with a stratified retinula. Distally the rhabdom is formed from four cells; proximally from another three (Hesse, 1901). The water bug *Lethocerus* has an acone eye, has stout ommatidia in which the rhabdoms move distally in the dark and proximally in the light (Walcott, 1971a). In the lower row are representative ommatidia with more typical ratios of length to width. The lepidopterans have eucones; the firefly (*Photuris*), an exocone. The butterfly has a photopic eye with a long slender rhabdom; the moth and firefly have scotopic eyes with short thick rhabdoms lying below long crystalline tracts, longitudinal migrations of pigment, and migratory nuclei in the retinular or pigment cells. Skippers, diurnal insects derived from sphingid moths, have a modified scotopic eye with short rhabdoms and long crystalline tracts, but nonmigratory pigment. *Photuris* has both a basal retinular cell and a distal retinular cell, the latter with its rhabdomere separate from the bulk of the rhabdom. (Butterfly, from Yagi and Koyama, 1963; moth and skipper modified from Yagi and Koyama, 1963, with information from unpublished electron micrographs of W. H. Miller; *Photuris,* from Horridge, 1969a). Axons of the photoreceptors (ax); basement lamina (b. l.); basal retinular cell (b. ret. c.); crystalline cone (cn.); cornea or corneal lens (cor. or cor. l.); corneal process of butterfly (cor. p.); corneal pigment cell (cor. p. c.); crystalline tract (c. t., consisting of processes of ret. c. or cn.); distal retinular cell (ret. c.); retinular cell nucleus (ret. c. n.); rhabdom (rhb.); secondary pigment cell (s. p. c.); Sempter's (cone) cell (Smp. c.); tracheols (tr.); tracheolar tapetum (tr. t.). (From T. H. Goldsmith and G. D. Bernard, in M. Rockstein, Ed., *Physiology of Insecta,* Vol. 2, Academic Press, New York, 1964.)

317

of the ommatidium. The retinula cells have the medial aspects of their plasma membranes thrown into regular arrays of microvilli. The latter, called rhabdomeres, contain the photoreceptive pigment rhodopsin on the P face leaflet of their membranes (Chi and Carlson, 1979). These rhabdomeres may be contiguous with those of other cells, forming a fused rhabdom; or may be separate, forming an open rhabdom; or may be arrayed in alternate or overlapping layers. The long axis of each rhabdomeric microvillus is usually parallel to the electric vector of linearly polarized light, enabling maximal light absorption. The rhabdom as a whole may contain multiple rhabdomeric orientations that collectively cover several light orientations. In cross-sectioned ommatidia the rhabdom area is seen to differ extremely among species, a dimension often relating directly to general sensitivity to light (Smola, 1976) (Fig. 8-16). The nocturnal cockroach has a large, fused rhabdom that is 16 times greater in area than that of the honeybee. The fly, which is active only in bright light, has an open rhabdom 2 times that of the bee, allowing the fly to maintain high acuity at rapid flight speeds.

The dimensions of the rhabdom together with its higher index of refraction compared to the surrounding cytoplasm permits the rhabdomere to function as an optical light guide. In this way, light tends to stay in and to propagate down the rhabdom, increasing the probability that it will be absorbed by the visual pigments (Carlson and Chi, 1979). Both the physical dimensions and refractive indices of the diopteric apparatus and the retinula cells must be considered in determining the physiological angle of acceptance and, ultimately, the light reaching the receptive areas of the rhabdom (Carlson and Chi, 1979; Horridge, 1975).

Additional structural features of the retinula cells are important in understanding their light reception efficiencies. In many species the retinal cells are of different lengths, so that some insects have a distal and others a proximal rhabdom. In still other species a single rhabdom runs the entire length (Fig. 8-15). The retinula cells are usually surrounded by primary and/or secondary pigment cells, and sometimes by trachea, that serve to absorb, scatter, or reflect light entering relative to the optical axis of the ommatidium (Carlson and Chi, 1979). Several mechanisms exist for the movement of pigments so that the compound eye can be light-adapted to reduce the amount of light reaching the rhabdom or dark-adapted to maximize the amount of light reaching the rhabdom, depending on ambient light conditions (Burkhardt, 1977; Horridge, 1975; Smola, 1976). (These mechanisms are discussed in Section 8.4.3.1.)

8.4.2. Transduction

The rhabdom of insect ommatidia, like the rods of vertebrate eyes, contains the light-sensitive pigment rhodopsin as 60–80 Å unordered particles as seen on the P face leaflet of the microvillar membrane (Chi and Carlson, 1979). Rhodopsin consists of a chromophore molecule, retinal (vitamin A alde-

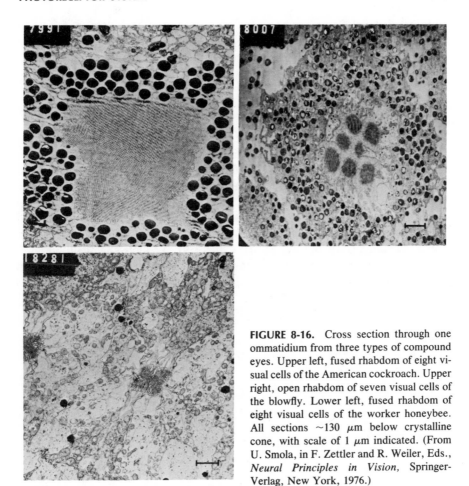

FIGURE 8-16. Cross section through one ommatidium from three types of compound eyes. Upper left, fused rhabdom of eight visual cells of the American cockroach. Upper right, open rhabdom of seven visual cells of the blowfly. Lower left, fused rhabdom of eight visual cells of the worker honeybee. All sections ~130 μm below crystalline cone, with scale of 1 μm indicated. (From U. Smola, in F. Zettler and R. Weiler, Eds., *Neural Principles in Vision,* Springer-Verlag, New York, 1976.)

hyde), conjugated by a Schiff-base linkage to a protein molecule called opsin. In *Drosophila* reared without vitamin A, the sensitivity to light is reduced by 2–3 log units, due to the reduction in visual pigment resulting from the lack of the necessary precursors for proper biosyntheses (Pak, 1979).

In the dark, the retinal is in its *cis* configuration and has the characteristic wavelength of maximum absorption (λ_{max}) of the "native" pigment. When light is absorbed, the *cis*-11-retinal isomerizes to *trans*-11-retinal with a concomitant generation of a receptor potential through some unknown mechanism. The *trans*-11 form can remain as a stable intermediate, and metarhodopsin of the latter can be photoisomerized back to rhodopsin by absorption of long-wavelength photons. Depending on the pH, a hydrolysis can occur, cleaving retinal and opsin. When accompanied by a loss of color, this hydrolysis is known as bleaching; however, in most of the insects studied, light causes the formation of a metarhodopsin that absorbs at longer wavelengths so that no bleaching occurs. In fact, the intermediate is even more colored than native rhodopsin. Since light and dark adaptation are correlated with this series of light-mediated chemical reactions, the dynamics of photoreceptor response are comparatively slow.

8.4.3. Feature Extraction

8.4.3.1. Intensity. Since insects (collectively) are active throughout the day and night, they are exposed to full sunlight and moonlight, which differ in light intensity 8 log units, or 100,000,000 times! Numerous adaptations of compound eyes have evolved for regulating the amount of light that actually reaches the light-sensitive rhabdom. In contrast, the ocelli usually have a convex lens with a rather wide angle of acceptance of light, so that nearly all the light passing through the lens is focused below the retinula cells (Patterson and Chappell, 1980; Carlson and Chi, 1979). As such, the ocelli function as intensity receptors over a wide range of wavelengths, as has been shown by ERG studies in many species (Fig. 8-17) (Hu et al., 1978; Patterson and

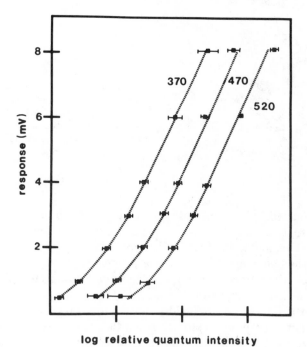

log relative quantum intensity

FIGURE 8-17. Amplitude of the ERG response of the ocellus as a function of the log relative quantum intensity. Mutants of *Drosophila melanogaster* which lack all ommatin screening pigments were used. Curves to the three wavelengths are arbitrarily displaced on intensity axis for comparison of similar curves. Intensities required to elicit a 0.5-mV response for 370, 470, and 520 nm are 1×10^{13}, 9×10^{12}, and 1×10^{12} quanta/cm² sec, respectively. Standard error bars are given. [From K. G. Hu et al., *J. Comp. Physiol.* **126**, 15 (1978).]

Chappell, 1980; Carlson and Chi, 1979). Recent studies have shown that the ocellar plexus of the dragonfly receives inputs from the compound eyes and wing receptor systems that facilitate the input from the retinular cells. Earlier behavioral studies have indicated that the ocelli have a modulating effect on locomotor activity in many species, but many details of the integration of this intensity-related input remain unknown (Goodman, 1975).

The compound eyes have evolved with several specializations for regulating the amount of light reaching the rhabdom. Like the retinula cells of ocelli, those of the ommatidia give increased response with increased intensity of light, but the spectral content is also very important (see Section 8.4.3.3). The ommatidia have several types of adaptation mechanisms for regulating the light received (Carlson and Chi, 1979). The ommatidia may be surrounded by a cellular palisade within which pigment granules migrate radially, or the cone cells and rhabdoms may actually move. The most common type of adaptation involves longitudinal pigment migration in the pigment cells in insects with clear zone eyes (Horridge, 1975). In some cases, pigment will migrate around the crystalline tracts or away from them so that a circular pupil as well as a longitudinal pupil mechanism exists (Fig. 8-18 summarizes some of the known pupil mechanisms).

In the light-adapted state, in which the intensity of available light exceeds that required for proper vision, the pigment migrates proximally to cover the maximum area surrounding the retinula cells, so that only light entering through the corneal lens of that ommatidium is effective. In this way each ommatidium is isolated from its neighbors. In the dark-adapted state, when the intensity of ambient light is much lower, the screening pigments migrate distally, surrounding only the lateral margins of the crystalline cone. These dynamics enable a clear zone beneath the cones, and some of the light entering one ommatidial lens is refracted laterally to the retinula cells of adjacent ommatidia. This increase in intensity of light reaching the retinula cells of each ommatidium occurs at the loss of image formation, since the physiological angle of acceptance is greater than the structural angle of acceptance. In fast, day-flying insects (e.g., true flies, dragonflies, bees, and some butterflies), good image resolution is needed, so the ommatidia are numerous but reduced in cross-sectional area, so that the structural acceptance angle of light is reduced to 2° or 3°. Thus, a smaller amount of light enters, and pigment migration for altering the physiological acceptance angle need only operate over a relatively narrow range of intensity. In crepuscular or night-active species like cockroaches and moths, larger ommatidia occur with larger structural acceptance angles, so that the only way to adapt to increased intensity is by mass pigment migration.

8.4.3.2. Form and Flicker Fusion.   How images are formed within the compound eye of insects is still not totally known (Carlson and Chi, 1979). Earlier studies led to classifying apposition or photopic eyes as those which could form an image directly below the cone within one ommatidium, and

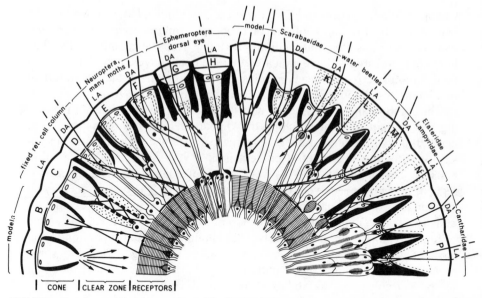

| CONE | CLEAR ZONE | RECEPTORS |

FIGURE 8-18. Ommatidia of eyes adapted for dim light vision by having a clear zone between cornea and receptor layer. The diversity lies in the presence and position of a second lens and especially in the mechanisms of adaptation to a range of intensity. Retinula cell nuclei are solid black, cone cell nuclei are open circles, significant pigment is solid black, inhomogeneous refractive index is shown by dashed zones, rhabdoms are cross-hatched, and light rays bear arrows. A. Even if light is scattered from the cone tip across the clear zone, there is an increased sensitivity by summation on the receptors. B. The cone tip can act as a pinhole so that a poor image forms on the receptors. This is not known to be functional, but can form artifacts in eye slices examined for images. C, D. Crystalline cone and fixed (retinula cell) light guide, found in some Lepidoptera. Adaptation is by pigment migration around the light guide. Hesperioidea are of this type but permanently DA, and with an excellent focusing of off-axis rays by a second lens within the cone. E, F. Typical neuropteroid type with a column of retinula cells that acts as a light guide and is surrounded by pigment in LA state. Neuroptera, some lower Coleoptera, many moths. Note movement between LA and DA states. G, H. Dorsal eye of Ephemeroptera. Fixed retinula cell column forms a light guide. Focusing is poor and pigment movement small, so the eye is probably suited for a particular ambient intensity. I. When the cornea is very thick, with internally projecting corneal cone, the inner curved surface of the cone acts as a second lens and contributes toward focusing on the receptor layer. J. DA state in many scarabaeoid beetles, with focusing by inhomogeneous crystalline cone, and cell movement as in E and F during adaptation. K, L. In beetles of the families Carabidae, Gyrinidae, Dytiscidae, and Hydrophilidae there is, in addition to the other neuropteroid mechanisms, a solitary distal rhabdomere (cross-hatched) which lies directly in the light path of a single facet. In the DA state, light is partially focused across the clear zone on the receptor layers. M, N. Beetles of the families Elateridae, Lampyridae, and Lycidae have inhomogeneous corneal cones, fixed crystalline tracts, and extensive pigment migration as shown. The DA eye is well focused, and in the LA eye each ommatidium operates separately by its light guide. O, P. Beetles of the family Cantharidae have a narrow clear zone crossed by a (cone-cell) crystalline tract. In the LA state the eye is apposition only, but in the DA state another set of rhabdomeres catch light that crosses the clear zone obliquely outside the light guide. (Horridge, 1974).

superposition or scotopic eyes as those that form images in one ommatidium with the light entering from several ommatidial lenses (Dethier, 1964). This interpretation has remained questionable, since numerous images can be formed by diffraction (Carlson and Chi, 1979). Additional complications arise from the wave guide properties of the rhabdoms and the relative positions of rhabdoms of different lengths (Carlson and Chi, 1979).

The patterns of light that fall onto the compound eye may be resolved by the optical system as distinct or not depending on the intensity of light, the rate at which it changes, the portion of the eye illuminated, and the spatial resolution of the system. Not only are there differences among species in the threshold response to light, but regions of the compound eye of an individual may differ (Goldsmith and Bernard, 1974). Behavioral studies have shown that, with increasing intensity of light, the ability to distinguish two points, that is, visual acuity or resolution, increases (Carlson and Chi, 1979). Rarely are patterns of light received as stationary, so that movement across the visual field usually accompanies form. Insect eyes seem particularly well developed for movement detection. Flying diurnal insects, such as the honeybee and true flies, have particularly rapid responses to changes in light, and flicker fusion (ff) frequency, may be as high as 300 pulses/sec. Such insects are said to have "fast" eyes in contrast to insects like *Periplaneta* and *Aeshna* larvae, which have slow eyes with ff of 20–60 pulses/sec.

Regional differences in the size of ommatidia and their spatial resolutions exist in many species (Burkhardt, 1977; Goldsmith and Bernard, 1974; Van Praagh et al., 1980). The locust has a fused rhabdom system in each of its ommatidia, and all the retinular cells of an ommatidium synapse at the first optic neuropile. Thus, the ommatidium is the functional optical unit. The resolution is some function of the dimension of the ommatidium and its angle of acceptance, and this relationship is characteristic for apposition eyes. The situation is distinctly different in flies, however, as seen in Fig. 8-19. Here there is an open rhabdom, so that individual retinula cells in different ommatidia are aligned on the same path and have their axons crossing over to a common site in the first optic neuropile (lamina ganglionaris) of the brain. Here the axons, each from six different ommatidia, converge onto several monopolar interneurons to form an optic cartridge, which is the functional unit of the neural superposition eye (Horridge, 1968). In the locust and other insects with apposition eyes, the resolution may well be a function of the angle of acceptance of the ommatidia, but in flies and other species with neural superposition eyes, the smaller acceptance angles and complex integration combine to give resolutions greater than 0.01° as indicated behaviorally (Carlson and Chi, 1979). Thus, neural superposition optics permits both good spatial resolution and enhanced sensitivity.

8.4.3.3. Spectral Response. It has long been known that insects can perceive the colors we see with an additional sensitivity to wavelengths of the

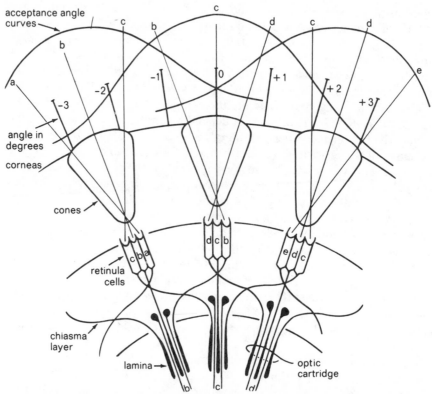

FIGURE 8-19. The relation between the optical axes of retinula cells and the connections of their terminals to second-order neurons of the optic lamina of the fly. Three ommatidia are inclined at an angle of about 2.5° to each other, with acceptance angle curves of sensitivity to light of a typical retinula cell on the axis as shown. The axes of the retinula cells, along the lines a, b, c, d, and e, are inclined to the axes of the ommatidia. The important experimental finding is that the connections of the ommatidia with the optic cartridges of the lamina, as effected by the axons of the retinula cells, bring together the axons of retinula cells that are pointing in the same directions. Note the inversion in the optical path and subsequent erection by the crossings of the axons (Bratenberg, 1966, in Horridge, 1968).

near ultraviolet beyond our visual range (Dethier, 1964; Goldsmith and Bernard, 1974). Behavioral studies have demonstrated that color vision is widespread among insects of several orders, but details of how this is accomplished by the CNS are only beginning to emerge (Kong et al., 1980). Early in this century it was shown that the honeybee had behavioral responses to colors, and recent intracellular recordings from retinula cells confirm the presence of three cell types with different, but overlapping, spectral responses (Fig. 8-20). Not all three types are found with the same frequency in the compound eye, so that even though the basis for trichromic color vision exists, it may well function differently from our own. Perhaps some butterflies are world champions in the spectral range in which they can see. They

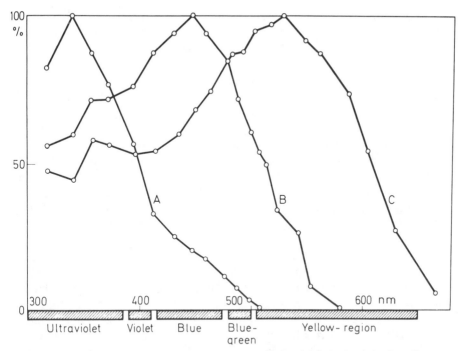

FIGURE 8-20. Three characteristic response curves of visual cells in the drone bee (Autrum and Wrehl, 1963). Later the same response types were found in the worker bee. Beneath the abscissa are indicated the three basic color regions described by Daumer (1956), and the regions of acute wavelength discriminability according to Kühn (1927). Note that the violet region lies directly beneath the point of intersection of the curves for UV and blue receptors, while the blue-green region coincides with the intersection point for the blue and yellow-green receptors. [From D. Burkhardt, *J. Comp. Physiol.* **120,** 33 (1977).]

have UV and red receptors whose max is 60 nm longer than those known for man and primates.

In flies, particularly *Calliphora* and *Drosophila,* studies of their open rhabdom eye have revealed that there are eight retinula cells in each ommatidium and three functional categories within that unit. Cells R1–6 contain rhodopsin and metarhodopsin and have $2\lambda_{max}$ of 470 and 500 nm (blue and green), cell R7 has a peak at 350 nm (UV), and cell R8 has a peak at 500 nm (blue) (Pak, 1979). These studies were based on genetic white-eyed mutants, so the normal eye pigments would not become a complicating factor. In the cockroach there are three UV- and five green-sensitive cells per ommatidium, with uniform distribution over the entire compound eye (Butler, 1971). In the locust there are two or more pigments per cell, and the ommatidium of the moth *Deilephila* has a UV-sensitive distal region, a green-sensitive medial region, and a proximal region most responsive to violet. The fused rhabdom in this ommatidium consists of nine retinula cells of differing lengths (Schlect et al., 1978).

In addition to having different photoreceptor pigments within the same ommatidium or even the same cell, many insects have secondary pigments that give characteristic bands or spots to the eyes and act as color filters, altering the wavelengths that can reach the rhabdoms below (Kong et al., 1980). In others (e.g., the honeybee), UV-sensitive cells are concentrated into the dorsal region of the eye (Burkhardt, 1977). In the owl fly, *Ascalaphus,* there is an entirely separate dorsal UV eye, a ventral non-UV eye, and screening pigments (Fig. 8-14, lower right) (Gogala, 1978). The dorsal UV-responsive eye, because of its visual pigment homogeneity, has proven to be a model system for studying the biochemistry and kinetics of the UV-absorbing visual pigment (Gogala, 1978). The ultimate survival value of this UV receptor relates to its being at or near 100% sensitivity as the insect flies in blue sky light which photoisomerizes the blue-absorbing metarhodopsin to the UV native pigment.

8.4.3.4. Polarized Light Response. Light waves vibrate in planes (their electric vector) at right angles to the direction in which they are traveling (their magnetic vector); thus, light is polarized. The vibrations in the electric vector are detected by a plane-polarized detector. Insect eyes are uniquely structured for such reception, and the rhabdomere is the polarization detector. Furthermore, it has been reported that the short and long twisted retinula cells of the honeybee are the basis of *e* vector discrimination (Wehner et al., 1975). The short polarization-sensitive cells would compare their UV photon catch against the long UV-sensitive cells which were twisted and thus not sensitive to *e* vector. In addition, UV light from the sky is least affected by changing atmospheric conditions; UV-sensitive cells on the dorsal surface of the honeybee eye are known polarized UV detectors. (In fact, all polarization-sensitive receptors have their λ_{max} in the UV.) Less than 1° of sky and three to seven ommatidia are sufficient for the bee to use this short-wavelength polarized light as a visual cue in its orientation in flights to and from the hive (Menzel and Snyder, 1974).

8.4.4. Adaptive Radiation of Photoreceptor Systems

Behavioral responses to colors, patterns, and movements have been analyzed in many insects. These studies, together with electrophysiological studies of retinula cells and the successive layers of neurons in the optic neuropile of the brain, are beginning to allow the determination of a functional link between receptor cell properties and the behavioral responses. Extremely sophisticated techniques for determining the optical properties of insect compound eyes are revealing additional features (Carlson and Chi, 1979; Horridge, 1975). Although we cannot yet account for visually elicited behavior on the basis of the receptors activated and the neural circuits involved, examples suggest much progress has been made. We shall consider two such examples and the design diversity of the compound eye that they represent.

The design of the compound eye is a compromise among the requirements for a relatively compact design of limited size and the capabilities for maximal spatial and temporal resolution of form, absolute sensitivity to light, and the ability to adapt to the range of light intensities encountered in the habitat of the insect. A thorough understanding of the resolution of form by the compound eye requires some knowledge of the principles of optics, together with a careful consideration of the morphology of individual retinula cells, ommatidia, and the overall shape of the compound eye. In a simple apposition eye, the ommatidium is the functional unit, and the smallest form that can be distinguished is determined largely by the diameter of the lens and the angle between the optical axes of adjacent ommatidia, called the interommatidial angle and denoted as $\Delta\phi$. In Fig. 8-18, if ommatidia a, b, and c represented adjacent ommatidia, the angle between the center line drawn through each would represent $\Delta\phi$.

Thus, a compound eye with many small facets would have greater resolution capabilities than one with fewer larger ommatidia. In some insects, different regions of the same compound eye may differ in $\Delta\phi$, so that maximum resolution may reside in only a portion of the eye. In addition to $\Delta\phi$, each ommatidium of an apposition eye has its own angle of acceptance of light, called $\Delta\rho$, which is morphologically fixed by the corneal lens, crystalline cone, and their arrangement to the receptor cells beneath. The pigment granules that can migrate according to the state of light adaptation further alter the light diffraction, therefore resulting in a physiological acceptance angle that differs from the morphological acceptance angle. An experimentally determined $\Delta\rho$ must, therefore, be specified at some level of light-adapted state. Finally, the overall shape of the compound eye, which differs greatly among species and even sexes of the same species, determines how many ommatidia with aligned optical axes point in the same direction.

A visual field composed of such aligned ommatidia may be of widely varying size, shape, and direction in its location relative to the central axis of the head. Compound eyes of insects commonly allow visual fields extending in virtually all directions. Where the visual fields of each compound eye intersect each other, a field of binocular overlap results in which depth determination based on stereoscopy is possible, enabling distance determination. In insects with superposition eyes, the rhabdomeres of retinula cells in *different* ommatidia are aligned in the same optical axis, so that the relationship of $\Delta\phi$, $\Delta\rho$, facet size, and the effects of light adaptation are much more complex (Fig. 8-19).

8.4.4.1. The Fly Eye. Male flies of many species pursue other flies, particularly females, in rapid zig-zag flights. This requires a photoreceptor system with high temporal and spatial resolution that can fixate a rapidly moving target and furnish the information required for rapid adjustments to the flight motor system. This optomotor system in the male house fly allows it to respond to movements of the female of up to 2500°/sec angular velocity with a delay in response of only 30 msec (Land and Collett, 1974). Furthermore, the male fly can track the moving female against the moving visual background of the environment that results from his own flight. The neuroanatomical basis for this resides in the organization of lobula plate neurons, which are quite different in the two sexes.

The large compound eye of the male house fly has a wide field of view extending both in front of the insect and to the rear. The region of binocular overlap in the dorsal–frontal and ventral–frontal regions of the head is small, less than 20°. The eyes of female house flies differ in shape and size and have binocular overlap in the ventral–frontal region. The compound eyes of both sexes consist of a large number of small facets that are very similar in size over most of the surface and have a $\Delta\phi$ of 10° (Fig. 8-14). The ommatidium is of the open rhabdom plan (Fig. 8-16), and retinula cells from six different ommatidia (each viewing a common point in space) converge into one optic cartridge that synapses with interneurons in the lamina ganglionaris (first optic neuropile). There are as many ommatidia as cartridges (about 3000) (Fig. 8-19). The eye is thus structured to operate in bright light with good spatial resolution, and it possesses a high flicker fusion frequency of 300 pulses/sec. A number of other features enhance these basic capabilities.

The photoreceptor cells of the fly eye consist of essentially two populations of cells (peripheral and central cells); most of these converge onto second-order neurons in the lamina ganglionaris, while others pass through the latter "ganglion" to synapse in several strata in the second optic neuropile. An extremely sophisticated and complex system exists. These visual interneurons are selectively responsive to form, motion, and intensity changes of stimuli projected from the peripheral retina in the compound eye. Some of these interneurons have a receptive field only a few ommatidia wide, and some have much greater receptive fields, so that one must understand the degree of synaptic convergence and lateral inhibition present in the neuropiles of the optic lobes (Strausfeld, 1976). As to lateral inhibition among laminar cells, there are a number of lateral-going plexes between retinal axons and their follower interneurons. In addition, contrast enhancement and edge effects could also be accomplished in nonneural ways based on the glial–neuronal compartmentalization of the neuropile, which would dictate the sources, sinks, and electrical resistance patterns in the neuropile.

Several interneurons of the fly visual neuropile have been studied electrophysiologically and injected with dye, so that their entire dimensions and terminations have been mapped (Fig. 8-21). One such identified neuron, the H-1 neuron in the lobula optic neuropile of the blow fly, shows a directionally selective motion detector (DSMD) response (Eckert, 1980). This cell responds by increased frequency of spikes to forward movement across a particular portion of the compound eye. Movement in the opposite direction

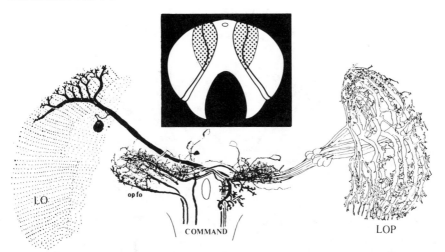

FIGURE 8-21. Diagram of visual interneurons from a male fly mapped with the aid of cobalt sulfide stain. The eight lobula plate vertical giant neurons (LOP) are shown on the right, and a single giant neuron is shown on the left, with the ommatidia with which it connects indicated by dots in the insert. (Modified from N. J. Strausfeld, in F. Zettler and R. Weiler, Eds., *Neural Principles in Vision,* Springer-Verlag, New York, 1976.)

inhibits its response. These responses, which are frequency, wavelength, and intensity dependent, increase over specific ranges until the wavelength reaches a value which is less than 2 times the interommatidial angle. At this point geometrical interference results and the response of the H-1 neuron reverses (Eckert, 1980). How this reversal affects further integration is not known, but this same pattern of reversal of responses occurs in the optomotor flight behavior of flies (Eckert, 1980). In addition, there are vertical and horizontal movement detector neurons in the optic neuropiles beneath the compound eyes, the stimulation of which (through lower-order neurons) results in changes in the thrust and torque forces delivered to the wings (Eckert, 1980). Probably the H-1 interneuron is distinctly involved in processing this movement information that regulates flight, but more detailed studies are needed to confirm it.

The complex integration in the fly visual neuropiles may enhance visual perceptions in a number of ways. Lateral inhibition increases the detection of edge contrast by one optic cartridge by inhibiting the responses of adjacent cartridges (Strausfeld, 1976). Some interneurons give faster-rising EPSPs than those of the primary visual cells that drive them, so that temporal enhancement also exists (Strausfeld, 1976). Thus, the detection of a female fly is accomplished by a photoreceptor system with geometrical features for high spatial resolution and with receptor cells with inherently high temporal resolution. Both of these inputs receive some enhancement and extensive processing by the CNS in ways yet to be fully determined. Designed to function best in bright light, this system suffers a severe loss of

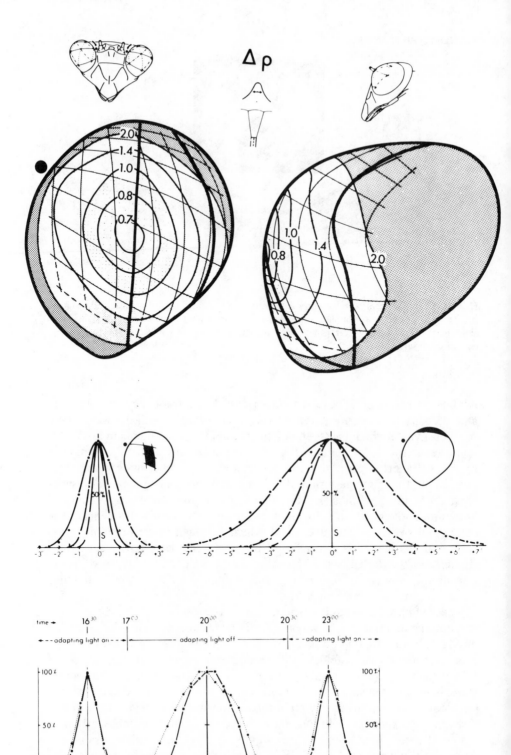

Δ ρ

resolution in dim light. The mating of flies thus occurs only in daylight hours, leaving the darker periods for species with different visual capabilities.

8.4.4.2. The Mantid Eye. The preying mantis must also detect rapidly moving flies and other insects, but for an entirely different purpose—that of feeding. In this case, the photoreceptor system must detect and follow an insect until it is within reach of the raptorial front legs (Rossel, 1979). The visual task is similar to that of the male fly pursuing its potential mate, but additional compromises have been imposed on the functioning of the mantis eye in meeting requirements for survival, so that the adapted structure and functioning is quite different.

The compound eyes of mantids are shaped quite differently from those of the house fly, and their positions at the vertices of a triangular-shaped head allow for vision in nearly all spatial directions (Fig. 8-22, upper). In the mantid *Tenodera australasiae* huge binocular fields extend 240° vertically with a maximum horizontal overlap of 35° in the frontal eye regions (Rossel, 1979). Unlike the eye of the house fly, that of the mantid exhibits large variations in the size of the facets and interommatidial angles across the compound eye (Fig. 8-22, upper). The decreased ommatidial angles of the larger fovea result in their being flattened. The ommatidial axes within this region are further tilted, so that they are no longer normal to the corneal surface. As a result of this tilting, the optical axes focus more toward the center of the fovea than do the anatomical axes.

The fovea, thus, contains ommatidia that are aligned with the center of this region. The acceptance angles ($\Delta\rho$) of the larger ommatidia in the fovea are reduced, so that the capacity for increased resolution resides in this specialized region. The angular sensitivity of the fovea is within ±3°, while that of the dorsal portion is more than twice this value (Fig. 8-22, middle). As with other apposition eyes, the mantis is equipped with screening pigments that migrate longitudinally so that a difference in angular sensitivity occurs with light adaptation (Rossel, 1979). During the daylight hours, the mantis eye is green, with a narrow fovea that can be light adapted in about half an hour. As dusk approaches, however, the fovea enlarges, and as pigment migrates in a circadian rhythm, the entire eye becomes black (Fig. 8-22,

---

FIGURE 8-22. Variation in acceptance angles of different regions of the compound eye of the praying mantis, *Tenodera australaisae*, and their relationship to adaptation state and time of day. Upper, the acceptance angle of an individual retinula cell Δp varies from 0.7° in the central foveal region (left) to 2.5° in the edge regions of the eye (right) for light-adapted cells. Under these conditions the acceptance angles and interommatidial angles have a one-to-one correspondence. Middle left, mean angular sensitivity of the foveal region (darkened in inset) decreases from dark-adapted state (±3°) to the light-adapted state (±1°). Middle right, mean angular sensitivity of the dorsal region (darkened in inset) decreases from dark-adapted state (±7°) to the light-adapted state (±3°). Below, two foveal cells show ±1° annular sensitivity during light periods, increasing to ±1.5°–2° during dark periods. [From S. Rossel, *J. Comp. Physiol.* **131**, 95 (1979).]

bottom). During this period, light adaptation requires 2 hr. Thus, the mantis's visual acuity in the foveal region differs between daytime and night.

The mantid responds to prey targets by moving its head to face the prey. Rapid jerky movements of the head called saccadic moments keep the image totally in the binocular field of the fovea (Lea and Mueller, 1977). This rapid movement limits the time of blurring of background image while the prey is tracked. In the mantis these movements occur at 400°/sec angular velocity with a duration of 116 msec. The mantid responds similarly to prey of different sizes, shapes, and directions of movement, so that all of these have equal value in releasing the strike (Rilling et al., 1959). Color and odor have no releasing effect. Nor do responses vary with experience, so learning is not involved. The most important feature of the prey target is rapid, jerky movements (Rilling et al., 1959).

Although no electrophysiology has been performed on mantis visual interneurons, jittery movement fibers (JMF) are well documented in predatory dragonfly larvae and in locusts (Frantsevich and Mokrushov, 1977; Rowell, 1971). These interneurons respond to steady movement with high rate bursts of 150–250 spikes separated by pauses. JMF habituate to repeated movement along the same trajectory, producing an inhibition that spreads laterally. JMFs of larvae of the dragonfly, *Aeshna,* are sensitive to target shifts of 0.5–3°/sec as well as slow movements of less than 10°/sec and have receptive fields in either portions or the entire eye of the opposite side.

Mantids have both protective coloration and form, so that during the daylight hours they can catch food while remaining relatively inactive and free from predation by birds. Mantids are also nocturnal, but as previously indicated, their visual acuity is much reduced in dim light. During the night, prey catching by visual fixation is a secondary occupation compared to finding a mate. Little is known about the role of vision in mating (Roeder, 1963). However, the mantid eye can obviously handle the tasks of catching flying prey in daylight and attempting to find and approach the female from behind at night.

8.5. CHEMORECEPTOR SYSTEM

The chemical senses of insects are extremely well developed and serve numerous key functions in their survival. Success in finding mates, food, and oviposition sites and in avoiding predation are often dependent on the senses of smell and taste in solitary species (Visser and Minks, 1982). For social insects, the ability to communicate alarm at the intrusion of enemies, follow trails to food sources, or recognize members of different castes depends on their chemical senses (Kaissling, 1971). The chemical tasks of an individual insect can be quite limited within a short lifetime of a few days. For example, a male moth needs only to find a female and mate before dying and the required sense of smell can be quite specialized (Seabrook, 1978). For others, like the female moth, chemoreception may be related to courtship, finding food, and locating a suitable site to lay eggs, so that the senses of smell and taste require a wider range of responsiveness.

Studying the chemical senses is hampered by the difficulties of isolating and identifying the minute quantities of compounds from insects and plants that serve as stimuli. The extremely small sizes of chemosensory cells have precluded intracellular recording, making detailed studies of the functional capabilities difficult. For these reasons, knowledge of the specializations in insect chemosensory systems is somewhat limited. Because so many key functions are played by chemoreception, its adaptive arena is probably the largest and most diverse.

Chemical senses are rather arbitrarily divided into olfaction (smell) and gustation (taste), based on the way in which stimulus molecules are contacted. The olfactory sense detects molecules in the gaseous phase that are mixed with the air that may be present in extremely small concentrations. Thus, the olfactory threshold is extremely low in insects, among the lowest recorded for any animal (Kaissling, 1971). The gustatory sense detects molecules in the liquid phase, for example, those dissolved on the surfaces of leaves or other insects. Here the concentration of stimulus is usually 3–5 log units higher than for olfaction (Dethier, 1976). The source of chemicals must actually be contacted by the gustatory sensilla, so that this is also called contact chemoreception. Some contact chemosensitive cells are also responsive to odors (Stadler and Hanson, 1975). Thus, both senses probably detect stimulus molecules in the same way, although little is currently known about the details of transduction in insect chemoreceptors (Futrelle, 1984; Dethier, 1976; Norris, 1981).

8.5.1. Structure and Distribution

8.5.1.1. Uniporous Sensilla. Uniporous sensilla occur as hairs, pegs, or papillae of various sizes on the surface of the insect (Zacharuk, 1980). They are usually multimodal; the characteristic structures of the numerous chemosensitive cells are shown in Fig. 8-1, middle. The cuticular part of uniporous sensilla may be long hairs of several hundred microns such as those surrounding the labellum of the house fly (Fig. 8-23, upper left), or shorter hairs such as those found on the tarsi (Fig. 8-23, upper right). These sensilla are particularly concentrated on the mouthparts of both adults and immatures, being most abundant on the maxillary and labial palpi (Fig. 8-23, middle left). Internal chemosensilla are found in the buccal cavity on the inner surface of the labrum (Fig. 8-23, middle right) and on the hypopharynx. Uniporous sensilla found on the tarsi of many insect species function in gustation and with those on the mouthparts serve to determine which foods are acceptable and which are not. The antennae also contain uniporous

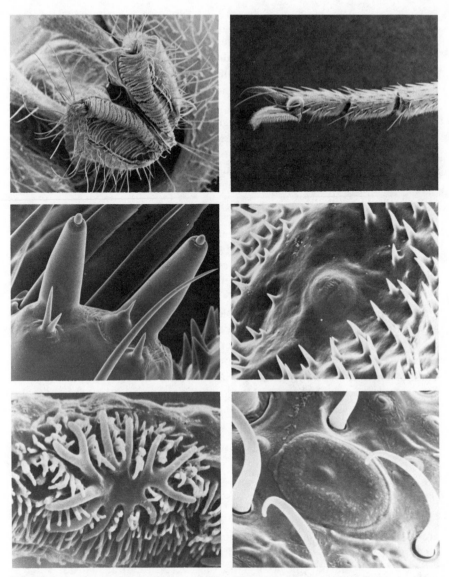

FIGURE 8-23. Insect chemosensitive sensilla. Upper left, labellum of the fly *M. domestica* bearing long uniporous hairs (100×). Upper right, foretarsus of *M. domestica* with ventral uniporous hairs (98×). Middle left, uniporous styloconic sensilla on the maxilla of the caterpillar *H. zea* (633×). Middle right, uniporous dome on epipharyngeal surface of the labrum of the caterpillar of *H. zea* (1725×). Lower left, multiporous branched sensillum on the antenna of the aphid *Calaphis betulaecoleus* (3200×). Lower right, multiporous plate on the antenna of the worker honeybee (8025×). (Scanning electron micrographs courtesy of D. Ave and L. D. Hatfield.)

sensilla, particularly on the distal and ventral surfaces, where they may be used to contact conspecifics during courtship or to tap possible sources of food or oviposition sites as the insect moves about its habitat.

8.5.1.2. Multiporous Sensilla. Multiporous sensilla occur in an enormous variety of shapes and sizes, primarily on the antennae, but to a lesser extent on the palpi and ovipositors (Altner and Prillinger, 1980; Zacharuk, 1980; Dethier, 1964). These sensilla often contain 20 or more chemosensory cells with highly branched dendrites that communicate with the outside world through the many minute pores in the cuticle (Fig. 8-1, lower). These sensilla are also multimodal, occasionally including thermo- and hygrosensitive cells along with the chemosensitive cells (Altner and Prillinger, 1980; Zacharuk, 1980). The antennae of many species, best studied in moths, are highly branched or have large surface areas that together with the large multiporous sensilla (Fig. 8-23, lower left) intercept odor molecules from the air and allow them to diffuse through the cuticular pores to the dendrites, where they presumably bind with receptor proteins in the membrane producing receptor potentials in a manner yet undetermined (Kaissling, 1971; Seabrook, 1978). Being contained within pits or depressions in the antenna may limit these sensilla's efficiency in capturing odors.

8.5.2. Feature Extraction

8.5.2.1. Threshold and Concentration–Response Curves. The sensitivity of the insect chemoreceptor system for a given stimulus compound can be determined by measuring either the response of a single receptor cell, which yields the absolute sensitivity of the system, or by measuring a behavioral response, which gives the relative overall sensitivity of the system (Kaissling, 1971). For the blow fly, *Phormia regina,* the number of impulses produced by stimulation of different tarsal hairs gives the concentration–response curves shown in Fig. 8-24. The concentration giving half-maximal response, which is used to compare the stimulating effectiveness of different sugars (Seabrook, 1978; Norris, 1981), indicates that the sugar-sensitive cells in different hairs respond in some characteristic manner for each stimulus group.

By presenting solutions to the tarsi of restrained flies, one can observe a lowering of the mouthparts, called the labellar response, to determine a behavioral threshold (Fig. 8-25). Table 8-3 compares the electrophysiological and behavioral thresholds of the blow fly to sugars. That the behavioral threshold is almost always lower than that of the receptor seems anomalous at first glance. However, the behavioral response results from the exposure of a large population of receptor cells, only a few of which must produce nerve impulses to elicit the labellar response. The threshold of the cells measured singly is expressed as the concentration that always gives a response. The probability of stimulation of the receptors with low thresholds

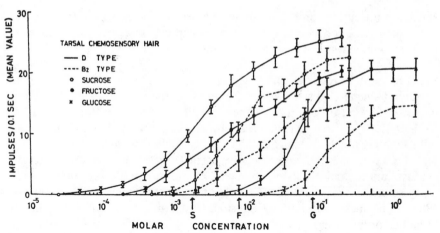

FIGURE 8-24. Mean concentration–response curves for tarsal chemosensory hairs of the blow fly. The arrows show behavioral acceptance thresholds for sucrose (S = 1.8 × 10⁻³ M), fructose (F = 8 × 10³ M), and glucose (G = 8 × 10⁻² M). [From A. Shirahashi and Y. Tanabe, J. Comp. Physiol. **92**, 161 (1974).]

in the population is greater in the behavioral experiment, so the observed behavioral threshold is lower.

Studies to determine the olfactory thresholds of insects in which the criterion of threshold response in a single cell is taken as that concentration that produces an impulse 20–50% of the time also yields receptor thresholds above those measured behaviorally (Davis and Sokolove, 1975). For the silkworm, *Bombyx mori,* 3×10^4 molecules/cm^3 of the pheromone molecule bombykol produced an impulse 35% of the time, while 650 molecules/cm^3 caused males to flutter their wings (Fig. 8-26). Here the number of receptor cells is even greater than in the previous example, so the probability of stimulating receptors with low thresholds in the behavioral experiment is

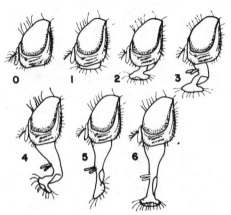

FIGURE 8-25. Labellar response of flies to tarsal stimulation by sugars. Arbitrary scale 0–6 used to quantify response magnitude. (Modified from Dethier, 1976.)

TABLE 8-3. Relationship of Sugar Receptor and Labellar Behavioral Responses in the Blow Fly with Concentration of Sugars

Sugar	Slope[a]	x intercept (M)	Number of Flies	Number of Hairs	Number of Concentrations	Behavioral Responses (M)
Fructose	16 ± 2	0.015 ± 0.01	11	111	4–10	0.0058
Sucrose	38 ± 7	0.025 ± 0.02	6	62	3–5	0.0098
Glucose	48 ± 2	0.264 ± 0.12	11	111	4–10	0.132
D-Arabinose	52 ± 7	0.210 ± 0.09	3	30	4	0.144
Maltose	50 ± 13	0.035 ± 0.02	3	30	3	0.0043

[a] The slope is expressed as impulses per second per logarithm unit.
Source: Modified from Dethier (1976).

also greater. A comparison of the sensitivities of insect gustatory and olfactory systems with those of humans reveal nearly equal capabilities in many instances (Kaissling, 1971; Seabrook, 1978). The olfactory thresholds of the honeybee for several compounds are within the same order of magnitude as those of humans, while the response of the fly to sugars is equal to or better than our own taste (Seabrook, 1978). These similarities exist for chemicals that can be perceived by both humans and insects, but there are many biologically important chemicals that are detected only by insects (Kaissling, 1971; Seabrook, 1978; Visser and Minks, 1982).

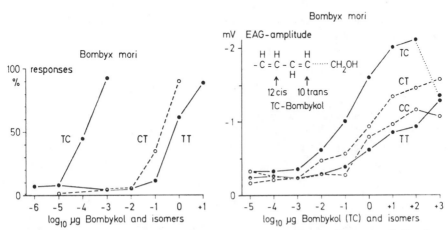

FIGURE 8-26. Behavioral and electroantennogram responses of male silkworm moth (*Bombyx mori*) to different isomers of its sex attractant. Left, behavioral thresholds for bombykol (TC) and two isomers. Right, maximum amplitude (quasistationary) of the summated receptor potential (EAG) to bombykol and its steroisomers (CT, C, and TT) (Kaissling, 1971). (Redrawn from Schneider et al., 1967.)

8.5.2.2. Specificity. Like other sensory systems of insects, the chemoreceptor system is composed of individual receptor cells that have different, but overlapping, responses (Kaissling, 1971; Seabrook, 1978). The specificity of a chemoreceptor cell can be determined by the stimulus response curves of all effective substances. The concentrations that give half-maximal responses can then be compared as indicators of the specificity of the cell (Kaissling, 1971). The specificities of the four chemosensitive cells of the labellar hairs of the blow fly have been described as sugar, cations, anions, and water (Dethier, 1976). The sugar-sensitive cell is also responsive to some amino acids and fatty acids, while the water receptor is inhibited by salts and reactivated by certain sugars at low concentrations (Stadler and Hanson, 1975). In other words, the specificity of a gustatory cell is described according to its optimal stimuli, with the range of additional effective stimuli often determined by the number of compounds that have been tested. The occurrence of sugar-, salt-, and water-sensitive cells in insects is widespread (Dethier, 1973, 1976; Stadler and Hanson, 1978; Ma, 1972). Gustatory-chemosensitive cells that are most sensitive to unique plant chemicals have evolved in caterpillars and several other species of phytophagous insects (Visser and Minks, 1982; Stadler, 1977).

Olfactory-chemosensitive cells have evolved with unique specificities where the olfactory tasks are highly specialized. For example, the male silkworm moth, *Bombyx mori,* responds differently to the various geometrical isomers of the pheromone molecule, bombykol (Fig. 8-26). Recently, a second pheromone component, bombykal, was shown to be detected by a second highly specific olfactory cell (Kaissling and Kasang, 1978). Such chemosensitive cells with limited spectra of response are called odor specialists (Ali, 1978; Dethier and Schoonhoven, 1968; Davis and Sokolove, 1975). Odor specialist cells probably occur widely among insects utilizing pheromones for communication as well as among those insects with specific behavioral tasks. An example of the latter would be a mosquito using a situation-unique compound like CO_2 to orient to a warm-blooded host.

The honeybee also possesses a CO_2 receptor that is important in the survival of the hive, and, in addition, a number of chemosensitive cells of the pore plate are responsive to a wide variety of odor compounds produced by flowering plants (Kaissling, 1971). These are representative of the odor generalist cells widespread among insects with varied olfactory tasks, especially of those species that have coevolved with plants. Even here the reaction spectra vary from a few to several compounds (Kaissling, 1971; Stadler, 1977).

8.5.3. Adaptive Radiation of Chemoreceptor Systems

Much of the adaptive uniqueness of insect chemoreceptor systems lies in their capabilities for detecting chemicals of special biological importance. Differing chemoreceptor systems have evolved to meet different chemically mediated behavioral tasks. The American cockroach, *Periplaneta americana,* which utilizes pheromones for aggregation and mating and feeds on a wide variety of foods, possesses a correspondingly wide range of chemoreceptor capabilities (Sass, 1978). Females of the summerfruit tortrix moth, *Adoxophyes orana,* have olfactory receptors that respond to odors from apple tree leaves, bark and fruit, as well as to their own pheromone; in contrast, males respond only to the pheromone (Den Otter et al., 1978). The female imported cabbage worm moth, *Pieris brassicae,* can detect a pheromone on the eggs laid by other females that deters her from laying her own eggs proximate to others, thus preventing excess competition for food (Behan and Schoonhoven, 1978). Bark beetles have an elaborate receptor system for pheromones and host tree volatiles (Payne, 1979).

8.5.3.1. Gustation in Flies. The first chemoreceptor system of insects to be studied in detail was that of contact chemoreception in flies (Dethier, 1976). Many species of flies, including *Phormia regina,* have tarsal and labellar chemosensitive cells that detect sugars, salts, water, and proteins. If the tarsi randomly contact an adequate stimulus, the labellum is lowered so that longer, more sensitive labellar hairs contact the source, giving a further sensory evaluation. With suitable labellar input, a fly spreads the labellar lobes, ejects saliva, and begins ingestion by sucking up the partially digested food through the interpseudotracheal papillae (Fig. 8-23, upper left). Constant monitoring of the ingested food is further provided by internal chemoreceptor cells in the pharynx. Ingestion continues until negative-feedback signals from internal stretch receptors on the foregut inhibit the motor output to the muscles which hold the labellum extended (Dethier, 1976). The decision by the CNS to accept or reject the sampled food may be accomplished within 150 msec after the first stimulation. Acceptance of the food is dependent on the frequency of spikes arriving from the tarsal and labellar chemoreceptors, with sugar and water cells having an excitatory effect and the salt cells having an inhibitory effect (Dethier, 1976). The frequency of these inputs is dependent on the concentrations of stimuli in the mixture. These inputs, together with the internal states, are integrated by the CNS in some unknown way to produce the final behavioral output of acceptance or rejection (Dethier, 1976).

Many factors contribute to the acceptance and rejection of food by the fly. For one thing, there are different classes of sugar receptors on the tarsi (Fig. 8-24) that occur in different frequencies (Shirahashi and Tanabe, 1974). In addition, the number of functional cells decrease with age; furthermore, the acceptance thresholds are 3–7 times lower and the rejection thresholds 3–4 times higher in freely moving flies compared to those restrained (vander Molen and vander Schaal, 1979). It is also known that repetitive stimuli seem to produce long-lasting altered activity within the CNS (Dethier, 1976; Hall, 1980). Unfortunately, these central excitatory and inhibitory states affect information processing in ways not yet understood (Dethier, 1976).

The nutritional state of the insect and its previous diet history are involved in a more indirect manner in determining the behavioral reaction of a fly to food (Dethier, 1976). The various interactions of salts and sugars on the separate chemosensory cells, in combination with the above factors, make it even harder to determine how environmentally significant stimuli are encoded and acted upon by the CNS. What is known is that the fly eats a variety of substances and rejects others with the use of only a few chemosensory cells. This constitutes a classic example of the parsimony of insect sensory systems.

8.5.3.2. Gustation in Caterpillars. Caterpillars possess two uniporous pegs on each maxilla (Fig. 8-23, middle left) and two uniporous domes on the epipharyngeal surface of the labrum that contain gustatory receptor cells (Fig. 8-23, middle right). The lateral and medial pegs of the maxilla each contain four cells, with different optimal stimuli in different species (Table 8-4). A given species usually possesses a salt cell, sugar cell, inositol or sugar alcohol cell, and a cell that responds to specific plant compounds (the deterrent cell). The uniporous dome contains a sugar cell, a salt cell, and a deterrent cell in the few species studied (Table 8-4). These two sets of gustatory cells provide the major sensory inputs regulating feeding in the caterpillar. The insect may obtain additional cues through olfaction, but usually it takes test bites and, if acceptable, continues feeding. It is believed that the cells of the maxilla control biting and the cells of the epipharyngeal sensilla control swallowing (Schoonhoven, 1972; Ma, 1972).

Plants and their insect herbivores have apparently coevolved, each developing strategies against the other. Plants produce allelochemicals that may have adverse effects on competing plants, pathogens, or insects (Visser and Minks, 1982; Rosenthal and Janzen, 1979). Not surprisingly, insects possess specialized chemoreceptor cells for detecting these chemicals and numerous enzymes, such as the mixed-function oxidases that can degrade many plant natural products (Rosenthal and Janzen, 1979). Caterpillars are an integral part of this evolutionary struggle and accordingly have developed specialized feeding habits. With 110 sensory cells, much fewer than the fly's several thousand, caterpillars can detect an amazing variety of plant compounds (Schoonhoven, 1972; Stadler, 1977; Hanson, 1983).

In some species, like *Pieris brassicae,* the imported cabbage worm, a cell specific for glucosinolate compounds found only in cruciferous plants occurs, and its input stimulates feeding (Ma, 1972). *P. brassicae,* a monophagous species, has become so well adapted to its host plant that it uses the host-specific compounds as feeding stimulants. On the other hand, many other species do not feed on cruciferous plants because glucosinolates are effective deterrents for them (Schoonhoven, 1972; Stadler, 1977). The deterrent cell in most species studied has a wide response spectrum that allows it to detect the compounds in the plants that the insect does not accept for

TABLE 8-4. Distribution of Gustatory Receptor Cells in Various Lepidopterous Larvae[a]

Insect	Medial Sensilla				Lateral Sensilla				Epipharyngeal			Reference
	1	2	3	4	1	2	3	4	1	2	3	
Manduca sexta	SALT/A		Alk	INOS	SALT/A	S/G	SA	INOS	SALT	SUC	D	DeBoer et al. (1977)
Pieris brassicae	SALT	SALT	D	MOG/I	AA	S/G	MOG	ANTH C	SALT	SUC	D	Ma (1972)
Mamestra brassicae					SALT	S	MOG/GLY	SALT				Wieczorek (1976)
Philosamia cynthia	SALT	CON	G	INOS	AA	S/G	G	INOS				Schoonhoven (1972)
Bombyx mori	SALT	SALT	D	WATER	SALT	S/G	G	INOS				Ishikawa et al. (1969)

[a] Abbreviations: A, acid; AA, amino acid; ANTH C, anthocyanins; CON, conessin; D, deterrents; G, glucose; GLY, glycosides; INOS, inositol; MOG, mustard oil glycosides; S, sucrose; SA, salicin.

feeding. Cutting off the appropriate sensillum can render unacceptable plants acceptable.

While the sensory responses of caterpillars with varying feeding habits have been recorded for a variety of plant saps and diets, the question remains as to how these diverse sensory inputs code for the specific plant sap with its mixture of chemicals. The maxillary cells have varying specificities, and their frequency of firing depends on the concentration of stimuli. However, ultimately all of these sensory inputs must be integrated in the CNS.

Like that of the fly, the caterpillar CNS appears to add the excitatory and inhibitory inputs to determine acceptance or rejection (Hanson, 1983). In caterpillars, however, the deterrent cells may adapt more slowly than the excitatory cells, so that a negative input lasts longer (Schoonhoven, 1972). The possibility that a plant-specific pattern of impulses is produced by the maxillary cells has been proposed by Dethier (1973). Figure 8-27 shows hypothetical concentration–response curves for a sugar, a salt, and a deterrent cell that are based on the responses of such cells recorded from caterpillars. If we assume that each receptor has a unique response spectrum which may or may not overlap with that of the others, and that no stimulus interactions occur, then the responses of these receptors to three different plants can be analyzed (Fig. 8-28). Here the magnitude of response for each cell is given for plants differing in concentrations of the compounds indicated in Fig. 8-27. Plant A has a high concentration of sugars and a low concentration of deterrent. The sugar receptor input dominates, and this plant would be acceptable. Plant B has a high concentration of organic acid, a possible

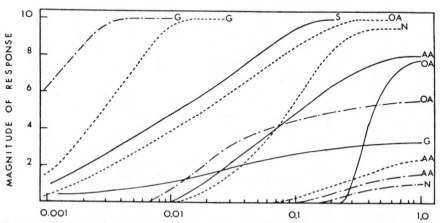

FIGURE 8-27. Concentration–response curves for three hypothetical receptors of a caterpillar: ——, a sugar receptor; ― ― ―, a salt receptor; —·—, a rejection or deterrent receptor. Each receptor is postulated as responding to four of five compounds: G, a glycoside; S, a sugar; N, a salt, OA, an organic acid; AA, an amino acid. The slopes of the curves are arbitrary but are based on the general characteristics of actual curves. [From V. G. Dethier, *J. Comp. Physiol.* **82**, 103 (1973).]

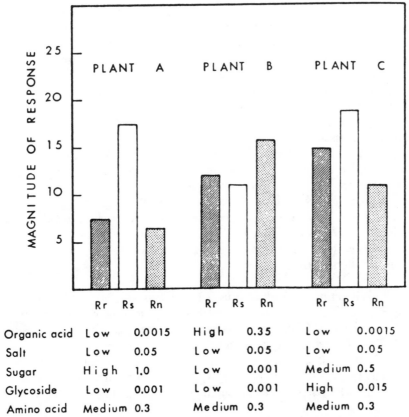

	Rr	Rs	Rn	Rr	Rs	Rn	Rr	Rs	Rn
Organic acid	Low		0.0015	High		0.35	Low		0.0015
Salt	Low		0.05	Low		0.05	Low		0.05
Sugar	High		1.0	Low		0.001	Medium		0.5
Glycoside	Low		0.001	Low		0.001	High		0.015
Amino acid	Medium		0.3	Medium		0.3	Medium		0.3

FIGURE 8-28. Three patterns of response of a sensillum made up of the three hypothetical receptors described in Fig. 8-27; R_r, repellent or deterrent receptor; R_s, sugar receptor; R_n, salt receptor. Each pattern represents the response to a different plant (A, B, C) in which concentrations of five compounds were set arbitrarily. Each vertical bar representing the combined magnitude of response to all constituents of the plant was generated by reading from the response curves in Fig. 7-27 the magnitude of response given by each receptor to each of the compounds at the concentrations shown. [From V. G. Dethier, *J. Comp. Physiol.* **82,** 103 (1973).]

deterrent, and a low concentration of sugar. All three cells respond, but the salt and deterrent responses predominate, and the plant is rejected. Plant C is intermediate to A and B, and it might be minimally accepted or rejected, but it would be for reasons entirely different from B. This idea of across-fiber patterning is consistent with the responses recorded from caterpillar receptors, and although simplified, still holds for more complex situations where possible interactions among stimuli may occur at the level of the receptors.

While how the brain interprets the gustatory inputs that regulate feeding behavior is not exactly understood, some progress toward understanding has been made by recording the responses of candidate receptor cells to a number of chemicals over a selected concentration range and then measuring the amount of feeding on a defined diet containing these same concentrations and chemicals. In *P. brassicae,* studying these parameters resulted in some interesting findings (Ma, 1972; Blom, 1978). By plotting the receptor cells, response to sucrose as sensory input, the amount eaten as behavioral output, and the processing as a black box, the investigator can simply look at the relationship between input and output. Predictably, over the mid-concentration range the sensory input will be coding for intensity linearly and the behavioral output will be linearly related (Fig. 8-29).

The situation is undoubtedly more complex than this. Ablation experiments coupled with feeding studies revealed that the sensory input from the cells of the epipharyngeal dome are given equal weight with the input from cells of both the right and left maxilla (Ma, 1972). The same approach with *Mamestra brassicae,* a relatively polyphagous species, did not give a linear relationship of sensory input to behavioral output, suggesting that inputs from other, perhaps undescribed, receptors are required or that processing is nonlinear (Blom, 1978).

8.5.3.3. Olfaction in Male Moths. In contrast to flies and caterpillars, lepidopterous males feed very little, their main task being to find a female and mate within a few days or weeks. This chemically mediated task has resulted in the evolution of the most sensitive and specific olfactory cells known in the animal kingdom, the pheromone receptors (Kaissling, 1971; (Mitchell, 1981). For males of many species of Noctuidae, Tortricidae, and Saturniidae, the volatile pheromone blends of compounds produced by the

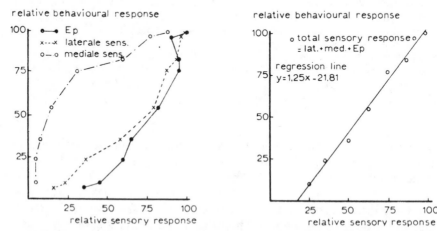

FIGURE 8-29. Correlation between relative sensory response of gustatory receptor cells and relative behavioral response of ingestion over the same concentration range of sucrose in the caterpillar *Pieris brassicae.* Left, response of each of the three sucrose-sensitive cells versus behavior. Right, responses of the three cells summated versus behavior. [From F. Blom, *Netherlands J. Zool.* **28,** 277 (1978).]

females of their species are sign stimuli that release upwind, anemotactic flight, allowing the male hopefully to reach the female and to copulate (Seabrook, 1978). This entire sequence can involve a sequential exchange of stimuli between sexes (see Chapter 14).

The pheromone receptor cells of the redbanded leafroller moth *Argyrotaenia velutinana* occur as two cells/sensillum on the male antenna (O'Connell, 1975) (see Chapter 14, Section 14.4.3). The female-produced pheromone of this species is a mixture of 9 parts (E)-11-tetradecenyl acetate [(E)-11-TDA] to 1 part (Z)-11-tetradecenyl acetate [(Z)-11-TDA] that elicits behavioral responses in males at 2 ng (Roelofs, 1978). The receptor cell with the largest spike amplitude (unit A) is stimulated by (Z)-11-TDA, resulting in a steep response slope and a higher maximum than is the receptor with the smaller spike amplitude (unit B) (Fig. 8-30). The opposite effect occurs with stimulation by (E)-11-TDA, that is, the B unit responds more than the A unit, which in fact is inhibited by (E)-11-TDA at concentrations below 1 μg (Fig. 8-30). The receptors show maximal responses to the pheromone components, but other structurally similar compounds are also stimulatory and inhibitory (Table 8-5). Pheromone receptors can be sufficiently specific to discern optical isomers (Chapman et al., 1978).

While the silkworm moth, *Bombyx mori,* has been in laboratory culture for centuries and has perhaps lost some of the genetic diversity of its wild counterparts, its pheromone receptor systems have been well studied (Kaissling, 1971). Its females also produce a two-component pheromone consisting of 164 ng of the alcohol (Z,E)-10,12-hexadecadien-1-ol (bombykol) and 15 ng of the aldehyde (Z,E)-10,12-hexadecadien-1-al (bombykal) (Kaissling and Kasang, 1978). Like the redbanded leafroller male, the *Bombyx* male possesses two receptor cells/sensillum on the antenna (Kaissling, 1971). One of these responds to bombykol, the other to bombykal. The response thresholds for the bombykol receptor have been calculated as a single molecule of pheromone, making these moth pheromone receptors the most sensitive olfactory receptors currently known (Kaissling, 1971; Seabrook, 1978).

In male saturniid moths, *Antheraea pernyi* and *A. polyphemus,* the region of the brain that receives pheromone olfactory cell inputs is enlarged, forming a macroglomerulus (Boeckh et al., 1977). The number of olfactory receptor cells, very large in male Lepidoptera, exceed 100,000 in *Bombyx* and *Antheraea,* with similar large numbers in the American cockroach. One antenna of *Periplaneta americana* contains more than 150,000 olfactory cells that converge onto less than 250 cells leaving the glomerulus (Boeckh et al., 1977). Intracellular recordings from the visual neuropile of flies show that convergence of 6:1 can result in significant reductions in noise and increase in sensitivity of the processing interneuron (Smola, 1976). If the olfactory convergence in male moths is similar to that of the cockroach, the convergence ratios of several hundred to one seem possible. This could render the increase in sensitivity of central neurons in the pheromone system even

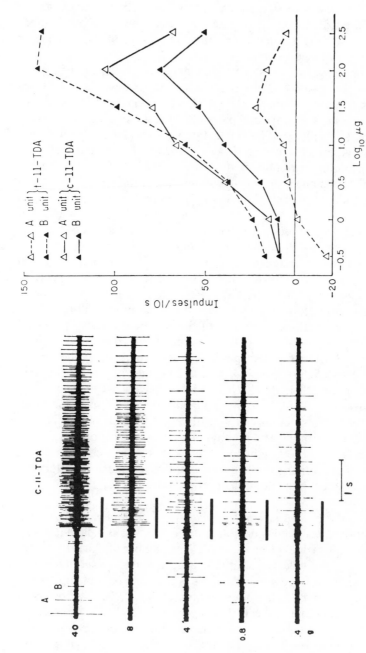

FIGURE 8-30. Response of olfactory receptor neurons of the male red-banded leafroller moth to isomers of the sex attractant; tetradecenyl acetate (TDA). Left, responses of A and B neurons to decreasing amounts of *c*-11-TDA. Note persistence of response following end of one second stimulation. Right, dose–response curves for the two receptors for *t*-11-TDA and *c*-11-TDA. (Reproduced from *J. Gen. Physiol.*, 1975, Vol. 65, p. 179, by copyright permission of The Rockefeller University Press.)

TABLE 8-5. Responses of Olfactory Receptor Neurons of the Redbanded Leafroller Moth

Sensillum	Unit	Attractant		Inhibitors					Synergists	
		(Z)-11-TDA	(E)-11-TDA	TriDA	(Z)-11-TDol	(Z)-11-TDF	DDA	(Z)-7-DDA	10-PDA	10-UDA
1	A	+	+	0	0	+	+	++		
	B	+	+	0	0	+	+	−		
2	A	++	+	−	−	+	−	−		
	B	++	++	+	0	++	0	0	+++	+
3	A	+++	0	0	−	+	−	0		
	B	+	+	−	0	−	−	+		
4	A					−			0	+
	B	+++	+++			+			+	
5	A	++	+			+	+	0	+	
	B	0	0			+	+	−	+	0

Key: $R \leq -6 = -$; $-5 \leq R \; 5 = 0$; $6 \leq R \; 50 = +$; $51 \leq R \; 100 = ++$; $R \geq 101 = +++$; Blank = not tested.

R response magnitude in impulses/10 sec.

* All compounds presented at an intensity of 31.6 µg/µl.

Source: Modified from O'Connell (1975).

more dramatic than that of the visual system. Indeed, it could account for some of the reduced behavioral thresholds when compared to receptor thresholds observed in adult moths.

In addition to the extreme sensitivity and the high selectivity of phero-mone receptor cells, their responses to single pulses of stimulus can produce fluctuations and responses that persist after the stimulation has ended. Such an increase in the temporal aspects of stimulation could be an important adaptation for a male moth flying upwind in a discontinuous odor plume. Indeed, the response features of insect pheromone receptors that have been studied in detail represent truly monumental achievements in adaptive strategy.

8.6. OTHER SENSORY SYSTEMS

Our knowledge of the sensory capabilities of insects for temperature, humid-ity, and geomagnetic reception is only recently beginning to include electro-physiological data. However, it has long been known that insects respond behaviorally to changes in temperature and humidity, often selecting opti-mum conditions when offered a choice (Dethier, 1964). The existence of such behavioral choices suggested that specific receptors may exist, and methodical ablations of various appendages led to the antenna as the princi-pal receptor site (Roth and Willis, 1952). In addition, the behavior of some species, such as mosquitoes and other biting flies and the buprestid beetle, *Melanophila acuminata,* that flies to forest fires, suggests that special sen-sory capabilities may exist. Nevertheless, for the vast majority of species, the roles of such detectors are not readily apparent (Ali, 1978).

8.6.1. Thermoreceptor System

Thermoreceptors do not usually provide the animal with the same kind of detailed information as other sensory systems. In general, the sense of tem-perature seems to be involved in aiding the animal in maintaining an optimal body temperature, and insects appear to possess temperature detectors (Ali, 1978), which, when combined with behavioral mechanisms, result in altering the body temperature.

Insects for which electrophysiological proof of temperature receptor functioning has been determined are shown in Table 8-6. The majority of species studied to date possess cold receptors that increase their rate of firing to decreasing temperature. These receptors respond with tonic in-creases in frequency to changing temperatures and to different tempera-tures.

Although cold receptors have been found on a variety of species, these insects nevertheless constitute a very small sample of the class Insecta. Cold receptors on the antennae of caterpillars can detect temperature changes

TABLE 8-6. Insects that Possess Temperature Receptors as Determined by Electrophysiological Recordings

Insect	Order	Sensillum Type	Location	Receptor Type	Reference[a]
Apis mellifera	Hymenoptera	Peg in pit	Antenna	Cold	
Locusta migratoria migratorioides	Orthoptera	Peg in pit	Antenna	Cold	
Periplaneta americana	Orthoptera	Peg without pores	Antenna	Cold	
Carausius morosus	Orthoptera	Peg without pore	Antenna	Cold, warm	
Manduca sexta (larva)	Lepidoptera	Multiporous peg (?)	Antenna	Cold	Dethier and Schoonhoven (1968)
Heliothis zea (larva)	Lepidoptera	Multiporous peg (?)	Antenna	Cold	
Pieris brassicae (larva)	Lepidoptera	Multiporous peg (?)	Antenna	Cold	
Dendrolimus pini (larva)	Lepidoptera	Multiporous peg (?)	Antenna	Cold	
Philosamia cynthia ricini (larva)	Lepidoptera	Multiporous peg (?)	Antenna	Cold	
Protoparce sexta (larva)	Lepidoptera	Multiporous peg (?)	Antenna	Cold	
Aedes aegypti	Diptera	Peg in pits	Antenna	Cold, warm	Davis and Sokolove (1975)
Triatoma sp.	Hemiptera	(?)	Mouthparts	Cold?	

[a] Consult Altner and Prillinger (1980) for specific references.

349

resulting from evaporative cooling of the air by moisture leaving the cut surfaces of leaves, which may serve as an indicator of food turgidity and quality (Dethier and Schoonhoven, 1968). In the locust, honeybee, and bug *Triatoma* cold receptors respond best in the range of 20°–30°C, the optimum for body functioning, while in the stick insect *Carausius* a linear response to decreasing temperature exists below 20°C.

A cold receptor in the cockroach shows variable impulse shape, with sensitivity of −155 impulses/sec/°C, and a very marked phasictonic response, with initial frequency exceeding 250/sec. This receptor is found within a small cone-shaped sensillum (Fig. 8-31) on the ventral surface of the antenna on a few segments and responds to both the rate of change of temperature and the instantaneous temperature (Altner and Prillinger, 1980). Other studies have shown that a cold receptor cell can exist together with a hexanoic acid–sensitive chemoreceptor cell in another type of sensillum that is double-walled and has longitudinal grooves (Altner and Prillinger, 1980).

The mosquito *Aedes aegypti* is notable in having a pair of thermoreceptive units located in a peg in a pit at the tip of the antennae. One of these cells is a warm receptor and the other a cold receptor (Fig. 8-32). The warm cell gives a phasic–tonic response to increasing temperature, with a maximum between 25° and 28°C. This cell also responds to different rates of change in temperature, the magnitude of response depending on the starting tempera-

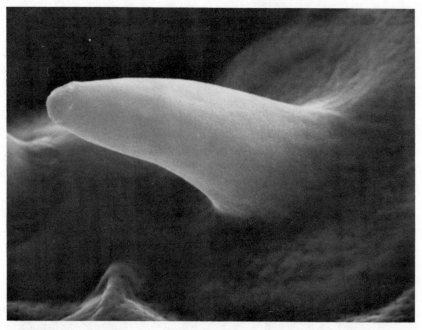

FIGURE 8-31. The sensillum from the antenna of the American cockroach that contains both cold and dry receptor neurons (10350×). (Courtesy of M. Sullivan.)

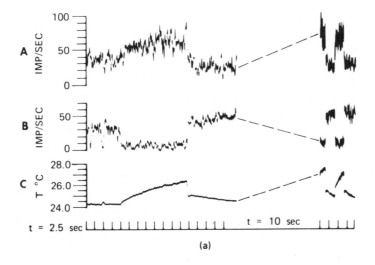

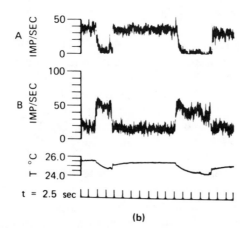

FIGURE 8-32.   Frequency responses of two cells simultaneously recorded from the same pit peg sensillum on the antenna of the mosquito *Aedes aegypti*. Two cells are differentiated by their relative spike amplitudes. Upper, line *A* is a plot of the instantaneous spike frequency of a warm-sensitive neuron and *B* of a cold-sensitive neuron. Line *C* is the plot of the temperature as measured by the thermistor adjacent to the antenna. Below, line *A* response of warm receptor to decreasing temperature; line *B* response of cold receptor to decreasing temperature. [Modified from E. E. Davis and P. G. Sokolove, *J. Comp. Physiol.* **96**, 223 (1975).]

ture. The cold receptor responds in a similar manner to decreasing temperatures. These cells did not respond to infrared radiation, to CO_2, or to other chemicals. Convection currents with thermal differences of 1–2° were demonstrated at distances of more than 40 cm from the human arm and can furnish adequate stimuli to aid in host location by the mosquito (Davis and Sokolove, 1975).

8.6.2. Hygroreception

A number of electrophysiological studies, which are mostly the same as those reported for temperature receptors in Table 8-6, have confirmed the presence of hygroreceptors on the antennae of caterpillars, honeybees, mosquitoes, locusts, bugs, and cockroaches. The cockroach has moist receptors that increase frequency in a phasic–tonic manner to increasing humidity and dry receptors that increase frequency to decreasing moisture of the air. Moist and dry receptor cells occur together in the same sensillum, and their antagonistic responses may be important in CNS integration (Altner and Prillinger, 1980). The situation is seriously complicated by the fact that humidity changes can also be accompanied by temperature changes. Both changes could cause ionic or structural changes of the receptor cells affecting transduction. Loftus (Altner and Prillinger, 1980) has described the dry receptor of *Periplaneta,* which may also have a dual modality as a temperature receptor.

8.6.3. Geomagnetic Reception

Honeybees and termites have been shown to have behavioral responses to magnetic fields, but no receptors have been identified by electrophysiology or by ablation (Martin and Lindauer, 1977; Gould, 1980; Goldman and O'Brien, 1978). In termites locomotor activity rhythms and feeding rhythms vary with geomagnetic fluxes, and screening with magnetic shielding induces changed feeding cycles (Becker and Gerisch, 1977).

Honeybees show four effects of magnetic fields:

1. In the waggle dance, bees convert the angle flown to a food source with respect to the sun into an angle with respect to gravity.
2. When the comb is turned to a horizontal position removing the usual gravity cue, bees either stop dancing or become disoriented. Eventually the bees can reorient, but canceling magnetic fields eliminates this reorientation.
3. A displaced swarm in an empty cylinder will build the comb in the magnetic orientation as it had in the parent hive.
4. Bees will set their circadian rhythms by daily fluctuations in the earth's magnetic fields, but strong fields will disrupt these rhythms. Transversely oriented magnetic material in the front of the abdomen has been proposed to be involved (Gould, 1980; Goldman and O'Brien, 1978).

8.7. CONCLUSIONS

This chapter has focused on the principles of operation common to all insect sensory systems. The unique relationships of these to the structure of insect cuticle and the way in which information about specific features of the external environment are detected by insect senses are known in varying detail for each system. These areas of current research, together with concurrent electrophysiological studies of individual sense cells and new noninvasive techniques, should provide many new details of sensory cell functioning in the near future.

Currently, we know practically nothing about the processing of sensory information by the CNS. Our knowledge of insect behavior indicates that many types of sensory inputs are simultaneously integrated by the CNS. Preliminary studies of the processing neurons in the cockroach olfactory glomerulus indicate that a uniquely high degree of within and across modality convergence takes place. Other recent investigations indicate that a plethora of small peptidic neuromodulators are produced and act throughout the insect nervous system. We can thus expect these neuropeptides to be focal areas for research in insect sensory information processing in the immediate future. One result of this work will be the description of many new sites, in biochemical terms, that will allow the design of new kinds of chemicals for insect control through their action of behavioral modification. This approach has much to offer in realizing a new level of sophistication in our continuing efforts to develop successful methods of insect control.

ACKNOWLEDGMENTS

The generous contributions of the following people are gratefully acknowledged: D. A. Ave, S. D. Carlson, V. G. Dethier, F. E. Hanson, L. D. Hatfield, and G. F. Shambaugh for reviewing the manuscript; D. A. Ave, L. D. Hatfield, and M. L. Sullivan for SEM and photographic expertise; K. Knighten for copyright correspondence; L. Fulton for patiently typing the manuscript; and D. L. Shankland and M. S. Blum for involving me in this project.

REFERENCES

M. A. Ali, Ed., *Sensory Ecology: Review and Perspectives*, Ser. A, Vol. 18, Plenum Press, New York, 1978.

H. Altner and L. Prillinger, *Int. Rev. Cytol.* **67**, 69 (1980).

M. Anderson and L. H. Finlayson, *Physiol. Ent.* **3**, 157 (1978).

K. Arikawa and K. Aoki, *J. Comp. Physiol.* **148**, 483 (1982).

V. G. Becker and W. Gerisch, *Z. Ang. Ent.* **84**, 353 (1977).

M. Behan and L. M. Schoonhoven, *Ent. Exp. Appl.* **24**, 163 (1978).

P. Belton, "An analysis of direction finding in male mosquitoes," in L. B. Browne, Ed., *Experimental Analysis of Insect Behavior*, Springer-Verlag, New York, 1974, p. 139.

D. R. Bentley, *Science* **174**, 1139 (1971).

G. D. Bernard and D. G. Stavenga, *J. Comp. Physiol.* **134,** 95 (1979).

F. Blom, *Netherlands J. Zool.* **28,** 277 (1978).

J. Boeckh, V. Boeckh, and A. Kühn, "Further data on the topography and physiology of central olfactory neurons in insects," in J. LeMagnon and P. MacLeod, Eds., *Olfaction and Taste VI,* Information Retrieval, London, 1977, p. 315.

D. Burkhardt and G. Schneider, *Z. Naturforsch.* **126,** 139 (1957).

E. Burkhardt, *J. Comp. Physiol.* **120,** 33 (1977).

R. Butler, *Z. Vergl. Physiol.* **72,** 67 (1971).

S. D. Carlson and C. Chi, *Ann. Rev. Ent.* **24,** 379 (1979).

O. L. Chapman, J. A. Klun, K. C. Mattes, R. S. Sheridan, and S. Maini, *Science* **201,** 926 (1978).

C. Chi and S. D. Carlson, *J. Morph.* **161,** 309 (1979).

E. E. Davis and P. G. Sokolove, *J. Comp. Physiol.* **96,** 223 (1975).

G. DeBoer, V. G. Dethier, and L. M. Schoonhoven, *Ent. Exp. Appl.* **21,** 287 (1977).

C. J. Den Otter, H. A. Schuil, and A. Sander-Van Oosten, *Ent. Exp. Appl.* **24,** 370 (1978).

V. G. Dethier, *The Physiology of Insect Senses,* Methuen, New York, 1964.

V. G. Dethier, *J. Comp. Physiol.* **82,** 103 (1973).

V. G. Dethier, *The Hungry Fly,* Harvard University Press, Cambridge and London, 1976.

V. G. Dethier and L. M. Schoonhoven, *J. Insect Physiol.* **14,** 1049 (1968).

K. Dumpert and W. Gnatzy, *J. Comp. Physiol.* **122,** 9 (1977).

H. Eckert, *J. Comp. Physiol.* **135,** 29 (1980).

A. Ewing, *Physiol. Ent.* **3,** 33 (1978).

M. B. Fenton and J. H. Fullard, *J. Comp. Physiol.* **132,** 77 (1979).

L. Frantsevich and P. Mokrushov, *J. Comp. Physiol.* **120,** 203 (1977).

J. H. Fullard, M. B. Fenton, and J. A. Simmons, *Can. J. Zool.* **57,** 647 (1979).

R. P. Futrelle, *Trends in Neurosci.* **7,** 116 (1984).

K. P. Gaffal, H. Tichy, J. Theib, and G. Seelinger, *Zoomorphologie* **82,** 79 (1975).

M. Gewecke and J. Philippen, *Physiol. Ent.* **3,** 43 (1978).

M. Gogala, "Ecosensory functions in insects (with remarks on Arachida)," in M. A. Ali, Ed., *Sensory Ecology Review and Perspectives,* Series A, Vol. 18, Plenum Press, New York, 1978, p. 123.

A. I. Goldman and P. J. O'Brien, *Science* **201,** 1026 (1978).

T. H. Goldsmith and G. D. Bernard, "The visual system of insects," in M. Rockstein, Ed., *Physiology of Insecta,* Vol. 2, Academic Press, New York, 1974, p. 165.

L. J. Goodman, in G. A. Horridge, Ed., *The Compound Eye and Vision of Insects,* Clarendon Press, 1975, p. 515.

J. L. Gould, *Amer. Sci.* **68,** 256 (1980).

M. J. R. Hall, *Physiol. Ent.* **5,** 17 (1980).

F. E. Hanson, "The behavioral and neurophysiological basis of food plant selection by lepidopterous larvae" in S. Ahmad, Ed., *Herbivorous Insects,* Academic Press, New York, 1983, p. 3.

G. A. Horridge, *Interneurons: Their Origin, Action, Specificity, Growth, and Plasticity,* W. H. Freeman, San Francisco, 1968.

G. A. Horridge, *The Insects of Australia,* Melbourne Universtiy Press, 1974, p. 5.

G. A. Horridge, "Optical mechanisms of clear zone eyes," in G. A. Horridge, Ed., *The Compound Eye and Vision of Insects,* Clarendon, Oxford, 1975, p. 265.

K. G. Hu, H. Reichart, and W. S. Stark, *J. Comp. Physiol.* **126,** 15 (1978).

T. Ichikawa and H. Tateda, *J. Comp. Physiol.* **149,** 317 (1982).

S. Ishilkawa, T. Hirao, and N. Arai, *Ent. Expt. Appl.* **12**, 544 (1969).

K.-E. Kaissling, "Insect olfaction," in L. M. Biedler, Ed., *Handbook of Sensory Physiology,* Vol. 4, *Chemical Senses,* Part 1, *Olfaction,* Springer-Verlag, New York, 1971, p. 351.

K.-E. Kaissling and G. Kasang, *Naturwissenschaften* **65**, 382 (1978).

Y. Katsuki and N. Suga, *J. Exp. Biol.* **37**, 279 (1960).

K. Klamring, B. Lewis, and A. Eichendorf, *J. Comp. Physiol.* **127**, 109 (1978).

K. L. Kong, Y. M. Fung, and G. S. Wasserman, *Science* **207**, 783 (1980).

A. Kühn, *Z. Vergl. Physiol.* **5**, 762 (1927).

M. F. Land and T. S. Collett, *J. Comp. Physiol.* **89**, 331 (1974).

J. Y. Lea and C. G. Mueller, *J. Comp. Physiol.* **114**, 115 (1977).

W. C. Ma, *Dynamics of Feeding Responses in Pieris brassicae Linn. as a Function of Chemo-sensory Input: A behavioral, ultrastructural, and electrophysiological study,* H. Veenman and Zonen, N. V., Wageningen, 1972.

H. Markl and J. Tautz, *J. Comp. Physiol.* **99**, 79 (1975).

H. Martin and M. Lindauer, *J. Comp. Physiol.* **122**, 145 (1977).

R. Menzel and A. W. Snyder, *J. Comp. Physiol.* **88**, 247 (1974).

A. Michelsen, "Hearing in invertebrates," in W. D. Keidel and W. D. Neff, Eds., *Handbook of Sensory Physiology,* Vol. 5, Part 1, Springer-Verlag, New York, 1974, p. 389.

A. Michelsen, *Amer. Scientist* **67**, 696 (1979).

L. A. Miller, *J. Insect Physiol.* **17**, 491 (1971).

L. A. Miller and J. Olesen, *J. Comp. Physiol.* **131**, 113 (1979).

E. M. Mitchell, Ed., *Management of Insect Pests with Semiochemicals,* Plenum Press, New York, 1981, 513 pp.

A. Moiseff, G. S. Pollack, and R. R. Hoy, *Proc. Nat. Acad. Sci., USA* **75**, 4052 (1978).

A. Morchen, J. Rheinlaender, and J. Schwartzkopff, *Naturwissenschaften* **65**, 656 (1978).

D. M. Norris, *Perception of Behavioral Chemicals,* Elsevier North-Holland Biomedical Press, Amsterdam, 1981.

R. J. O'Connell, *J. Gen. Physiol.* **65**, 179 (1975).

J. Olesen and L. A. Miller, *J. Comp. Physiol.* **131**, 121 (1979).

W. L. Pak, "Study of photoreceptor function using *Drosophila* mutants," in X. O. Breakefield, Ed., *Neurogenetics: Genetic Approaches to the Nervous System,* Elsevier North-Holland, New York, 1979, p. 67.

J. Palka, R. Levine, and M. Schubiger, *J. Comp. Physiol.* **119**, 267 (1977).

J. A. Patterson and R. L. Chappell, *J. Comp. Physiol.* **139**, 25 (1980).

T. Payne, "Pheromone and host odor perception in bark beetles," in T. Narahashi, Ed., *Neurotoxicology of Insecticides and Pheromones,* Plenum Press, New York, 1979, p. 27.

A. V. Popov, *Tr. Uses. Entomol. Obshchest (Horae Soc. Entomol. Rossicae)* **54**, 182 (1969).

R. J. Pumphrey and A. F. Rawdon-Smith, *Proc. Roy. Soc. Lond. Ser. B.* **121**, 18 (1936).

J. Rheinlaender and A. Morchen, *Nature* **281**, 672 (1979).

S. Rilling, M. Mittelstaedt, and K. D. Roeder, *Behavior* **14**, 164 (1959).

K. D. Roeder, *Nerve Cells and Insect Behavior,* No. 4, *Harvard Books in Biology,* Harvard University Press, Cambridge, 1963.

K. D. Roeder, *J. Insect Physiol.* **12**, 843 (1966).

K. D. Roeder, *J. Exptl. Zool.* **194**, 75 (1975).

K. D. Roeder and A. E. Treat, *Amer. Scientist* **49**, 135 (1961).

W. L. Roelofs, *J. Chem. Ecol.* **4**, 685 (1978).

G. A. Rosenthal and D. H. Janzen, Eds., *Herbivores: Their Interactions with Secondary Plant Metabolites.* Academic Press, N.Y., 1979, 718 pp.

S. Rossel, *J. Comp. Physiol.* **131,** 95 (1979).

L. M. Roth and E. R. Willis, *J. Morph.* **91,** 1 (1952).

C. H. F. Rowell, *Z. Vergl. Physiol.* **73,** 167 (1971).

H. Sass, *J. Comp. Physiol.,* **128,** 227 (1978).

P. Schlect, K. Hamdorf, and H. Langer, *J. Comp. Physiol.* **123,** 239 (1978).

L. M. Schoonhoven, "Plant recognition by lepidopterous larvae," in H. F. van Emden, Ed., *Insect/Plant Relationships, (Symposia of The Royal Entomological Society of London, Number Six),* Blackwell Scientific, Oxford, 1972, p. 87.

J. Schwartzkopff, "Mechanoreception," in M. Rockstein, Ed., *Physiology of Insecta,* Vol. 2, 2nd ed., Academic Press, 1974, p. 509.

J. Schwartzkopff, *Ann. Rev. Psychol.* **28,** 61 (1977a).

J. Schwartzkopff, *J. Comp. Physiol.* **120,** 11 (1977b).

W. D. Seabrook, *Ann. Rev. Ent.* **23,** 471 (1978).

A. Shirahashi and Y. Tanabe, *J. Comp. Physiol.* **92,** 161 (1974).

U. Smola, "Voltage noise in insect visual cells," in F. Zettler and R. Weiler, Eds., *Neural Principle in Vision,* Springer-Verlag, New York, 1976 p. 194.

E. Stadler, "Sensory aspects of insect-plant interactions," in J. S. Packer and D. White, Eds., *Proc. 16th Int. Congress Ent.,* Ent. Soc. Amer., College Park, MD, 1977, p. 211.

E. Stadler and F. E. Hanson, *J. Comp. Physiol.* **104,** 97 (1975).

E. Stadler and F. E. Hanson, *Physiol. Ent.* **3,** 121 (1978).

N. V. Strausfeld, "Mosaic organizations, layers, and visual pathways in the insect brain," F. Zettler and R. Weiler, Eds., *Neural Principles in Vision,* Springer-Verlag, New York, 1976, p. 260.

N. Suga, *J. Insect Physiol.* **12,** 1039 (1966).

N. Suga, *J. Auditory Res.* **8,** 129 (1968).

J. Tautz, *J. Comp. Physiol.* **118,** 13 (1977).

J. Tautz, *J. Comp. Physiol.* **125,** 67 (1978).

G. Theophilidis and M. D. Burns, *J. Comp. Physiol.* **131,** 247 (1979).

H. Tischner, *Acustica* **3,** 335 (1953).

J. P. van Praagh, W. Ribi, C. Wehrhahn, and D. Whittmann, *J. Comp. Physiol.* **136,** 263 (1980).

J. N. vander Molen and A. W. J. vander Schaal, *Chemical Senses Flavour* **4,** 341 (1979).

J. H. Visser and A. K. Minks, *Proc. 5th Int. Symp. Insect Plant Relationships,* Wageningen, 1982.

R. Wehner, G. D. Bernard, and E. Geiger, *J. Comp. Physiol.* **104,** 225 (1975).

K. Weise, *J. Comp. Physiol.* **78,** 83 (1972).

H. Wieczorek, *J. Comp. Physiol.* **106,** 153 (1976).

C. A. G. Wiersma and L. L. M. Roach, "Principles in the organization of invertebrate nervous systems," in E. R. Kandel, Ed., *Handbook of Physiology (The Nervous System),* Vol. 1, Part 2, American Physiological Society, Bethesda, MD, 1977, p. 1089.

R. S. Wilcox, *Science* **206,** 1325 (1979).

K. Wise, *Naturwissenschaften* **56,** 575 (1969).

R. Y. Zacharuk, *Ann. Rev. Ent.* **25,** 27 (1980).

9

NERVOUS SYSTEM: FUNCTIONAL ROLE

JOHN L. EATON

Department of Entomology
Virginia Polytechnic Institute and State University
Blacksburg, Virginia

CONTENTS

SUMMARY

The function of an insect's nervous system is to receive and integrate information about changes in its external and internal environment and to coordinate and control organ systems in response to the changes. Taking a broader view, the ability of the nervous system to control complex behaviors demonstrates its integrative capacity. Inputs to the nervous system are via the sensory receptors, and outputs are via skeletal and visceral muscles and the endocrine system. The relationship of the nervous system to the endocrine system is very complex, and the role of sensory input is poorly understood, especially the ability of hormones to influence the activity of the nervous system and thereby change the insect's behavior.

Much of what is known about insect nerve function comes from the study of specific problems. Regarding coordination of skeletal muscles, a picture has developed of pattern generator systems including the oscillating interneurons that pattern the outputs of groups of excitatory and inhibitory neurons to antagonistic muscle groups. The frequency of these oscillating systems may be modulated directly or indirectly by input from sensory receptors which frequently form part of a proprioceptive feedback loop. Ethological concepts have proved useful for the study of complex behaviors, many of which are innate. However, learning, which has been well demonstrated in many hymenopterous insects, is also involved in some complex behaviors. Attempts to reveal learning mechanisms through physiological studies have been less successful, and considerable controversy exists on this subject.

9.1. INTRODUCTION

The function of the insect nervous system may broadly be defined as the reception and integration of information concerning changes in the insect's external and internal environment and the coordination and control of the organ systems in response to the changes. Inputs to the nervous system are through the sensory receptors. Outputs are through the two distinct, but often related, motor and neuroendocrine pathways. Because of the relative simplicity of the insect nervous system, insects have often been chosen as research animals by developmental and neurobiologists. This has resulted in certain areas of insect nerve function receiving far more study than others. However, an effort has been made to provide the reader with a broad, but admittedly incomplete, coverage of the function of the insect nervous system.

One may approach this subject from either the neurobiological point of view of examining the mechanisms at work in the nervous system or from the entomological point of view of examining the control of the various insect activities by the nervous system. The latter approach has been taken here. The reader should be aware that the activities of insects are highly integrated with one another and that the subdividing of the topic treated here has been done to make it more comprehensible.

9.2. NERVOUS SYSTEM CONTROL

9.2.1. Control of Motor Activities

The motor activities of insects are those mediated via the muscular system. The basic pattern of motor output often results from the activity of endogenous pacemakers, groups of interneurons, within the central nervous system (CNS). This basic pattern is modified by inputs from peripheral sensory receptors.

In insects the execution of one or more motor activities depends on the state of the animal's readiness or arousal. Although absent in some insects, such as the stick insect *Carausius,* in most endogenous circadian pacemakers are the major regulators of the daily periodicity of motor activities. These genetically coded endogenous circadian pacemakers are located in the CNS and modulated by the insect's environment. In spite of intensive research on circadian rhythms over the last 25 years, the precise nature of the endogenous circadian pacemakers remains obscure.

In the cockroaches *Leucophaea maderae* and *Periplaneta americana,* the clock controlling locomotor rhythms is believed to be in the optic lobes. In *Drosophila* spp. the eclosion clock is in the anterior end of the pupa, and in pupal *Hyalophora cecropia* and *Antheraea pernyi* it is in the brain. The major factors modulating the periodicity of circadian pacemakers are light and temperature. In *L. maderae* and *P. americana,* changes in photoperiod are detected by the compound eyes. In pupal *H. cecropia* and *A. pernyi,* rhythms are reset by the direct effects of light on the brain (Saunders, 1974; Beck, 1980). While circadian rhythms exert a dominant effect on the daily periods of insect motor activities, other factors also influence the insect's readiness to respond to stimuli during those periods of activity.

The causes of arousal or readiness to respond to stimuli are not well understood in insects (Miller, 1974a,b). Increased arousal may be the result of all sensory inputs acting through specific interneurons and affecting many motor systems, for example, the increased locomotor activity of starved insects. It may also be the result of specific forms of input that produce increased motor activity. For example, in the mating behavior of insects, specific signals such as mating calls or pheromones may stimulate increased motor activity or changes in motor activity. It has been suggested that these effects could be mediated through a command interneuron with a highly selective input and a less selective output, which would "arouse" the insect by affecting the thresholds for activity in motor systems. The movement detector systems or orthopterans, in which specific interneurons are excited by small objects in the visual field, may be an example of such a system (Fraser, 1971; Fraser and O'Shea, 1980).

Changes in motor activity may also be evoked by hormones, as in the initiation of eclosion and adult behaviors in giant silkmoths or of mating behavior by a *corpus allatum* hormone in adults of *Schistocerca* (Truman and Riddiford, 1974).

The level of an insect's response to stimuli may also be depressed under certain conditions. For example, mating, molting, and cryptic behavior may result in insects failing to respond to stimuli that would otherwise be effective.

9.2.2. Control of Locomotion

The major modes of locomotion in insects are walking and flying. In both, centrally generated programs produced by pattern generators are thought to control the basic pattern of activity in motor nerves supplying the muscles of the legs and the wings, while input from peripheral sensory receptor systems on the appendages provides a feedback system for precise control of appendage movement. Additional sensory input to the locomotor systems comes from receptors on the head and elsewhere.

9.2.2.1. Walking. In the most frequently observed gait used by walking insects, the legs form alternating tripods, with the prothoracic and metathoracic legs of one side and the mesothoracic leg of the opposite side contacting the substrate at any given time. Thus, the stepping of the legs of an individual segment is 180° out of phase. Other gaits using all six legs are known, and many insects may walk on four legs, for example, mantids, grasshoppers, and danaid butterflies. In some cases, as in swimming insects, water striders, and lepidopterous larvae (prolegs), the legs of the segment step synchronously.

The neural control of walking involves coordination within and between segmental ganglia as well as the incorporation of information from leg proprioceptors regarding leg position and stresses. In slow walking, phasic reflexes appear to be important in controlling stepping. However, in rapid running, stepping appears to be under central control (Miller, 1974a; Hughes and Mill, 1974; Fourtner, 1976).

The reflex control of slow walking has been studied in the metathoracic leg tibial extensor muscle of *Schistocerca*. This muscle is innervated by fast, slow, and inhibitory axons. In slow walking, only the slow and inhibitory axons are active. The activity of the slow axon is influenced by input from the femoral chordotonal organ and by tarsal sensilla. The common inhibitor neuron is also affected by inputs from the tarsal sensilla. During fast walking, the extensor muscle slow axon fires alternately with slow axons to flexor muscles, and the inhibitory axon fires just before contraction of the flexor muscle. Presumably the inhibitory axon activity accelerates relaxation of the extensor muscle. Abolition of tarsal input reduces activity in the inhibitory axon and causes the slow extensor axon to continue firing during flexion. This indicates that phasic sensory input from leg proprioreceptors helps control the amplitude and timing of motor output by interacting with a central program.

The fast axon of the metathoracic tibial extensor muscle is active only during jumping and kicking. It is known that the activity of slow axons to extensors and flexors increases just before the jump and the firing of the fast axon produces the jump. In *Schistocerca,* courtship kicking is controlled the same way as jumping, but defensive kicking involves only the slow axons. In *Cyrtacanthacris,* whose metathoracic legs are armored, fast axons are also involved in defensive kicking (Miller, 1974a).

In cockroaches, bursts of nonoverlapping motor spikes of patterned activity have been recorded from motor neurons of partially deafferented preparations (Fig. 9-1). This has been interpreted as a central program that can pattern output in the absence of phasic sensory input. The activity patterns recorded from these preparations differ from those of free-walking cockroaches, especially at high stepping frequencies. Such deafferented prepara-

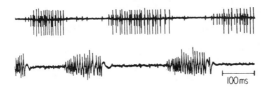

FIGURE 9-1. Reciprocal burst activity in the levator motoneurons 5 and 6 (top trace) and depressor motoneuron D8 (bottom trace) during rhythmic leg movements. Spikes in the top trace were recorded extracellularly from nerve 6Br4. Axons in this nerve innervate the coxal levator muscles. The large- and intermediate-amplitude spikes are from motor axons 6 and 5, respectively. The small spikes are from motor axon 3. The bottom record is an electromyograph from the coxal depressor muscles 177D,E. Each spike is the extracellularly recorded junctional potential elicited by an action potential in motoneuron D8. The variation in amplitude of the spikes is due to facilitation of the EJPs and slight movement of the recording electrodes. [From K. G. Pearson and C. R. Fourtner, *J Neurophysiol.* **38,** 33 (1975).]

tions also show a lower maximum frequency, 15 Hz, compared to normal running animals, 20 Hz (Miller, 1974a).

As in locusts, phasic reflexes exert important effects on the central program of slowly walking cockroaches but are less important in running cockroaches. It has been shown that inputs from trochanteral campaniform sensillae supply a positive feedback to the slow depressor axon during leg retraction. Loading the leg causes this reflex action to increase the thrust of the leg and also slows the stepping frequency. This action suggests that the reflex is important in adjusting to an increased load on the leg as in climbing. Inhibitory fibers may also be under phasic control. Normally, three common inhibitors of coxal depressor muscles fire only during elevation and probably aid relaxation of the depressors. Similarly, leg protractor neurons are inhibited so long as the leg campaniform sensilla are active. Defferentiation causes a reduced stepping frequency and a slowdown of depressor activity between elevator bursts. This suggests that input from campaniform sensilla is important in controlling the activity of the common inhibitor neurons (Miller, 1974a; Pearson and Duysens, 1976).

In coordinating stepping movements between segments, both peripheral reflexes and central interneurons appear important. In intact and deafferented cockroaches, bursts of spikes occur in the mesothoracic levators just after those in the metathoracic levators, thus requiring a central interneuron. However, in denervated preparations, a second mesothoracic burst of spikes to the levators may occur before the second metathoracic burst. This suggests that peripheral sensory input would normally suppress this activity (Miller, 1974a; Hughes and Mill, 1974).

Presently, it is believed that patterns for each leg are generated by endogenous oscillator interneurons and that there is a reciprocal coupling between the oscillators of a segment as well as between those of adjacent segments. The intrasegmental oscillators operate antagonistically in normal walking and synchronously when limbs of segments are moved together. The intensity of the coupling between segments can vary, and increasing the degree of coupling reduces the time for information transfer from one segment to another. Each oscillator is believed to be influenced by afferent inputs from the leg it is controlling. This afferent input has little effect on the activity of adjacent oscillators (Hughes and Mill, 1974).

Four different nonspiking identified* interneurons that appear to be involved in pattern generation have been observed in the cockroach metathoracic ganglion. These interneurons influence either extensor or flexor motor neurons and two have been subjected to extensive study (Fourtner, 1976).

Interneuron I is an intraganglionic neuron lying on the ipsilateral side of the ganglion, whose soma is on the contralateral side of the ganglion

---

* Identified neurons have identical morphologies and can be found in the same location in different preparations of the same species.

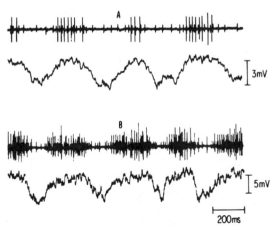

FIGURE 9-2. Intracellular recording from interneuron I during rhythmic leg movements. A and B are two different preparations. Top traces, extracellular recordings from nerve 6Br4; bottom traces, intracellular recordings from interneuron I. In A, the small depolarizations occur at the same time as an increase in activity in motor axon 3 (small spike) and weak bursts of activity in motor axon 5 (large spike). In B depolarizations are larger and motor axon 6 (largest spike) is required. Note the occurrence of IPSPs during the repolarizations phase following the third and fourth depolarizations in B. [From K. G. Pearson and C. R. Fourtner, *J. Neurophysiol.* **38,** 33 (1975).]

(Fourtner, 1976). Its membrane potential oscillates at the frequency of stepping, its depolarizing phase coincides with the bursts of spikes in the flexor motoneurons, and its hyperpolarizing phase corresponds with bursts in the extensor motoneurons (Fig. 9-2). When neuron I is stimulated in a resting animal, the flexor motoneurons are activated. Increasing the stimulus intensity recruits additional slow flexor motoneurons. Stimulating interneuron I with a depolarizing pulse during the normal hyperpolarizing phase resets the rhythm of leg movements.

Interneuron II also oscillates during leg movements. However, it is 180° out of phase with interneuron I, hyperpolarizing during flexion and depolarizing during extension. Neither the ability of interneuron II to reset the rhythm nor its morphology have been determined. A second interneuron type (morphology also unknown), referred to as a *command integrating neuron,* has also been recorded in the metathorax. Depolarization of this interneuron produces alternating bursts in both flexor and extensor motor neurons.

Based on this and other information, a model for the central pattern generator of the cockroach metathoracic leg has been proposed (Fig. 9-3). In this model, nonspiking interneurons I and II form part of a rhythm pattern generator that exhibits excitatory and inhibitory influences over the extensor motor neurons. There is also a possible reciprocal inhibitory interaction between interneurons I and II that would facilitate their oscillations. The

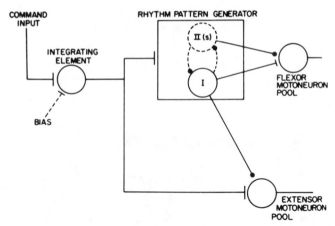

FIGURE 9-3. Hypothetical model of the central mechanism producing rhythmic leg movements in the cockroach. The broken lines suggest very tentative proposals since there is no direct evidence that interneuron II is part of the rhythm pattern generator, nor is there evidence that interneurons I and II are linked via reciprocal inhibition. The bias input in the integrating element is input from segmental afferents and other central neurons. Lines ——| designate excitatory connections and ——● designate inhibitor connections. (From C. R. Fourtner, in R. M. Herman, S. Grillner, P. S. G. Stein, and D. G. Stuart, Eds., *Neural Control of Locomotion,* Plenum Press, New York, 1976.)

command integrating interneuron interacts between command input and the rhythm pattern generator. In this system, because of the delay incurred in the rhythm pattern generator, the extensor motor neurons would be activated first, followed by the flexor motor neurons. Once started, the system would continue to oscillate so long as command input is maintained. A system of interneurons with similar properties has also been observed in the locust metathoracic ganglion (Burrows, 1979).

9.2.2.2. Flying.   The movement of the insect wings is accomplished either by direct or indirect flight musculature. The nerves controlling the flight musculature may show a synchronous or one-to-one relationship to muscle contraction and the rhythm of the flight motor is under neurogenic control (Chapter 8). Alternatively, the relationship of nerve impulses to the flight musculature is asynchronous and the rhythm of the flight motor is under myogenic control. The frequency of wing beat in insects with myogenic rhythms usually exceeds 100 Hz. In myogenic (asynchronous) fliers, the wing-beat frequency is determined by mechanical conditions involving the thoracic exoskeleton and the fibrillar flight muscles. Myogenic rhythms are found in members of the orders Thysanoptera, Psocoptera, Heteroptera, Homoptera, Hymenoptera, Apocrita, Diptera, and Coleoptera. Due to their large size, Heteroptera and Coleoptera, while retaining the myogenic mechanism, have wing-beat frequencies below 100 Hz (Pringle 1974, 1976).

In most insects, the loss of tarsal contact with the substrate or the "tarsal reflex" causes the unfolding of the wings and subsequent flight. Some insects must be in the proper physiological condition for the tarsal reflex to be effective. For example, Lepidoptera must have a high enough thoracic temperature for flight initiation. In cockroaches (*Periplaneta*) and locusts (*Schistocerca*), the wings are opened by the direct flight muscles. In *Oncopeltus,* the wing base must be released by moving the prothoracic lobe before a group of nonfibrillar muscles contracts to open the wings and the asynchronous flight muscles are activated. In *Calliphora,* the pleural wall must first be stiffened to permit proper wing articulation before the tergosternal muscles can contract, producing a simultaneous jump and wing opening. This is immediately followed by activation of motor nerves to the flight motor, causing flight to begin. Although some insects will fly on a tether in still air, most require antennal sensory stimulation to maintain flight. This can be input from additional sensory receptors (Johnston's organ) and from sensory hairs on the head. Visual information is also important. In bees and muscid flies, moving stripes are sufficient to maintain flight in still air (Pringle, 1974).

In neurogenic (synchronous) fliers such as *Schistocerca,* the flight motor pattern controlling wing beat is produced by two reciprocally active groups of motoneurons. These motoneurons are believed to be activated by coupled interneuronal oscillators (pattern generators) which synapse with the motoneurons causing them to produce bursts of action potentials at the wing-beat frequency (Burrows, 1973). The presence of these oscillating interneurons has been inferred from several studies. It is known, for example, that the motoneurons of deafferented preparations show patterned excitatory postsynaptic potential (EPSP) activity at the flight frequency in the complete absence of motor spikes. This activity is presumably produced by interneurons. It is also known that larval locusts, whose wings cannot be moved, produce patterned spikes in wing muscles. Furthermore, the wing-beat frequency of adults increases with maturation, suggesting the maturity of a pattern generator system.

Candidate pattern-generating interneurons have also been inferred from electrophysiological studies of motoneurons of the left and right forewing and in ipsilateral tergosternal elevator muscles of locusts. These motoneurons show a dual pattern of EPSP activity, which appears to be related to both ventilation and wing beat frequencies. One phase of the activity pattern is a slow depolarization which coincides with expiration and is a part of the ventilatory rhythm. Superimposed on the depolarizing phase of the slow rhythm is a fast rhythm of EPSPs occurring at about the wing-beat frequency of 20 Hz (Fig. 9-4). These slow and fast rhythms are known to occur in 30 flight motoneurons and ventilatory motoneurons. Presently, it is believed that a pair of mesothoracic interneurons contact and may be generating the rhythms in both flight and ventilatory motoneurons (Burrows, 1973, 1976).

FIGURE 9-4. Common synaptic inputs to the tergosternal motoneurons of the fore wings and hind wings. (a) Ipsilateral meso- and metathoracic tergosternal motoneurons are depolarized together in both slow and fast rhythms. One complete depolarizing phase of the slow rhythm is shown. [From M. Burrows, *J. Exp. Biol.* **63**, 713 (1973).]

Studies of the motoneurons to spiracular opener and closer muscles have shown that they are under control of these interneurons.

It is not so clear that these interneurons are flight pattern generators. The fast rhythm does occur in flight motoneurons and can cause spikes in the motoneurons under flight conditions. Furthermore, a rhythmic modulation of the spike frequency in a wing elevator muscle (=flight frequency) is tied to the ventilatory rhythm. These data suggest that one pair of interneurons may serve a dual function in flight and ventilation. A similar system of interneuron pattern generators for flight has also been implicated in dragonflies (Simmons, 1977). These results are consistent with the conclusion that the flight rhythm is generated by interneuronal pattern generators (Miller, 1974b).

Peripheral control of wing-beat amplitude frequency is exerted by sensory receptors on the wing. A well-studied example is a stretch receptor at the base of each wing of locusts which monitors wing elevation. This receptor discharges with increased frequency at the peak of wing elevation. The receptor system is believed to influence wing beat through monosynaptic excitatory contacts with wing depressor motoneurons and inhibitory contacts with wing elevator motoneurons. Each receptor makes ipsilateral contacts with motoneurons of both wings, but not with contralateral wings (Pringle, 1976). Wing depression receptors are also known. These sensory receptors evoke excitatory potentials in ipsilateral depressor motoneurons. They also exert contralateral effects, presumably via interneurons, on wing motoneurons of the same or the adjacent segment.

Changes in the speed of forward flight or power are controlled through input from receptors on the antennae and sensory hairs on the head (Gewecke, 1974; Miller, 1974a). Input from these receptors influences both wing-beat amplitude and frequency in locusts, flies, and bees. In *Locusta*, stimulation of mechanoreceptors on the antennal pedicel resulting from bending of the antennae reduces flight speed. Stimulation of wind-sensitive hairs on the frons and vertex increases flight speed. Apparently, flight speed is controlled via the interaction of these two receptor systems.

Control of power and direction can be achieved by direct means through modifications of the pattern generator or by indirect means. Power can be adjusted by recruitment of fresh motor units or by changes in the amplitude of the wing stroke (Lepidoptera) or angle of incidence of the wing (*Schistocerca*). Turning and correcting for yaw can be achieved in unilateral changes in the frequency of the motor rhythm or by unilateral changes in the spike-burst length in motoneurons. Unilateral changes in phase are also used in turning. In Lepidoptera, a delay is seen in the firing of a basalar muscle motoneuron to the wing on the inside of the turn relative to the same motoneuron on the outside of the turn.

Indirect methods may also be used to control direction. In asynchronous fliers such as *Calliphora,* tonically active muscles control wing beat-frequency, amplitude, angle of incidence, and unfolding. The legs and the abdomen may also be used for steering, as in *Schistocerca* (Miller, 1974a).

9.2.2.3. The Cockroach Evasive Response. In cockroaches the evasive running behavior elicited by puffs of air detected by cercal sensory receptors has long been thought to be mediated, in part at least, by the giant fibers' interneurons of the ventral nerve cord. Recent studies show that cercal stimulation with wind puffs produces trains of spikes in giant fibers with frequencies of 200–350 Hz. Studies of the giant fiber–motoneuron pathway using higher stimulus frequencies showed that stimulation of single giant fibers with 15 pulse trains at 340 Hz produces spikes in identified motoneurons demonstrating that both small and giant fibers appear to be involved in the evasive response. Furthermore, the cercal–giant fiber system also has been shown to convey information about wind direction, thus permitting the cockroach to determine the source of stimulation and turn away from it (Ritzman and Camhi, 1978; Camhi, 1980).

9.2.3. Control of Feeding

Studies of neural control of feeding have largely concerned the role of external and internal sensory receptors in providing information to regulate food intake. Two species, *Phormia regina* and *Locusta migratoria,* have been studied extensively in this context (Barton-Browne, 1975).

Most insects show increased locomotor activity after food deprivation. Termed *locomotor preingestion activity,* this is the first of a series of events controlled by the nervous system and related to feeding. The pattern of preingestion locomotion may contain random elements but is also likely to be directed by information from antennal olfactory receptors. In *Phormia,* once the fly locates a food source, stimulation of tarsal chemosensory hairs provides information to CNS interneurons to initiate final orientation to the food source, terminate locomotion, and promote the extension of the proboscis to probe the food. This is termed *nonlocomotor preingestion behavior.* This series of events requires transfer of information from the thoracic ganglion to the subesophageal ganglion. It is not known whether the brain is involved in this behavior.

Electrophysiological studies of elements of proboscis extension support the results of behavior studies. The threshold for proboscis extension due to tarsal stimulation with sugars is high immediately after feeding and declines with time. This relationship is apparent in electrophysiological studies where activity in tarsal sugar receptors and in motoneurons innervating the proboscis extensor–adductor complex are recorded simultaneously. In flies starved for 60 hr, the first two spikes in the stimulated sugar receptor must occur with a 5-msec interval to elicit spikes in the proboscis motoneuron (Fig. 9-5). After 72 hr of starvation, the motoneuron activity was elicited by spikes 25 msec apart, indicating a decrease in the threshold of the tarsal sugar receptor system (Getting, 1971).

Stimulation of tarsal sugar receptors alone is insufficient to maintain proboscis extension. This is apparently due to both peripheral and central adaption mechanisms. Studies with single hairs have shown that the interspike interval of the sugar receptor increases to more than 20 msec in 1–13 sec after stimulation, depending on sugar concentration, indicating adaptation at the sensory receptor. These levels of activity are insufficient to maintain proboscis extension. When a single hair was allowed time for recovery from adaptation by stimulating it at 1-min intervals, proboscis extension still was not maintained, suggesting that central adaptation also occurs. Changing the pathway by stimulation of a different hair restored the response, thus confirming central adaptation and not sensory adaptation or motor fatigue as the cause.

The threshold for proboscis extension is also influenced by spatial summation. The simultaneous application of sugar solutions to two tarsal sugar receptor sensilla results in synergism, presumably through central summation of the motoneuron response. Similarly, stimulation of two sensilla with sugar and salt solutions produces central interaction in the form of both summation and inhibition with respect to proboscis extension (Barton-Browne, 1975; Dethier, 1976).

Stimulation of the chemosensory hairs on the labellum leads to the next

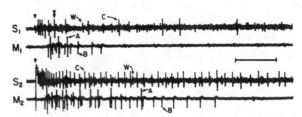

FIGURE 9-5. Electrophysiological records of sucrose stimulation of a single labellar sensillum and the resultant motor activity in the extensor–adductor muscle complex of a water-satiated fly starved 72 hr. S1 and M1, receptor and motor activity, respectively, in response to 40 mM sucrose. The two spike sizes in the sensory records represent activity in the sugar (C) and water (W) receptors. Muscle activity in the extensor–adductor complex also consists of two spike sizes, large (A) and small (B). The single arrow at the beginning of both sensory traces marks the instant solution contacts with the sensillum. The double arrow in S1 marks the last sugar spike for which the sensory interspike interval (ISI) is less than 20 msec. Note the cessation of motor activity in M2 despite a constant sensory ISI of less than 20 msec. Time mark, 100 msec. [From P. A. Getting, *Z. Vergl. Physiol.* **74**, 103 (1971).]

step in the feeding process of *Phormia,* the spreading of the labellar lobes. The electrophysiology of this system has been studied by recording from furcal retractor muscles and their motor nerves. The importance of spatial summation in reducing the response threshold has been shown. Studies reveal that simple increases in the number of hairs stimulated with sugar solutions are not adequate to cause opening of the labellar lobes. Rather, it is the combination of hairs which is important. Stimulation of labellar hairs generally only causes opening of the ipsilateral lobe.

Opening the labellar lobes brings the interpseudotracheal papillae into contact with the sugar solution, and input from these papillae via the subesophageal ganglia and the brain activates the sucking pump so ingestion can begin. This initiates ingestion behavior in both the tsetse fly and in *Calliphora.* The sucking pump of these flies is powered by the reciprocal action of ventral and dorsal dilator muscles inserting on the pump. Ventral and dorsal stretch receptors in the epithelium of the anterior pharyngeal wall also are excited by actual or simulated pumping. The labro-pharyngeal nerve conveys afferent information from the stretch receptors and motor activity to the muscles of the pump.

A model for pumping has been proposed. In it information from intrapseudotracheal papillae causes a central interneuron(s) to be activated and in turn activates the motoneuron to the ventral pump dialator muscle (Fig. 9-6). Fluid is sucked into the ventral pharnyx and the ventral stretch receptors are activated. Afferent information from the ventral stretch receptor excites the

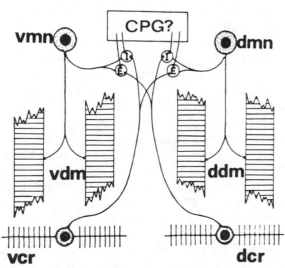

FIGURE 9-6. Model for control of cibarial pumping. CPG, central pattern generation; E, excitatory synapses; I, inhibitory synapses; dmn, dorsal motoneuron; vmn, ventral motoneuron; ddm, dorsal dialator muscle; vdm, ventral dialator muscle; dcr, dorsal cibarial pump receptor; vcr, ventral cibarial pump receptor. [Modified with permission from M. J. Rice, *J. Insect Physiol* **16,** 277 (1970).]

dorsal motoneurons and inhibits the ventral motoneurons. Thus, as the dorsal dialator contracts, the ventral dialator relaxes. Stimulation of the dorsal stretch receptor reverses the cycle and it begins again. This model provides a means for adjusting pumping rate to changes in food volume and viscosity and perhaps permits monitoring the volume of food ingested (Dethier, 1976).

Additional information regarding control mechanisms suggests that a central pattern generator may be an important part of the pump system in *Phormia* and likely in other insects. Severing the labro-pharyngeal nerve of flies blocks afferent information from stretch receptors. If the labral receptors of such flies are stimulated, the motoneurons still produce bursts of activity, but of irregular duration. This suggests that a central pattern generator may generate the basic pumping pattern and that feedback from the stretch receptors is needed to maintain it (Barton-Browne, 1975; Dethier, 1976).

Cessation of feeding can be produced by any of three mechanisms: long-term adaptation of peripheral receptors, a response generated by receptors in the digestive tract, and internal action of the food material or a metabolite. In *Phormia* sensory feedback from presumed stretch receptors in the foregut and the abdominal body wall is known to play a major role in stopping feeding in satiated flies. This action has been demonstrated by transection of the recurrent nerve which innervates the foregut of the fly (Dethier, 1976). Similarly, transection of the nerves to abdominal stretch receptors of the fly also induces hyperphagia. Recordings from the recurrent nerve and the median abdominal nerve show that levels of sensory input from foregut receptors and abdominal receptors increase with increasing distension of the gut, supporting the hypothesis that their inputs are important in controlling food intake (Barton-Browne, 1975; Dethier, 1976).

In *Locusta*, anterior foregut stretch receptors and sensory adaptation appear important in termination of feeding. In addition, a hormone released by the *corpora cardiaca* causes pore closure on palpal chemoreceptors, blocking sensory input for up to 2 hr after feeding. Hormones have also been implicated in prevention of food-seeking behavior. These behavioral reactions have been produced by simulating feeding through injecting agar into the foregut or *corpora cardiaca* homogenates into the hemocoel. That injection of blood from recently fed insects will delay the beginning of the next feeding in *Locusta* further suggests a hormonal control over feeding behavior (Bernays and Chapman, 1974). In *Rhodnius prolixus* and *Aedes aegypti*, abdominal stretch receptors are important in terminating feeding (Barton-Browne, 1975).

Determination of food palatability is yet another aspect of the neural control of feeding and affects both feeding and meal size. It is determined peripherally by the sensory receptor systems. Apparently the acceptability of a material is checked at several stages in the process leading to ingestion. For example, in *Phormia,* stimulation of the labellar hairs with L-arabinose will elicit proboscis extension and opening of the labellar lobes. L-Arabinose is not, however, acceptable to labellar papillae, and ingestion does not occur (Dethier, 1976). Similarly, stimulation of oral chemoreceptors of locusts by deterrent materials from plants reduces meal size and deters subsequent feeding on plants containing the deterrent compounds (Barton-Browne, 1975).

9.2.4. Control of Ventilation

Ventilation, or the movement of air through the tracheal system, involves neural control of the pumping movements of the body, which produce expiration and inspiration, and of the opening and closing of the spiracles (Miller, 1966) (see section 9.2.2.2).

9.2.4.1. Control of Abdominal Pumping. In most insects where active ventilation of the tracheal system occurs, the hardening of the head and pterothorax has relegated this function to the abdomen. Electrophysiological recordings from isolated abdominal nerve cords or lateral nerves of abdominal ganglia reveal bursts of activity that correspond in frequency to intact ventilatory nerves, suggesting a centrally generated rhythm. These studies have also shown that the embryonic third abdominal ganglion contains the most anterior and important pacemaker for ventilation.

In locusts and dragonflies, larval abdominal pumping movements are produced by alternate excitation of antagonistic expiratory and inspiratory segmental muscles. In cockroaches and other insects, only expiratory muscles are present, and inspiration is the result of abdominal elasticity. The ventilatory system of the locust is best known and will serve as an example.

In *Schistocerca* each abdominal ganglion has two lateral branches (N1 and N2) and a median nerve. In resting locusts ventilation takes place in the dorsoventral plane. Branch N2 innervates dorsoventral expiratory muscles of the same segment. The inspiratory muscles are innervated by the median nerve of the preceding segment. Thus, two ganglia innervate the antagonistic muscles of one segment. In stressed locusts, the ventilation pattern changes to longitudinal telescoping movements of the abdomen. Under those conditions, N1 innervates longitudinal expiratory muscles and N2 innervates inspiratory muscles. Recordings from peripheral nerves and from muscles indicate that some muscles have two slow axons. Some evidence of inhibitory axons is also seen (Lewis et al., 1973).

Recordings from N2 and the median nerve of resting locusts show synchronous bursts of activity that do not overlap. Lesion experiments, which isolate the abdominal and thoracic CNS or remove the brain, reduce, but do not abolish, the burst frequency.

Intersegmental coordination has been observed by simultaneous recordings from homologous nerves from four segments. These studies show that the third abdominal ganglion generally is the initiator of expiratory and inspiratory bursts. Recordings from abdominal connectives also reveal bursts of spikes phase-locked to abdominal expiratory bursts. The left and right connectives seem to be independent, suggesting that each connective contains an interneuron that extends through the abdominal nerve cord and is activated in the third abdominal ganglion.

A model consisting of three interneurons and the expiratory and inspiratory motoneurons has been proposed to describe the control of abdominal ventilation in the locusts (Fig. 9-7). In it, interneuron 1 is a burst-producing interneuron(s) in the metathoracic ganglion which receives excitatory and inhibitory input from command fibers which respond to carbon dioxide, oxygen, and other factors at other sites in the CNS. Interneuron 1 inhibits interneurons 2, the coordination interneurons that extend the length of the abdomen in each of the two connectives. Interneurons 2 synapse with a small interneuron 3, which in turn sends branches to excite expiratory moto-

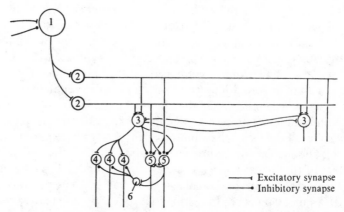

FIGURE 9-7. A model which can account for some features of the coordinated ventilatory activity in the abdomen of an intact locust. Only a few of the motoneurons of one segment are shown. Cell I is a burst-producing interneuron, or group of interneurons, in the metathoracic ganglion which receives excitatory and inhibitory input from anterior centers. It produces bursts of impulses of relatively constant duration and these establish the inspiratory phase; interbursts of variable duration establish the expiratory phase. Bursts of impulses produced by cell I inhibit the auto-activity of two coordinating interneurons, one in each connective (2). These cells run the length of the abdomen and in each ganglion they synapse with a small interneuron (3) which sums their activity and distributes it to expiratory motoneurons of both sides (4). Thus cells 2 drive cells 4 via cell 3, which also strongly inhibits the inspiratory motoneurons (5). Cells 2 also directly but weakly inhibit the inspiratory motoneurons. When inhibition by cell 3 is lifted, the inspiratory motoneurons fire as a result of their own endogenous activity and there is weak positive coupling between them. Simultaneously their activity inhibits the expiratory motoneurons via an interneuron (6). The expiratory motoneurons too may be positively coupled to each other but this has not been indicated. The pattern is repeated in each segment. In addition, further connections between neighboring cells 3 are shown and these account for the coordination of ventilation when cells 1 and 2 are inactive (e.g., after removal of GIII). When not driven by cells 2 they are capable of endogenous burst formation. Their short bursts determine the expiratory strokes which are separated by long periods of firing by the inspiratory motoneurons. [From Lewis et al., *J. Exp. Biol.* **59**, 149 (1973).]

neurons and other branches to inhibit motoneurons. After inhibition by inter-neuron 3 declines, the inspiratory motoneurons produce an endogenously generated burst to the inspiratory muscles and simultaneously inhibit the excitatory muscles. Interneuron 3 is also capable of endogenous burst pro-duction, and coupling between these interneurons could explain the coordi-nation of ventilation when the nerve cord is severed behind the metathoracic ganglion (Lewis et al., 1973).

9.2.4.2. Control of the Spiracles. Innervation of thoracic spiracles occurs via the median nerve of the segmental ganglion or of the next anterior gan-glion. The axons of the neurons, one or two, supplying the closer muscles divide into right and left branches, which follow the transverse nerves. The closer muscle of each spiracle accordingly receives the same pattern of motor impulses. Sensory nerves running from the area of the spiracles into the CNS have also been described. A similar pattern is evident in the abdo-men, with the cell bodies lying in the ganglion of the segment anterior to the spiracle and the axons following the connectives to the next segment before exiting in the branches of the median nerve (Miller, 1966).

Spiracle opening and closing may be controlled separately and sometimes locally by chemical means. The spiracles may also be controlled by a central mechanism that coordinates their movements with ventilatory pumping of the body.

The mechanisms regulating opening and closing have been studied in spiracle 1 of *Schistocerca,* which has both opener and closer muscles. The closer muscle receives two motor axons from the prothoracic ganglion. The opener muscle also receives two motor axons from the prothoracic ganglion and a third motor axon from the mesothoracic ganglion. Normally the spira-cle opens slowly and stays open longer than it is closed. During expiration the prothoracic opener motor nerves often discharge at the same time as the closer motor nerves. Their activity continues into the inspiratory phase of ventilation, and the opener motor nerve also becomes active. This may maintain a state of tetanus in the opener, causing the valve to open wider when the closer relaxes. Increasing CO_2 levels in the brain, prothoracic ganglion, or mesothoracic ganglion stimulate activity in the opener nerve. In this system the spiracle opening and closing appears due to the action of the closer muscle, and the degree of opening depends on the opener muscle which responds to CNS stimulation.

Some spiracles like spiracle 2 of *Schistocerca* have only closer muscles. This spiracle is innervated by two axons referred to as fast and slow, and both power and control are exercised through the closer muscle. In spiracles of this type, CO_2 acts as a peripheral control mechanism by directly causing the closer muscle to relax and permit the spiracle to open.

Synchronization of spiracular opening and closing with abdominal pump-ing provides a mechanism for unidirectional ventilation of the tracheal sys-tem. *Schistocerca* again provides a model. Here, synchronized activity of the two thoracic spiracles results from input from mesothoracic ganglion interneurons. Probably these interneurons are somehow coupled to those coordinating abdominal pumping movements described above (Miller, 1966).

9.2.5. Control of Sound Production

Different sound-producing mechanisms for intraspecific communication have evolved in different insects. The best known and most studied are found in the Orthoptera and Heteroptera. The neural control of sound production is best known from studies of representative Orthoptera. In this group, sounds are produced by three mechanisms. In the suborder *Ensifera* (Tettigoniidae and Grylloidea), a "file" produced on the undersurface of the wing is moved across the edge (plectrum) of the opposite wing. This action produces a sound upon closure of the wings and also, in some species, upon opening of the wings. In the suborder *Caelifera* (Acridoidea), sound is produced by rubbing a row of pegs on the inner edge of the hind femur against a vein on the elytra. In this group, sound is produced by both upward and downward movement of the femur. The third mechanism (in some acridids) produces sound by opening and closing of the hindwings (Elsner and Popov, 1978). The first two mechanisms are most widespread and best known.

Electromyograms have been extensively and successfully used to study motor coordination during stridulation. Studies of the motor units of single muscles have shown that each is characterized by a discrete type of activity. Three types of activity have been identified. They are phasic (1–3 spikes/stroke), bursting (the activity lasting for the duration of the stroke), and phasic–tonic (continuously active units which vary in activity according to the song pattern). In both crickets and grasshoppers, the motor units of a given muscle have always been activated in the same sequence "in all species and in all types of behavior in which they are involved (stridulation, walking, flight)" (Elsner and Popov, 1978). In sound production, a high degree of muscular coordination results in intense sounds and less coordination produces weak sounds. The degree of coordination varies according to the sound pattern produced.

Electromyographic study of intermuscle coordination has shown that the patterns produced are quite complex and do not simply reflect stridulatory movements at the muscle level. In crickets, on first examination, the wing movement appears to be produced by the action of antagonistic opener and closer muscles. The opener muscles are activated near the end of the elytra's inward motion and the closer muscles activated toward the end of its outward motion. These same muscles function as wing depressors or elevators in flight. Close examination of the activation of the opener muscles from pulse to pulse within a chirp shows that they are activated at progressively later and later times. Thus, the activity of the openers does not reflect the duration of the opening and closing phase. This lack of correspondence between muscle activity and stridulatory movement is also present in grasshoppers. It is particularly complex in slowly (1.1–14 Hz) stridulating species. In fast-stridulating species (50 Hz), muscle activity does tend to reflect stidulatory movements.

In crickets the two elytra operate as a functional unit. A high degree of contralateral coordination exists due to electrical coupling of motoneurons. This system is probably driven by a common interneuron(s).

In grasshoppers sound is produced by two metathoracic legs rubbing against the edge of the elytra. These movements are also strongly coupled. There is, however, a phase lag with one side leading the other. Through time the lead may shift from one leg to the other, especially between song sequences (Elsner and Popov, 1978).

There is little doubt that central program generators are of fundamental importance in insect sound generation. The pro- and mesothoracic ganglia (crickets and tettigoniids) and the metathoracic ganglia (grasshoppers) are the probable locations of central oscillators. At the intrasegmental level, studies with *Gryllus campestris* have shown that pulses within chirps are produced by the alternating activity of mesothoracic motoneurons to wing opener and closer muscles. The opener motoneurons initiate the sequence, and 3–7 msec after the first spike appears in it, the closer motoneuron hyperpolarizes and then depolarizes. After the second and succeeding depolarization in a chirp, an IPSP appears in the opener motoneuron in the interpulse interval. This reciprocal activity continues until the end of the chirp. A model proposed to explain the pulse pattern suggests that the opener and closer motoneurons are coupled via interneurons which inhibit both closers and openers. A chirp-timing slow oscillator starts the activity by exciting the opener motoneuron. Afterward, the reciprocal activity of the opener and closer motoneurons is produced through interneurons which inhibit both openers and closers. This system could also be important in terminating the chirp as well as the development of refractoriness in the opener motoneuron.

Insofar as control of chirp generation is concerned, evidence indicates a network of interneurons which undergo alternating activity. Interneurons have been observed which show bursting-spike activity during chirps and hyperpolarization between chirps. Other interneurons are hyperpolarized during chirps and produce bursts of spikes between chirps. The exact relationship of these units to chirp production remains to be determined.

While studies of stridulation indicate that some elements of control occur at the level of single ganglia, these networks are not autonomous. Input from areas both anterior and posterior to the thorax are known to exert significant effects on stridulatory behavior. For example, the presence of a spermatophore, signaled via proprioreceptors in the genitalia of male crickets, serves to induce calling and courtship songs. Similarly, the rivalry song, which is detected by tympanic organs, is induced by the calling of nearby conspecific males. Severing the nervous system of crickets at the neck connectives or between the meso- and metathoracic ganglia blocks calling, but after a few weeks "spontaneous" calling returns. This has led to the suggestion that calls can be produced autonomously. However, if the connectives on both sides of the thorax are severed, "spontaneous" calling does not occur even after electrical stimulation.

Studies in which the brain has been subjected to electrical stimulation suggest that the central body and the *corpora pedunculata* are involved in song control. In contrast to the findings that songs could be generated in thoracic ganglia separated from the brain by severing cervical connectives, brain interneurons have been recorded whose descending activity in the ventral nerve cord of *G. campestris* corresponds to song patterns. A correspondence between chirp and respiratory rhythms has also been observed.

From these data it is evident that both peripheral input and brain activity affect stridulatory behavior. The exact relationships are yet to be explained (Elsner and Popov, 1978).

9.2.6. Control of Ecdysis and Eclosion

Neural control of ecdysis and eclosion may be divided into two phases. The first, initiation, occurs in response to environmental stimuli and/or endogenous circadian rhythms. Accordingly, insects (crickets and locusts) preparing for ecdysis may show increased locomotor activity until a suitable site is located, when presumably it is terminated by the appropriate sensory input. In other insects, initiation is under the control of an endogenous circadian clock. In the pupae of the moths. *Antheraea pernyi* and *Manduca sexta* and of the fly *Drosophila pseudoobscura,* the clock is believed to be in the head, probably the brain, and produces a diel rhythm of eclosion (Truman, 1980a). Therefore, the timing of the initiation of ecdysis or eclosion is determined by the response of the nervous system to suitable exogenous or endogenous stimuli. This response results in release of the prothoracicotropic neurohormone, which initiates the succeeding events in ecdysis and eclosion (see Chapter 13, Section 13.5.5).

The second phase of neural control of ecdysis and eclosion concerns the neural control of the body movements necessary to shed the old cuticle (Truman, 1980a). Ecdysis and eclosion are complex processes, and the insect is very vulnerable at this stage. It should not be surprising then that precise motor programs have been developed for shedding the old cuticle.

Neural control of ecdysis has been studied in crickets and locusts. In the cricket, *Telogryllus oceanus,* ecdysis has been divided into four phases: a preparatory phase, an ecdysial phase, an expansional phase, and an exuvial phase. Each phase can be further subdivided into a set of motor programs. A motor program is a series of contractions in a set of muscles which accomplish a phase of the movements required for ecdysis. These motor programs must occur in a set sequence to produce a successful ecdysis. One example in the crickets is the motor program controlling withdrawal of the forelegs from the old exocuticle. Here, bilaterally homologous muscles undergo alternating contractions at about 3-sec intervals to withdraw the legs. These coordinated, patterned movements suggest the existence of a central pattern generator.

Superimposed on the phases and the motor programs is a division of ecdysial behavior into "bouts" of activity separated by quiet "interbouts." These periods of activity and rest vary in frequency and do not seem to be under rigid control.

Adult eclosion of saturniid and sphinx moths may also be divided into phases. In these moths, patterns of motor activity associated with escape from the pupal cuticle are induced by the action of eclosion hormone on the nervous system (see Section 13.5.5, Chapter 13). The three phases of *H. cecropia* are preeclosion behavior (rotatory movements of the abdomen), eclosion behavior (peristaltic contractions of the abdomen), and wing inflation behavior. In isolated *H. cecropia* abdominal nervous systems, the first two behaviors are prepatterned into the nervous system and released by eclosion hormone. Thus, application of the hormone initiates bursts of activity in the nervous system which correspond first to preeclosion behavior and then to eclosion behavior (Reynolds, 1980).

Each abdominal ganglion contains preeclosion behavior pattern-generating neurons which respond to eclosion hormone. However, the patterns generated by the more anterior ganglia occur at a higher frequency than those of posterior ganglia. This suggests that the anterior ganglia drive preeclosion behavior. In eclosion behavior, motor patterning varies with the site of hormone application. Ganglia 3 and 4 produce an abnormal response when treated with eclosion hormone. This suggests that, under normal circumstances, the sixth abdominal ganglion drives the pattern generators in the other ganglia during eclosion behavior.

The fact that two different sites drive preeclosion and eclosion suggests these two behaviors may be independent and further that their temporal organizations depend on the relative latencies of their responses to eclosion hormone. In aged isolated abdomens, the ability to show preeclosion and eclosion behavior declines with age sequentially. Thus, some abdomens which will not produce preeclosion behavior do produce eclosion behavior. Similarly, low concentrations of eclosion hormone will induce preeclosion but no eclosion behavior. Addition of more hormone will induce eclosion behavior. These studies further support the concept of two essentially independent pattern generators (Truman, 1980b).

9.2.7. Control of the Digestive Tract

Movements of the digestive tracts of insects are the result of intrinsic myogenic rhythms and neural control systems which may override the myogenic system to regulate the force and duration of contractions and also operate the valves which regulate the movements of food through the gut. The stomodeum (esophagus, pharynx, crop) and the anterior region of the mesenteron are innervated by fibers from the stomatogastric nervous system which includes the frontal, hypocerebral, and ingluvial or ventricular ganglia. The brain is also involved in controlling gut movements. For example, in *Phormia,* as mentioned in Section 9.2.3, the nerves controlling the extrinsic dorsal and ventral dialators of the sucking pump originate in the brain (Rice, 1970). Similarly, in locusts, foregut muscles are innervated by tritocerebral neurons which pass through the stomatogastric nervous system (Evans, 1980). The frontal connectives joining the brain and the frontal ganglion also provide circumstantial evidence for the brain's role in controlling movements of the foregut and receiving sensory input from it.

Generally, the frontal ganglion is considered the center of activity in the stomatogastric nervous system. It receives sensory input from the esophagus and pharynx. In *Oryctes,* it contains about 80 interneurons and motoneurons, plus a sensory neuropile. Interneurons or motoneurons pass from the frontal ganglion through recurrent nerves to the hypocerebral ganglia where the interneurons synapse with motoneurons and the motoneurons continue to muscles of the foregut. The recurrent nerve may continue posteriorly to the ingluvial ganglia (Orthoptera) or the ventricular ganglion (Lepidoptera). Here branches extend to the posterior areas of the foregut, the proventricular region, and the anterior midgut (Evans, 1980; Miller, 1975).

Experiments in which various nerves in the stomatogastric nervous system have been severed indicate that peristaltic movements of the foregut are largely under myogenic control and that the stomatogastric nervous system serves to coordinate the strength and type of movements of the foregut by overriding the intrinsic myogenic rhythms.

The proctodeum and possibly the posterior region of the midgut are innervated by branches of the proctodeal nerves which arise in *Periplaneta* as branches of the sixth abdominal ganglion and in *M. sexta* larvae from the terminal abdominal ganglion (Miller, 1975; Reinecke et al., 1973). In the cockroach, the proctodeal muscles receive a polyneuronal innervation from the proctodeal nerves. Stimulation of these nerves produces gut contractions at frequencies proportional to stimulus rate, up to 50 Hz (Evans, 1980). Studies of the nerve endings on the hindgut show neurosecretomotor endings which may indicate the presence of either B type aminergic or peptidergic transmitter substances. Application of exogenously applied materials suggests that either glutamate or the peptide proctolin may be synaptic transmitters in the hindgut.

9.2.8. Control of Circulation

In insects the dorsal vessel is the major organ involved in the circulation of hemolymph. Rhythmic contractions of the posterior region of the dorsal vessel or the heart provide the force for the movement of hemolymph through the heart and into the anterior aorta. In a few insects, the heart is not believed to be innervated and its pulsations are believed to have myogenic origin (larval *Tipula*, larval and pupal *Anopheles*).

In those insects whose dorsal vessels are innervated, fundamental patterns of contraction are thought to be myogenic with neural or hormonal modulation (Evans, 1980). Nerve fibers to the dorsal vessel may come from one or two locations: the lateral cardiac nerves and/or the segmental ganglia (Jones, 1977).

The lateral cardiac nerves of *Periplaneta* arise from the hypocerebral ganglion of the stomatogastric nervous system. The hypocerebral ganglion also sends a short median fiber to the aorta. Lateral branches from the hypocerebral ganglion pass to the *corpora cardiaca–allata* complex and then extend posteriorly as lateral cardiac nerves, parallel to the aorta and toward the heart. About 40 axons in each branch of the nerve appear to be both neurosecretomotor and nonneurosecretory. Similar lateral cardiac nerves have been reported from other orthopterans (*Blaberus, Melanoplus,* and *Locusta*), Odonata, and Hemiptera. Fibers which innervate the aorta from the stomatogastric nervous system are also known from Diptera, Lepidoptera, and Hymenoptera (Jones, 1977).

Segmental cardiac nerves (Orthoptera, Lepidoptera, Diptera, and Hymenoptera) arise from the segmental ganglia, often as branches of the unpaired median nerves. The branches of these segmental nerves may innervate either the alary muscles (alary nerves) or the heart (segmental cardiac nerves).

Cardiac ganglion cells, whose cell bodies are in the lateral cardiac nerves of *Periplaneta,* are thought to act as motorneurons for the heart. Neurosecretory neurons in the lateral cardiac nerves have their cell bodies where the segmental nerves join the neurons of lateral cardiac nerves. Both neurons show spontaneous activity. The segmental nerve contains one nonneurosecretory and two neurosecretory fibers. One of the neurosecretory fibers does not synapse with the heart but apparently releases its contents into the hemolymph near the lateral cardiac nerve. The other neurosecretory neuron synapses with heart muscle. The nonneurosecretory fiber also appears to synapse with the cardiac ganglion cells, but not the heart muscle.

In the cockroach, the heart is innervated directly by two types of intrinsic cardiac neurons which are in turn modulated by the CNS. Nevertheless, complete denervation of the heart of *Periplaneta* does not stop the heartbeat, suggesting that it is myogenic with a superimposed neural control system. On the other hand, in most insects stimulation of the segmental nerves causes cardioacceleration (Evans, 1980). By contrast, in the moth *Manduca sexta,* there is evidence that the heart may respond to heating of the thorax with increased amplitudes of pumping. In addition, transection of the abdominal nerve cord at the second abdominal ganglion decreases the amplitude of heart contractions and suggests that these effects are mediated via the segmental nerves to the heart (Heinrich, 1971). It has also been noted that hemolymph circulation is aided by the ventral diaphragm and by accessory pulsatile organs, but no information is available on the neural control of these organs (Miller, 1975).

9.2.9. Control of the Organs of Reproduction

To describe the neural control of insect reproductive organs, the motor activity of the muscles of both external and internal genitalia must be considered. Specific descriptions of neural control of the genitalia muscles are very scarce.

Both exogenous factors (pheromones and other sensory stimuli) and endogenous factors (hormones) influence the readiness for and initiation of copulatory behavior. It is known, for example, that the subesophageal ganglion of cockroaches and other insects is a center for inhibiting copulation. Apparently, a hormone from the *corpora cardiaca* acting on the subesophageal ganglion can release this inhibition. Patterned activity in the phallic nerve of male cockroaches results, which causes rhythmic twisting movements similar to those seen during copulation (Miller, 1975).

In both male and female insects, visceral muscles of the internal genitalia may aid sperm or spermatophore movement and storage. Similarly, the glands associated with the genitalia are often musculated, but details of their innervation and neural control are lacking.

Muscles are also important in the movement of the egg during oviposition. In *Locusta,* the sixth abdominal ganglion controls the muscle involved in oviposition. Sensory input must also be important as evidenced by the sensory receptors found on ovipositors of some flies (Rice, 1976). Decapitation induces oviposition in *Mantis religiosa* and *Aedes aegypti,* suggesting the existence of a control system analogous to that controlling copulation in male insects (Miller, 1975).

9.3. INTERACTIONS WITH THE ENDOCRINE SYSTEM

The importance of the neurosecretory cells of the brain as a site for production and release of hormones concerned with molting and metamorphosis is well known (see Chapter 13). It is also recognized that neurosecretory cells are distributed in smaller numbers in the other segmental ganglia and that neurosecretions from the brain and the ventral ganglia serve diverse functions. The secretions and functions include glandotropic neurohormones, which affect endocrine glands; morphogenetic neurohormones, which guide ontogenesis; myotropic neurohormones, which affect kinetics of skeletal and visceral muscle; metabolic neurohormones, which influence metabolism; chromotropic neurohormones, which affect color change; and ethotropic neurohormones, which affect the function of the nervous system (Fraenkel, 1980). In addition, the function of the nervous system may also be affected by secretions of the endocrine glands.

9.3.1. Control of Molting and Metamorphosis

Neural control of the processes of molting and metamorphosis involves the action of at least three classes of neurohormones. They are glandotropic neurohormones [brain hormone(s) or PTTH], ethotropic neurohormone (eclosion hormone), and morphogenetic neurohormone (bursicon).

Considering the vast literature that exists regarding the endocrinology of molting and metamorphosis, surprisingly little is known about the role of the nervous system in sensing conditions that cause the release of the brain hormones initiating these phenomena. It is known that in *Rhodnius* stimulation of stretch receptors in the abdomen increases activity of nerves to the *corpora cardiaca,* which presumably leads to the release of the brain hormone. In *Locusta* stimulation of pharyngeal stretch receptors is thought to bring about brain hormone release. External stimuli may also be involved. For example, in bee larvae contact with the cocoon induces pupation. In *M. sexta,* brain hormone release occurs when the larva reaches a critical body weight. In other insects, crowding or disturbance has been shown to reduce growth and molting, suggesting that a suitable sensory input is required for normal development (Wigglesworth, 1972; Ishizaki and Suzaki, 1980) (see Chapter 10, Section 10.11.2).

The release of the brain hormone(s) is followed by a series of events, some of which involve the CNS and may be initiated by it. Brain hormone directly affects the *corpora allata* and the prothoracic glands, stimulating the release of juvenile hormone and ecdysone, which guide the production of the new epidermis (see Chapter 13). When the new cuticle is complete, release of the eclosion hormone initiates the process of shedding the old cuticle. This release of eclosion hormone is precisely timed by an environmentally entrained "clock" in the brain (Reynolds and Truman, 1980). The action of eclosion hormone on the nervous system has been discussed previously (Section 9.2.9). Following molting, the third neurohormone, bursicon, is released from the CNS and causes tanning of the new cuticle (Tobe and Stay, 1981). The control of the release of these hormones is not well understood, but undoubtedly some type of nervous control is involved.

9.3.2. Neuroendocrine Effects on Internal Organ Systems

The products of the neurosecretory cells of the brain pass down the axons of the first and second *corpora cardiaca* nerves to the *corpora cardiaca.* Similarly, the segmental neurosecretory cells send axons to segmental neurohemal organs which may provide release sites. Intrinsic cells of the *corpora cardiaca* may produce neurohormones, as may neurohemal organs as well. In addition, many axons arising from the median nerve contain neurosecretory granules and innervate skeletal muscles and visceral muscles.

The products of the neurosecretory cells and neurohemal organs are only beginning to become known. Extracts of the *corpora cardiaca* have resulted in the isolation of numerous unidentified materials, some of which are presumably neurohormones, that have a variety of myotropic and metabolic effects including cardioacceleration, hyperglycemic effects, and adipokinetic effects (Kammer and Rheuben, 1981).

Of special interest are several materials whose structures are known and which have received considerable recent attention. They are the biogenic amines—5-hydroxytryptamine (5HT), dopamine, and octopamine—and the peptide proctolin. 5-Hydroxytryptamine and octopamine, found in both the CNS and in neurohemal organs, affect a number of processes. Both 5HT and octopamine appear to modulate the activity of skeletal and visceral muscle. Furthermore, octopamine may influence the metabolic rate of skeletal muscle and does affect the activity of light organs. In addition, octopamine injected into insects also raises the level of excitation of the insect and its levels increase in the hemolymph during flight (Kammer and Rheuben, 1981). Dopamine and 5HT influence the activity of salivary glands. All three biogenic amines are active in the CNS, where they serve as neurotransmitters or modulators of activity. It has been suggested, although the evidence is not strong, that 5HT's effects follow a diel rhythm (Evans, 1980). Thus, insect activity could be differentially modulated throughout the day.

Proctolin is produced and secreted by the terminal abdominal ganglion. Recent immunological studies suggest that it is also distributed in other segmental ganglia, the brain, and peripheral nerves. It affects the activity of visceral muscle—including heart, rectal, and foregut muscles of *P. americana*—and skeletal muscle, for example, the extensor tibia of *Schistocerca*. It has been described as either a neurotransmitter or a modulator in the CNS and peripherally (Miller, 1983).

9.3.3. Hormonal Effects on Behavior

During insect development the nervous system also grows. It adds neurons and, particularly in the case of Endopterygota, undergoes vast structural changes related to changing functions. These changes are influenced by the action of hormones, which in several cases also induce temporary or permanent changes in behavior of the insect (see Chapter 10).

9.3.3.1. Effects on Developmental Behavior. *Schistocerca* larvae approaching the molt show decreased activity. This change correlates with an increase in interneuron activity in thoracic ganglia and a decrease in motoneuron activity. In intermolt larvae, these same effects can be produced by treating the CNS with blood from molting larvae high in ecdysone and 20-hydroxyecdysone. Similarly, the decline in juvenile hormone (JH) levels in the mature larvae of *Mimas tiliae* is believed to cause the cessation of feeding behavior and the initiation of positive geotaxis associated with seeking a pupation site. Implantation of active *corpora allata* to serve as a JH source

reverses this behavior. Hormones have also been implicated in the premeta-morphosis behavior of *M. sexta* (Riddiford and Truman, 1974).

The initiation of eclosion behavior by the action of eclosion hormone on the CNS of lepidopterous larvae has already been discussed (Section 9.2.9). This same hormone apparently "turns off" pupal behavior and "turns on" adult behavior. In this case, the loss of pupal behavior is accompanied by a loss of certain "pupal" interneurons and muscles and the activation of adult nerve centers (Riddiford and Truman, 1974). Thus, in these cases, the CNS loses the ability to direct one set of functions and assumes the ability to direct another set, sometimes permanently.

9.3.3.2. Effects on Reproductive Behavior. For certain grasshoppers and locusts with a reproductive dispause to express mating behavior also re-quires an increase in the titer of JH. In *Locusta* this is apparently brought about by activation of C-type neurosecretory cells of the *pars intercerebralis* under proper environmental conditions. The JH, in turn, apparently acti-vates CNS nerve centers that direct mating behavior. This system controls the onset of mating behavior so that it only occurs under the proper environ-mental conditions.

The control by the *corpora allata* and JH of sexual receptivity in female insects shows considerable variation. In some, the *corpora allata* are not necessary and in others they are essential. In the cockroach, *Byrsotria fumi-gata,* allatectomized females do not produce sex pheromone but are other-wise capable of mating. On the other hand, allectectomy of the acridid *Gomphocerus* prevents onset of female receptivity. Treatment with JH ana-logs reverses this condition. Although there is some variability, a similar situation seems to exist in certain cockroaches and certain Diptera. In some species, products of the neurosecretory cells may contribute to bringing about behavioral changes associated with female receptivity (Riddiford and Truman, 1974).

Hormonal effects, possibly from the neurosecretory hormones, on the nervous system may affect the initiation of oviposition behavior of orthop-terans, hemipterans, coleopterans, and lepidopterans. In the bug, *Iphita,* injection of hemolymph from females ready to oviposit into females in which the eggs have not matured induces premature oviposition. Similar effects are induced by injection of extracts of *corpora cardiaca* from ovipositing fe-males into virgin females of *Schistocerca.*

9.3.3.3. Effects on Migratory and Diapause Behavior. Insects that migrate or enter diapause cease feeding and reproductive behavior during these peri-ods. In the case of migration, locomotor behavior greatly increases; during diapause, the insect become quiescent. In both cases, environmental stimuli such as photoperiod, temperature, and food availability cause the patterns of activity expressed by the nervous system to undergo a drastic change. These effects on the CNS are believed to be caused by changing hormone levels. Possible involvement of hormones from the neurosecretory cells of the *pars intercerabralis* and from the *corpora allata* and elsewhere has been suggested for *Melolontha melolontha, Leptinotarsa decemlineata, Locusta migratoria, Danaus plexippus,* and *Oncopeltus fasciatus* (Rankin, 1978) (see Chapter 12, Section 12.6).

The migration of *O. fasciatus* is one of the best-studied examples (Rankin, 1974, 1978). In these bugs, photoperiod and food availability interact to produce changes in the level of JH in the blood. A short photoperiod induces an adult reproductive diapause and subsequent migratory flights. A long photoperiod and high temperature inhibit flight and stimulate reproductive activity. Experimental studies suggest JH production declines as days shorten in the fall, bringing on reproductive diapause and southward migration. In the spring, longer days and higher temperatures permit JH production to increase, and this results in migratory flights followed by mating and ovarian development.

In insects that migrate, environmental stimuli apparently provide the cues for changing the level of neuroendocrine and endocrine activity. These changes, in turn, cause changes in the activity of the CNS and other organ systems. The changes in the activity of the CNS are reflected as changes in behavior.

In considering neural involvement in diapause, discussion will be limited to photoperiodically induced diapause of insect larvae, pupae, and adults. The events surrounding this type of diapause closely parallel those described above for migration.

In larvae and pupae, diapause is usually induced by exposure of the developing insect to short photoperiods, although in some species, such as the European corn borer, *Ostrinia nubilalis,* and the pink bollworm, *Pectinophora gossypiella,* periodic low temperatures have also been shown to induce diapause (Beck, 1980). Continuous exposure to long days prevents diapause in insects with a photoperiodically induced diapause. The best evidence to date indicates that the changes in photoperiod associated with diapause induction are detected by extraocular receptors located in the insect head and presumably the brain (Chippendale, 1977). The eyes and the ocelli do not appear to be directly involved.

In larvae, exposure to short photoperiods in the egg and/or the larval stage initiates diapause. In the rice stem borer, *Chilo suppressalis,* the larval *corpora allata* secrete JH, and blood titers of JH remain high during diapause. These high levels of JH are thought to inhibit the brain–prothoracic gland system and inhibit development. A similar control over larval diapause apparently also exists for the southwestern corn borer, *Diatraea grandiosella.* High JH titers were found in diapause-induced European corn borer larvae and codling moth larvae. Some researchers believe that it plays a role in diapause induction (Beck, 1980; Chippendale, 1977) (see Chapter 13, Section 13.5.6).

Termination of larval diapause is enhanced by long-day photoperiods and increasing temperatures. Little information is available on the ways in which these environmental parameters are detected, but apparently they lead to the release of PTTH from the brain neurosecretory cells and the subsequent production of ecdysone. In the European corn borer, the ileum of the hindgut has been suggested as the source of hormonelike material, procto-done, which stimulates the brain neurosecretory cells to produce and/or release PTTH. This hypothesis is not universally accepted.

In pupal diapause, short-day effects on preceding larval stages are de-tected by means of unidentified extraocular photoreceptors. Their effect is to stop the production of PTTH and JH. In *Hyalophora cecropia* the absence of PTTH prevents activation of the prothoracic glands and the release of ecdy-sone, which stops the insect's development toward the adult stage (Beck, 1980). Indeed, the importance of the absence of ecdysone in pupal diapause is revealed by injections of ecdysteroids, which initiate development. Diapause termination is, in turn, often correlated with long days and rising temperatures. These factors exert an effect on the neuroendocrine system of the brain which must result in PTTH production and/or release. On the other hand, debrained pupae remain in a permanent state of diapause (Beck, 1980).

Diapause in adult insects, which is associated with a suppression of repro-ductive activities and gonadal development, may be induced by changes in photoperiod or by temperature. A similar reproductive diapause is present in migrating insects as previously indicated in this section. These changes are believed to be due to stimulation of the neuroendocrine system.

Adult insects entering diapause also show significant behavioral changes, indicating that the nervous system is affected by the diapause state. Feeding, locomotion, and sensory responses may be different during diapause behav-ior. Colorado potato beetles, *Leptinotarsa decemlineata*, normally posi-tively phototactic and active feeders, become negatively phototactic and cease feeding. Several other beetles exhibit a similar behavior as they enter diapause. The convergent ladybird beetle, *Hippodamia convergens,* exhibits both migration and diapause behavior in response to the long photoperiods of midsummer. In the spring, short photoperiods and warmer temperatures produce a migration from diapause sites to feeding and reproduction sites. The photoperiodic changes result in changes in the hemolymph JH levels produced by the brain which affect these behavioral changes (Rankin, 1978; Rankin and Rankin, 1980).

9.4. INSTINCTIVE CONTROL OF COMPLEX BEHAVIORS

This chapter has dealt primarily with neural control of portions of specific activities revealing, as far as possible, the specific neurons, muscles, and endocrine glands involved and their relationships to one another. The per-formance of complex behaviors by insects is, however, dependent on the ability to assemble various specific activities into groups that permit the insect to complete the behavior involved. Although some insects' behaviors may be learned, in the vast majority of cases specific behaviors are controlled through genetically determined instinctive or innate behaviors.

The concepts and terms of ethology provide a useful means of explaining these behaviors. A specific behavior may consist of one or several "fixed-action patterns" (FAPs). The expression of the FAPs depends on a combination of factors, including developmental state and motivational and release-controlling factors (Markel, 1974). Examples of FAPs are found in grooming movements of beetles and cockroaches and nest construction activities of several insects. The characteristics of these activities are that they are highly stereotyped and initiated by "releasers" or "sign stimuli" (Markel, 1974; Matthews and Matthews, 1978). Furthermore, the same nerves and muscles are active in the same sequence each time the behavior is performed. This implies that they are initiated by the same external or internal stimuli each time and, once initiated, proceed to completion without additional input. For example, insects groom a missing appendage as if it were present, thus feedback is not required. Also, the same motions will be used each time the missing appendage is groomed.

As indicated above, the expression of FAPs depends on developmental state. Thus, adults do not express larval FAPs and vice versa. The importance of hormones in the expression of behavior was discussed in Section 9.3. The proper releasers or sign stimuli must also be present. This has led to the concept that for each FAP there is an "innate releasing mechanism" (IRM). The IRM is the mechanism for summing all the internal and external inputs and determining the thresholds for release of the FAP. Thresholds are dependent on the motivation of the insect. A deprived insect will have a lower threshold than a satiated one. Since a specified behavior pattern often consists of a sequence of FAPs, for each to proceed, the prior step must be completed and so signaled by its IRM.

In most cases, the completion of a behavior sequence is terminated by means of a feedback loop. An example of such a loop is the input provided by gut stretch receptors to terminate feeding (see Section 9.2.3).

9.5. LEARNING

Learning has been defined as "any enduring or relatively permanent change in behavior that occurs as a result of experience or practice" (Markel, 1974). Memory is a requisite for learning, since previous experiences must be encoded or stored as a part of the learning process. Studies of insect behavior have resulted in the definition of three types of learning. The first is habituation, which is loss of responsiveness to stimuli that are harmless or unavoidable. Habituation is distinguished from sensory or neural adaptation or motor fatigue by its increased duration. A second type of insect learning is

associative learning. It is defined as the "ability to form associations between previously meaningless stimuli and reinforcements such as reward
and punishment." The training of honeybees to associate colors with the
presence of nectar is an example of associative learning. The third type of
learning is latent learning. It differs from associative learning in that no
obvious reinforcement exists. An example of latent learning is the ability of
bees and wasps to learn the location of their nest through the recognition of
landmarks (Markel, 1974).

Attempts to uncover learning mechanisms through physiological studies
have met with only limited success, and a considerable amount of controversy surrounds the subject. In some of the earliest studies, cockroaches
were shown to be capable of learning to avoid shocks to any of their six legs.
This shock-avoidance training was carried out for 45 min and lasted from 15
min to 24 hr or longer after the end of the training period. It was also
reported that training can be transferred from one ganglion to another during
the avoidance training. Furthermore, by progressive removal of portions of
the nervous system, preparations that contain only a single ganglion were
shown to be capable of being trained. However, attempts to train split ganglia have been unsuccessful (Eisenstein, 1972).

In another series of experiments with *Schistocerca,* shock-avoidance
training produced an increase in the spike frequency in the motor nerve to
coxal adductor muscles each time the spontaneous discharge in the muscle
fell below a set threshold. These results have been questioned by other
workers who were unable to repeat them (Miller, 1974). Others have criticized the whole concept of the study of isolated preparations on the grounds
that such preparations may behave differently than intact preparations and,
therefore, they should be prefaced by behavioral studies of whole animals
(Menzel et al., 1974).

It has been suggested that administration of the protein-synthesis inhibitor, cyclohexamide, decreased learning in cockroaches. However, these
studies have been criticized based on the lack of adequate controls and the
secondary effects of cyclohexamide. Similarly, changes in acetylcholine esterase and γ-aminobutyric acid have been reported during leg learning in
cockroaches. These studies have also been criticized because of the difficulties in accurately measuring levels of enzymes and transmitters in small
tissue samples and inability of others to repeat them (Miller, 1974a). Recent
studies have also failed to demonstrate the ability of the cockroach leg
preparation to achieve a memory duration greater than 15 min, and it has
been suggested that it is an unsuitable subject for studies of memory
(Willner, 1978).

9.6. CONCLUSIONS

Advances in techniques such as intracellular recording from interneurons,
iontophoretic injection of dyes to visualize neurons, and new methods for

the identification of chemical messengers have contributed to major advances in our knowledge of the function of the insect nervous system. They have also caused a reassessment of the view that, because it contains a small number of neurons, the insect nervous system is simple and correspondingly easy to understand. It is now known that the connections made by neurons with one another and the mechanisms for information transfer between them are highly complex. Nevertheless, many challenging problems remain to be studied, among which the following are particularly significant: (1) the mechanisms involved in changing the level of arousal and in modulating activity in response to daily changes in environmental conditions are poorly understood and impact on many insect activities; (2) newly discovered and yet-to-be discovered chemical messengers including neurotransmitters, neuromodulators, and neurohormones must be studied in order to reveal their role in function of the insect nervous system; and (3) the relationship of the nervous system and the endocrine system, particularly the pathways involved in the control of hormone release by the nervous system and the effects of the endocrine system on the nervous system, also need more study. Increasing knowledge of the function of the insect nervous system will undoubtedly be of considerable significance in comparative neurobiology and may be of major importance in insect control as well.

ACKNOWLEDGMENTS

I would like to thank J. Camhi, A. E. Kammer, T. A. Miller, and J. W. Truman, for providing preprints and other useful information. Thanks also to D. E. Mullins and D. G. Cochran for their critical reviews and to M. E. Erickson for typing the manuscript.

REFERENCES

L. Barton-Browne, "Regulatory mechanisms in insect feeding," in J. E. Berridge and V. B. Wigglesworth, Eds., *Advances in Insect Physiology,* Vol. 2, Academic Press, New York, 1975, pp. 1–116.

S. D. Beck, *Insect Photoperiodism,* Academic Press, New York, 1980.

E. A. Bernays and R. F. Chapman, "Regulation of food intake by acridids," in L. Barton-Browne, Ed., *Experimental Analysis of Insect Behavior,* Springer-Verlag, New York, 1974, pp. 48–59.

M. Burrows, *J. Exp. Biol.* **63,** 713 (1973).

M. Burrows, "Neural control of flight in the locust," in R. M. Herman, S. Grillner, P. G. S. Stein, and D. G. Stuart, Eds., *Neural Control of Locomotion,* Plenum Press, New York, 1976, pp. 419–438.

M. Burrows, *J. Neurophysiol.* **42,** 1108 (1979).

J. Camhi, *Scientific Amer.* **243,** 158 (1980).

G. M. Chippendale, *Ann. Rev. Ent.* **22,** 121 (1977).

V. G. Dethier, *The Hungry Fly,* Harvard University Press, Cambridge, Mass., 1976.

E. M. Eisenstein, "Learning and memory in isolated insect ganglia," in J. E. Treherne, M. J. Berridge, and V. B. Wigglesworth, Eds., *Advances in Insect Physiology,* Vol. 9, Academic Press, New York, 1972, pp. 112–181.

N. Elsner and A. V. Popov, "Neuroethology of acoustic communication," in J. E. Treherne, M. J. Berridge, and V. B. Wigglesworth, Eds., *Advances in Insect Physiology,* Vol. 13, Academic Press, New York, 1978, pp. 229–355.

P. D. Evans, "Biogenic amines in the insect nervous system," in M. J. Berridge, J. E. Treherne, and V. B. Wigglesworth, Eds., *Advances in Insect Physiology,* Academic Press, New York, 1980, pp. 318–473.

C. R. Fourtner, "Central nervous control of cockroach walking," in R. M. Herman, S. Grillner, P. S. G. Stein, and D. G. Stuart, Eds., *Neural Control of Locomotion,* Plenum Press, New York, 1976, pp. 519–537.

G. S. Fraenkel, "Forward and overview," in T. A. Miller, Ed., *Neurohormonal Techniques in Insects,* Springer-Verlag, New York, 1980, pp. i–xv.

C. H. Fraser Rowell, *Z. Physiol.* **3,** 167 (1971).

C. H. Fraser Rowell and M. O'Shea, *J. Comp. Physiol.* **137,** 233 (1980).

P. A. Getting, *Z. Vergl. Physiol.* **74,** 103 (1971).

M. Gewecke, "The antennae of insects as air-current sense organs and their relationship to the control of flight," in L. Barton-Browne, Ed., *Experimental Analysis of Insect Behavior,* Springer-Verlag, New York, 1974, pp. 100–113.

B. Heinrich, *J. Exp. Biol.* **54,** 153 (1971).

G. M. Hughes and P. J. Mill, "Locomotion: Terrestial," in M. Rockstein, Ed., *The Physiology of Insecta,* Vol. 3, Academic Press, New York, 1974, pp. 335–379.

H. Ishizaki and A. Suzaki, "Prothoracicoptropic hormone," in T. A. Miller, Ed., *Neurohormonal Techniques in Insects,* Springer-Verlag, New York, 1980, pp. 245–276.

J. C. Jones, *The Circulatory System of Insects,* Charles C. Thomas, Springfield, Ill., 1977.

A. E. Kammer and M. B. Rheuben, "Neuromuscular mechanisms of insect flight," in C. F. Herreid and C. R. Fourtner, Eds., *Locomotion and Energetics in Arthropods,* Plenum, New York, 1981, pp. 163–194.

G. W. Lewis, P. L. Miller, and P. S. Mills, *J. Exp. Biol.* **59,** 149 (1973).

H. Markel, "Insect behavior: Functions and mechanisms," in M. Rockstein, Ed., *The Physiology of Insecta,* Vol. 3, Academic Press, New York, 1974, pp. 3–148.

R. W. Matthews and J. R. Matthews, *Insect Behavior,* Wiley, New York, 1978.

R. Menzel, J. Erber, and T. Masuhr, "Learning and memory in the honeybee," in L. Barton-Browne, Ed., *Experimental Analysis of Insect Behavior,* Springer-Verlag, New York, 1974, pp. 195–217.

P. L. Miller, "The regulation of breathing in insects," in J. W. L. Beament, J. E. Treherne, and V. B. Wigglesworth, Eds., *Advances in Insect Physiology,* Vol. 3, Academic Press, New York, 1966, pp. 297–354.

P. L. Miller, "The neural basis of behavior," in J. E. Treherne, Ed., *Insect Neurobiology,* Elsevier, New York, 1974a, pp. 359–430.

P. L. Miller, "Rhythmic activities and the insect nervous system," in L. Barton-Browne, Ed., *The Experimental Analysis of Insect Behavior,* Springer-Verlag, New York, 1974b, pp. 114–138.

T. A. Miller, "Insect visceral muscle," in P. N. R. Usherwood, Ed., *Insect Muscle,* Academic Press, New York, 1975, pp. 545–606.

T. A. Miller, "The properties and pharmacology of proctolin," in R. Downer and H. Laufer, Eds., *Invertebrate Endocrinology,* Vol. I, *Endocrinology of Insects,* A. R. Liss, Inc., New York, 1983, pp. 101–107.

K. G. Pearson and J. Duysens, "Function of segmental reflexes in the control of stepping in cockroaches and cats," in R. M. Herman, S. Grillner, P. S. G. Stein, and D. G. Stuart, Eds., *Neural Control of Locomotion*, Plenum Press, New York, 1976, pp. 519–537.

K. G. Pearson and C. R. Fourtner, *J. Neurophysiol.* **38**, 33 (1975).

J. W. S. Pringle, "Locomotion: Flight," in M. Rockstein, Ed., *The Physiology of Insecta*, Vol. 3, Academic Press, New York, 1974, pp. 433–476.

J. W. S. Pringle, "The muscles and sense organs involved in insect flight," in R. C. Rainey, Ed., *Insect Flight Symposia of the Royal Entomological Society of London*, No. 7, Blackwell Scientific, London, 1976, pp. 3–15.

M. A. Rankin, "The hormonal control of flight in the milkweed bug," in L. Barton-Browne, Ed., *Experimental Analysis of Insect Behavior*, Springer-Verlag, New York, 1974, pp. 317–328.

M. A. Rankin, "Hormonal control of insect migratory behavior," in H. Dingle, Ed., *Evolution of Insect Migration and Diapause*, Springer-Verlag, New York, 1978, pp. 5–32.

M. A. Rankin and S. Rankin, *Biol. Bull.* **158**, 356 (1980).

J. P. Reinecke, B. J. Cook, and T. S. Adams, *Int. J. Insect Morphol. Embryol.* **2**, 177 (1973).

S. E. Reynolds, "Integration of behavior and physiology in ecdysis," in M. J. Berridge, J. E. Treherne, and V. B. Wigglesworth, Eds., *Advances in Insect Physiology*, Academic Press, New York, 1980, pp. 475–595.

S. E. Reynolds and J. W. Truman, "Eclosion hormones," in T. A. Miller, Ed., *Neurohormone Techniques in Insects*, Springer-Verlag, New York, 1980, pp. 169–215.

M. J. Rice, *J. Insect Physiol.* **16**, 277 (1970).

M. J. Rice, *Aust. J. Zool.* **24**, 353 (1976).

L. M. Riddiford and J. W. Truman, "Hormones and insect behavior," in L. Barton-Browne, Ed., *Experimental Analysis of Insect Behavior*, Springer-Verlag, New York, 1974, pp. 286–296.

R. E. Ritzman and J. M. Camhi, *J. Comp. Physiol.* **125**, 305 (1978).

D. S. Saunders, "Circadian rhythms and photoperiodism in insects," in M. Rockstein, Ed., *The Physiology of Insecta*, Vol. 2, Academic Press, New York, 1974, pp. 461–533.

P. Simmons, *J. Exp. Biol.* **71**, 141 (1977).

S. S. Tobe and B. Stay, "Neurosecretions and hormones," in W. J. Bell and K. G. Adiyodi, Eds., *The American Cockroach*, Chapman and Hall, New York, 1981, pp. 305–342.

J. W. Truman, "Eclosion hormone: Its role in coordinating ecdysial events in insects," in M. Locke and D. S. Smith, Eds., *Insect Biology in the Future*, Academic Press, New York, 1980a, pp. 385–401.

J. W. Truman, "Organization and release of stereotyped motor programs from the CNS of an insect," in C. Valverde-Rodriguez and H. Arechiga, Eds., *Comparative Aspects of Neuroendocrine Control of Behavior, Frontiers of Hormone Research*, Vol. 6, S. Karger, Basel, 1980b, pp. 1–15.

J. W. Truman and L. M. Riddiford, "Hormonal mechanisms underlying insect behavior," in J. E. Treherne, M. J. Berridge, and V. B. Wigglesworth, Eds., *Advances in Insect Physiology*, Vol. 10, Academic Press, New York, 1974, pp. 297–352.

V. B. Wigglesworth, *The Principles of Insect Physiology*, Chapman and Hall, London, 1972.

P. Willner, *Animal Learn. Behav.* **6**, 249 (1978).

10

BEHAVIOR AND PHYSIOLOGY

WALTER R. TSCHINKEL
Department of Biological Science
Florida State University
Tallahassee, Florida

CONTENTS

SUMMARY

The physiological causes of behavior can be somewhat arbitrarily divided into behaviors released by specific agents and those associated with more general changes in physiological state. Hormones, the most important specific agents causing behavioral changes, can be subdivided into three somewhat overlapping groups according to their type of action: (1) behavioral releasers, which act directly on the nervous system to cause the appearance of preprogrammed behaviors; (2) switches of behavioral state, which unmask the capacity for certain behaviors whose release then depends on subsequent stimuli; and (3) developmental hormones, which cause the development of new capacities in the insect's nervous system, usually paralleling morphological development. Ecdysteroids and juvenile hormone are the most important hormones of this third type. The hormonal causes of behavior are usually themselves under higher, often extrinsic, control. Two common extrinsic controlling factors are time of day (circadian) and season (often photoperiod), but a variety of other extrinsic cues are also used.

Feeding behavior is a prime example of those associated with general physiological state. Preingestion and ingestion behavior and meal size are regulated by a variety of positive- and negative-feedback cues which cause the insect to seek out food and to feed to a certain state of repletion.

Examples of behaviors affecting physiology are discussed. One, thermoregulatory behavior, regulates the rates of practically all biological reactions as well as many physical processes. Insects accomplish thermoregulation primarily by choosing the most favorable temperature in a heterogeneous microenvironment. In those insects that do not use heat generated by muscular activity, the chief extrinsic source of heat is solar radiation.

In general, the behavioral causes of physiological changes are poorly documented and have received little attention. In certain insects, the behavioral aspects of crowding lead to profound physiological changes, such as phase change in aphids or inhibition of pupation in tenebrionid beetles. Among some social wasps, behavior associated with dominance rank causes changes in their reproductive physiology and behavior.

10.1. INTRODUCTION

While most people are reluctant to believe that humans and closely related animals are living machines, they have little difficulty in adhering to an automaton view of insects. Perhaps for this and other reasons, the insects lend themselves unusually well to the mechanistic explanation of behavior. Consequently, the study of behavioral development in insects is in a rather primitive state, inhibited by the view that insect behavior is "genetically fixed," and that environmental input during development plays no role in the outcome.

Behavior is often the final effector in a developing sequence of physiological and morphological changes, and in this chapter we assume that behavioral development has already taken place. Each set of behaviors is associated with an appropriate internal state; the absence of that state is mirrored in the absence of the set of behaviors. The first part of this chapter, therefore, deals with the physiological correlates of various kinds of behavior, or, by implication, the physiological causes of behavior. Behavioral causation within the nervous system is treated in Chapter 9.

Being accustomed to making a complete separation between the animal and its environment, we have, until relatively recently, tended to overlook the fact that behavior is not only caused by the physiology and morphology of the animal but may also feed back to alter that physiology and even morphology. Behavior and physiology are thus not linked in a unidirectional manner but are two aspects of a reciprocal control system that assures proper conditions for the animal's growth and reproduction. The best examples of such reciprocal causation come from studies of avian reproduction in which it has been shown that the entire chain of behaviors and physiological events—from ovarian development through nest-building, copulation, oviposition, and brooding—was a reciprocating series of effects of hormones on behavior followed by an effect of behavior on hormones. Although there is no equally sophisticated work on insects, the second part of this chapter deals, often speculatively, with how insects' behavior can act on their physiology.

10.2. THE PHYSIOLOGICAL CAUSES OF BEHAVIOR

Broadly speaking, we can divide the physiological causes of behavior into two convenient categories:

1. Those behaviors released by discrete, readily recognized triggers called *hormones*. Hormones are usually released in small quantities by identifiable organs and themselves play no role as metabolites or internal homeostatic compounds.

2. Those behaviors associated with more complex and less readily defined changes in "physiological state." These changes, often homeostatic shifts, are caused by development, nutrition, changes in physical conditions, and so on.

In reality, this dichotomy does not always hold up, and there is much mixing of these behavioral causes. Hormones can cause physiological changes as well as behavioral ones, and the former can in their turn cause the latter. The exact chain of causation has been dissected out in relatively few cases.

10.3. HORMONALLY CAUSED BEHAVIOR

Hormones originate either in endocrine glands (e.g., ecdysone, juvenile hormone) or in the nervous system, in which case they are called *neurosecretory* hormones. To qualify as a hormone, an agent must be secreted in small quantities by an identifiable gland or neuron; be carried by the circulatory system; and exert a physiological, behavioral, or developmental effect on the target tissue(s) or organ(s), including possibly the nervous system. The life of the insect—its development, maintenance and reproduction—involves a tightly orchestrated sequence of hormonal releases and responses which coordinate important events. While there are no horizontal studies of hormonally induced behaviors throughout the life of any single insect, vignettes from various insects allow putting together a plausible whole.

We can distinguish, although somewhat arbitrarily, three kinds of hormonal control over behavior: (1) Hormonal *releasers*—the hormone acts directly on the nervous system to release the behavior, without significant releaser effects from environmental input; (2) hormonal *switchers of behavioral state*—the hormone switches on a capacity for new behaviors, but the behaviors are only expressed on receipt of the appropriate sensory cues or environmental conditions; and (3) hormonal *developmental agents*—the hormone brings about the development of new behavioral states or capacities. These types are discussed in the following sections.

10.4. HORMONAL RELEASERS OF BEHAVIOR

From the point of view of hormonal causation of behavior, releasers represent the tightest link between the behavior and the hormones and would appear to constitute a relatively simple system. At a glance, this "straightforward" relationship is evident, for example, in the pupa–adult eclosion behavior of the saturniid *Hyalophora cecropia*. This behavior consists of three distinct phases that have been readily classified as (1) preeclosion, (2) quiescence, and (3) eclosion (Truman, 1978). Although the ability of an eclosion hormone to *rapidly* trigger these phases of ecdysial behavior in the adult moth is treated in Chapter 11 and will not be discussed further in this section, it should be emphasized that the apparent "simplicity" manifested by this behavioral repertoire is illusory. Indeed, the ecdysis of adult insects appears to be characterized by considerable complexity, a fact that becomes evident if this behavior is subjected to detailed analytical scrutiny.

10.4.1. Complexity of Ecdysial Behavior

The individual behavioral elements of ecdysis in the cricket *Teleogryllus oceanicus* have been analyzed in considerable detail and give an appreciation of the true complexity of the behaviors set into motion by eclosion hormone. The entire sequence is divided into four major phases: (1) a preparatory phase during which the cuticle is split, loosened, and anchored to the substrate; (2) an ecdysial phase during which the insect extracts itself from the cuticle (similar to the eclosion behavior in silkworms); (3) an expansional phase during which the new adult structures (wings, legs, etc.) are inflated and properly folded and tanned; and (4) an exuvial phase during which the exuvium is eaten. Each of these four phases could be further subdivided into specific *motor programs* resulting in particular muscular contractions and therefore behaviors. Forty-eight programs have been identified for *T. oceanicus,* and these occur in a highly coordinated, controlled sequence (Fig. 10-1). Each motor program typically occurs in *bouts,* rather than continuously, with coordination among the bouts of simultaneous motors making the entire organization more complex.

A final point should be made on stereotypy and peripheral feedback. So far, we have dealt with the behavior and motor program as though, once released, it always occurred in identical form with or without sensory feedback. In reality, this is not so, and the motor program played out by the deafferented CNS or otherwise modified pharate adults is not identical to the normal programs. Thus, while the programs may be centrally prepatterned, some depend on *sensory input* to determine the details of how and when the resulting behavior is expressed. The degree to which this is true varies among the motor programs. Organization into phases seems (at least in *T. oceanicus*) to be relatively fixed. Nevertheless, in other insects, completion of one behavioral phase (e.g., escape from the cocoon) is often required for commencement of the next, and prevention of a phase will inhibit the following program. Overall, then, we must conclude that control of ecdysial behavior is the result of central motor programs interacting to varying degrees with sensory feedback. Eclosion hormone sets this entire chain of cause–effect and feedback into motion, probably by acting only on the first step. A num-

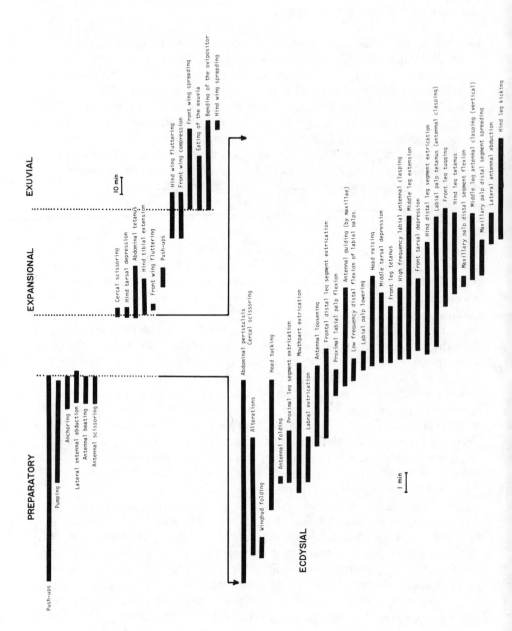

ber of authors have used the metaphor of "orchestration of behavioral events," and this analogy with the movements (phases), instruments (motor programs), and notes or measures (bouts) seems to give the proper air of magnificence to ecdysial behavior (Reynolds, 1980).

10.4.2. Other Examples of Releaser Hormones

A number of behaviors seem to be directly triggered by a hormone, usually a neurohormone. The examples that follow show the scope of such releasers.

10.4.2.1. Copulation. In adult male cockroaches and mantids, the motoneurons that innervate the phallic musculature are in a state of inhibition from the subesophageal ganglion. Severing the nerve cord anywhere between the subesophageal and terminal abdominal ganglia results in a lifting of this inhibition and activates a highly ordered motor program that brings about the coordinated movements of the phallomeres. In mantids, the female's well-publicized flair for devouring the courting male's head and prothorax removes the subesophageal ganglion and therefore the inhibition, resulting in vigorous copulatory movements. Typically, headless males copulate more effectively.

An extract of the *corpora cardiaca* (CC), when either injected or applied directly to the nerve cord, mimics the effect of severing the nerve cord. After a similar time lag, a motor program can be recorded in the phallic nerve that is like that produced by decapitation. The extract is able to elicit phallomeric movements that are quite complex and coordinated and are judged to be "probably" like those occurring during copulation.

Preliminary evidence suggests that the presumed hormone is not a polypeptide. Since how it is released, or whether it is released from the CC at all, is unknown, the evidence for the hormonal nature of this factor is incomplete.

10.4.2.2. "Calling" Behavior. In many moths the female attracts the male for copulation by releasing a sex pheromone from an abdominal gland. Pheromone is released by the eversion of the gland and protrusion of the terminal abdominal segments. This is usually referred to as "calling" behavior, although no sounds, only chemicals, are emitted. For most species such behavior occurs only at certain times of the day and is thus controlled by photoperiod (see Section 10.7.1). For example, virgin polyphemus moths "call" only at night.

FIGURE 10-1. The sequence of motor programs in a typical ecdysis of *T. oceanicus*. The entire 4-hr ecdysis is divided into four phases as shown in the top part of the figure (time calibration, 10 min), but the ecdysial phase is shown below on an expanded time scale (calibration, 1 min). Horizontal bars represent durations, time of recruitment, and termination for the 48 motor programs that can be observed externally. [Modified from J. R. Carlson, *J. Comp. Physiol.* **115**, 299–317 (1977).]

As to more proximal causes of calling, removal of the *corpora allata* (CA) of virgin polyphemus moths has no effect on calling, but when the CC are removed or denervated, the proportion of calling is reduced from 90% (controls) to about 20%. Reimplantation of CA–CC complexes does not restore calling behavior, perhaps because neural connections are necessary. However, blood injected from calling into noncalling females causes the latter to call even in circumstances in which they normally do not. Blood from noncalling females has no such effect. Thus, a *blood-borne factor* releases calling behavior, perhaps by direct action on the CNS, but its source is not clearly established.

Note that the cardiacectomy did not abolish calling behavior in about 15% of the individuals. If this were not the result of technical problems, it could be interpreted in two ways. Possibly the virgin moth population was heterogeneous—that the calling behavior is under CC control in some but not in others. Alternately, perhaps the behavior was not simply under the control of the CC and some other factor(s) acted in an additive or interactive manner with it, the quantitative importance of these factors varying among individuals. Such problems of complete interpretation are quite common but seldom mentioned in reviews. Ignoring them makes conclusions simpler and easier to comprehend, but the real world may actually be as complicated as the results indicate.

10.4.2.3. Oviposition. Egg-laying behavior may also be controlled by a releaser hormone from the CC. In the moth *Hyalophora cecropia* and other Lepidoptera, as well as the blood-sucking bug *Rhodnius prolixus,* virgin females lay some eggs, but this egg-laying rate increases dramatically upon mating. This is not an all-or-none effect, but an increase over baseline rate.

In the female cecropia moth, removal of the CC–CA complex abolishes the increased oviposition rate in response to mating, while allatectomy alone has no effect. As in calling behavior, reimplantation of CC–CA complexes does not restore the oviposition rate. However, injection of blood from ovipositing females increases oviposition rate in base-rate females, indicating the presence of a blood-borne factor. As in calling behavior, evidence that this factor comes from the CC is incomplete.

On the other hand, in *R. prolixus,* an extract of the *pars intercerebralis* cells applied to the oviducts of virgin or mated females brings on violent and rhythmic contractions of the ovarian sheaths and oviducts. Whether the responsible factor functions as a hormone or whether it normally causes oviposition via the observed muscular contractions was not established. The relationship between the *pars intercerebralis* and the CC makes it tempting to conclude that the *Rhodnius* evidence supports the conclusions relating factors affecting the oviposition in the moth.

Compounds present in the seminal ejaculates of male insects may constitute hormonal releasers for inseminated females. Male crickets transfer the enzyme prostaglandin synthetase during copulation. Prostaglandin stimulates oviposition in several species and may stimulate the smooth muscles of the oviducts directly. If this is correct, it represents a releaser not involving the nervous system directly.

These examples illustrate some of the properties of hormonal releasers of behavior and the situations in which they occur. In the more complete cases we know that the hormone acts directly on the CNS to release the prepatterned motor program, which may play itself out with (usually) or without further environmental, sensory, or physiological input. In the other cases the direct action on the CNS is a presumption. They are included here partly because of the lack of evidence to the contrary and partly because they seem analogous to the more complete cases.

10.4.3. Properties of Neurohormonal Releasers

Most of the neurohormonal factors characterized from vertebrates and insects have proven to be polypeptides or proteins. It is likely that because many of the hormonal releasers seem to be neurohormones, most will turn out to be polypeptides as well.

Neurosecretory control over behavior combines the advantage of general, nonlocalized action of endocrine systems with the rapid response and neural integration of the nervous system. We might expect such neuroendocrine control over behaviors that require rapid, general effects throughout the nervous system in response to specific stimuli, but in reality the reason a behavior comes under neuroendocrine control is not at all clear.

10.5. HORMONAL SWITCHES OF BEHAVIORAL STATES

The second type of hormone–behavior relationship identified in Section 10.3 is one in which the hormone causes the capacity to carry out a certain behavior (change in behavioral state), but the actual release of the behavior depends on factors such as environmental conditions or other sensory input. In other words, an existing behavioral program is unmasked, or switched on, and becomes responsive to further stimuli. While we are distinguishing such hormones from those that influence the development of the behavior, it seems likely that these two types will show intergradation when more information is available.

In looking for hormonal switchers of behavioral states, it is wise to concentrate on the major transitions in the insect's life. For example, when a juvenile insect undergoes metamorphosis, adults exhibit a suite of behaviors (sex, reproduction, flight, etc.) not characteristic of immatures. Other profitable transitions to examine might be the changes from unmated to mated state, from the feeding to the nonfeeding state in preparation for each molt, from nonmigratory to migratory, or from aggregative to dispersal. A number of these have been shown to involve hormones in some measure.

10.5.1. Adult Behavioral State

Certainly, one of the most dramatic behavioral changes in an endopterygote's life occurs at the time of pupal–adult ecdysis. Evidence that this is
actually a case of switching on a formed behavioral program comes from
experiments with pharate adults that have been experimentally freed from
their pupal cuticles (peeled). Prior to eclosion the pharate adult behaves like
a pupa, showing very few behaviors characteristic of the adult, if the pupal
cuticle is peeled. However, the typical adult behaviors are all established
within a short time after adult eclosion. According to Truman (1978), the
eclosion hormone, in addition to releasing preeclosion and eclosion behavior, throws the switch that turns on the entire assortment of typical adult
behaviors, while at the same time turning off the pupal behaviors.

To establish that there is switching, it is necessary to show that adult
behaviors can indeed be released experimentally in pupae. A number of
investigators have observed low levels of adult behavior in peeled pharate
adults: for example, peeled silkworms show low levels of wing flapping,
walking, leg extension, and leg grasping on the substrate. The neural activity
associated with adult behaviors such as flight or, for that matter, elements of
eclosion behavior can be recorded from developing adults still within the
pupal cuticle and long before ecdysis. In many cases, especially early in the
developmental stage, the neural activity is too weak to cause postsynaptic
potentials and muscular contractions, but it is interesting to speculate that
this neural activity might be important to behavioral development. In any
case, the fact that such behaviors or neural activity *can* occur indicates that
the neural organization needed for these behaviors is already developed
before adult eclosion but is *inhibited* from being expressed.

At least in crickets there is some evidence that the inhibitory center is
located in the CNS and is neural in nature. Larval crickets show moto-
neuronal output typical of singing and flight, although they lack the wings
and other structures necessary to express this behavior. When certain areas
of the brain are lesioned, such larval crickets can be induced to "sing"
prematurely.

After the switch is thrown on adult behaviors, these programs can respond to the proper sensory input or environmental conditions (Section
10.7). Thus, singing by male crickets will now occur at the proper circadian
phase (photoperiodic input) when the male is carrying a spermatophore in its
genital pouch (input from stretch receptors in genital pouch). Likewise,
walking, flight, and reproductive behaviors will all occur under appropriate
cues.

10.5.2. Appearance of Sexual Receptivity in Females

A second dramatic example of behavioral switching is the development of
sexual receptivity or activity in adults. Hormonal control of sexual behavior
seems to be quite heterogeneous (Barth and Lester, 1973).

In female insects gonadal function is independent of sexual behavior and possibly even oviposition. Gonadectomized insects show normal sexual and even ovipositional behavior under appropriate conditions. On the other hand, in many female insects the CA are not essential for female receptivity, although they do affect secretion of pheromone.

As a case study of an insect in which JH may play a role, the mosquito *A. aegypti* is instructive (Fig. 10-2). When the female mosquito first emerges from the pupa, she is refractory to insemination for 40–50 hr. While she may

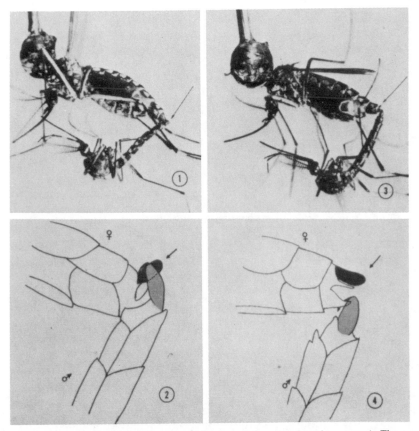

FIGURE 10-2. (A) Copulation by a sexually mature female of *Aedes aegypti*. The arrow indicates the cerci of the female extending from beneath the claspers of the male. This copulation resulted in insemination of the female. (B) Illustration of the degree of genital contact characteristic of copulation by a sexually receptive virgin female. The male claspers (shaded) fully engage the female cerci (dark). (C) Coupling without copulation by a sexually refractory female of *A. aegypti*. The arrow indicates the cerci of the female free and well above the claspers of the male. This coupling did not result in sperm transfer or insemination. (D) Illustration of the superficial genital contact typical of coupling by young refractory virgin females or previously inseminated females. The male claspers (shaded) do not touch the female cerci (dark). Copulation and insemination are impossible in this position. [Reprinted with permission from R. W. Gwadz et al., *Biol. Bull.* **140**, 201–214 (1971).]

copulate, no sperm is transferred. Allatectomy extends the refractory period to 12 days, and reimplantation of CA into allatectomized females restores their ability to mate at the normal time. However, implantation of extra CA into newly emerged intact females does not shorten the refractory period. On the other hand, treating newly emerged females with topically applied juvenile hormone analog reduces the refractory period from 44 to about 20 hr.

10.5.3. Sexual Activity in Males

While in most male insects sexual behavior does not seem to be under endocrine control, in some species of locusts and grasshoppers it does. Because they demonstrate the operation of both quantitative and interactive effects of hormones, rather than the usually hoped-for all-or-none and additive effects, these studies are worth discussing in some detail. Table 10-1 summarizes the available information on the effects of allatectomy, neurosecretory cells (NSC) cautery, and therapy on the level of sexual activity in locusts and grasshoppers. From these data, we can draw some tentative conclusions.

First, within the Acrididae, hormonal control over male sexual behavior is a *labile* trait, which even shows different degrees of control between geographic varieties of the same species. This variation probably represents an adaptation to regional conditions. Second, control by the CA is one of *degree* and varies from complete through partial to none. When there is partial control by the CA, such as in *Locusta migratoria migratorioides,* the CA may be under tropic control from the neurosecretory cells (C cells) in the brain. Third, even between the phases of the same variety of locusts, the CA–NSC plays a role only in the gregarious phase, allatectomy having no effect on the solitarious phase. However, the fact that allatectomized gregarious and intact solitarious males have about the same level of sexual activity implies that the CA's role is to enhance the base level of sexual activity caused by the direct action of the NSC hormone. Unfortunately for this hypothesis, NSC cautery and replacement have not been carried out on solitarious locusts. Nonetheless, we can pose this working hypothesis: (1) when hormonal control over sexual behavior evolves, it does so via the NSC cells directly; and (2) when an enhancement of sexual activity and its regulation evolves, the direct action of the NSC products is supplemented with a tropic action on the CA whose product in turn increases sexual activity.

Hormonal control of sexual activity tends to evolve in situations in which sexual activity and reproduction need to be turned on and off and back on at least once in the life of an individual. For example, hormonal control seems to be characteristic of those species showing seasonal reproductive diapause.

An added complication in swarm locusts is that at least two pheromones appear to bring all males in a group to maturity simultaneously (see Section

TABLE 10-1. Relationship of Neuroendocrine System to Level of Sexual Behavior in Locusts and Grasshoppers

Species	Allatectomy	Level of Male Sexual Behavior		
		CA Reimplantation	NSC Cautery	NSC Reimplantation
Crowded *Schisto-cerca gregaria,* immature	None	Normal after 10 days	—	—
Crowded *Schisto-cerca gregaria,* mature	Reduced	—	—	—
Crowded *Noma-dacris septemfas-ciatus*	None	Normal		
Crowded *Locusta migratoria*				
var. *cinerascens*	None	?	None	?
var. *marutensis*	Reduced	?	—	—
var. *migrato-rioides*	Reduced	Normal	None (C cells)	Reduced (?) PI+3CA none
Isolated *L. m. migratorioides*	Normal	Normal	—	—
Oedipoda miniata	Reduced	Normal	—	—
Gomphocerus rufus	Normal	—	Normal	—
Enthystira brachy-ptera	Normal	—	—	—

Source: M. P. Pener, in L. Barton-Browne, Ed., *Experimental Analysis of Insect Behavior,* Springer-Verlag, Berlin, 1975.

14.4.1.1). If such pheromones also affect the level of sexual behavior, they might mediate the control by the CA and NSC. Decisive evidence is missing.

10.5.4. The Switch to Postmating Behavior

In a large variety of insects, stimuli associated with mating terminate sexual receptivity in the female and initiate responsiveness to ovipositional stimuli. We have already dealt with the hormone that mediates the response to ovipositional cues in the cecropia moth. How is the cecropia female switched to the responsive state in which additional stimuli bring about the release of the hypothetical "oviposition hormone"?

As noted previously, mated females lay eggs at a much higher rate than do virgin females (Fig. 10-3A,B). The ovipositional state is therefore not en-

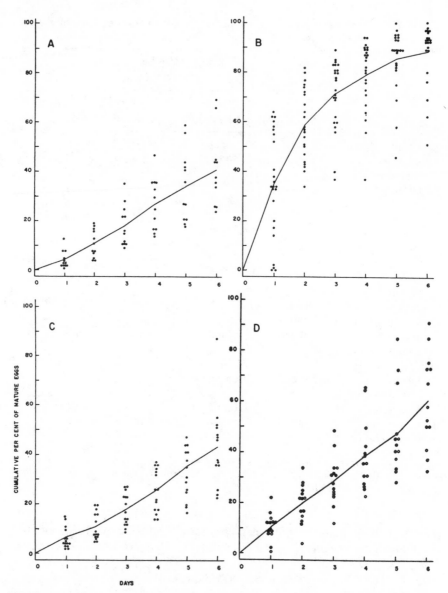

FIGURE 10-3. The oviposition patterns of female cecropia moths. (A) Virgin females; (B) females successfully mated to normal males; (C) females successfully mated to castrated males; (D) females from which the CA and CC have been extirpated before mating with normal males. Open circles indicate females with reimplanted CA–CC complexes. [Reprinted with permission from J. W. Truman and L. Riddiford, *Biol. Bull.* **140**, 8–14 (1971).]

tirely "off" in virgins, but proceeds at a baseline rate. When virgin cecropia are mated to males whose testes have been removed, they show the virgin-egg-laying rate (Fig. 10-3). Such castrated males transfer a spermatophore, but it lacks sperm. However, additional studies that apparently correlated the higher ovipositional rates of mated females with the presence of a sperm-filled bursa copulatrix (the pouch that receives the sperm during insemination) are seriously flawed by the absence of *both* (1) analyses of variance of the ovipositional rates and (2) sham-operated controls. Significantly, analyses of variance ultimately demonstrated that experimental females oviposited at rates *higher* than virgin but *lower* than mated females. Furthermore, the same can be said for the apparent inhibition of oviposition of mated females from which the CC–CA complex had been removed, and again the possible effects of surgical trauma could not be measured because no sham-operated controls had been carried out.

The lesson is threefold. First, without careful attention to controls, confounding effects make interpretation uncertain. Second, in our eagerness for answers, we should not be blind to the fact that most biological phenomena are causally affected by more than a single factor. In cecropia, the roles of such additional factors are not very dramatic, but they are there. In other species, such complexity of causation becomes more obvious; a case follows this paragraph. Third, we often cannot easily determine by eye which treatments result in larger or smaller values. This problem becomes acute as the mean values approach one another and/or the variances increase. Such judgments must be made with the aid of statistics, a tool that has yet to make major inroads into physiology.

A detailed analysis of factors affecting both oogenesis and oviposition in the larch bud moth, *Zeiraphera diniana,* has been provided. In this moth, egg production (number of eggs produced in a female's lifetime) is under an all-or-none switch associated with viable sperm transferred during mating, but oviposition (eggs laid) is under more complex control. There is evidence that the number of eggs produced is independent of the proportion laid and that the processes are under separate control. The minimum requirement for oviposition and oogenesis in *Z. diniana* is adequate space and nutrition (Fig. 10-4). Without these, neither occurs, although each female contains an average basic stock of about 37 eggs (Fig. 10-4, group 1). If space and sugar water are provided, egg production rises to an average of about 70 eggs and about 1.5% of them are laid (Fig. 10-4, group 2). If a larch twig (oviposition site) is now provided in addition, egg laying is raised to 20%, but egg production is unaffected (Fig. 10-4, group 3). Mating raises oogenesis and oviposition, but the interaction of factors is similar. With space and nutrition, but without a larch twig, mated females produce a mean of 95 eggs and lay 20% of them (Fig. 10-4, group 4). With larch present, mating to a sterile male (no sperm in spermatophore) does not increase egg production over unmated, but increases the proportion laid to about 54% (Fig. 10-4, group 5). Finally, with a larch twig and normal mating, egg production rises to a mean of 150, of

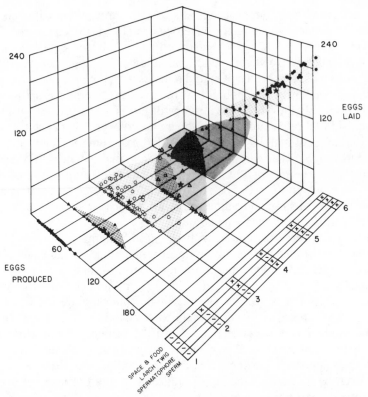

FIGURE 10-4. The effects of various experimental treatments on the egg production and egg laying of individual larch bud moths. Experimental treatments are shown on the Z axis and are: (1) unmated, no space, no food, no larch twig (oviposition site); (2) unmated, space and food present, no larch twig; (3) unmated, space and food, larch twig present; (4) mated to normal male, space and food no larch twig; (5) mated to sterile male (spermatophore present, but without sperm), space and food, larch twig present; (6) mated to normal male, space, food, larch twig all present. The shaded planes enclose the individual values and the stars indicate the means. (Data from G. Benz, *J. Insect Physiol.* **15,** 55–71, Copyright 1979, Pergamon Press, Ltd.)

which 94% are laid. Note that the larch twig has no effect on oogenesis if the female is uninseminated. Therefore, the interaction of larch with insemination is greater than the effect of larch alone.

Nutrition, oviposition site, and mating thus interact in a complex way to control oogenesis and oviposition. Stimulation of oogenesis by these factors is an additive effect, but stimulation of oviposition is a multiplier effect (Fig. 10-5). The oviposition rate without stimulation other than space and nutrition is taken as the base rate, and the values in the boxes indicate the multiplication of the base rate attributed to each factor. In addition, note that the effect of two simultaneous factors is 1.66 times greater than either factor alone. Factors thus interact *synergistically,* that is, their effects are not simply additive but are greater in concert than alone.

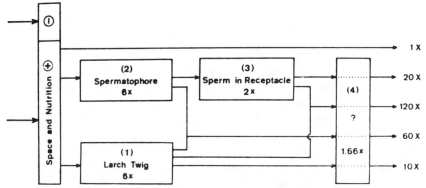

FIGURE 10-5. The average effects on oviposition of the larch bud moth by the factors (1) larch twig, (2) spermatophore or seminal fluid, and (3) presence of viable spermatozoa in the *receptaculum seminis*. These are considered as multiplier units, each acting with the specific factor indicated below the name. Space and nutrition are taken as a deblocking unit. If it is positive, a small stimulus is forwarded, leading to the deposition of the basic number of eggs. If one or several factors are acting, the basic number, X, will be multiplied by the corresponding factors. All combinations are possible. In order to arrive at the number experimentally found, a fourth multiple (4) quantifying the interaction (1.66) was inserted. (Reprinted with permission from G. Benz, *J. Insect Physiol.* **15,** 55–71, Copyright 1969, Pergamon Press, Ltd.)

This is a complicated picture, but it is probably closer to the real world than what seems intuitively obvious. Let us complicate matters still further by drawing attention to the uncertainty associated with all our statements as a result of the variability of the data. Our conclusions are actually only *statistically* true. Any given individual may deviate rather widely from them, for these actually apply to populations of animals. For example, while unmated females with space, nutrition, and larch twigs matured an average of 63 eggs and laid an average of 12, the actual values ranged from about 20 to 145 for production and 0 to 120 for laying (Fig. 10-4). Clearly, then, there are individuals in this group whose production and laying are not distinguishable from those of individuals in the comparable mated group or, for that matter, from the average performance of that group. It is only the means for the populations that are different. The "truth" for an individual may thus deviate considerably from that of a population of individuals. This kind of variation is, to a greater or lesser degree, a basic characteristic of all organisms and biological data. Unravelling the many complex sources of such variation can be a long and difficult task and is often considered not worth the effort, but in our struggle to extract average truths, we should never forget that the price of ignoring the variation is the distortion of reality.

Data on several other Lepidoptera indicate that, in general, the effect of mating is to switch to increased oviposition, but the full response usually depends on additional sensory and physiological inputs. The change from receptivity to various kinds of mated behavior (termination of receptivity, oviposition, searching for oviposition sites, maternal care, etc.) is a common result of mating throughout the Insecta, but the actual stimuli bringing it about vary. Thus, in the cabbage white butterfly and certain cockroaches the stimulus occurs when a stretch receptor on the genital pouch is filled with a spermatophore. A glass bead in the genital pouch has the same effect as mating. In bedbugs, the stimulus is the presence of sperm in the upper reproductive tract of the female. In *Drosophila,* house flies, and mosquitoes, it is a substance secreted by the male sex accessory gland and contained in the spermatophore. In the cecropia and larch bud moths, and in other Lepidoptera as well, it is the presence of sperm and/or a spermatophore in the *bursa copulatrix* and/or spermatheca. In the blood-sucking bug *Rhodnius,* it is also the sperm in the spermatheca.

Evolution has clearly exploited any of the stimuli associated with insemination as signals telling the female that she has been mated. These stimuli are used either alone or in combination, quantitatively or more or less qualitatively. The signal never consists of all of the stimuli resulting from mating, but only a *token* of the total. This parallels other realms of signaling within and between animals.

The case of the mosquito, *A. aegypti,* is of particular interest because the effects of mating on oviposition and the termination of receptivity have been clearly separated and because the latter seems to be under relatively simple control. In this mosquito, the male accessory glands add a material termed *matrone* to the spermatophore, which terminates receptivity in the female. Recall that receptivity in this species is associated with the posture of the female's cerci, and this posture is under control of the last abdominal ganglion. Severance of the ventral nerve cord does not prevent mating in mature virgins and does not prevent refractoriness in matrone-injected females. The action of matrone is thus not neurally mediated through the brain, but it is not possible to conclude that it acts directly on the terminal ganglion, since other hormonal mediators could be involved. In any case, a signal derived from mating terminates receptivity, and its action is coordinated by, if not directly on, the terminal ganglion. Whether other insects terminate receptivity in an equally simple manner, it is too soon to tell. In certain cockroaches, mated females remain receptive, but mating shuts off the secretion of sex attractant and therefore they no longer attract males for mating.

10.5.5. Other Cases of Switching of Behavioral States

Three further examples illustrate the variety of behaviors affected by switching and the diversity of the switching mechanisms.

The milkweed bug, *Oncopeltus fasciatus,* is one of many species of insects that migrate north- and southward in spring and fall, respectively. For the spring population moving northward, this migration seems to be mostly prereproductive, a fact that suits the generalization that insects respond to migratory stimuli primarily during the prereproductive, postteneral period.

Once reproduction comes into full swing, migration is supposed to be inhibited. Because of the involvement of the CA and JH in reproduction and the either/or nature of the reproduction–migration states, a number of authors have suggested that JH may control both. In the milkweed bug, migratory behavior is measured in the laboratory as the proportion of a test group that shows bouts of tethered flight longer than 30 min. When this measure is used, there is a relationship between JH level and flight activity, flight being maximal at intermediate JH levels.

JH level can be increased by application of JH or decreased by starvation, but it has not been possible to allatectomize this species successfully. When adults are starved, a drop in JH level and an increase in flight follows, but with longer starvation, JH level drops to near zero, and flight decreases to control level or below. Treatment of such long-starved bugs with a JH analog increases their migratory flight. These data were interpreted as follows: Oogenesis and oviposition occur at high JH titers, and such titers prevent migratory behavior. Starvation results; a drop in JH prevents oogenesis and switches the animal to the migratory mode, causing it to seek greener pastures.

While it is clear that JH stimulates flight, how this relates to the seasonal control of migratory flight is much less clear. Milkweed bugs on long-day photoperiods show a rapid, postteneral rise of JH, while those on short-day treatments show a slower rise. The level of flight activity associated with each intermediate JH titer is much higher in the short-day than the long-day group. Thus, flight and JH titer are correlated, but the difference between the long- and short-day groups cannot be explained by JH titer alone. At the very least, the sensitivity of migratory behavior to JH stimulation is affected by photoperiod. The mechanism of these effects must await further work.

Endocrines have also been shown to play a role in the cyclical short-range migrations of the female cockchafer, *Melolontha melolontha*. The female emerges in open fields and flies to the forest edge to feed. When her eggs are matured, she returns to the open field to oviposit, and then returns once more to the forest to feed. In this way, she may go through three cycles of feeding, migration, and oviposition. During the prefeeding flight, the female is attracted by dark silhouettes while in flight, but during the oviposition flight she flies in the reverse compass direction, even if displaced to unfamiliar countryside. Postoviposition flight direction is also maintained in unfamiliar countryside. CA from preoviposition females transplanted into prefeeding females causes the latter to reverse their direction of flight to that characteristic of the preoviposition females.

The reverse transplantation of CA (from prefeeding to preoviposition females) causes no change, nor does the sham-operated and normal controls. The CA of preoviposition females can cause the males to undertake the reverse flight to the open fields, even though they never do this normally. The CA of preoviposition females also prevent further oogenesis when transplanted to prefeeding females.

From the point of view of behavioral switching, the interesting feature is that the migrations are primarily accomplished by a reversal of the sense of direction and that this is linked to the CA. Nevertheless, there must be some regulation of when the beetles take flight which may not be CA linked.

The control of queen-rearing behavior in worker honeybees illustrates a hormonelike effect of a pheromone. As long as a healthy queen is present in a honeybee colony, the workers do not rear additional queens (except during swarming season) and worker ovaries remain undeveloped. Should the queen die or be removed, the worker behavior changes rapidly. They begin to construct special cells, transfer larvae to them, and rear replacement queens. Their ovaries also undergo development, and some may lay eggs. Suppression of queen-rearing behavior and ovarian development in workers by the queen is the result of pheromones produced in her mandibular glands and in her abdomen (queen substance and queen scent). The action of the queen's pheromone is thus hormonelike, even though it is extrinsic to the workers, and it accomplishes the switching of worker behaviors. The pheromones are passed around the colony by food sharing, but their mode of action is not well understood (see Section 14.4.1.1).

Social insects may provide another special case of hormonal influences on a shift in behavior. Workers of most social Hymenoptera (ants, wasps, bees) show a gradual shift in behavior with age, young workers performing mostly innest duties such as brood and queen care, shifting later to building, guarding, and defense. Only during the last portion of their lives do they take up duties outside the nest, primarily foraging. There is preliminary evidence that this shift is affected by JH in honeybees. Considering the association of JH with reproduction, this hypothesis seems plausible, but more evidence is needed. In any case, this is not an all-or-none switch in behaviors, but a shift in the balance of behaviors over time. Furthermore, this trend in behavior in honeybees can be reversed to some degree and is generally quite flexible (see Section 14.3.5.4). Endocrine control on such subtle effects is difficult to study.

10.6. HORMONES INFLUENCING BEHAVIORAL DEVELOPMENT

Hormones bringing about the development of *new behavioral programs* belong to the third type of hormonal causes of behavior (Section 10.3.1). Typically, the hormone initiates (and possibly steers) a long series of developmental, morphogenetic, and physiological cause–effect events, which culminate in the development of a new behavioral capacity. As in most development, there is a varying amount of environmental influence and modification of development, and the release of the final behavioral program is under some degree of control by extrinsic factors or perhaps further intrinsic ones.

Because so many events come between the release of such a hormone and the final behavior, and so many other factors steer and influence the outcome, there are practically no case studies specifically on the development of a behavior under hormonal influence. What does exist is a body of literature on the major developmental hormones of insects (ecdysteroids, JHs, and the various neurosecretory tropic factors), whose emphasis is morphogenetic and physiological, but in which associated behaviors are commonly noted. In fact, many of the new behavioral programs that appear at various times throughout the life cycle of an insect are associated with the development of new structures, and the development of the new neural programs is but an aspect of changing behavioral needs associated with morphogenetic change. Clearly, wings and genitalia develop under the hormonal milieu of metamorphosis, but they would be useless structures to an adult insect if its nervous system did not contain programs for their use. In almost all cases, we see the appearance of the new programs only as correlates to morphogenesis and have no idea what role the hormones play in their development.

We have used many of the behaviors as examples in our discussions of hormonal releasers (Section 10.4) and hormonal switching of behavioral state (Section 10.5). What we were discussing were only the final or semifinal stages in a long chain of events set into motion by a developmental hormone. For example, we used ecdysial behavior of the cecropia moth as an instance of direct causation of behavior. However, the neural programs released by the eclosion hormone as well as the synthesis of eclosion hormone itself developed under the influence of ecdysteroids in conjunction with low levels of JHs hormones that bring about metamorphosis. Actual release of the eclosion hormone and behavior is then caused by a reponse to photoperiod interacting with some unknown cue of developmental readiness, a responsiveness that itself must have developed under hormonal influence. The same is true with adult behavioral states. Because of the complexity of events intercalated between the hormonal release and the behavioral outcome, we can only make rather weak statements about this third type of hormonal causation. It is the habit of science to dissect such long and complex causation chains into their individual components to understand them, but this dissection is still awaited for hormonal development of behaviors.

Interestingly enough, in some cases the same hormones initiating development of behavior can also have switching or direct-releaser effects, illustrating the parsimonious nature of evolution. For example, the burst of ecdysteroid secretion that initiates apolysis also appears to act directly on the CNS of locusts to enforce the premolt quiescence typical in this and many other insects. In the hornworn larva, the initial burst of ecdysteroid appears to release the prepupation wandering that brings the larva to its site of pupation. Whether the ecdysteroid acts as a releaser or a behavioral switching hormone in this case is not clear, but it is most probably not acting as a developmental hormone. The release of cocoon-spinning behavior also seems to represent a more immediate causal link of ecdysteroid to a behavior.

10.7. HIGHER CONTROL OF HORMONAL CAUSES OF BEHAVIOR

So far, we have focused on the internal causes of behavior or the physiological cues that release, change, or bring about the development of a new behavior pattern. However, a very large variety of internal control mechanisms, including hormones, evolve to come under control of one or more of a large variety of external environmental stimuli.

Let us view the larch bud moth (Section 10.5.4) again briefly from the point of view of *extrinsic controlling factors*. The factors food, space, larch twigs, and mated state are actually all extrinsic factors that are somehow transduced into an internal stimulus. The transduction may involve sensory, nutritional, or pheromonal processes, but their effect must eventually be integrated, presumably by the brain, into an action on the secretion of the hypothesized oviposition hormone that turns the ovipositional response on or off.

The larch bud moth also illustrates two other common and important properties of such extrinsic control. First, the actual stimulus used is not the entire complex phenomenon, but only a token property of the whole. The second property is the multifactorial nature of extrinsic control. The larch bud moth case illustrates that not only does each single factor (food, space, larch twig, mated state) have less than full control, but the factors often interact with one another in a nonadditive way. The number of factors and the proportion of control by each varies widely among insects.

Perhaps because it represents an important aspect of an insect's adaptation to local conditions, extrinsic control over hormones appears to be rather labile and to evolve with relative ease. This lability is witnessed by three observations. (1) Extrinsic control is very common, perhaps nearly universal if all hormones of each species are considered. (2) Different environmental stimuli control the same hormone in different species. Thus, for example, ecdysteroid release may be extrinsically controlled by a variety of stimuli including photoperiod, circadian time, feeding, crowding, space, and so on, while the "oviposition hormone" may be controlled by food, space, oviposition site, mating (sperm, spermatophore, male accessory gland substances), and/or other stimuli. (3) The strength of control, or even its presence or absence, varies within the geographic races of a single species (e.g., *Locusta migratoria*, Section 10.5.3). Similar variation has been described for CA control over female receptiveness in the female cockroach *Leucophaea maderae*.

10.7.1. Circadian Control over Hormones

Practically all insects show responsiveness to time of day. As with internal stimulators, they do not respond to a general melange of stimuli associated with time of day but to one or a small number of very specific properties such as the timing of the light–dark transition or the dark–light transition. The transition sets an "intrinsic circadian clock" that somehow estimates the time of day and is reset every 24 hr. Thus, a female silkworm moth deprived of this 24-hr cue by being kept in constant darkness would still "call" at the approximately "correct" time of day, but the time of calling would gradually drift or become arrhythmic without any timing signal. Of the extremely wide variety of behaviors and hormones under circadian control, a few examples will suffice.

Upon completion of adult development, the timing of the release of eclosion hormone from the CC in the cecropia moth is under control of a circadian timer, probably in the brain. When brains were transplanted from pupae to isolated abdomens, the abdomens initiated ecdysial behavior on a schedule characteristic of the treatments received by the brains. It appears that the brain contains a photoreceptor and clock that time the release of eclosion hormone.

Eclosion of adult *Drosophila* from cultures in constant darkness is random throughout the 24-hr cycle, but a very short flash of light synchronizes the emergence of individuals at a particular phase every 24 hr. In saturniid moths, oviposition as well as the calling behavior (Section 10.4.2) are under at least partial circadian control via the brain clock and hypothesized hormones.

Circadian control is often superimposed over some developing capacity such as the ability to eclose or lay eggs. This is referred to as "gating" because if the capacity develops to the threshold for release after a certain time of day, the behavior or hormonal release is prevented until the proper time on the following day. Eclosion and prepupation wandering behavior are examples of such "gated" behaviors. Gating is an interactive phenomenon involving extrinsic factors and intrinsic hormonal and developmental processes.

Light may not be the only circadian timing cue of importance. In crickets, for example, it was found that temperature fluctuation under constant light could serve as an entraining cue to singing behavior in the males. A number of other entraining cues have also been indicated. It seems likely that in real life the insect uses such cues in a multifactorial, interactive manner, as the larch bud moth used the factors food, space, larch twigs, and mated state.

10.7.2. Seasonal Control over Hormones

Conditions for growth and/or reproduction are favorable only during particular seasons, and each insect species must ensure that it synchronizes its growth and reproduction with the correct season, be it spring, summer, fall, wet or dry, or even winter in a few cases. Because growth and reproduction are under endocrine control, it is not surprising that a wide array of species have evolved seasonal control over their endocrine systems. In most cases the token signal of season is photoperiod. Because photoperiod (day length) and its pattern of change vary in a predictable way with season, they are better indicators of season than is temperature or some other seasonal cue. Nevertheless, temperature often interacts with day length as a seasonal indicator in photoperiodically controlled systems.

Commonly, an insect escapes from seasonally unfavorable conditions in one of two ways:

1. Escape in time through a physiological shutdown called *diapause*. This shutdown can affect the whole animal and carry it through the unfavorable period in an inactive state (hibernation—winter, estivation—summer) or it may affect only certain organ systems, such as the reproductive system and reproductive behavior (reproductive diapause). For example, in the adult Colorado potato beetle, *Leptinotarsa decemlineata*, short-day photoperiod results in low JH titers, which cause reproductive diapause and, by an unknown mechanism, suppression of sexual behavior.

A number of lepidopterous larvae and pupae have been shown to undergo seasonal diapause mediated by a photoperiodic response of JH or ecdysteroid or both. Most cases of diapause show obvious parallel changes in behavior, but control of the behavior itself is unknown at present.

2. Escape in space by *migration*. Migration, which moves the insect from an unfavorable area to a more favorable one, can be stimulated by seasonal cues. Many butterfly species migrate south in the fall. The best known of these is the monarch, which overwinters in spectacular roosts in the central Mexican highlands. The annual migrations of ladybird beetles between the Central Valley of California and the Sierra Nevada provide another example. For the long-range migrators such as the butterflies, the relationship of migration and reproduction and their respective control are not yet clear. Some Lepidoptera, such as the sulfur *Phoebis sennae* or the skipper *Urbanus proteus* lay eggs as they migrate south, while others, such as the monarch, appear reproductively inactive. Perhaps we should expect large variation in the endocrine mode of control of migration and reproduction.

10.7.3. Other Extrinsic Factors

If we accept that calling behavior in polyphemus moth females is internally controlled by a "calling hormone," this species presents an unusual example of extrinsic control. Not only does calling take place only during certain hours of the night, but the female must be exposed to oak leaves or their odor before calling takes place (see Chapter 11). The odor component, (E)-2-hexenal, can be substituted for oak leaves, although, because dose–response relations are unknown, we cannot say this is the only cue of importance.

In insects such as mosquitoes and blood-sucking bugs, reproduction is strongly cyclic because of the need for blood meals to nourish the developing eggs. We might expect that the secretion of the gonadotropic hormones, and thus the control of gonadal activity and reproductive behavior, would come under control of stimuli emanating from feeding. In *Rhodnius,* stretch receptors in the gut bring about the postfeeding release of ecdysteroids in larvae, and a similar mechanism might operate on gonadotropic hormone during the adult reproductive phase.

10.8. SUMMARY OF HORMONAL CAUSATION OF BEHAVIOR

The main points of hormonal causation of behavior are summarized in Fig. 10-6, which focuses on the development of behaviors as prepatterned CNS programs and their switching and release by hormonal and extrinsic factors.

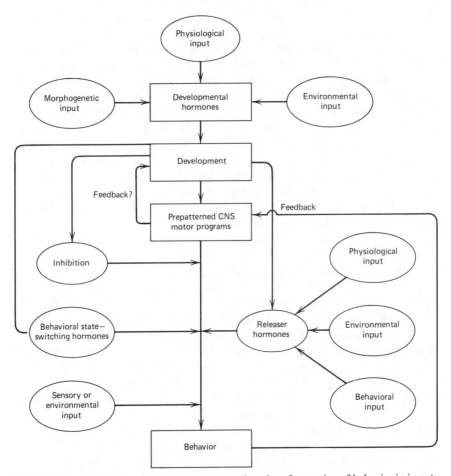

FIGURE 10-6.   Schematic diagram of the hormonal modes of causation of behavior in insects.

10.9. HORMONALLY REGULATED BEHAVIOR THROUGHOUT LIFE CYCLE

A discussion of the hormonal regulation of behavior through the life of an insect must be a composite picture, spliced together from the vignettes discussed, a good deal of extrapolation, and a fair dose of plain speculation. Obviously, there is no single species of insect for which the sequence to be discussed has been demonstrated, and in many cases it is certain that the patterns will vary among species.

The insect begins its active life by hatching from the chorion of the egg, a behavior probably released by eclosion hormone or a similar neurosecretory hormone. The larva feeds and grows until it has gained enough weight to molt. Thereupon, a pulse of ecdysteroid brings on apolysis and probably premolt quiescence as well as premolt gut-voiding. When the new cuticle is fully formed and when environmental conditions are appropriate (correct circadian time, lack of crowding, other physical cues), eclosion hormone is released from the subesophageal ganglion, releasing eclosion behavior and possibly posteclosion behaviors such as eating the cast cuticle, swallowing air, or postmolt quiescence. This sequence of hormones and behaviors is repeated for each larval molt until the time for metamorphosis has arrived. When the metamorphic molt is due, the burst of ecdysteroid that initiates the molt is not accompanied by high JH levels and thus results in the development of pupal or adult cuticle.

Significantly, this hormonal secretion results in the appearance of new metamorphosis-associated behaviors such as cocoon spinning, prepupal wandering, and pupation-chamber excavation. Presumably, these have been developed during the larval period under hormonal influence. In addition, the usual premolt behaviors are also released (gut-voiding, cessation of feeding, quiescence). With present knowledge, it is difficult to say whether each hormone plays a releaser, switching, or developmental role. Whatever the case, when pupal development is complete and the proper conditions of photoperiod, circadian time, and other physical conditions are met, eclosion hormone is once again secreted from the subesophageal ganglion, releasing ecdysial behaviors. It probably also brings on the switch to pupal behaviors, such as these are. Because JH is low during the pupal stage, adult apolysis follows the next burst of ecdysteroid. Morphogenesis is presumably paralleled by the development of appropriate adult behaviors, and we must presume that this requires a certain hormonal milieu. Since pupae are so inactive, this behavioral development must occur mostly within the nervous system, but we know little of this process.

When adult morphogenesis and behavioral development are ultimately complete, eclosion hormone is once again secreted, this time from the CC. The resulting ecdysis is followed by the switching on of the previously developed adult behaviors, beginning with postecdysial behaviors such as expansion of wings, legs, genitalia, and postecdysial wandering where it exists. In species that enter adult life with developed gonads, sexual behavior and receptivity may be immediately apparent as well. In others, the gonads develop under the influence of JH (and perhaps ecdysteroids), and sexual behavior and receptivity are not switched on until gonadal development is complete. In such cases, JH may itself play a role in the appearance of these behaviors. Neurosecretory releaser hormones such as the "calling hormone" may be important during this phase. The end result of sexual hormone is mating, naturally. After mating, sexual behavior is generally turned off, at least in females, and replaced with such postmating behaviors as oviposition, perhaps released by an "oviposition hormone" which may also bring about such associated behaviors as brooding or provisioning.

Under some conditions, gonadal development may be delayed as a result of a delay in JH secretion. Such delays may be associated with migratory behavior or diapause. In such cases, JH secretion and gonadal development commence when conditions are once again favorable. JH secretion in such cases has come under environmental control.

In insects that reproduce in distinctive cycles, feedbacks of various kinds regulate the gonadal and receptivity hormones so that gonadal development alternates with sexual behavior or quiescence.

10.10. BEHAVIOR ASSOCIATED WITH GENERAL PHYSIOLOGICAL STATES: FEEDING BEHAVIOR

If the approach of suppertime is contrasted with the period after the dessert plates are cleared, it becomes obvious even to the nonspecialist that hunger and feeding are inextricably linked in a feedback loop. Hunger leads to feeding; feeding results in negative feedback that terminates this behavior and determines meal size. Feeding in many insects follows the same general lines and has been extensively studied.

For analysis, feeding behavior can be divided into three phases:

1. Preingestion locomotion. This may be oriented or apparently random.
2. Nonlocomotory preingestion behaviors. These often involve the positioning of the animal or movements of its mouthparts in preparation for feeding.
3. Ingestion behavior. This brings the food into the animal's alimentary system and includes chewing, sucking, or whatever the animal does to ingest its food.

Experimental analysis of these phases has taken two approaches. In the first, the behavior of a cohort of animals is tested before and after feeding. This has the effect of confounding aging with the result of feeding, although this objection can be partly overcome by going through several cycles of feeding and hunger. In the second approach, one cohort is fed, while another cohort of equal-aged animals serves as an unfed control. The responses of the dependent measure in the two groups are then compared.

10.10.1. Prefeeding Locomotion

In general, deprived insects are *more active* than fed ones. This difference increases the probability of contact with food or stimuli leading to food. Measured prefeeding locomotion appears to be usually random, but in those species showing oriented locomotion, there are often changes in the response to light, gravity, or other orienting cues upon deprivation. While species from diverse orders show increased locomotion in response to deprivation, by far the best studied have been the locust, *Locusta migratoria,* and the blow fly, *Phormia regina*.

Prefeeding locomotion can consist either of spontaneous locomotion or of locomotion resulting from increased reactivity of the animal to external stimuli. Either or both may change with deprivation, and the change may be by different mechanisms. Experiments do not always distinguish between these.

Single blow flies housed in an actograph (measures activity) show increased activity with deprivation. This activity drops to a low level soon after feeding, only to rise again as the time since the last meal increases. If two blow flies are parabiotically paired so that one is riding upside down on the back of the other, the activity of the walking fly is reduced when the inverted fly is fed, indicating that at least part of the cause of postfeeding inactivity is *blood-borne*. Injection of glucose solution or water into deprived flies also decreases activity, suggesting hemolymph changes are involved. If the crop duct is ligated before feeding so that the fly cannot take fluid into the crop, the fly becomes active again very soon after feeding. A number of factors were not found to be important to the decline of activity: increased weight, crop distension, hemolymph potassium, labellar input, and collapse of abdominal air sacs. Interactions of factors have not been tested, and most of the available evidence seems to link inactivity with changes in the hemolymph composition, perhaps the continued presence of sugar. This is supported by the observations on parabiotic flies and the fact that the crops of flies fed dilute sugar solution empty into the midguts more rapidly than the crops of those fed more concentrated solutions, while the dilute-fed flies become active earlier than the concentrated fed.

Locusts given free access to food feed in discrete, well-spaced meals. Their spontaneous activity declines gradually after a meal until just before they take their next one. This decline is due in part to hormonal material released from the CC, the release resulting from signals from the stretch receptors stimulated by distension of the foregut upon feeding. Thus, filling the foregut with nonnutritive agar or injecting homogenates of CC into deprived locusts has the same effect as feeding.

The marching activity of locust hoppers, a social activity that results in marching in groups, increases with food deprivation for up to 4 hr. Marching is, however, not simply spontaneous activity, but includes a strong element of reactivity since it requires social interactions among hoppers. The physiological mechanism of this increased marching is poorly understood.

Deprivation also increases the *responsiveness* of locusts to the odor of grass. This responsiveness is reduced by feeding in proportion to the size of the meal. Similarly, deprived individuals of many species of insects become more responsive to and are attracted to water. It is thought that this response increases the probability of finding food, since many feed on wet materials. A number of species change their response to orienting cues such as light or gravity upon deprivation in such a way that their chances of finding food are increased.

10.10.2. Nonlocomotor Preingestion Behavior

The insect, having arrived at the food through one or several of the previously described locomotory activities, must now stop moving, orient properly to the food, and bring its mouthparts into contact with it. In many insects (e.g., butterflies and many Diptera), this behavior is mediated through receptors on the tarsi. If only the tarsus on one side is stimulated, the insect will turn toward that side. When tarsal chemoreceptors on both forelegs are stimulated, the insect stops walking and lowers its mouthparts into contact with the food so that feeding behavior can commence. The *threshold* for this mouthpart extension varies with the state of hunger and has been used extensively to study the regulation of nonlocomotor preingestion behavior, especially in the flies *Calliphora* and *Phormia*. The threshold for proboscis extension in fed flies is often several-hundred-fold that of deprived flies, but returns to a low value after another period of deprivation. The maximum threshold is reached within about an hour of feeding, and its magnitude depends directly on the volume and concentration of sugar solution ingested.

By what mechanism is this change in responsiveness to food brought about? Possibilities include neural, hormonal, hemolymph composition, stretch reception of gut filling, or combinations of these.

If the guts of flies are ligated just behind the proventriculus so that the sugar solution cannot enter the midgut, there is no effect on the rise in threshold. Likewise, threshold is not affected in flies given a sucrose enema to fill the midgut.

Experiments indicate that sugar solution solely in the midgut does not affect threshold and that a full crop alone is not responsible for the elevated threshold. Apparently, all that is needed is the presence of sugar solution in the *foregut*. The fly could detect this either by chemosensation or by distension of the foregut. Peristalsis or controlled enlargement of the foregut increases neural activity in the recurrent nerve, which activity responds to concentration in the same way that threshold elevation does. Sectioning the recurrent nerve causes hyperphagia (overeating), indicating that feedback from the foregut probably plays an important role in determination of tarsal thresholds to sugar. Injection experiments and others have shown that hemolymph changes in sugars probably play little role.

In many insects, feeding brings about the release of material from the CC, and at least in some cases this release raises the threshold of maxillary chemoreceptors. It may do the same for tarsal receptors, although this has not been established. Such hormonal release may form a link between foregut stimulation and tarsal threshold elevation. Although the mechanism may be entirely neural, this seems unlikely with present evidence.

Regulation of chemosensory thresholds other than tarsal ones is less clear and more heterogeneous. Labellar threshold shows a response to feeding in some insects, but not in others. In *Locusta* feeding raises the thresholds of the chemoreceptors on the maxillary palpi. Stretch receptors on the foregut feed back to the brain via the posterior pharyngeal nerve and bring about the release of material from the CC. The CC hormone causes the closure of pores in the sensilla on the maxillary palps, and this closure may be responsible for the rise in threshold, although the evidence is circumstantial. Thus, locusts palpate but do not feed for 1–2 hr after feeding, a period during which the pores are closed. The interval between meals is about 1–2 hr.

10.10.3. Determination of Meal Size

Many insects offered food in excess stop feeding after a characteristic meal size. Meal size is presumably determined when *inhibitory* stimuli arising from the intake of food exceed the *excitatory* cues that stimulate ingestion. On a crude level, the intuitive notion that meal size is related to the length of deprivation is borne out in many species. In blow flies, for example, the volume of fructose solution imbibed increases with the length of deprivation and depends as well on the volume taken in the previous meal. Meal size, however, is not as simple as filling a passive bag to a certain level at intervals. Thus, for example, the volume and rate of ingestion of various sugars by blow flies not only varied positively with concentration up to the limit set by viscosity but also depended on the type of sugar as well. Some type of chemosensory input thus plays a role in terminating feeding. The major control in blow flies, however, seems to be *neural* feedback, for section of the recurrent nerve or the median abdominal nerve leads to pronounced hyperphagia. If both these nerves are sectioned, the flies lose all control over feeding and continue to feed long after they have already burst (ugh!). Recording from either of these nerves shows increasing input to the brain as distension of the foregut and crop progress during feeding. The exact route this neural feedback takes in terminating feeding is not entirely clear.

Besides an increase in neural negative feedback, other factors play a passive or modifying role in meal-size determination. All sugar-sensitive receptors of blow flies are known to adapt, and such adaptation probably plays some role in terminating feeding. These receptors include those on the tarsi, labellum, interpseudotracheae, and pharynx. Since some adapt more rapidly than others, their importance to feeding termination probably varies.

Stimulation of the tarsal or labellar sugar receptors seems to produce an excitatory state in the CNS lasting up to 45 sec. During this time, the flies search more intensely and extend the proboscis to lower-quality stimuli. The intensity of this state increases with sucrose concentration and may contribute to the determination of meal size long after the chemoreceptors are adapted.

The mechanism determining meal size in blow flies is proposed to be as follows. Feeding is terminated when the negative influence of the neural input from the foregut and crop stretch receptors overcomes the excitatory influence from other sources. The level of excitatory influence to be overcome in the CNS is determined by input from the chemoreceptors, which subsequently may adapt. The higher the level of the excitatory state, the greater the inhibitory feedback must be to terminate feedback. This explains why meal size increases with more stimulating food and why nerve section results in hyperphagia.

We should not expect control of meal size to be the same in all species. In deprived migratory locusts, meal size increases rapidly up to about 3 hr deprivation and more slowly up to 6 hr. It also varies according to the palatability of the food offered. Meal size is positively correlated with the rate of ingestion and negatively correlated with the amount of food still in the crop at commencement of feeding. The amount of food in the rest of the gut plays no role. Section of the postpharyngeal and recurrent nerves results in larger-than-normal meals and indicates that the brain receives negative feedback from stretch receptors on the crop. Since these stretch receptors are located at the anterior end of the crop and the crop fills from back to front, it is possible that the feedback from the crop is not graded, but all-or-none. Supporting this possibility is the fact that this negative feedback limits meal size only when the locust is feeding on highly palatable food. In other words, it becomes important only when the maximum possible meal is taken. Other factors, such as chemoreceptors and excitatory CNS states, must limit meal sizes of less palatable foods.

Osmotic pressure of the hemolymph may be one such factor. The osmotic pressure of hemolymph increased 24% by the end of a 15-min meal, mostly as a result of water loss during salivation. Injection of hyperosmotic fluids into locusts before feeding brought about a reduction in meal size in proportion to the osmotic excess over base level. At present, it is unclear how osmotic pressure changes, CC hormones, and chemoreceptor input interact with, and are integrated with, the negative feedback from the gut-stretch receptors.

The information available on the mechanisms controlling meal size in insects other than blow flies and locusts indicates considerable variation in the details of control (Barton-Browne, 1975).

10.11. EFFECTS OF BEHAVIOR ON PHYSIOLOGY

Traditionally, insect behavior has been considered a product of physiological events. The reverse has rarely been considered. Therefore, any discussion of the behavioral causation of physiological changes among insects suffers from a scarcity of information and an absence of any organizing principles.

Simply to partition our speculations, let us arbitrarily divide the physiological effects into two extremes: (1) general effects without easily identifiable target organs or processes and (2) specific effects in which the change caused by the behavior can be localized to a single organ or process or a small number of these. Let us also divide the behaviors in question into two categories: (1) that of the individual acting upon itself and (2) that of other individuals acting upon a test individual. The former is a feedback system within the individual, while the latter amounts to environmental input. Mechanistically, neither of these introduces any new phenomena but merely call attention to the possible position of a behavior in the total spectrum of cues or causes. These two dichotomies, while not always sharp or easy to apply, give us four combinations of behavior and physiological effect (Table 10-2). Naturally, we are likely to be lumping widely disparate kinds of phenomena together.

10.11.1. General Effects: Behavioral Thermoregulation

Because behavior is the most common "first response" in the process of adaptation, it often brings the animal into a more propitious environment and results in general physiological change. We shall limit ourselves to only one such response, behavioral thermoregulation. The effect of body temperature

TABLE 10-2. Examples of Physiological Effects of Behavior

Behavior in Question	Physiological Effect	
	Specific	General
Behavior of self	Pupation after dispersal in tenebrionid beetles Adipokinetic effect Heartrate acceleration	Thermoregulation
Behavior of others	Effects of crowding in locusts, aphids, tenebrionid beetles	General effects of crowding: mortality, nutrition, vitality, growth, activity
	Ovarian development in social wasps, bees	

is of obvious physiological importance because it affects the rates of most chemical and physical processes.

As a group, insects are ectotherms; that is, their body temperature is dependent on, although not necessarily equal to, the environmental temperature. Nevertheless, many insects are able to regulate the temperature of their bodies, or parts of them, to a surprising degree, and the behaviors are often quite specific ones. They do this in two major ways: (1) by behavioral exploitation of the *temperature differences* in their microenvironments and (2) by using the heat produced by the flight muscles during exercise. The latter may include behavioral as well as physiological mechanisms and constitutes a type of homeothermy. We shall deal only with the first type of thermoregulation, basing discussion on the thorough review of Casey (1981).

The body temperature of an insect (or any object) is determined by the rate of heat gain and the rate of heat loss. When the rate of gain equals the rate of loss, the body temperature will remain constant (steady state). Heat can be gained and lost by conduction, convection, and radiation. Practically, for insects, heat is gained by radiation (mainly sunlight) and lost by convection, evaporative cooling playing little role. For flying insects, muscular action is also a source of heat.

Since heat exchange takes place only at the surface of an object or insect, the rate of gain or loss is related to body size. Thus, all other factors being equal, as an insect increases in size, its rate of warming or cooling will decrease because the rate of heat transfer is related to the square of the dimension (surface area) while the mass to be warmed or cooled is proportional to the cube. This also causes the equilibrium temperature to increase with size. Figure 10-7 shows this effect for small and large individuals of the desert locust warmed by sunlight.

At a constant volume, rates of heat exchange are affected by shape because of the changing ratio of surface to volume. A sphere is the most

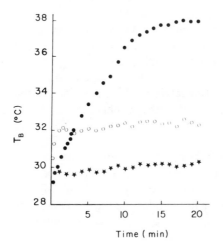

FIGURE 10-7. Effect of solar radiation on the internal temperature (T_B) of green solitarious hoppers of *S. gregaria*. Open circles represent first instar; solid circles represent fifth instar; stars indicate air temperature. The small hopper reaches a lower equilibrium temperature and does so more rapidly. (Reprinted with permission from Stower and Griffiths, 1966.)

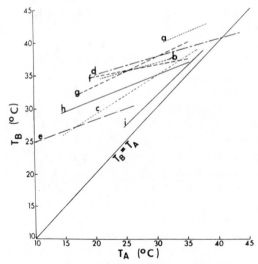

FIGURE 10-8. The relation of body temperature (T_B) to air temperature (T_A) in various insect species while relying on behavior patterns to control body temperatures. (a–c) Locusts, *Schistocerca gregaria, Locustana pardalina,* and *Psoloessa delicatula;* (d) the desert cicada *Diceroprocta apache;* (e) the syrphid fly, *Syrphus* sp.; (f, g) the dragonflies *Libellula* spp. and *Pachydiplax longipennis;* (h, i) the sphingid caterpillars. (Reprinted with permission from T. M. Casey, in B. Heinrich, Ed., *Insect Thermoregulation,* Wiley-Interscience, 1981.)

compact geometric shape and has the lowest surface-to-volume ratio and therefore the lowest rate of heat exchange. As the shape becomes more elongate or flattened or lobed, the ratio of surface to volume increases, and the rate of heat exchange increases with it. Elongate insects therefore cool and warm more rapidly than compact ones of the same volume and also have lower equilibrium temperatures. For insects of the same shape and volume, dark surface color increases the rate of exchange and equilibrium temperature and the rate at which it is reached.

Insect temperatures are also affected by a number of environmental properties. Obviously, the rate of warming increases with the intensity of the solar radiation. Convective exchange is proportional to a power of the wind velocity, as long as the wind exceeds 30 cm/sec. Below this, convection is essentially independent of wind speed. Convective exchange is also related directly to some fractional power of the body size, depending on shape. Large size and more compact shapes result in lower rates of convective exchange.

Whether or not an insect thermoregulates can be determined by plotting its body temperature (T_b) against the air temperature (T_a). If the insects do not regulate at all, the slope of this line will be equal to 1, while if they regulate perfectly, the body temperature will be independent of the air temperature and will have a slope of 0. Figure 10-8 shows this relationship for a number of insect species. While most regulate to some degree, none do so perfectly, and larvae of *Manduca sexta* do not regulate at all.

All of the data in Fig. 10-8 are for insects regulating behaviorally only, without substantial metabolic contribution from flight. The excess in body over air temperature is achieved by the behavioral exploitation of temperature differences of the insect's microenvironment. Put another way, the air temperature is often not a good measure of the actual or potential thermal environment because it does not take radiative heat into account.

The thermal microenvironment is measured by placing models of the same shape, color, posture, size, and other physical characteristics (perhaps a dead individual) in the microhabitats and measuring their equilibrium temperatures. The range of these "true" environmental temperatures then represent the selection of temperatures available in that habitat. When "true" environmental temperatures are compared to the actual body temperature of nonflying insects in the same microhabitat, the two are found to be equal. That is, the body temperature of the model and the live insect are the same. Any observed temperatures in excess over the environmental temperature are due to physiological causes.

Exploitation of thermal heterogeneity can be illustrated by comparing the measured body temperature of two species of grasshoppers from different habitats against the predicted range of body temperatures available to each, based on the range of microhabitat temperatures available (Fig. 10-9). In *Eritittix simplex,* the range of temperatures available is small because the habitat is moist and evaporation keeps the substrate temperature near the air temperature. On the other hand, in *Psoloessa delicatula,* the habitat is higher and a large temperature heterogeneity results from sunlight warming the substrate. In Fig. 10-9 the air temperature and the calculated range of possible body temperatures are shown for each species as a function of time of day. It can be seen that both actual body temperature functions fall within the predicted range. While *E. simplex* can be said not to regulate because it has no choice, it is clear that *P. delicatula* is maintaining its body temperature near the lower end of the range of body temperatures available to it.

Having framed the problem and established the criteria for behavioral thermoregulation, it is now appropriate to survey the kinds of behavior that insects use to exploit their thermal environments for thermoregulation.

Because the main source of heating above air temperature is sunlight, and because few insects are spherical, changes in body orientation and posture can be used to maintain body temperature within a favorable range. Many insects, when their body temperatures are below the range for normal activity, orient the aspect of the body with the largest surface area perpendicular to the sun's rays. Such "basking" behavior maximizes heat gain by exposing more body surface to the sun. Basking is widespread among insects and has been reported for cicadas, dragonflies, butterflies, caterpillars, tenebrionid beetles, and others.

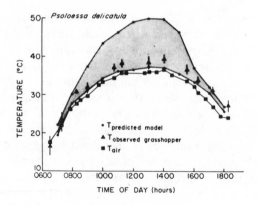

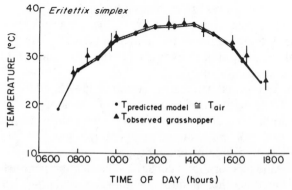

FIGURE 10-9. Temperature relations of *P. delicatula* and *E. simplex* in their natural environments in relation to the time of day. Upper curve connecting solid circles represents predicted maximum body temperature (T_b) achievable by the grasshoppers; lower curve connecting solid circles represents the predicted minimum body temperature achievable by the grasshoppers. (Reprinted with permission from T. M. Casey, in B. Heinrich, Ed., *Insect Thermoregulation*, Wiley-Interscience, 1981.)

When body temperature approaches the upper "set point," the behavior of the animal changes, so that now it presents its smallest aspect toward the sun, thus minimizing heating and possibly facilitating heat loss by the shaded portions of its body. Many insects from diverse orders show such behaviors.

Dragonflies show a variety of postures and body orientations that relate to ambient temperatures (Fig. 10-10). They bask when the air is cool and face directly into the sun or adopt the so-called "obelisk" posture when their body temperature nears the upper set point.

The wings of some insects play an important role in basking. Butterflies, among the most conspicuous baskers, hold their wings in one of two ways depending on the species. The dorsal baskers hold their wings out flat and present the dorsal aspect to the sun; lateral baskers, their wings tightly folded dorsally, present their lateral aspect to the sun. In either case, when

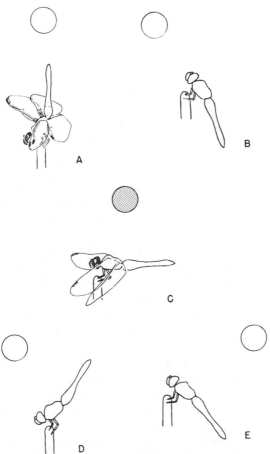

FIGURE 10-10. (A, B) Postures in libellulid dragonflies that minimize exposure to the sun. (C, D, E) Postures that maximize exposure to the sun. Stippling indicates sun above or behind plane of paper. Typical wing postures shown in A and C. [Reprinted with permission from L. M. May, *Ecol. Monogr.* **46,** 1–32 (1976). Copyright © 1976, the Ecological Society of America.]

the body temperature approaches the upper "set point," both kinds of baskers direct dorsally folded wings toward the sun.

By partial shading of the wings of dorsal baskers, it can be shown that only the wing bases are important for thoracic heating. Wings that have been cut off at the base and placed about 1 mm from the thorax are still as effective in heating the thorax. The most likely explanation is that the sun warms the air under and between the overlapping wings, and this warm air in turn warms the thorax. Basking dragonflies over a warm substrate also capture warm air under the downward and forward bent wings.

In some butterflies, the wing bases that are exposed to the sun during basking are dark, contributing significantly to heating. In *Natholis iole* the degree of dark coloration is higher in spring and fall than in summer, and the basking dark forms reach thoracic temperatures up to 3°C higher than the light forms under identical conditions.

Flightless animals also use body orientation for control of body temperature. Since caterpillars are elongate animals, a caterpillar changing its orientation from perpendicular to parallel to the sun's rays may reduce by 10-fold the area heated by the sun. The parallel orientation, most likely to occur during the heat of the day when the sun is overhead, brings the caterpillar into a vertical orientation, which in turn maximizes cooling by the wind. On cool days, most of the caterpillars remain horizontal, thus maximizing heating by the sun and minimizing cooling by the wind.

Insects that live on the ground have special problems because, especially when dry, the ground may reach high temperatures through solar heating. Moreover, the existence of a boundary layer near the ground reduces heat transfer by wind, while reflected radiation may intensify the heating. On sunny days, ground temperatures exceed air temperatures in midmorning, so many insects, including flying insects, use the ground to warm up early in the day. As body temperature approaches the maximum for normal activity, insects may burrow into the soil or enter animal burrows to reach cooler zones. They may also seek or shuttle in and out of shade, as within shade, body temperature will follow air temperature. Other insects, for example, many tenebrionid beetles and locusts, show a distinctive behavior called "stilting" by which they extend their legs to increase the distance between their ventral surfaces and the ground. This behavior can decrease their body temperature by several degrees because it raises their bodies higher in the boundary zone, exposing them to cooler temperatures and higher wind velocity.

Later in the day, when the ground is warmer than the air, some insects (locusts and some dragonflies) exhibit "crouching" behavior to raise their body temperature. They are warmed by increased contact with the warm ground and reduction of convective heat loss because the body is lower down in the boundary layer where wind velocity is less.

All but the smallest insects show metabolic heating of the thorax during flight. When air temperatures are near the upper limit for activity, this metabolic heat contribution can raise thoracic temperatures to damaging levels. Many insects are thus reluctant to fly, or fly only for a few seconds, on very hot days. Insects such as locusts and dragonflies, which do fly at high air temperatures, spend a greater proportion of their flight time gliding. This shift from powered flight to gliding reduces metabolic heat production and may have thermoregulatory importance as well, but it should be noted that updrafts, and therefore conditions for soaring flight, are also associated with the heat of the day.

Figure 10-11 shows the full range of behaviors and movements used for thermoregulation by a single species, the desert locust, *Schistocerca gregaria,* in the course of a day. The torpor of the night gives way to warmup

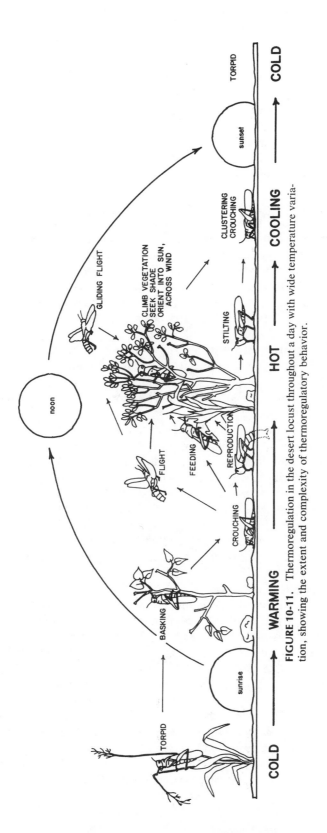

FIGURE 10-11. Thermoregulation in the desert locust throughout a day with wide temperature variation, showing the extent and complexity of thermoregulatory behavior.

behavior until mid- or late morning after which feeding, flight, and reproductive activities occur. As the day gets hot, behaviors are directed at preventing overheating. In the evening, as air temperatures in this desert habitat are dropping rapidly, activity is extended by behaviors that exploit the warm spots in the microenvironment. Torpor returns some time after sundown.

10.11.2. Specific Effects of Crowding

A number of insects show a physiological response to crowding. This is illustrated by *Zophobas rugipes,* a tenebrionid beetle often found in dense populations in deposits of bat guano in Central America. Notwithstanding the crowded conditions that larvae encounter in the field, in crowded laboratory cultures these larvae are inhibited from pupating by the crowding. Isolation from such crowded situations results in the initiation of metamorphosis within 3–4 days, with the larval–pupal molt occurring in about 12 days. Unless isolated, larvae die of "old age" after 10–18 months in such lab cultures.

The only attribute of crowding shown to be important in *Z. rugipes* was the *tactile stimulation* resulting from larvae crawling over and past other larvae. When isolated larvae were stimulated mechanically by simulated larvae in a "stimulatorium," they failed entirely to pupate, but did so when the stimulation ceased (Fig. 10-12). The sensory input resulting from tactile stimulation is somehow transduced into the inhibition of metamorphosis.

Since under crowded conditions the larvae continue to molt and grow—usually more than doubling their weight over that at the time they are first capable of pupating upon isolation—crowding inhibits only the transition to the pupal stage, perhaps larval–pupal apolysis. This transition is controlled physiologically by the ecdysteroids and JHs (Chapter 13). Because crowded larvae still molt, the secretion of ecdysteroids is probably not inhibited. Preliminary evidence suggests that the inhibition is not through the CA and JH either, because allatectomy does not overcome the effect of crowding, nor does injection of JH overcome the effect of isolation. Moreover, measurement and bioassay of the larval CA show these to be inactive.

In natural populations, crowding is also common. However, since these populations are not confined, larvae that have attained sufficient age and weight disperse from the crowded sites. Having found an undisturbed place, they prepare a pupation site and pupate. Each larva's own behavioral response to crowding (dispersal) thus results in removal from the inhibitory tactile stimulation and allows metamorphosis to take place (Tschinkel, 1981).

In effect, by dispersing when it does, each larva "chooses" the age and body weight at which it will pupate. This choice has far-reaching consequences for both its fitness and longevity as an adult. These parameters are both low for very early pupation at low weights, rapidly reach a maximum, and then decline with still later pupation and greater body weight. The ulti-

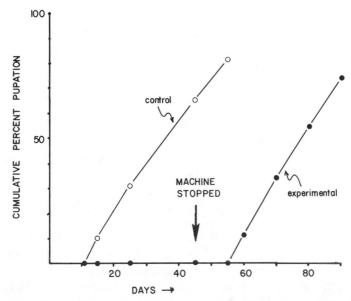

FIGURE 10-12. The effect of tactile stimulation on pupation of the tenebrionid beetle *Zophobas rugipes*. Isolated, mature larvae were stimulated 240 times per day by light chains dragged over them (experimental group) or were protected from contact with the moving chains (control). Tactile stimulation entirely prevented pupation in the experimental group, but pupation proceeded normally once stimulation ceased. [Modified from W. R. Tschinkel and C. D. Willson, *J. Exp. Zool.* **164,** 81–86 (1971).]

mate cause of these physiological effects is the timing of a behavioral event—dispersal.

In aphids some aspect of crowding, possibly behavior, results in developmental changes. A typical life cycle of such aphids follows: The spring populations are produced parthenogenetically and viviparously by usually wingless (apterous) founding females who are themselves hatched from overwintered, sexually produced eggs. Throughout the summer, reproduction continues to be essentially parthenogenetic and viviparous, and the aphids form population clones. As crowding in these clones increases, some apterous females produce winged (alate) offspring, which disperse to start their own clones elsewhere, still parthenogenetically. Their own daughters tend to be mostly apterous but can respond to crowding by producing alate daughters. With approaching fall (short days, plant senescence) the parthenogenetic females produce sexually reproducing males and females, which mate and lay eggs that overwinter.

Aphid life cycles are thus characterized by a number of developmental switches or polymorphisms (male–female; apterous–alate; sexual–parthenogenetic; viviparous–oviparous). While most of these seem to be under physical environmental control, the change from apterous to alate daughter production by parthenogenetic females is the result of crowding. Lees (1966) provided evidence that the tactile stimulation resulting from crowding, rather than chemical, visual, or other cues, causes the switch to alate production. In another species of aphid it was possible to cause alate production by forcing another aphid to walk over the test aphid 10 time in a 2-min period. Since a wide variety of unrelated insects can produce the effect in otherwise isolated aphids, tactile stimulation seems the likely cause.

Once again, the endocrine system provides a likely causal link between tactile stimulation and alate production. The CA of developing aphid larvae that will become apterous are larger and have larger nuclei than those that will become alates. Crowding of the mother may somehow influence the CA of the daughters developing in her body, causing them to become alates as a result of the higher levels of JH. Unfortunately, since application of JH to aphids has thus far produced equivocal results, confirmation of this hypothesis awaits further evidence.

Several species of Old World locusts, including the fabled plague locusts of biblical renown, show a wide variety of responses to crowding, often acting across one or even two generations. Increased density affects the size, weight, and vitality of hatchlings; body coloration of hoppers and adults; and activity level, behavior, rate of development, instar number, adult size, degree of sex dimorphism, fecundity, and fertility. Many of these effects are quantitative, so that varying degrees of crowding result in a continuum of morphologically and physiologically different locusts (phases), whose extremes are called either "solitarious" (reared in isolation for at least three generations) or "gregarious" (reared crowded for three generations) (Uvarov, 1966). There is some indication that the endocrine system may be involved in determining some of these phase characters. The properties of crowding that cue the endocrine system probably include behavior. For example, the change from nonaggregative to aggregative behavior caused by crowding in hoppers can be brought about in isolated hoppers by tactile stimulation with fine wires. If these changes in gregariousness have an endocrine basis, the tactile stimulation may act ultimately on the endocrine system.

Population density has a wide array of effects on a large number of species of insects, and there are probably many as yet undetected effects of crowding. At least some of these are probably the result of the effects of behavior on physiology.

10.11.3. Dominance Rank and Reproduction in Social Insects

Social and subsocial insects provide fertile ground for a search for effects of the behavior of one individual on the physiology of another. All social insects show at least some division of reproductive labor. That is, some individuals (or a single individual) become specialized to lay most or all of the eggs, while the remainder of the colony carry out colony maintenance and brood rearing. In the more highly evolved social insects, this reproductive–nonreproductive dichotomy is fixed during development into morphological castes (e.g., queen–workers), but this is not so in a number of social insects. Rather, colonies are formed by the association of similar founding individuals, with the characteristic division of reproductive labor appearing subsequently. In at least one such social insect, the paper wasp, *Polistes annularis,* preliminary evidence indicates that the division of labor in egg-laying is the result of a struggle for *dominance* among the founding individuals.

In many species of *Polistes,* colonies are annual, producing foundresses in the fall. After overwintering, small groups of these foundresses found the new colonies in the spring. The nests are built of cells formed from chewed-up plant fibers (paper) and are placed in sheltered locations. The foundresses are a representative sample of overwintering queens and, at the beginning, are not distinguishable from one another, either by their behavior, morphology or ovarian development. There ensues a struggle for dominance in which queens exhibit aggressive and submissive behaviors toward one another, and it soon becomes possible for the observer to rank all of the foundresses into a linear *dominance hierarchy,* with an alpha wasp at the top and with each of the other individuals submitting to all above her and dominating those below her. Such a hierarchy is stable over a long period of time, and if the alpha wasp is removed, the second wasp in the hierarchy will assume her position.

It can be observed that higher-ranking individuals lay more eggs, the alpha wasp often being the only wasp to lay eggs, and that rank is correlated with the degree of ovarian development (Fig. 10-13). These differences in ovarian development are caused by differences in JH secretion by the CA. Dominant wasps have larger, more active CA than do subordinate wasps (Röseler et al., 1980). The question is, what controls the CA and therefore the JH level? How do rank, behavior, and JH level interact? Since all of these are correlated, cause and effect cannot be determined without experimentation.

In the first experiments, ovaries from *P. gallicus* wasps of known behavioral rank were removed, and it was found that their rank and behavior were entirely unaffected. Ovarian development is, therefore, not the cause of rank. Treatment of wasps of small nests of known hierarchy with JH and JH analogs in a preliminary series of experiments demonstrated that compared to the controls the JH-treated wasps showed increased ovarian development, as expected. More interesting, the dominance relations of the treated nests were upset, showing an increase in aggressive interactions and a number of rank displacements of queens by JH-treated subordinates. Therefore, JH levels are somehow linked to aggressive and dominance behavior as well as to increased rank and ovarian development. The experiments do not settle the question of whether dominance behavior causes JH levels or JH levels cause dominance behavior. It is also possible that rank is not deter-

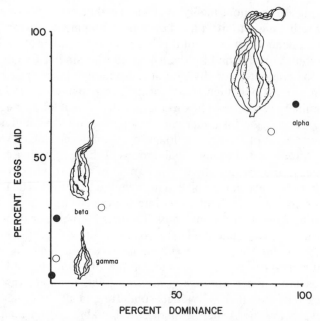

FIGURE 10-13. Relationship among dominance, ovarian development, and eggs laid by individual paper wasps. *Polistes gallicus*. Percent of times an individual showed behavioral dominance over a nestmate is plotted against the proportion of the colony's eggs laid by that individual. Rank is indicated for individuals. Drawings show degree of ovarian development for each rank. Data for two nests (open and closed circles). [Taken from L. Pardi, *Physiol. Zool.* **21**, 1 (1948).]

mined simply by behavioral interactions but involves pheromones or an interaction of behavior and pheromones. This is an interesting and fertile area. Physiological and hormonal causes of dominance rank are common among mammals, but their existence among insects is yet to be established.

A somewhat similar system may be operating among certain primitively social halictid bees, such as *Lasioglossum zephyrum*. Here, although the colonies are founded by a single female, the workers she rears are morphologically indistinguishable from the queen, except by their less-developed ovaries. There are, of course, behavioral differences, and it has been suggested that the differences between workers and queens are the result of "behavioral control" by the queen. Removal of the queen results in her replacement by the next ranking female, whose ovaries in turn enlarge. Although simple observations cannot distinguish between strictly behavioral and pheromonal control, JH treatment of *L. zephyrum* workers results in increased ovarian development, indicating once again that the queen's behavior (or pheromone) may be acting on worker JH. This hypothesis awaits further investigation.

10.12. CONCLUSIONS

One of the implicit underlying principles uniting behavioral physiology is that much of behavior is "hard-wired" into the nervous system and is released by a wide variety of mechanisms under a broad array of external conditions. Behavior is a critical interface between an organism's physiological needs and its ecology. Therefore, behavioral physiology will continue to be an important interface between physiology and ecology, in part because it identifies the extrinsic factors to which behaviors are tuned and thus contributes information on adaptiveness.

A central question is how the "hard-wired" behaviors come to be there, that is, the process of behavioral development. The few studies that have approached this question have investigated primarily the anatomical changes in the developing nervous system, assuming a link between neural anatomy and behavior. While this expectation is probably fulfilled on a gross level, behavioral development undoubtedly proceeds through modes other than the gross anatomical ones. The field is largely unknown territory and awaits clever and dedicated pioneers.

The study of behavioral releasers and, to a lesser extent, switches of behavioral state, are primed for the contribution of more examples to demonstrate the diversity or universality of the mechanisms involved. The number of case studies is still small, but the road has been clearly marked.

For the lover of *terra incognita,* the place to be is the effect of behavior on physiology. There are few uniting principles and very few recognized examples, yet, in many cases, it seems likely that behavior must have an influence back on physiology, closing the feedback loop. Novel phenomena undoubtedly await the explorer who ventures into this virgin territory.

REFERENCES

R. H. Barth and L. J. Lester, *Ann. Rev. Ent.* **18,** 445 (1973).

L. Barton-Browne, *Adv. Insect Physiol.* **11,** 1 (1975).

T. M. Casey, "Behavioral mehanisms of thermoregulation," in B. Heinrich, Ed., *Insect Thermoregulation,* Wiley-Interscience, New York, 1981.

A. D. Lees, *Adv. Insect Physiol.* **3,** 207 (1966).

S. E. Reynolds, *Adv. Insect Physiol.* **15,** 475 (1980).

P. F. Röseler, I. Röseler, and A. Strambi, *Insectes Soc.* **27,** 97 (1980).

J. W. Truman, *Horm. Behav.* **10,** 214 (1978).

W. R. Tschinkel, *Anim. Behav.* **29,** 990 (1981).

B. Uvarov, *Grasshoppers and Locusts,* Vol. 1, Cambridge University Press, 1966.

11

REPRODUCTIVE SYSTEMS

S. J. Berry
Department of Biology
Wesleyan University
Middletown, Connecticut

CONTENTS

SUMMARY

Reproductive strategies of insects are extraordinarily diverse, reflecting the full range of complexities one might expect of the most successful of all animal groups. Nurture of the developing embryo illustrates all the strategies developed by other groups of animals including oviparity, ovoviparity, viparity, and subsequent nourishing with glandular secretions. More unusual nutritive mechanisms include the devouring of the maternal tissues by the larvae. Although sexual reproduction is the general mode, parthenogenesis is also common, and chromosomal mechanisms of sex determination may vary in different species. External genitalia usually include an intromittent organ in the male and a receptive organ in the female, but either or both may be duplicated or absent. Ovaries and testes are formed from a basic tubular structure which is modified in different ways.

The protein yolk of eggs consists mainly of peptides from the fat body transferred through the blood to the ovary. The nucleic acid content of the eggs may be synthesized using the gamete chromosomes as template or the chromosomes of trophic cells. Seminal fluid and various noncellular coatings of the eggs are synthesized by accessory glands.

Various neural mechanisms and accessory structures and secretions are involved in the location of receptive partners and the release of copulatory behavior. The role of hormones in the regulation of growth and development has been extensively studied, but many of the features of the "fine-tuning" of reproductive mechanisms by hormones are still poorly understood.

11.1. INTRODUCTION

Reproduction is the substrate upon which evolution must operate, because the main force of selection pressure is brought to bear on reproductive structures and strategies. It would be logical to suppose that the largest and most diverse groups of animals, the arthropods, and particularly the insects, should evolve an extraordinary array of reproductive mechanisms. This is indeed the case, but it is only possible in a single chapter to emphasize the most generalized descriptions of major trends in the evolution of insect reproduction. Whenever possible, some of the range of solutions to particular problems will be described. In some cases, the range of these solutions can be conveyed by examining a single order or family of insects. Regrettably, it will not be possible to cover in detail the rich array of imaginative and sometimes bizarre reproductive mechanisms found in nature.

11.1.1. Nurture

A simple example of the diversity of mechanisms is the provision of "prenatal" care for the developing embryo. All possible permutations of mechanisms for the provision of nutrients and mechanical protection seem to be represented. The most common mode is oviparity, in which a self-contained egg is produced which contains substantial nutrient supplies enclosed in a protective chorion. These eggs hatch outside the female reproductive tract in the familiar manner of a chicken egg. Ovoviviparity is less common and involves the retention of a shelled egg within the reproductive tract until hatching or just before hatching. In both of these situations the nutritional needs of the embryo are contributed by the ooplasm, the egg cytoplasm. Viviparity, by contrast, involves not only the retention of the egg within the reproductive tract, but the provision of substantial nutrient supplies for the developing "fetus." In the cockroach suborder Blattaria, all three mechanisms are represented. *Blatta orientalis* deposits an egg case soon after it is formed, while *Blattella germanica* carries the egg case internally until just before the eggs hatch, and *Diploptera punctata* nourishes the naked developing larvae in a brood sac. A particularly noteworthy example of viviparity is found in the tsetse fly, *Glossina palpalis,* in which the larvae are maintained, one at a time, and nourished by the secretions of glands in the wall of the uterus. The secretion of these glands is released at the apex of a papilla directly into the mouth of the larva.

Eggs provide a proportion of the nutrition of the early larvae of some species of eusocial insects (bees, wasps, ants, and termites) and also some crickets. In some cases these eggs may be conventional oocytes, at least potentially fertilizable, but most are specialized trophic eggs. Trophic eggs generally have little or no chorion and are produced either by the reproductive female or by nonreproductive workers. They are produced specifically for nutritional purposes and are sometimes consumed by adults, but generally by larvae. Some trophic eggs consist of several fused, incompletely differentiated oocytes and have been described as a "trophic omelette."

Perhaps the ultimate example of maternal provision of nutrition is found in the cecidomyiid flies, where the eggs begin development in the reproductive tract, and the larvae consume the maternal tissues. They eventually "hatch" by rupturing the empty shell, which is all that remains of the mother. Another feature of reproduction in this group is that it occurs in immature females by a process referred to as paedogenesis. Although the somatic tissues of the female are immature, the ovaries are functional, and the unfertilized oocytes begin parthenogenetic development. Paedogenesis does not always require parthenogenesis, however, as the psychid Lepidoptera demonstrate. The adult males of these groups are easily recognizable as winged moths, while the females inhabit silken bags (bagworms) and resemble pupae. The females show no trace of wings, antennae, compound eyes, or other structures associated with adult Lepidoptera. The male alights on the bag and copulates with the female, which never emerges from this structure.

11.1.2. Sexuality

Sexual reproduction is the general rule in insects, although many exceptions and modifications are observed. As already mentioned, parthenogenesis is common in some species of flies and in aphids, wasps, and Lepidoptera. Cyclical and facultative parthenogenesis occurs in aphids, cynipid wasps, and cecidomyiids and involves production of individuals that reproduce either sexually or asexually. When the fungus on which the cecidomyiids feed dries up, they convert from wingless, paedogenetic, and parthenogenetic forms to winged adults that reproduce sexually. Aphids and cynipid wasps reproduce asexually during the spring and summer and then revert to sexual reproduction in the fall. Sexual reproduction allows the cecidomyiids to disperse to new sources of food, while in the aphids and cynipids it allows them to produce eggs adapted for winter survival.

In the more highly evolved social insects, reproduction is limited to a small number of individuals, often one queen and a small number of males (drones). Differentiation of primary and secondary sexual characters is suppressed among the remaining population, and suppression of sexuality is maintained by pheromones secreted by the reproductives. In some cases, loss of the reproductively differentiated adult will release the workers from the block and allow sexual differentiation to occur.

In the Hymenoptera, males are haploid and are produced by unfertilized eggs. The parthenogenetic production of males, termed *arrhenotoky,* is facultative in the sense that the female lays both fertilized and unfertilized eggs. *Thelytoky,* the production of females from unfertilized eggs, is common among the phasmids and may be obligate in some species where males are rare or nonexistent. Thelytokous females, which are diploid, are produced by eggs that fail to undergo meiosis or by fusion of a polar body with the oocyte.

11.1.3. Sex Determination

Sex determination is comparable to that for vertebrates, with the female being the homogametic sex (XX) and the male the heterogametic (XY). Exceptions are the Lepidoptera and Trichoptera, in which, like the birds, the males are homogametic (ZZ) and the females heterogametic (WZ). Sex-determining mechanisms are not as simple as they first appear. Autosomes as well as sex chromosomes appear to be involved in determination, and in some cases a Z0 genotype may produce a male, while in others, it results in a female phenotype. If a sex chromosome is lost from a mitotic spindle during embryonic divisions, mosaic mixtures of cells of both sexes can exist in the same individual. If the mosaic cells include the gonads, a gynandromorph is produced. A number of other unusual variations in chromosome mechanics have been observed in different insect species. In some of the scale insects, the paternal set of chromosomes is heterochromatic in the somatic cells of the males. The result of this chromosome behavior is that phenotypic traits are inherited from maternal grandfathers, but not from fathers.

There are at least two well-documented examples of phenotypic reversal of sexual genotype. The "transformer" mutation in *Drosophila* converts genotypic females into individuals whose somatic tissues show typical male characteristics and which attempt to copulate with females. The gonads appear to be rudimentary, nonfunctional testes that, however, contain cells resembling trophocytes. Feminization of genotypic male larvae of the mosquito, *Aedes stimulans*, can be accomplished by raising them at the elevated temperature of 27°C. Ovaries induced by this procedure are functional, and male accessory glands are replaced by a spermatheca.

11.2. EXTERNAL GENITALIA

The organs specialized for copulation and oviposition vary greatly even between species and, because they are readily visible, have been adopted by taxonomists as prime features for distinguishing between related species (Fig. 11-1). The lack of "fit" of copulatory organs may be a major barrier to hybridization and thus may effectively separate species that are still genetically compatible.

11.2.1. Male Genitalia

The mesodermal tubules that will form the *vas deferens* attach to the ninth abdominal segment where normally a single ejaculatory duct forms. In some primitive forms, including the earwigs, the two tubes do not fuse with the single ejaculatory duct but are connected with paired external structures. In some of these species, one of the paired penes is greatly reduced in size, and most insects, including the more advanced earwigs, have a single penis located in the midline. The major segment of the penis is the aedeagus, which is often withdrawn into the body cavity but is everted during copulation. Other modifications of structures on the ninth segment may include clasping organs that assist in initiating and maintaining copulation.

11.2.2. Female Genitalia

In the mayflies, the most primitive arrangement of reproductive organs is maintained. There are no accessory glands in either sex, and the oviducts open directly to two genital pores behind the seventh segment. The males are equipped with short, paired penes. Most other insects have a single common ovipositor associated with the eighth or ninth abdominal segment. In the apterygotes and some of the winged insects, the ovipositor is a simple opening for both copulation and for the deposition of eggs. The structure and elaborateness of the ovipositor is determined by the site of egg deposition. The most common site is the surface of plants or shallow cracks and crevices of various kinds. For these sites, the ovipositor is a relatively short tube from which the eggs emerge coated with colleterial secretions. The ovipositor of *Drosophila* has sharpened ends that penetrate the surface of fruit,

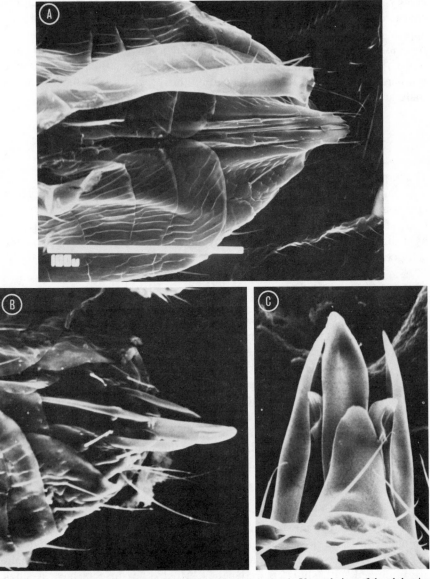

FIGURE 11-1. External Genitalia of *Trichogramma minutens*. A. Ventral view of the abdomen of a female. The ovipositor of this tiny wasp is used to deposit eggs into the eggs of various Lepidoptera. The ovipositor is hinged at the tip and folded back against the abdomen. (500×) B. The tip of the abdomen of a male, ventral view. (850×) C. Dorsal view of the male genitalia. (2000×)

while the ovipositors of some of the predatory wasps are long and sharp to allow penetration of the body of the insect prey. Possibly the most spectacular ovipositors belong to the ichneumonid wasps, which are parasitic on the larvae of Lepidoptera and woodboring beetles. One genus, *Thalessa*, parasitizes the wood-boring larva of the pigeon horn-tail, *Tremex columba*, and the ovipositor may be up to 15 cm in length. The ovipositor of stinging insects such as bees is modified by the addition of poison glands and reservoirs that evacuate the venom through the hollow sting.

The final abdominal segments are often modified to form various structures that aid in the process of egg deposition. Specialized "valves" formed by the terminal abdominal segments of grasshoppers function as shovels to excavate hollows in the soil into which the eggs are deposited. Pressure for digging and later oviposition is developed by inflation of air sacs and swallowing air to extend the length and volume of the abdomen.

11.3. INTERNAL ORGANS OF REPRODUCTION

11.3.1. The Structure of the Ovary

Ovaries of insects are composed of strings of developing oocytes, in which the most mature oocytes are located nearest to the common oviduct and the others are progressively less mature as one moves toward the germinal epithelium where division of the germ cells occurs (Fig. 11-2). Maturing oocytes are encased in a continuous, single layer of follicle cells, if any, and follicle cells form continuous strings within the separate oviducts. The mature, chorionated oocytes are moved along the separate oviducts to a common oviduct where fertilization occurs and where various colleterial secretions may be added.

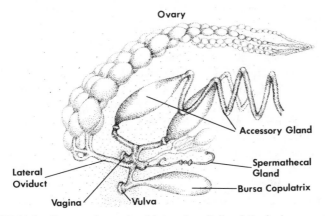

FIGURE 11-2. Ovary and associated internal genitalia of *Hyalophora cecropia*.

Most eggs contain relatively large amounts of stored material which support the early development of the embryo. This stored material consists of yolk proteins, carbohydrates, and lipids that are synthesized outside the oocyte and are contained in membrane-bound vesicles. Two major classes of stored compounds are nucleic acids and ribonucleoproteins. If these compounds are synthesized by trophic cells that remain attached to the oocyte, the ovary is classed as meroistic. Meroistic ovaries are further subdivided into polytrophic, in which the trophic (nurse) cells are included within the follicle with the oocyte, and telotrophic, in which the oocyte is attached by a long cellular process to a syncytial mass of trophic tissues at the distal end of the ovary (see Fig. 11-3).

The panoistic mode of oocyte construction, probably the most primitive, is found in the oldest families of insects, such as the cockroaches. The set of maternal chromosomes, which will also contribute to the zygote chromosome complement, is the site of synthesis of ribosomal, transfer, and maternal messenger RNA that will support the early development of the embryo. The lampbrush configuration of chromosomes, which is thought to be a manifestation of this synthesis in sea urchins and vertebrates, can be demonstrated in the developing oocytes in panoistic insect ovaries but not in meroistic. In some species, large accumulations of extrachromosomal DNA are found in the nucleus. The best-understood example of this phenomenon is the Giardini body of the water beetle *Dytiscus,* which may contain over half the nuclear DNA of the oocyte. This DNA seems to be enriched for the ribosomal locus but appears to contain other chromosomal regions as well.

Providing the developing oocyte with enough ribosomes to complete the formation of the pregastrulation embryo is a problem faced by all animals that produce eggs larger than somatic cells. This problem is exacerbated by the fact that until shortly before gastrulation, embryos produce none of their own ribosomes, but produce large amounts of new protein. Ribosomal RNA makes up more than 90% of the RNA of all cells, including eggs. Constructing a giant cell such as an egg could thus be expected to require many more ribosomal genes to serve as template than are required for a normal somatic cell. The Giardini body could help to supply ribosomal genes but is confined to a few species of insects and generally disappears at an early stage of oogenesis. Panoistic follicles contain no accessory cells such as nurse cells to provide extra template. The chromosomal ribosomal region is used as a template to produce thousands of copies of the ribosomal genes in the form of small, double-stranded circular DNAs which are no longer part of the chromosome. This "amplified" ribosomal DNA is gathered into accessory nucleoli retained within the oocyte nucleus and transcribed to produce ribosomal RNA. Selective amplification of the ribosomal genes has been adopted for oogenesis in amphibians and probably other vertebrates, as well as the less evolutionarily advanced insects. Amplified ribosomal DNA has been studied particularly extensively in crickets (see review by Cave, 1982).

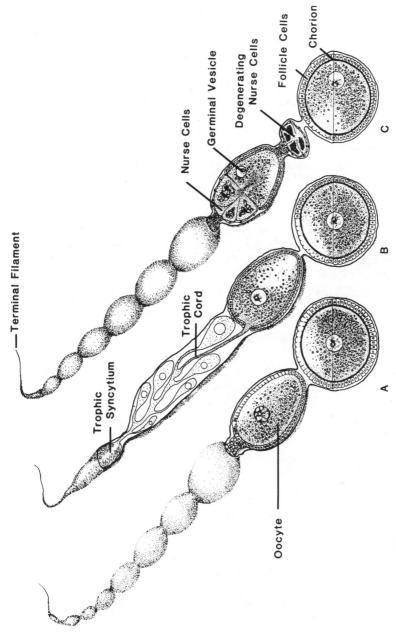

FIGURE 11-3. Three major types of insect ovary. A. Panoistic, in which no trophic cells are involved. B. Telotrophic, in which trophic cords connect the oocytes with the trophic syncytium. C. Polytrophic, in which trophic nurse cells are included in each follicle.

445

By contrast, in meroistic ovaries, the oocyte genome is segregated from the genome of the sister trophic cells whose synthetic activity will provision the egg with RNA. Thus, the chromosomes that will contribute to the fertilization process can be maintained in a quiescent state, while transcription for provisioning of the oocyte utilizes chromosomes that will not be utilized in the embryo. One advantage of the polytrophic mechanism is that the genome of the trophic cells can be replicated many times to supply template to support the synthesis of the enormous number of transcripts stored in the oocyte. This duplication of the relevant genes, such as those for ribosomal RNA, can be accomplished without the specific amplification necessary for the production of panoistic oocytes. The nurse cells in the follicle of the fruit fly have been estimated to contain 1500 times as much DNA as the diploid cells, and in the cecropia silkmoth direct measurements reveal 37,000 times as much DNA (see Fig. 11-4).

Elements synthesized in the trophocytes are transferred to the oocytes through cytoplasmic connectives called ring canals or fusomes. The mechanism of transport has recently been shown to be migration down an electrical field maintained along the length of the follicle (Woodruff and Telfer, 1973). The transport of labeled RNA can be seen in autoradiograms of the ring canal region (see Fig. 11-5).

11.3.2. Formation of Germ Cells

Both sperm and eggs are derived from primordial germ cells set aside very early in the development of the embryo. The area of ooplasm where the presumptive germ cells will form can be identified by differential staining properties in many species before the egg is fertilized. This specialized cyto-

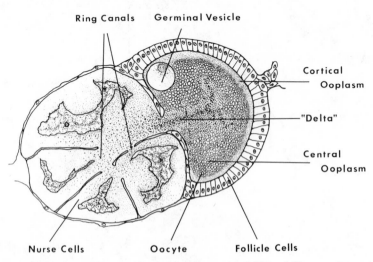

FIGURE 11-4. Diagram of a developing ovarian follicle of *H. cecropia*.

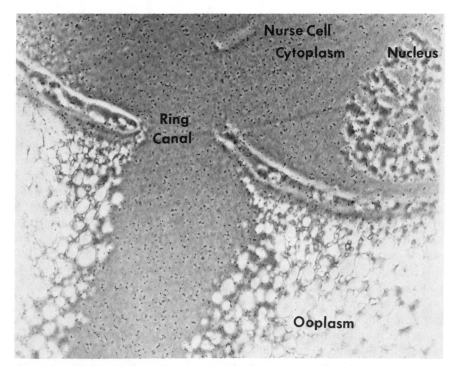

FIGURE 11-5. RNA in transit from nurse cells to the ooplasm. In this light micrograph, radioactive polyuridine has been annealed to the complementary polyadenylic acid regions of mRNA. The small, dark black dots are silver grains activated by disintegration of tritium in the labeled polyuridine.

plasm, often located at the posterior end of the egg, can be identified at the fine structure level by the presence of large, dense granules which are not found in the general ooplasm. When this area of the ooplasm is colonized by nuclei, a small group of large cells with rather slow mitotic activity is formed. These cells, referred to as pole cells, will later migrate to the interior of the embryo and eventually populate the gonads whose nongerminal tissue is constructed by cells derived from mesoderm. The mesodermal elements will also contribute the follicle cells in the ovarian follicles.

The pole cells share some unique and unexplained characteristics with the primoridal germ cells of nearly all animals. As indicated above, germ cells arise from specialized egg cytoplasm and remain aloof from the developing somatic tissues until the construction of the gonads is complete. When the gonads are prepared to receive them, germ cells migrate along the surface of other developing organs from their position in the uncleaved yolk, extraembryonic membranes, or the surface of the embryo, to the gonads. Meiosis, which may include some rather extraordinary changes in chromosome morphology, is restricted to the germ cells, the only cells that normally exhibit genetic programs that lead to the construction of eggs and sperm.

One exclusive germ cell mechanism that has been observed particularly well during the early development of certain midge embryos is chromosome elimination. One example is found in the development of the gall midge, *Mayetiola destructor,* described by Bantock (1970). The nucleus formed by the fusion of the egg and sperm pronuclei contains 40 chromosomes. At the fifth nuclear cleavage stage, the nuclei that have migrated into the region of the egg cytoplasm containing the polar granules retain their 40 chromosomes, while the others abruptly reduce the number of chromosomes to 6–8. The change in the number of nuclei from 16 to 28 precludes a mechanism involving lack of a round of replication and dividing up of the chromosomes already present. If a ligature is placed around the embryo so that it produces a constriction that prevents nuclei from colonizing the pole plasm, a normal embryo forms, but the gonads lack any eggs or sperm, and the individual is sterile. If the ligature is removed after all the nuclei have reduced their chromosome content to 6–8, these nuclei can populate the pole plasm, but still no germ cells are formed. The result in this admittedly extreme case is that the cells of the "germ line" retain 40 chromosomes from generation to generation, while those of the "somatic line" are reduced to 6–8 chromosomes. The somatic line can produce all the differentiated cells of the larva, pupa, and adult with the exception of eggs and sperm. Two possible conclusions are, first, that one function of the polar granules is to prevent chromosome reduction and, second, that unique information concerning the formation of germ cells is lost during the reduction. Future research on this problem may determine whether this loss of chromatin is a general mechanism, simply exaggerated in the gall midges, and whether significant amounts of unique genetic information are destroyed in somatic cells.

In meroistic ovaries, the germ cells will form both oocytes and trophic cells. This relationship is very clear in panoistic ovaries where incomplete cytoplasmic division leaves the presumptive oocyte attached to the presumptive nurse cells by strands of cytoplasm (ring canals). The number of nurse cells varies according to species. In some of the earwigs, one nurse cell is associated with each oocyte, while in moths and flies 7–15 nurse cells are present, and 31 nurse cells are found with each oocyte in some wasps. Since the number of nurse cells is determined by synchronous divisions of the germ cell, and one daughter cell is reserved as the oocyte, the number of cells per follicle can be expressed as 2 raised to the power of the number of division cycles minus 1. For instance, in the cecropia silkmoth, three division cycles ($2^3 = 8$) produce seven nurse cells ($8 - 1$) and one oocyte. After germ cell division, the nucleus of the oocyte becomes synthetically inactive, while the nuclei of the presumptive nurse cells embark on an extensive program of RNA and DNA synthesis as well as some protein synthesis. When developing follicles are examined, the potential oocyte can be identified in the cluster of sister cells because the nucleus contains synaptonemal complexes. These complexes are thought to represent chromosomal pairing typical of cells entering meiosis rather than mitosis. These clusters are produced in the most distal area of the ovary, called the germarium, and migrate into the next region, the vitellarium, where the oocyte beings to accumulate yolk while the nurse cells enlarge by endomitosis.

Oogenesis is less well defined in the telotrophic ovary because, although the nutritive cord seems analogous to the ring canals in function, the relationship to individual trophic cells is less certain. In some species the cell boundaries of the trophic cells break down and form a trophic core which contains no plasma membranes but a large number of nuclei. There is direct autoradiographic evidence for the transfer of nucleic acid from the trophic cells to the oocyte via the trophic cord.

The bulk of protein and lipid reserve stored in the ooplasm is synthesized in the fat body and taken up from the blood by pinocytosis at the surface of the oocyte. To reach the surface, blood proteins must pass through channels between the follicle cells (see Fig. 11-6). These channels, which are very wide during the period of rapid yolk accumulation, are obliterated after yolk accumulation is complete. All the proteins in the blood are represented in the yolk, including foreign proteins such as ferritin, which can be injected into the blood and then identified by its density in electron micrographs. While the process is not selective in the sense that particular proteins are excluded, certain proteins are nevertheless concentrated. Vitellogenins, for instance, are 20- to 30-fold more concentrated in the ooplasm of cecropia oocytes and make up 80–90% of the yolk protein. In most of the species examined vitellogenin is synthesized by the fat body, but only in females. However, there are exceptions to this generalization. There are reports in the literature of vitellogenin synthesis in the ovary and even by males of a few species. Vitellogenins may be modified in the process of being taken up from the blood and are often referred to as vitellins when extracted from the oocytes. Vitellogenins and vitellins generally have two or more subunits with a cumulative molecular weight of about 10^5.

Some of the minor proteins found in the yolk are supplied by the follicle cells rather than the fat body. Because of the origin of the yolk proteins outside the oocyte and the pinocytotic mechanism of accumulation, they are contained in membrane-bound vesicles. Other proteins, not associated with the yolk, are probably synthesized by the ooplasm or the trophic cells. Glycogen in the cytoplasm is thought to be polymerized by oocyte enzymes, while the lipids are assumed to accumulate by pinocytosis.

A noncellular vitelline membrane surrounds the oocyte, which increases in width as oogenesis progresses. The chorion, or egg shell, is formed on top of the vitelline membrane by the altered secretory activity of the follicles once vitellogenesis is complete (see Fig. 11-7). The chorion is constructed in several layers, each with its characteristic structure, and more than 20 different proteins are involved in the construction of the entire shell. Scanning electron microscopy has revealed elaborate and precise structures that may vary over different portions of the shell, such as the micropile, which permits sperm entry, and various respiratory structures. The structure and

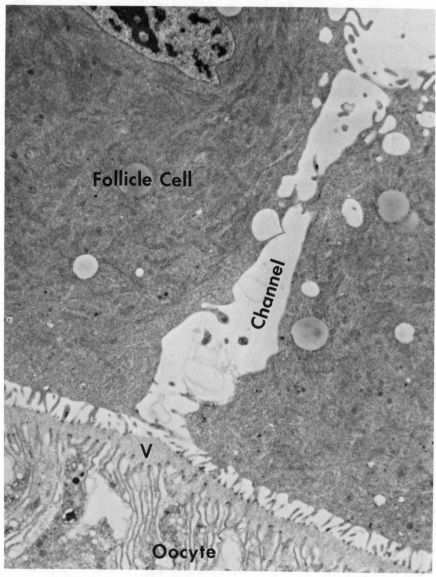

FIGURE 11-6. Electron micrograph of the surface of a developing oocyte of *H. cecropia*. Microvillae are seen extending down from the follicle cells and up from the oocyte to the forming vitelline membrane (V). A typical open channel from the hemocoel to the surface of the oocyte is plainly visible.

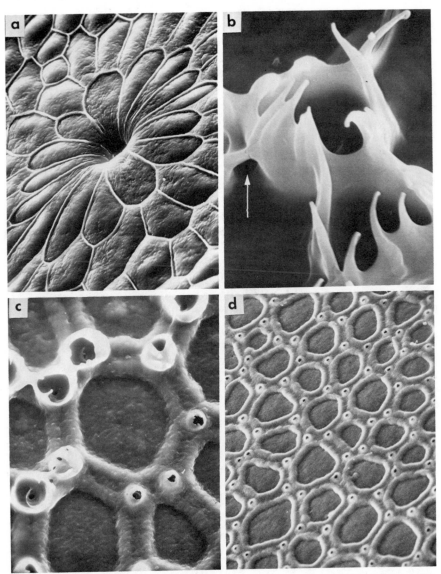

FIGURE 11-7. Scanning electron micrographs of structures on the surface of the chorion of *A. polyphemus*. a. Fifteen petal-shaped follicle cell imprints surround the opening of the micropyle (×875). b. Tall aeropyles. Tiny tunnels (arrow) appear between contiguous aeropyles (×2450). c. Transition zone between tall (left) and short (right) aeropyle zones (×875). d. Zone of short aeropyles. The aeropyles appear at the junctions of follicle cell imprints with an average of six surrounding each cell imprint (×440). (Reproduced by permission of F. C. Kafatos and Springer Verlag.)

synthesis of the chorion of several different insect eggs have been studied in considerable detail (Kafatos et al., 1977).

Oocytes may be destroyed in the ovary in response to a number of stimuli. In some Lepidoptera the first few oocytes that mature during the pupal–adult transformation routinely break down and are resorbed. Starvation or failure to copulate can also stimulate the destruction of mature oocytes. It has been suggested that the "shelf-life" of mature oocytes is limited, and if fertilization does not occur within a specific period of time, the oocyte is destroyed. The actual destruction is accomplished by the follicle cells. After assisting in provisioning of the oocyte, the follicle cells can activate new genetic programs resulting in either the synthesis of the chorion or the destruction of the oocyte.

Many insects require the services of symbiotic organisms in order to digest particularly fibrous foods such as wood or to synthesize required dietary constituents. "Infection" of the next generation by symbionts is accomplished in different ways, but commonly, bacteria are released from the maternal mycetomes and incorporated into the eggs. In the Curculionidae, symbionts are found in the primordial germ cells, while in many ant species they pass through the body of the follicle cells into the ooplasm. A number of specialized mechanisms, including receptive regions in the follicle cell investment and wandering mycetocytes which carry bacteria into the oocyte, ensure that the progeny are equipped with the required symbionts.

11.3.3. Structure of the Testis

The testis is formed from a tubule similar to the basic ovarian tubular structure (see Fig. 11-8). In some more primitive wingless insects and beetles this single tubule is the form of the adult testis. More often the testis is divided into subdivisions, each of which opens separately into the *vas deferens* via a short *vas efferens*. Generally the tubules are surrounded by a connective tissue capsule, the peritoneal sheath. The *vas deferens* widens into the seminal vesicle where mature spermatozoa accumulate, and generally further along its course the accessory glands enter. The *vasa deferentia* finally fuse to form a single ejaculatory duct that opens to the outside through the intromittent organ, the penis. In some Lepidoptera, the testes fuse secondarily to form a single testis, but with two sperm ducts retained.

Spermatogonial cells are found at the distal end of each tubule in close association with a giant apical cell. Connections with the apical cell are subsequently lost, and one or more spermatogonia establish cysts that migrate toward the *vas efferens*. Spermatogonia divide 6–8 times to produce 64–256 diploid spermatocytes. Haploid spermatids are formed by the meiotic division of the spermatocytes and mature to form spermatozoa. The cells in each cyst remain synchronized during their development, and cytoplasmic bridges reminiscent of those formed between nurse cells and oocytes are maintained for a short period. The pressure resulting from the

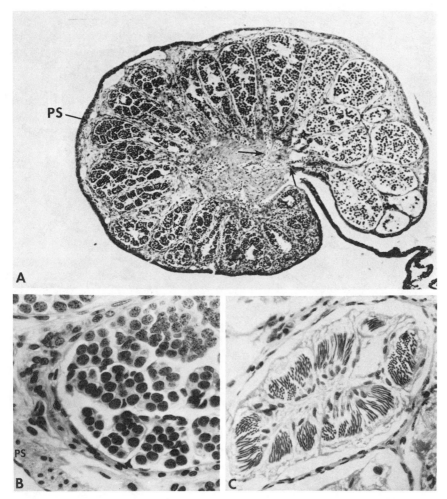

FIGURE 11-8. Testis of *Hyalophora cecropia*. a. Section of the entire testis showing the segmental arrangement of the tubules within the peritoneal sheath (PS). Arrows indicate the short *vasa efferens* which connect the tubules with the *vas deferens* seen approaching the testis from the lower right (×50). b. Higher magnification of the peritoneal sheath and the tip of one tubule. Differences in the state of chromosome condensation are seen in different clusters of spermatocytes (×400). c. A tubule containing nearly mature spermatids. "Trophic cells" are visible at the tips of some of the bundles of spermatids (×300).

formation of more cysts pushes the more mature bundles of sperm toward the *vas efferens,* producing a graded series of maturing cysts with the most mature at the *vas efferens* (see Fig. 11-9).

Other nongerm cells, associated with the cysts, are often referred to as trophic cells in analogy to the Sertoli cells of the vertebrate testis. No nutrient function for these cells has yet been definitely established.

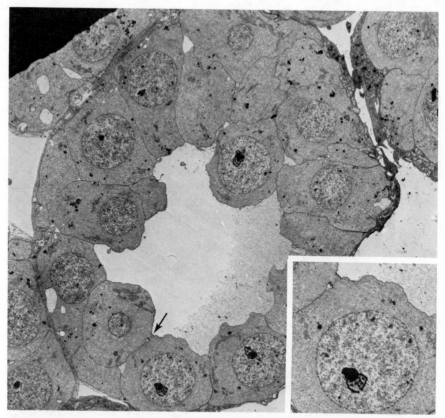

FIGURE 11-9.   Electron micrograph of a developing cyst in the testis removed from a pupa of *H. cecropia* just prior to completion of adult development. The inset is an enlargement of the area indicated by the arrow and clearly indicates a cytoplasmic bridge between two differentiating spermatocytes.

Differentiation of sperm follows the general pattern of vertebrate spermatogenesis: sperm cytoplasm is reduced, chromatin condensed, the mitochondria fuse to form one or a few nebenkern structures, and the typical 9 + 2 motile apparatus forms in the flagellum (see Fig. 11-10). As might be anticipated, there are some unusual structural departures whose significance is not readily apparent. A few examples of modification of the motile apparatus include the single central fiber in the sperm of the mosquito *Culex,* three central fibers in one species of fungus gnat, and seven central fibers in the caddisfly, *Polycentropus.* In some treehopper species, the flagellum divides into four branches, each of which contains two or three of the outer doublets. Among the distinctive features of the sperm of *Sciara* are 70 outer doublet fibers.

Two types of sperm, designated as eupyrene and apyrene, are produced

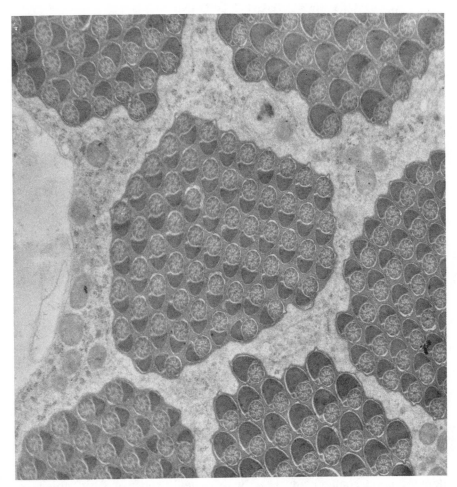

FIGURE 11-10. Bundles of developing sperm in the testis of *Drosophila*. The darker trapezoidal objects are the sperm nuclei, the wheel-shaped structures of the motile apparatus. (Electron micrograph courtesy of Professor B. I. Kiefer.)

by many species of Lepidoptera. The eupyrene sperm are involved in fertilization, while the apyrene sperm lack nuclei and thus serve no genetic function. The nuclei of apyrene sperm disappear by autolysis in late stages of maturation, but differences in the fine structure of the two types are detectable as early as the secondary spermatocyte stage. The apyrene sperm may be more motile than eupyrene under some conditions, and it has been suggested that their motility may assist in moving the nucleated sperm in the female reproductive system.

11.4. ACCESSORY REPRODUCTIVE ORGANS

11.4.1. Female Accessory Organs

The vagina, which is the common opening of the oviducts, serves as the copulatory duct as well in many species. In certain groups, notably the higher Lepidoptera, a separate copulatory opening, the vulva, leads to a sperm reception chamber, the *bursa copulatrix*. The *bursa*, in turn, is connected to the vagina by a short canal. If there is a single genital opening, the wall of the vagina is expanded into a pouch, the spermatheca. The *bursa copulatrix* and the spermatheca both serve as storage organs for sperm and release single sperm to the vagina where fertilization occurs as the oocytes pass down the reproductive tract. In ovoviviparous and viviparous species, a brood pouch or uterus, respectively, forms as an expansion(s) of the vaginal wall. Because they are of ectodermal origin, the vagina, spermatheca, *bursa,* and other pouches have a cuticular lining.

Secretory cells are associated with the walls of the spermatheca in some insects, and in others a separate glandular tubule, the spermathecal gland, empties into the spermatheca. The functions of spermathecal secretions have not been studied extensively. Some observations suggest that attraction of the sperm to the spermatheca and alteration of the beat frequency of the flagellae may be functions of the spermathecal secretions.

The cuticular lining of the *bursa copulatrix* may be modified to form one or more toothlike structures, the signa, which may aid in mechanical disruption of the spermatophore. Secretions of the *bursa* may also aid in digestion of the spermatophore.

11.4.2. Colleterial Glands

Most insects produce protective coatings for the newly laid eggs. The Lepidoptera simply coat individual eggs with a gluelike substance that hardens on contact with the air and attaches the eggs to the substrate on which they are laid. This attachment ensures that the larvae will hatch out in the vicinity of a supply of food such as a plant leaf. More familiar are the egg cases of the praying mantis, which are constructed of polymerized protein foam. Aquatic insects often produce egg cases of a gelled substance, but the most complex of these structures may be the tough ootheca of the cockroach, which consists of several layers of tanned protein formed into complicated egg chambers with elaborate respiratory apparatus and release valves.

The source of all these protective devices are the accessory or, more precisely, the colleterial glands, which also branch off the vagina. In the Lepidoptera, the colleterial glands consist of a pair of tubules subdivided into a proximal sac, in which secretion is stored and a tanning agent secreted, and a thinner duct responsible for synthesis and secretion of the major protein components. The colleteria of the cockroach present an interesting biochemical bilateral asymmetry. The left gland consists of a series of tubules that produce a secretion containing structural protein, an oxidase, and calcium oxalate, while the tubules on the right side secrete structural protein plus protocatechuic acid, a substrate for the oxidase. These secretions are mixed and extruded into the brood pouch where the ootheca is formed (see Chapter 4, Section 4.6.1).

11.4.3. Male Accessory Organs

The seminal vesicles are formed as modifications of the *vas deferens* and store mature sperm (see Fig. 11-11). The sperm appear to be immotile while in the seminal vesicle and the *vas deferens*. Some components of the seminal fluids appear to be added during the storage period in the vesicles.

Accessory glands comparable to the colleterial glands also open into the *vas deferens* or the ejaculatory duct. Those glands, which are associated

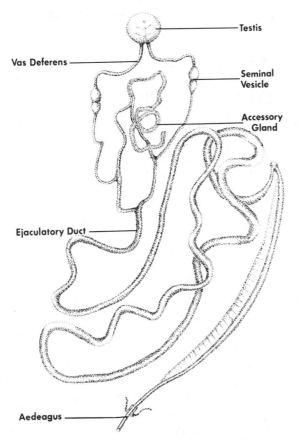

FIGURE 11-11. Diagram of the internal genitalia of *Manduca sexta*. The two testes in this species are fused in the midline, as is typical in many Lepidoptera.

with the ejaculatory duct, are ectodermal in origin and designated as ectadenia; those associated with the *vas deferens,* the mesodenia, are mesodermal.

The accessory glands may consist simply of secretory cells within the walls of the ejaculatory duct, as in the house fly, *Musca domestica,* but more commonly they are typical tubular glands. Like the colleterial glands, the male accessory glands provide a noncellular matrix for the gametes. These coatings for the sperm may vary from seminal fluids and gels to elaborate cuticular spermatophores, comparable to the oothecae of the female. A second function of the accessory gland secretions is the production of a mating plug which coagulates in the genital opening of the female and may either prevent loss of seminal fluid or serve as a barrier to further copulation.

Accessory gland secretions also have functions that depend on both their physiological properties and their biochemical constituents. For example, in many species sperm motility is activated by the seminal fluid. It has been further suggested that sperm activation may depend on augmentation of energy-supplying compounds derived from the seminal fluid. Seminal fluid causes vaginal swelling in fruit flies and mosquitoes, which may mechanically prevent further successful copulation, and in some species contains compounds (e.g., matrone in mosquitoes) that seem to evoke behavioral modifications that lead to avoidance of copulation (see Section 10.5.4).

Seminal fluid also may stimulate egg maturation and oviposition either by eliciting hormonal activity or by increasing the nutrient supply. Boggs and Gilbert (1979) fed male butterflies radioactive amino acids and allowed them to mate with females that were not treated with radiolabeled compounds. When eggs and unfertilized oocytes were extracted, they contained detectable amounts of radioactive compounds which could only have come from the sperm or seminal plasma. Since the unfertilized oocytes were labeled, penetration by radioactive sperm could not be responsible for the labeling, so it seems likely that radioactive amino acids or proteins were salvaged from the seminal fluid and taken up by the developing oocytes.

11.5. MATING

An enormous literature has been generated relative to the mating behavior of insects. As is the case for most groups of animals, mating raises two general problems: recognition and location of receptive individuals, sometimes at considerable distances, and the release of copulatory behavior.

Although visual acuity appears to be poor for most insects, visual recognition can serve as a stimulus. Some males are simply attracted to moving objects of the correct approximate size (e.g., house flies), but many are able to discriminate pattern, color, shape, or behavioral patterns. For example, males of the greyling butterfly *Eumenis semele* will approach flying objects of the correct size. If a dummy is towed at the end of a fishline, the attention of the males will be attracted. If the dummy is towed with an undulatory motion, they will be most strongly attracted.

As another example, although females of the butterfly *Pieris napi* occur as both yellow and white morphs, males prefer females of the same white hue over the yellow morph. If yellow females are painted white, they become as attractive as the white morphs. Other visual stimuli include the familiar flashes emitted by fireflies in which the patterns of light emission are species specific.

However, males are not always the sex to receive the visual stimulus. In several families of flies, the males swarm, often carrying a captured insect or a white balloon or bubble of silk to attract the female's attention. Females approach these swarms, enter, and select a mate, with the choice often depending on the object carried by the male. This is but one example of sexual selection, a major research topic of the last decade (Blum and Blum, 1979).

Sound is also employed in locating females. For example, the frequency of sound produced by the wing beat of female mosquitoes attracts males of the same species. The males can also be attracted by striking a tuning fork of the correct frequency, but other species of males with a different wing-beat frequency are not attracted. Many orthopterans produce extremely complex "songs" using the stridulatory apparatus associated with the legs. The frequency, shape of the pulse, and temporal pattern are all important characteristics, defining not only the emitting species but the physiological status of the emitter. Thus, one song may serve to maintain spacing or initiate swarming, another to attract females, and a fourth to stimulate or reinforce copulatory behavior. The use of sound signals is not confined to mosquitoes and orthopterans; it is employed by additional Diptera, Hemiptera, Coleoptera, and other groups as well.

Vibration frequency on the surface of water as well as in air can serve as a signaling device. Adult male water striders (*Gerris remigis*) produce high (80–90 waves/sec) and low (30–10 waves/second) frequency oscillations by moving their legs, while females produce only low-frequency signals. The low-frequency oscillations serve as an orienting signal in this and other species, while high-frequency signals inhibit attempts to copulate with the sender. Blinded males, attracted to females, were prevented from copulating with them by computer-generated high-frequency oscillations.

One of the most active fields of research into sex attraction mechanisms involves volatile chemicals released by one sex that are perceived by members of the opposite sex. These substances, once referred to as ectohormones and now classified as sex pheromones, will be discussed in detail in Chapter 14. Pheromones generally consist of complex mixtures and in social insects are employed for a number of activities not directly related to reproduction, such as generating alarm reactions and trail marking.

A demonstration of the interaction of pheromones and hormones in controlling reproductive activity is provided by the elegant studies of Riddiford and her co-workers (Riddiford, 1974; Sasaki et al., 1983) on the giant silkmoth, *Antheraea polyphemus*. The newly emerged adult female is not sexually receptive until exposed to fresh oak leaves or to a volatile product of the leaves, (E)-2-hexenal. Impinging of these molecules on the antennal receptors leads to activity in the CNS which causes the release of "calling" behavior. The calling posture involves protrusion of the terminal abdominal segment to expose pheromone-releasing glands. Calling is probably primarily released by electrical activity in the ventral nerve cord, which must be intact for the response to be initiated. Some hormonal component may be involved, because homogenates of active *corpora allata* or blood from calling or ovipositing females will elicit calling in isolated abdomens. Curiously, removal of the *corpus allatum* will not prevent the initiation of calling by a virgin female. Thus, the role of hormones in calling behavior remains to be defined.

Exposure of the tip of the abdomen of the calling female causes the release of a plume of female pheromone, in this case, a mixture of (E,Z)-6-11-hexadecadienal and (E,Z)-6-11-hexadecadienyl acetate. The pheromone interacts with receptors on the antennae of the male and stimulates scanning behavior via the CNS. There is some debate over how effective the gradient of female pheromone is in the actual location of the female, but males seem to be able to locate calling females over considerable distances, perhaps more than 1.6 km, by a process that appears much more efficient than random scanning.

Once the female is located and copulation completed, the presence of sperm in the *bursa copulatrix* stimulates it to release a humoral factor that stimulates oviposition. Implantation of the emptied *bursa* of a mated female into a virgin female stimulates oviposition but does not elicit calling. Transplantation of the *bursa* from one virgin female to another does not stimulate either calling or oviposition in the recipient, indicating that some factor released by an "activated" *bursa* is responsible for activation of oviposition (see Fig. 11-12).

In the case of giant silkmoths, the success of this elaborate mechanism is critically dependent on precise timing of the emergence behavior of both sexes. Giant silkmoths have poorly developed mothparts and do not feed, which appears to restrict the lifetime of the adults to 5–10 days. In some species, the male must fly actively for at least a day before it can respond to the female pheromone. The delayed responsiveness on the part of the newly emerged male may serve to reduce sibling mating, because the males from one batch of eggs may have dispersed several miles from the area where they developed. Females are relatively sedentary and thus stand a good chance of attracting nonsibling males located "downwind" from their site of emergence. The time of emergence of both males and females is determined by a temporal "window," which occurs at twilight and allows the vulnerable, newly emerged adult to expand and dry its wings during the night. The final refinement of this elaborate network of control mechanisms is the slightly accelerated development of males as opposed to females, which results in the earlier emergence of the males, allowing them to disperse for 2–3 days before any females are available for copulation. The

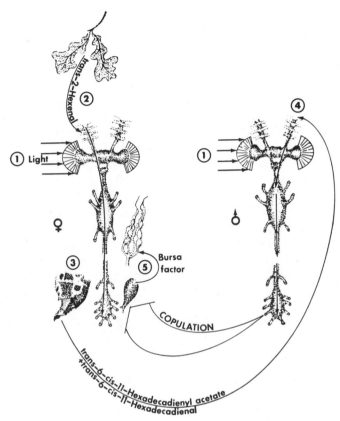

FIGURE 11-12. Summary of the hormonal control of the reproductive behavior of female silkmoths.

environmental cues for the entrainment of these behaviors are photoperiodic.

Many sex attractants also function as aphrodisiacs by releasing or reinforcing copulatory activity, but other types of compounds also possess aphrodisiac activity. For example, some male crickets and cockroaches produce a substance from the metanotal or Hancock's glands which induces the female to mount the male, to feed on the secretion, and thus orient in a copulatory attitude. The secretion may have nutritive value, but it also seems to have a behavior-modifying effect on the female CNS. Similar aphrodisiac activity is associated with a secretion of the hair-pencil butterfly (*Danaus gilippus*), named for a pair of thin glands at the tip of the abdomen. When the glands are brushed over the body of the female, she is induced to land and remain immobile long enough to allow copulation to occur. The tranquilizing effect appears to be produced by the secretion from the hair pencils.

The superior size of the female and the nutritional demands of egg maturation often make copulation a dangerous prospect for males. Indeed, courtship in the mantids has been described as more of a sneak attack than courtship. It involves the patient stalking of the female by the male, generally from the rear. When the male reaches striking range, he springs upon the female, and if she is feeding or recently fed, a long period of copulation ensues. Often, however, the female attacks and feeds on the male. The head is usually devoured first, resulting in more vigorous copulatory movements by the male. Though he may eventually be completely devoured, insemination usually takes place. This results because the feeding female destroys the subesophageal ganglion, which normally inhibits copulatory activity. Copulation by the headless male is still possible because this activity is autonomously encoded in the lower centers. Surgical decapitation of even an immature male results in the release of copulatory behavior and ovipositional movements by the female.

Many species of predatory flies utilize nuptial gifts of captured insects to distract the attention of the female during copulation. A curious sequence of modifications of this behavior has evolved. Some species wrap the offering in silk or blow a balloon around it. The next logical step is for the male to devour all but the cuticular shell of the offering and to present this wrapped in silk or encased in a balloon. The obvious final refinement is found in species that attract the female with empty balloons presented as a distraction during copulation.

11.6. COPULATION

The insertion of sperm or a spermatophore directly into the female reproductive tract is the most common mechanism of insemination. Insemination may be quite distinct from fertilization, since the sperm may be stored for up to several months before they are actually employed to fertilize eggs.

Indirect copulation is found among more primitive insects. The springtails (Collembola) possess no intromittent organs, so the male secretes a stalked spermatophore at the top of which is a droplet of fluid containing sperm. In most species these free-standing spermatophores are deposited in large numbers in an area where a female will be likely to blunder into them with her vulva. In one species, *Sminthurides aquaticus,* the two sexes lock antennae, and the male maneuvers the female into a position where her vulva comes in contact with the spermatophore. The bristletails (Thysanura) also deposit sperm packets on the ground or on silk threads where the female will come in contact with them and inseminate herself. The male often increases the probability of insemination by tangling the female or tying her down with silk in a position from which she will inseminate herself in an effort to escape.

A much less subtle strategy is the "traumatic insemination" practiced by the bedbugs (Hemiptera, Cimicidae). The female has no external genital receptacle to accommodate the intromittent organ of the male. There is, instead, a small notch located asymmetrically on the margin of the fourth abdominal segment with specialized epidermal cells underlying it. This notch, the organ of Berlese, is the target for the sharp, semicircular barb on the penis. The barb actually penetrates the cuticle of the female and allows passage of the sperm into the pouch formed by the underlying cells. The sperm, released from the organ of Berlese into the blood, migrate to the spermatheca where they are stored until the female is engorged with a blood meal. Sperm are then released again into the blood and penetrate the walls of the oviduct to reach the oocytes. It is difficult to construct a coherent hypothesis concerning the selective advantage of such a mechanism, but it is widespread among members of the family Cimicidae.

The most common mode of insemination involves the eversion of an intromittent organ (penis) which is erected by hydrostatic pressure. The penis is then inserted into the vagina or vulva and the sperm transferred to the spermatheca, or *bursa copulatrix*. In addition to the copious seminal fluids mentioned earlier, the spermatophore itself may also provide nutritive substances utilized for the maturation of eggs. In many species, the spermatophore is dissolved by proteolytic enzymes, and presumably the compounds produced by this process can be utilized by the oocytes. However, whereas the female neuropteran *Sialis* removes and eats the empty spermatophore, *Rhodnius* and cockroaches expel and discard it.

11.7. HORMONAL CONTROL OF REPRODUCTIVE ACTIVITY

The hormonal control of insect development is covered elsewhere in this volume (Chapter 13). The natural occurrence of gynandromorphs, in which cells of male and female phenotypes can literally exist side by side in the same individual, led early investigators to the generalization that endocrine control was absent in the regulation of insect reproductive physiology. This view is an obvious oversimplification, but sex-determining hormones such as estradiol and testosterone have still not been demonstrated to play a reproductive role in insects. The regulation of reproductive activity is accomplished mainly by ecdysones and juvenile hormones, whose chemical structures are described in Chapter 13, and to some extent by as yet uncharacterized peptides.

11.7.1. Juvenile Hormone and the Ovary

In *Rhodnius* juvenile hormone triggers vitellogenesis. If a last-instar larva is allatectomized after a blood meal but before a critical period has elapsed, the oocytes do not begin to accumulate protein. If active *corpora allata* are implanted, or if the allatectomized animal is parabiosed to an adult of either sex, the oocytes commence the accumulation of yolk. The role of juvenile hormone thus changes from the maintenance of the status quo during molting to stimulation of differentiative activity in the ovary.

The juvenile hormone-regulated activity can perhaps best be illustrated among ovoviviparous and viviparous cockroaches. In ovoviviparous species the maturation of the oocytes is suspended while the brood pouch is distended by an ootheca. This suspension is achieved by neural inhibition of the secretory activity of the *corpus allatum*. In some species, if the nerve cord of a pregnant female is severed, the release of juvenile hormone is resumed and stimulates the maturation of oocytes even though the brood pouch remains distended. A glass bead of the proper shape and size inserted into the brood pouch will similarly cause a suspension of oogenesis, demonstrating that no hormonal factors from the embryos need be postulated (Engelmann, 1970). In a few species, the regulatory mechanisms may be more complex because the ovary seems to be refractory to juvenile hormone up to the twentieth day of pregnancy, and neither interrupting the nerve cord nor injection of juvenile hormone will stimulate the resumption of vitellogenesis.

Accessory reproductive structures may also require juvenile hormone for activation. In the viviparous cockroaches, a proteinaceous fluid secreted by glands in the brood sac constitutes the major source of nutrients for the developing larvae. Secretion of this "milk" is stimulated by injected juvenile hormone and ceases if the *corpora allata* are extirpated. Secretion of the cockroach ootheca by the colleterial glands is also juvenile hormone dependent. Indeed, so precise is the juvenile hormone requirement that segments of the colleterial tubules have been used as a quantitative assay for juvenile hormone titer.

The role of juvenile hormone in the initiation and maintenance of reproductive organs is not consistent, even within taxonomic groups. For example, in the tobacco hornworm, *Manduca sexta,* vitellogenesis is initiated after the emergence of the adult moth and juvenile hormone is required for this process. On the other hand, in other species of moths, such as the silkmoths, removal of the *corpora allata* has no effect on vitellogenesis or colleterial functions, both of which are initiated during pupal life. Large quantities of juvenile hormone are found in the blood of *H. cecropia* males and, though much of it is concentrated in the male accessory glands, there is no evidence that it is transferred to the female at copulation or has any function in the development or activity of the male reproductive organs.

11.7.2. Ecdysone and the Ovary

Since the prothoracic glands in many species degenerate before vitellogenesis is initiated, it was assumed that ecdysone was involved in the early stages of oocyte determination, but not in vitellogenesis and later maturation. The discovery that the ovary itself can be a source of ecdysone has forced a reexamination of the original assumptions. It seems clear that in the mosquito *A. aegypti* and in blow flies, ecdysone is required for egg production. In locusts and cockroaches, the synthesis of ecdysone by the ovary has been demonstrated, and the hormonal titers in ovaries and blood fluctuate in parallel, probably because the hormone "leaks" into the blood (Bullière et al., 1979). Some authors believe that, in a number of species, ecdysone antagonizes the vitellogenesis-stimulating action of juvenile hormone.

11.7.3. Hormonal Control of the Testis

Hormonal regulation of testicular function has not been examined as extensively as the subject would seem to warrant. Ecdysone appears to stimulate spermatogonial divisions in some species including *Rhodnius prolixus* and locusts. Addition of ecdysone accelerated the conversion of spermatogonia in the G_1 to S and G_2 to M phases of the cell cycle. Experiments by Kambysellis and Williams (1971) suggest that, in the cecropia silkmoth, a blood-borne factor distinct from the molting hormone may be the active agent. If the intact testis is incubated in blood obtained from recently formed diapausing pupae, no acceleration is observed. Ecdysone also has no effect when added to this preparation or to naked cysts removed from the testis. If blood from an injured pupa, or from one stored for a few months, is substituted, the naked cysts will respond but the intact testis also requires ecdysone. These results are interpreted as indicating that ecdysone may serve as a vehicle to facilitate penetration of the peritoneal sheath by a blood factor christened "MF" (macromolecular factor).

Juvenile hormone has been found to antagonize the effects of ecdysone on spermatogonial divisions. Allatectomy stimulates precocious spermatogonial divisions. Implantation of active *corpora allata* in most cases inhibits the divisions, and the few cases in which it has been demonstrated to stimulate division seem attributable to prothoracotropic effects either of juvenile hormone or of neurosecretions carried along with the *corpus allatum*. Injected juvenile hormone analogs most often suppress sperm development (Dumser, 1980).

11.8. CONCLUSIONS

Insect reproductive mechanisms reflect the sort of diversity that one should expect from one of the most ancient, and perhaps the most successful, groups of animals. Many mechanisms parallel those developed by their only serious rivals, the vertebrates, and others are pushed to much greater extremes. The uterine gland of the tsetse fly is an obvious analogue of the mammary gland, for example. It is equally clear that analogous structures represent separate solutions, arrived at in complete isolation from one another. At this juncture, it is difficult to trace the evolution of biochemical control mechanisms, particularly reproductive hormone–target interactions, because we are still in the earliest stages of unraveling the molecular interactions involved. The spectacular progress in defining general mechanisms of steroid and peptide hormone control of vertebrate, particularly mammalian, reproduction has not found a parallel in studies of insect reproductive endocrinology. It seems clear that interaction of steroids with chromatin and cyclic nucleotide mediation of peptide hormone signals may be as applicable to insects as to vertebrates.

One reason for the lag in probing insect mechanisms may be the very diversity of the situations to be examined and the occasionally contradictory results encountered. Furthermore, the clinical imperative to unlock the secrets of insect reproductive mechanisms is lacking. There are, however, compelling reasons, both practical and theoretical, to study insect reproduction. Control of reproductive capacity is an important key to pest control. The centrality of insects, particularly *Drosophila,* to genetic research makes determination of the role of ecdysone and juvenile hormone on the regulation of the genome of pressing importance.

REFERENCES

C. R. Bantock, *J. Embryol. Exp. Morph.* **24,** 257 (1970).

M. S. Blum and N. A. Blum, Eds., *Sexual Selection and Reproductive Competition in Insects,* Academic Press, New York, 1979.

C. L. Boggs and L. E. Gilbert, *Science* **206,** 83 (1979).

D. Bullière, M. Bullière, and M. de Reggi, *Wilhelm Roux's Arch.* **186,** 103 (1979).

M. D. Cave, "Morphological manifestations of ribosomal DNA amplification during insect oogenesis," in R. C. King and H. Akai, Eds., *Insect Ultrastructure,* Vol. I, Plenum Press, New York, 1982, pp. 86–117.

J. de Wilde and A. de Loof, "Reproduction," in M. Rockstein, Ed., *The Physiology of the Insecta,* Vol. I, 2nd ed., Academic Press, New York, 1973, p. 12.

J. B. Dumser, *Ann. Rev. Ent.* **21,** 341 (1980).

F. Engelmann, *The Physiology of Insect Reproduction,* Pergamon Press, Oxford, 1970.

F. Engelmann, *Adv. Insect Physiol.* **14,** 49 (1979).

F. C. Kafatos, J. C. Regier, G. D. Mazur, M. R. Nadel, H. M. Blau, W. H. Petri, A. R. Wyman, R. E. Gelinas, P. B. Moore, M. Paul, A. Efstratiadis, J. N. Vournakis, M. R. Goldsmith, J. R. Hunsley, B. Baker, J. Nardi, and M. Koehler, "The eggshell of insects: Differentiation-specific proteins and control of their synthesis and accumulation during development," in W. Beerman, Ed., *Biochemical Differentiation in Insect Glands,* Springer-Verlag, Berlin, 1977, p. 45.

M. P. Kambysellis and C. M. Williams, *Biol. Bull.* **141,** 527 (1971).

R. A. Leopold, *Ann. Rev. Ent.* **21,** 199 (1976).

L. M. Riddiford, "The role of hormones in the reproductive behavior of female wild silkmoths," in L. Barton-Browne, Ed., *Experimental Analysis of Insect Behavior,* Springer-Verlag, Heidelberg and New York, 1974, p. 278.

M. Sasaki, L. M. Riddiford, J. W. Truman, and J. K. Moore, *J. Insect Physiol.* **29,** 695 (1983).

R. I. Woodruff and W. H. Telfer, *J. Cell. Biol.* **58,** 172 (1973).

12

INTERMEDIARY METABOLISM

STANLEY FRIEDMAN
Department of Entomology
University of Illinois
Urbana, Illinois

CONTENTS

SUMMARY

As we continue to enlarge our understanding of metabolic processes in insects, we find that the phrase *plus ça change, plus c'est la même chose* is entirely apropos. Comparative biochemistry teaches us that, with little exception, the methods used to regulate reaction sequences are similar across the animal kingdom; and, in the case of insects, we find that we can even analogize the major organ responsible for synthesis and detoxification, the fat body, with the mammalian liver. Muscle metabolism in insects also markedly conforms to that of vertebrates, so it can be said that our understanding of insect metabolism has been greatly enhanced by studies on vertebrate systems. However, the flow of information has not all been undirectional. This is exemplified by the classical studies of David Keilin on the cytochromes in the flight muscle of the wax moth, *Galleria mellonella,* which have led to a general appreciation of the role of these heme proteins in oxidative activity in living organisms.

A chapter concerned with intermediary metabolism is, necessarily, a summary of some of the major aspects of this subject. This chapter has concentrated, for the most part, on pathways that are of greatest importance to the processes embodied in a general treatise on insect physiology, and has skimmed the surface of reactions that have not been carefully studied in insects.

ABBREVIATIONS

Abbreviations used in this chapter are as follows:

AMP   adenosine monophosphate; adenosine 5′-phosphoric acid
cAMP   3′,5′-cyclic AMP
ADP   adenosine 5′-diphosphate
ATP   adenosine 5′-triphosphate
UDP   uridine 5′-diphosphate
UTP   uridine 5′-triphosphate
PPi   inorganic pyrophosphate
Pi   inorganic phosphate

NAD   nicotinamide adenine dinucleotide
NADP   nicotinamide adenine dinucleotide   pyridine nucleotide
       phosphate
FMN   flavin mononucleotide
FP   flavoprotein
FAD   flavin adenine dinucleotide

12.1. MOVEMENT OF MATERIALS ACROSS GUT WALL

The insect gut wall serves a dual function. It is a mechanical barrier, physically limiting the movement of high-molecular-weight materials from the lumen of the digestive tract into the hemolymph, and, as well, a metabolically active tissue. In this latter role, it secretes enzymes into the gut lumen (varying according to species and diet), which change complex foodstuffs into simple compounds. These are then handled by other enzymes located within the cells of the gut wall or they diffuse unchanged across the wall into the hemolymph. Thus, polysaccharides, proteins, large class-size lipids, and all other manner of dietary constituents will not enter the blood and provide sustenance unless the specific hydrolases, esterases, and glucosidases are either present on the gut wall as membrane-bound entities or secreted into the gut lumen. The interesting methods used to effect these changes and movements are detailed in Chapters 1 and 2 and are outlined here.

In the case of carbohydrate, it appears that simple sugars are produced in the gut lumen from polysaccharides and pass into the hemolymph by diffusion. This is facilitated in some insects by other tissue activity which removes the diffusing molecules from the blood.

The movement of amino acids is somewhat more complicated, since the blood of many insects contains high concentrations of certain of these compounds. Protein is, for the most part, split to its constituent amino acids in the lumen, and at least in the case of locusts (*Schistocerca gregaria*) and cockroaches (*Periplaneta americana*), passage across the gut wall is assisted by an active (H_2O transport) process which creates a positive diffusion gradient. In some flies (e.g., larvae of *Musca domestica*) the γ-glutamyltransferase cycle, a system found also in mammals, has been invoked. There is scant evidence for either of these strategies and both deserve further investigation.

Work done on lipid movement indicates a much more active participation of synthetic enzymes localized in the cells making up the gut wall. Complex compounds are hydrolyzed in the gut lumen to free fatty acid and glycerol or monoacylglycerol, and certain of these recombine within the cells of the midgut to form diacylglycerol. As a result, free fatty acids, glycerol, and diacylglycerol are released into the hemolymph. Once in the blood, the diacylglycerol is bound to specific protein carriers for transport to storage sites.

It is astonishing how little is known at this time of the processes whereby compounds, both useful and toxic, are introduced to the metabolically active interior of the insect. However, as will be seen, knowledge of the methods governing the utilization of many of these same compounds is equally lacking.

12.2. LEVELS OF STUDY OF INTERMEDIARY METABOLISM

Investigations of whole animal rates of oxidation of foodstuffs under varying conditions are of little worth except as guides to utilization capability since they provide no insights into the ways in which compounds are metabolized once they move across the gut wall. For an understanding of mechanism, the systems must be dissected and the reaction sequences described.

In recent years, studies of insects have begun to focus on the regulation of metabolic processes rather than the processes themselves. The questions of how an animal maintains a given level of some metabolic source and of how that source is brought into use at a rate comcomitant with need are important not only to the biochemist but also to those whose interests are concerned with the limits within which a given insect can function. Thus, an understanding of metabolic shifts in reaction to stress is a significant tool for both the quantitative ecologist examining resource partitioning and the control specialist trying to establish infective and lethal doses of biological or chemical reagents.

In general, process rates are governed at particular sites within a reaction sequence. There are a number of ways in which the reaction rate of any enzyme in a pathway can be changed (pH, concentration of substrate or product, etc.), but the control of a system is usually attained by more subtle and highly selective methods. The means wherby regulation of specific enzymes may be effected can be categorized as (1) modification of the enzyme by noncovalent binding of a compound to some site other than the catalytic one (allosteric regulation) and (2) making a reversible chemical change in the structure of the enzyme (inactive to active form or vice versa by phosphorylation, adenylation, etc.).

The complexities of regulation are such that end products of a sequence may behave as allosteric inhibitors, substrates may act as activators, and metabolities of other interconnected pathways may play roles as either activators or inhibitors. To highlight this, the actions of various compounds that specifically affect certain key enzymes in the important metabolic pathways will be mentioned in our description of those reaction sequences. The indirect and still obscure actions of some of the metabolic hormones will be treated in Section 12.6.

Although every cell in the insect is responsible for the synthesis and utilization of most metabolically significant compounds, major studies have only been carried out on two tissues: fat body and flight muscle. The most obvious reasons for this are (1) the internal body volume occupied by these tissues makes up a major portion of the total tissue mass contained in the hemolymph; (2) the fat body is the major synthetic and storage organ in the body; and (3) the indirect flight muscles are, generally speaking, the most actively metabolizing tissue in the body. It is quite difficult to obtain pure preparations of other tissue in amounts great enough to make quantitative comparisons of enzymic activity, but in a few cases this has been done, and some interesting variations on the generally accepted themes of metabolic transformation have emerged.

12.3. THE FAT BODY AS A METABOLIC SOURCE

12.3.1. Morphology

The major organ involved with the synthesis of macromolecules and the metabolism of nutrients and other compounds crossing the midgut wall is the fat body. It represents between 40 and 60% of the total body weight of some mature endopterygote larvae and is, in a sense, the insect equivalent of the mammalian liver: It is responsible for the synthesis, storage, and release of important metabolites; it contains the major detoxification enzymes; and because of its various capabilities, it can be considered to be the major tissue through which the rates of all large-scale substrate-dependent functions are modulated. (A good example of this, to be detailed later in this chapter, is the uptake of lipid and its synthesis and storage as triacylglycerol in the locust fat body and, in the same tissue, under the influence of a hormone, its conversion to diacylglycerol and release for use by flight muscle.)

The organization of the fat body varies in different insects and stages of development, but it may be described as a relatively loosely tied agglomeration of two or three cell types, of which the major one, the trophocyte, is specialized for the important activities mentioned above. The urocyte, involved in uric acid storage, and the mycetocyte (not found in most insects), the repository for symbiotic microorganisms, are not thought to play roles in the highly regulated, carefully balanced activities that determine the metabolic state of the animal.

The fat body is in many species a loose structure, irregularly distributed throughout the body, even surrounding the brain. Its close investment of many organs permits storage compounds to be moved quickly from source to sink, an important attribute in insects, which, having "open" circulatory systems, measure circulation rates in relatively long time periods. It may show different morphological characteristics in different places in the body (locust) and may exhibit functional differences over time and between sexes (*Bombyx mori*). Finally, in endopterygotes it may persist from larva to adult (mosquitoes) or undergo histolysis and reformation from imaginal tissue (*Drosophila* and other higher Diptera).

12.3.2. As Source of Carbohydrate Synthesis

With few exceptions, nutritionally derived high-molecular-weight carbohydrate has not been found in insect blood, and it is assumed that most carbohydrate enters the hemolymph from the gut as glucose and/or whichever other monosaccharides are produced in the breakdown of polysaccharide by gut-wall enzymes. Unless utilized directly by a metabolizing tissue, the sugar moves to the fat body, where in most insects it is converted to one of two major storage carbohydrates—trehalose, a circulating form, or glycogen, a tissue-bound form. These two compounds are related through common intermediates in their formation, are interconvertible for transport purposes in time of metabolic need, and, through enzymic action, are immediate sources of glucose.

α,α-Trehalose (α-D-glucopyranosyl-α-D-glucopyranoside), a nonreducing disaccharide,

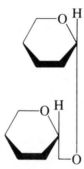

has been found in large amounts (20–170 mM) in the blood of a great number of species. Although present in members of orders ranging from Orthoptera to Diptera, thereby attaining the title of "*the* insect blood sugar," it is absent in various stages of different species [e.g., *Phormia regina* (Diptera) early larva] and, perhaps, from whole orders such as Dermaptera. In those insects, glucose may take the place of trehalose as a circulating source of carbohydrate.

The concentration of trehalose, although varying among species, has been shown to be constant and highly regulated within any given developmental stage. It is only present in small amounts in tissue and appears to function as the form in which carbohydrate is transported through the blood from storage sites to sites of utilization. It has at least two advantages over glucose for this purpose:

1. It is a large molecule and has a lower diffusion rate across gut cell membranes, resulting in a negligible loss from the body under normal conditions.

2. As a disaccharide, its structure makes for a lower osmotic load in the blood per unit of glucose transported.

When it reaches a cell with a need for sugar, it is cleaved to glucose in a reaction catalyzed by the hydrolytic enzyme, trehalase, which is found almost everywhere in the body:

$$\text{Trehalose} + H_2O \rightarrow 2 \text{ glucose}$$

The major site of trehalose synthesis is the fat body, although the synthetic enzymes are found in other tissues, and the compound may be produced in small amounts in gut and muscle in some insects. Its synthesis has been detailed in a number of species and is noteworthy because it is highly regulated. Glucose, entering the blood across the midgut wall, is moved into the fat body, where conversion takes place through the following energy-requiring steps:

Rxn 1   2 Glucose + 2 ATP → 2 glucose 6-P + 2 ADP   (hexokinase)

2   Glucose 6-P → glucose 1-P   (phosphoglucomutase)

3   Glucose 1-P + UTP →
        UDP-glucose + PPi (UDP-glucose pyrophosphorylase)

4   UDP-glucose + glucose 6-P →
        UDP + trehalose 6-P (trehalose 6-P synthase)

5   Trehalose 6-P → trehalose + Pi   (trehalose 6-P phosphatase)

Total: 2 Glucose + 2 ATP + UTP →
                trehalose + 2 ADP + UDP + PPi + Pi.

Regulation of trehalose production is, in some insects, almost certainly involved with its level in the blood. It is itself an end-product inhibitor of the reaction sequence, and the rate at which it is moved out of the fat body into the blood is instrumental in setting the level of inhibition of production. Movement into the blood is, of course, dependent on its rate of removal from the blood by those tissues utilizing it as a source of glucose.

The character of the inhibition of production may vary from species to species, becoming very complex in certain insects that depend on carbohydrate as the sole source of flight energy. A good example of this is found in the calliphorid flies in which the enzyme, trehalose 6-P phosphatase (Rxn 5), has a second active site which normally hydrolyzes glucose 6-P at a low rate but is allosterically activated by trehalose. Since glucose 6-P is a substrate in the reaction sequence leading to trehalose (Rxn 2), any activation of that hydrolytic site will decrease trehalose production. Thus, as the quantity of trehalose in the system increases, the glucose-P site on the phosphatase is activated, which increases the rate of glucose 6-P hydrolysis and leads, finally, to a decrease in trehalose synthesis.

Trehalose levels may also be regulated by reactions involved in glycogen synthesis and degradation, since UDP-glucose is an intermediate in the production of both compounds.

Glycogen, the storage polysaccharide ubiquitous in animal cells, is of great importance to insects. Although varying in amount in different tissues and stages of development, in carbohydrate-utilizing insects it is the basic energy source for flight and other activity. When present in a tissue in which there is a sudden need for glucose (e.g., fly flight muscle), it is converted directly (see Section 12.4.1); when it must be transported through the blood from the fat body, it is degraded to glucose 1-P by fat body glycogen phosphorylase (see Section 12.4.1 and 12.5.3) and reconstructed as trehalose (see above, Rxns 3–5).

Glycogen, which has been examined in considerable detail in blow fly flight muscle, possesses molecular weights ranging to 100 million. The outer chains are short, but in the presence of the proper enzymes, the compound can be completely cleaved to glucose.

Insect synthesis of glycogen is known only to the extent that, in both fat body and flight muscle, the addition of glucose by an $\alpha(1\rightarrow4)$ glycosidic linkage to a glycogen primer is catalyzed by UDP-glucose-glycogen transglycosylase (glycogen synthetase):

$$\text{UDP-glucose} + (\text{glucose})_n \rightarrow \text{UDP} + (\text{glucose})_{n+1}$$

No substantial information is available concerning the synthesis of 1,6 branch points in insects.

Glucose 6-P activates the synthetase, and in different insects the activation is different, affecting [honeybee (*Apis mellifera*) larva] or not affecting (*Hyalophora cecropia* larva) the K_m for UDP-glucose. This information, coupled with the preceding discussion of trehalose synthesis, make it obvious that glucose 6-phosphate is central to the control of production of both trehalose and glycogen. However, interactions between these syntheses in any given tissue depend to a great extent on the spatial relationships between the enzymes engaged in these processes, so that little can presently be said about the *in vivo* situation. In this context, it is also unwise to speculate about preferential utilization of UDP-glucose by one or the other system, although studies on cecropia larval fat body enzymes indicate that trehalose 6-P synthase has a lower K_m for UDP-glucose than does glycogen synthetase.

As an aside, it should be pointed out that there is a third major storage compound in insects, the $\beta(1\rightarrow4)$-linked *N*-acetylglucosamine polymer, chitin. As the principal structural component of the insect exoskeleton, it is produced by epidermal cells through reactions involving the synthesis of UDP–*N*-acetylglucosamine and its incorporation into chitodextran polymer (thus lengthening the chain). The enzyme catalyzing this reaction, chitin synthase, is incorporated into the epidermal cell membrane, and it is believed that the polymer is produced on the outside surface of the membrane from material transported through it (see Chapter 4). However, this high-molecular-weight source of glucose is not totally lost to the insect, since histological studies indicate a decrease in endocuticular thickness in starving insects. Our knowledge of how this takes place is limited to investigations of molting insects and the production and presence of chitinolytic enzymes in molting fluid. Perhaps there is much to be learned from intrastadial studies on the turnover of components of this polysaccharide.

12.3.3. As Source of Lipid Synthesis

As was stated earlier, complex lipids, after digestion by lipases and esterases, enter the midgut wall as glycerol and free fatty acids. Studies on locusts and cockroaches have shown that in the cells of the gut wall some of the free fatty acids may be rebuilt into glycerol phosphatides and acylglycerols and released into the hemolymph as diacylglycerols. These are bound to lipoprotein carrier molecules, rendering them soluble, and moved to the fat body for resynthesis and storage. Much of the glycerol passes unchanged across the gut wall into the hemolymph and from there to the fat body.

The fat body is the major site of *de novo* synthesis of free fatty acids and of di- and triacylglycerols. Almost all lipid synthesized and stored in the fat body is present as triacylglycerol, diacylglycerol, and free fatty acid. It is subsequently released and utilized by other tissues. The diacylglycerol found in the hemolymph has been shown to contain fatty acid moieties that differ from those which make up the stored triacylglycerol, so it has been assumed that it is not derived from simple hydrolytic cleavage of the triacyl ester. Recently, that assumption has been substantiated by the demonstration in locust fat body of a lipase that degrades triacylglycerol to 2-monoacylglycerol, and another enzyme, a monoacylglycerol acyl transferase, that synthesizes *sn*-1,2-diacylglycerol from 2-monoacylglycerol. Since the same *sn*-1,2 configuration is found in blood diacylglycerol, it is now believed that triacylglycerol is stored in one compartment in fat body and degraded and resynthesized into different diacylglycerols in a second compartment for release to the blood.

The pathway of synthesis of free fatty acids has been shown to follow that generally occurring in vertebrates, although the insect experiments have been done with complex systems and are so undetailed (e.g., malonate, HCO_3^-, and NADPH have been shown to be required) that no regulatory processes have been established with certainty. In vertebrates and other eukaryotes, the enzymes involved with fatty acid synthesis are found in the cytoplasm and comprise a complex in which a protein-(Acyl Carrier Protein) bound (thioester linkage) malonyl moiety combines with an enzyme (β-ketoacyl-ACP synthase) bound acyl group to produce a protein-bound (thioester linkage) acyl compound increased in length by two carbons and CO_2. The reactions are as follows:

1. Formation of a protein-bound moiety:
   a. Acetyl CoA + HCO_3^- + H^+ + ATP → malonyl CoA + ADP + Pi (acetyl CoA carboxylase).

    b. Malonyl CoA + ACP → malonyl ACP + CoA (ACP-malonyl transferase).

2. Formation of enzyme-bound acyl groups:

    a. Acetyl CoA + ACP → acetyl ACP + CoA (ACP-acyl transferase).

    b. Acetyl ACP + β-ketoacyl-ACP synthase → acetyl β-ketoacyl-ACP synthase + ACP.

3. Reaction between products 1b and 2b (catalyzed by β-ketoacyl-ACP synthase).

    Malonyl ACP + acetyl β-ketoacyl-ACP synthase → acetoacetyl ACP + CO_2 + β-ketoacyl-ACP synthase

4. Reduction of acetoacetyl ACP to butyryl ACP and formation of enzyme-bound acyl group:

    a. Acetoacetyl ACP + NADPH + H^+ → D-β-hydroxybutyryl ACP + $NADP^+$ (β-ketoacyl-ACP reductase).

    b. D-β-Hydroxybutyryl ACP → crotonyl ACP + H_2O (enoyl-ACP hydratase).

    c. Crotonyl ACP + NADPH + H^+ → butyryl ACP + $NADP^+$ (enoyl-ACP reductase).

    d. Butyryl ACP + β-ketoacyl-ACP synthase → butyryl-β-ketoacyl-ACP synthase + ACP.

5. Addition of protein-bound malonyl moiety to enzyme-bound acyl group:

    Butyryl (C_4) β-ketoacyl-ACP synthase + malonyl ACP (see Rxn 1) → β-ketohexanoyl (C_6) ACP + CO_2 + β-ketoacyl-ACP synthase

6. Repeat reaction sequence 4 at C_6 level, and so on.

Thus, acetyl CoA provides the two carbon moiety at the methyl end of the chain (Rxn 3) and malonyl CoA all the other carbons in the fatty acid. The enzyme β-ketoacyl-ACP synthase specifies the chain length limit of the fatty acid produced, and, finally, the fatty acid is released from the protein as free fatty acid by the action of a thioesterase.

Although fatty acid biosynthesis occurs in the fat body cytoplasm, it must be supplied with acetyl CoA from a mitochondrial source. The movement of the acetyl moiety from the mitochondrion may take place in one of two ways: by transfer across the mitochondrial membrane as a carnityl derivative after pyruvate decarboxylation (see Section 12.4.3) or, more certainly, through mitochondrial production of citrate from acetyl CoA and oxaloacetate (see Section 12.4.2), the citrate then moving into the cytoplasm to be converted into acetyl CoA and oxaloacetate by the cytoplasmic enzyme, ATP-citrate lyase:

Citrate + ATP + CoA → acetyl CoA + oxaloacetate + ADP + Pi

The acetyl CoA is used in lipogenesis and the oxaloacetate is returned to the mitochondrion for recycling after conversion to malate by cytoplasmic malate dehydrogenase (see Section 12.5.2).

In a different vein, the high concentrations of malic enzyme:

$$\text{Pyruvate} + CO_2 + NADPH + H^+ \rightarrow \text{L-malate} + NADP^+$$

in the fat body cytoplasm of both *Locusta migratoria* and *Periplaneta americana* suggest that the oxaloacetate needed to increase the rate of citrate production at the onset of lipogenesis is derived from this source, the malate crossing the mitochondrial membrane and being converted within it to oxaloacetate by the mitochondrial malate dehydrogenase which catalyzes the reaction:

$$\text{L-Malate} + NAD^+ \rightarrow \text{oxaloacetate} + NADH + H^+$$

The fat body mitochondrion is much more functionally diverse than the mitochondrion from flight muscle (see Section 12.4.2) and, as such, has an inner membrane that is permeable to a much greater array of compounds.

It is claimed that long-chain fatty acids can be synthesized within the mitochondria of *Drosophila melanogaster* by a reversal of the β oxidation (energy-generating) pathway of fatty acid degradation described in Section 14.4.3. This mitochondrial preparation is said also to be capable of chain elongation of preexisting fatty acids.

Unsaturated fatty acids (only monoenoic) are formed in either of two ways in insects: in locust fat body microsomes, by enzymic desaturation of C_{16} or C_{18} acids at the 9 position with the concomitant reduction of pyridine nucleotides; in *Drosophila*, by β-α dehydration of C_{10} or C_{12} β-hydroxy acids, resulting in C_3-enoic acid, with subsequent elongation to C_{16} or C_{18} bringing the double bond to the 9 position.

12.3.4. As Source of Amino Acid Synthesis

The high and varied concentrations of the diverse amino acids in insect hemolymph derive from endogenous synthesis as well as ingested and digested protein. The different metabolic activities of specific tissues are responsible for the differential use of absorbed products of protein digestion, so that, in those insects examined in detail, the blood levels of some amino acids seem to reflect only their presence in or absence from dietary sources, whereas levels of others (e.g., tyrosine and proline) appear to be highly regulated.

For all practical purposes the common protein-derived amino acids are present in insect blood as the L-enantiomers. However, in the hemolymph of specific insects the D-forms of some and derivatives of others, such as tyrosine phosphate and GABA occur. Free amino acids and certain derivatives have also been described in salivary and defensive secretions.

Although most amino acids found in insects derive from ingested food, only some are really essential to the diet. The required amino acids are the same 10 necessary for mammalian success: arginine, histidine, leucine, isoleucine, lysine, methionine, phenylalanine, threonine, tryptophan, and valine. Specific insects have been shown to need others for optimum growth and development, but, for the most part, all but the above amino acids can be synthesized from glucose at a rate sufficient to satisfy need (see Chapter 2, Section 2.5.2).

Highly regulated amino acids, such as proline and tyrosine, appear to have their major sources of synthesis and/or control located in the fat body. For example, tyrosine, which is produced by the hydroxylation of phenylalanine, is sequestered with other amino acids in the fat body of calliphorid fly larvae and built into a high tyrosine (12%), high phenylalanine (10.7%) storage protein named "calliphorin." The protein is released and stored in the blood through larval life. It is then moved into the cuticle where it may be used as a matrix of aromatic amino acids for the reactions leading to hardening and darkening of the adult cuticle (see Chapter 4).

Proline, in those insects that utilize it for energy (see Section 12.4.4), is not only obtained from exogenous sources but is synthesized in the fat body from alanine through the following series of reactions:

Rxn 1 Alanine + α-ketoglutarate $\rightarrow$
         pyruvate + glutamate (alanine-α-ketoglutarate aminotransferase)
     2 Pyruvate + CO_2 + NADPH + H^+ $\rightarrow$
                        L-malate + $NADP^+$ (malic enzyme)
     3 L-Malate + NAD^+ $\rightarrow$
                        oxaloacetate + NADH + H^+ (malic dehydrogenase)
     4 Oxaloacetate + acetyl CoA (from fatty acids) + H_2O $\rightarrow$
                        citrate + CoA (citrate synthase)
     5 Citrate $\rightarrow$ isocitrate   (aconitase)
     6 Isocitrate + NAD^+ $\rightarrow$ α-ketoglutarate
                        + CO_2 + NADH + H^+ (isocitrate dehydrogenase)
     7 Glutamate + ATP + NADH + H^+ $\rightarrow$ glutamate γ-semialdehyde
                        + ADP + Pi + NAD^+ (glutamate kinase and dehydrogenase)
     8 Glutamate γ-semialdehyde $\rightarrow$
                        Δ^1-pyrroline 5-carboxylate + H_2O (spontaneous)
     9 Δ^1-Pyrroline 5-carboxylate + NADPH + H^+ $\rightarrow$
                        proline + $NADP^+$ (pyrroline 5-carboxylate reductase)
    10 $NADP^+$ + NADH $\rightarrow$
                        NADPH + NAD^+ (pyridine nucleotide transhydrogenase)

Total: Alanine + acetyl CoA + ATP + NADPH + H^+ $\rightarrow$
                        proline + CoA + ADP + Pi + $NADP^+$.

The aforementioned enzymes have not been isolated, but, using radioactively labeled precursors, the general scheme has been established in the tsetse fly, *Glossina morsitans*. Measurements of the specific activities of certain fat body enzymes of *Leptinotarsa decemlineata,* the Colorado potato beetle, have shown that here, as well, malic enzyme and malic dehydrogenase are more likely utilized in the production of oxaloacetate from pyruvate (Rxns 2,3) than is the more common enzyme, pyruvate carboxylase, which catalyzes the reaction

$$\text{Pyruvate} + CO_2 + \text{ATP} + H_2O \rightarrow \text{oxaloacetate} + \text{ADP} + \text{Pi}$$

This and other information obtained from fat body and flight muscle (see Section 12.4.4) indicate that proline may play a major role in this insect also.

12.3.5. As Source of Protein

Protein synthesis takes place in every cell of the body during periods of growth and development of tissue. Since these processes underlie metamorphic changes in insects, attention is beginning to focus on the possibility that it is at the levels of transcription and/or translation that the developmental hormones exert their control (see Chapter 13).

Most of the synthetic studies have been carried out on whole larval *Tenebrio* (Coleoptera) extracts and the fat body and silk glands of various larval Lepidoptera. The posterior portion of the silk gland in *Bombyx mori* is responsible for the synthesis of the silk protein, fibroin, composed preponderantly of a few amino acids (45% glycine, 29% alanine, 10% serine), and it has been used as a source material for relatively pure messenger RNA (mRNA), from which fibroin can be specifically synthesized in a cell-free system. However, the synthetic process is more significant in its general aspects, so it will be described with information derived from a variety of sources.

The steps in synthesis (translation) follow rather closely those in other eukaryotes. In the first step, specific amino acids are coupled by ester linkages to their corresponding transfer RNAs (tRNAs) in a two-step reaction catalyzed by specific enzymes (aminoacyl-tRNA synthetases):

1. Amino acid$_1$ + ATP + synthetase$_1$ $\rightarrow$ aa$_1$-AMP-synthetase$_1$ + PPi.
2. aa$_1$-AMP-synthetase$_1$ + tRNA$_{aa_1}$ $\rightarrow$ aa$_1$-tRNA$_{aa_1}$ + AMP + synthetase$_1$.

As is well known, the triplet code is degenerate (more than one codon per amino acid), and, in a number of eukaryotes including insects (*Bombyx* silk gland), more than one specific tRNA and aminoacyl-tRNA synthetase have been found for each amino acid. The isoenzymes do not react equally with each isoaccepting (homologous for a given amino acid) tRNA, so development may be regulated by synthesis taking place within a framework of varying amounts of different tRNAs and isoenzymes. These variants could quantitatively specify the proteins synthesized from different mRNAs carrying codons specific for one or another tRNA.

The second step involves the formation of the protein synthesis initiation complex. This consists of the conjoining of a 40S ribosomal subunit, the mRNA, GTP, methionine-tRNA$_{met}$ (isoacceptor tRNA methionine, specific for initiation), and three initiation factors (different specific proteins). The complex combines with a 60S ribosomal subunit, at which time the initiation factors are liberated and GTP is hydrolyzed. At this point the initiator tRNA$_{met}$ is bound to the initiator codon on the mRNA, which, with the 40S and 60S subunits, is the active synthetic unit. Formation of this unit has been demonstrated in *Tenebrio* larval cell-free systems.

(This is another place at which regulation may take place i.e., specification of the initiation factors can determine whether the peptide chain will be begun.)

The third step is the elongation of the polypeptide. This consists of:

1. Specific binding of a second acylated tRNA: aa$_2$-tRNA$_{aa_2}$, to the site containing the aa$_2$ triplet on the mRNA next to the already occupied initiator codon site. (At any time, there are two active sites on the ribosome: the acceptor, A, site, to which the new amino acyl tRNA$_{aa}$ is bound, and the peptidyl, P, site which will transfer its peptidyl moiety (or to begin with, the initiator amino acid) to the newly bound amino acid.

2. Peptide bond formation between the peptide (or initiator amino acid) from the P site and the newly bound aa$_2$-tRNA$_{aa_2}$ on the A site with concomitant release of the initiator amino acid—or peptide—less tRNA from its attachment to the codon at the P site. (Activities at these sites control the rate of polypeptide elongation as the ribosome moves along the mRNA and the message is translated.)

3. Movement of the ribosome the length of one codon along the mRNA in the 5'–3' direction, with the simultaneous movement of the peptidyl-tRNA$_{aa_2}$ from the A site to the P site.

4. Binding of aa$_3$-tRNA$_{aa_3}$ complex to the A site on the messenger codon for aa$_3$ and repeat of step 2.

These reactions require the participation of GTP and two enzymes: transferase 1, which is involved in the binding of aminoacyl-tRNA$_{aa}$ to ribosomes, and a second enzyme (transferase 2, translocase) which is responsible for the movement of the peptidyl-tRNA$_{aa}$ from the A site to the P site, opening the A site for further binding. All of these enzymic activities have been examined in more than one insect species and shown to be quite similar to one another.

It is easily seen that the complex reactions leading to polypeptide synthesis can be regulated at any of a number of levels, particularly with respect to the action or inaction of one or another of the enzymes which confer specificity upon the synthetic process.

The production of RNA (transfer and messenger), patterned from activated portions of the chromosome (DNA available for transcription), is another and perhaps more significant level at which control may be asserted. Interest in the RNA polymerases responsible for synthesis (transcription) of tRNA and mRNA has increased with recognition that primary effects of developmental hormones might be exerted at the level of the genome and its immediate products. Although the transcription process has not been detailed in insects, a number of ongoing studies are concerned with the RNA polymerases specific for synthesis of each of the three major RNA species (polymerase I: ribosomal RNA; II: hnRNA (messenger precursor); III: 5S and pre-4S RNA (tRNA precursor). It has recently been found, for example, that insect RNA polymerase IIIs (involved with tRNA synthesis) have certain physical properties that distinguish them from other eukaryotic polymerase IIIs.

12.4. MUSCLE AS A METABOLIC SINK

12.4.1. Utilization of Carbohydrate: Glycolysis

Anatomical descriptions of insect muscles specialized for different functions reveal such divergence in number and location of organelles and tracheoles (see Chapter 6) that it is prudent to inquire into the possibility that the metabolic pathways from which energy is derived for contraction may also be distinct.

There are too few investigations concerned with skeletal (tubular) muscle to permit general comment on the carbohydrate storage capacity of this tissue. However, the relatively high level of glycogen phosphorylase activity in both locust and hemipteran (*Lethocerus cordofanus*) femoral muscle (Table 12-1) catalyzing the reaction:

$$(\text{Glucose})_n + \text{Pi} \rightarrow (\text{glucose})_{n-1} + \text{glucose 1-P}$$

indicates that glycogen must be used as a major source of energy.

In tubular muscle the classical Embden–Meyerhof pathway of glycolysis seems to be followed:

1. From glycogen stored in muscle:

$$(\text{Glucose})_n + \text{Pi} \rightarrow (\text{glucose})_{n-1}$$
$$+ \text{glucose 1-P} \quad (\text{glycogen phosphorylase})$$
$$\text{Glucose 1-P} \rightleftharpoons \text{glucose 6-P} \quad (\text{phosphoglucomutase})$$

2. From trehalose entering muscle from blood:

$$\text{Trehalose} \rightarrow 2 \text{ glucose} \quad (\text{trehalase})$$

TABLE 12-1. Activity (μmol min^{-1} g fresh muscle^{-1} at 25°C)[a]

Insect	Muscle	GP	HK	PFK	Cyt GPDH	Mito GPDH	LDH	PDH	CS	SDH	IDH	TGL	DGL	MGL	CPT
Locust	Flight	8.0	8.0	13.0	124.0	33.0	1.6	1.7	244.0	20.0	26.0	0.07	0.6	—	3.6
(*Locusta migratoria*)	Femoral	20.0	2.3	16.0	33.0	25.0	53.0	—	18.0	2.0	9.0	—	—	—	—
Cockroach	Flight	30.0	18.0	19.0	216.0	48.0	1.5	6.7	185.0	57.0	55.0	0.02	0.1	—	0.1
(*Periplaneta americana*)	Coxal	3.0	1.3	4.0	35.0	9.0	—	—	—	0.3	3.0	—	—	—	—
Waterbug	Flight	1.0	4.0	6.0	51.0	8.0	1.3	0.5	105.0	27.0	28.0	0.07	0.9	0.7	3.5
(*Lethocerus cordofanus*)	Femoral	12.0	0.05	—	13.0	0.9	59.0	—	—	—	—	—	—	—	—
Cockchafer	Flight	14.0	2.3	13.0	103.0	36.0	4.4	60.0	109.0	30.0	45.0	0.03	0.3	1.9	0.1
(*Melolontha melolontha*)															
Honeybee	Flight	4.0	29.0	20.0	257.0	44.0	1.5	1.5	345.0	—	113.0	0.04	0.3	0.4	0.2
(*Apis mellifera*)															
Silver Y Moth	Flight	2.0	50.0	41.0	—	110.0	—	—	—	—	—	0.04	0.5	8.3	1.1
(*Plusia gamma*)															
Tortoiseshell butterfly	Flight	8.0	4.8	14.0	—	24.0	—	2.8	280.0	—	55.0	0.06	0.5	—	0.9
(*Vanessa urticae*)															
Tsetse fly	Flight	1.8	2.3	—	29.0	4.0	1.5	40.0	74.0	—	20.0	—	—	—	—
(*Glossina austeni*)															
Flesh fly	Flight	57.0	17.0	—	270.0	100.0	2.5	3.0	141.0	80.0	72.0	0.04	0.3	—	0.3
(*Sarcophaga barbata*)															
Blowfly	Flight	54.0	14.0	46.0	300.0	110.0	1.7	—	—	—	—	—	0.2	—	0.2
(*Phormia terranova*)															

[a] Abbreviations: GP, glycogen phosphorylase; HK, hexokinase; PFK, phosphofructokinase; cytGPDH, cytoplasmic L-glycerol 3-P dehydrogenase; mitoGPDH, mitochondrial L-glycerol 3-P dehydrogenase; LDH, lactic dehydrogenase; PDH, proline dehydrogenase; CS, citrate synthase; SDH, succinate dehydrogenase; IDH, isocitrate dehydrogenase (NAD$^+$); TGL, triacylglycerol lipase; DGL, diacylglycerol lipase; MGL, monoacylglycerol lipase; CPT, carnitine palmitoyl transferase; —, not recorded.

Source: Adapted from Crabtree and Newsholme (1975).

3. From glucose produced from trehalose or entering from blood:

$$\text{Glucose} + \text{ATP} \rightarrow \text{glucose 6-P} + \text{ADP} \quad \text{(hexokinase)}$$

Then:

Glucose 6-P $\rightleftharpoons$ fructose 6-P   (glucose phosphate isomerase)

Fructose 6-P + ATP $\rightarrow$ fructose 1,6-diP + ADP   (6-phosphofructokinase)

Fructose 1,6-diP $\rightleftharpoons$ D-3-phosphoglyceraldehyde
$\quad\quad\quad\quad\quad$ + dihydroxyacetone P   (fructose diphosphate aldolase)

Dihydroxyacetone P $\rightleftharpoons$
$\quad\quad\quad\quad\quad$ D-3-phosphoglyceraldehyde   (triosephosphate isomerase)

2 D-3-Phosphoglyceraldehyde + 2NAD$^+$ + 2Pi $\rightleftharpoons$
2 D-3-phosphoglyceroyl P + 2NADH + 2H$^+$ (glyceraldehyde P dehydro-
$\quad\quad\quad\quad\quad\quad\quad\quad\quad\quad\quad\quad\quad\quad\quad\quad\quad\quad$ genase)

2 D-3-Phosphoglyceroyl P + 2ADP $\rightleftharpoons$ 2 D-3-phosphoglycerate
$\quad\quad\quad\quad\quad\quad\quad$ + 2ATP   (phosphoglycerate kinase)

2 D-3-Phosphoglycerate $\rightleftharpoons$
$\quad\quad\quad\quad\quad\quad$ 2 D-2-phosphoglycerate   (phosphoglyceromutase)

2 D-2-Phosphoglycerate $\rightleftharpoons$ 2 phosphoenolpyruvate + 2H$_2$O   (enolase)

2 Phosphoenolpyruvate + 2ADP $\rightarrow$ 2 pyruvate + 2 ATP
$\quad\quad\quad\quad\quad\quad\quad\quad\quad\quad\quad\quad\quad$ (pyruvate kinase)

From glycogen, the overall reaction to pyruvate is

$$(\text{Glucose})_n + 3\text{ADP} + 3\text{Pi} + 2\text{NAD}^+ \rightarrow$$
$$(\text{glucose})_{n-1} + 2 \text{ pyruvate} + 3\text{ATP} + 2\text{NADH} + 2\text{H}^+ + 2\text{H}_2\text{O}$$

From glucose, the reaction is

$$\text{Glucose} + 2\text{ADP} + 2\text{Pi} + 2\text{NAD}^+ \rightarrow$$
$$2 \text{ pyruvate} + 2\text{ATP} + 2\text{NADH} + 2\text{H}^+ + 2\text{H}_2\text{O}$$

Muscle associated with walking and jumping generally contains a relatively small number of mitochondria compared to flight muscle, indicating a
different source of energy or, at the very least, a markedly different rate
capability. Most muscle of this type is "white", although within single leg
muscles both "red" and "white" fibers may be present. (Such fibers are
distinguished from one another by differences in color, just as are vertebrate
fibers. However, vertebrate fiber color depends on myoglobin and cytochrome, and there is some question as to whether color is an indication of
metabolic activity. Insects, having no myoglobin, must derive the color of
their darker fibers from a higher cytochrome content, suggesting a greater
mitochondrial concentration.) In "red" muscle, which contains more mitochondria, energy is obtained from aerobic metabolism, and fatigue takes

longer to set in than in "white" fibers. From this observation, it might be expected that a preponderantly "white" muscle secures its energy, to a great extent, from anaerobic sources. And, indeed, Table 12-1 provides evidence of high lactic dehydrogenase and comparatively low oxidative enzyme activity in locust femoral muscle.

The reoxidation of NADH, reduced during glycolysis, takes place in "white" muscle by the formation of lactic from pyruvic acid:

$$2 \text{ Pyruvate} + 2\text{NADH} + 2\text{H}^+ \rightleftharpoons 2 \text{ L-lactate} + 2\text{NAD}^+$$

The lactic acid then either diffuses from the muscle into the blood to be reoxidized in the fat body or is reoxidized slowly by the muscle itself over the extended period of time that it is at rest.

The case is different for flight muscle, which is predominantly "red" or aerobic. The use of carbohydrate by this tissue type has been most extensively studied in *Phormia regina* (Diptera), in which carbohydrate is a major energy source. In this insect [as in others that derive energy from the oxidation of lipids or amino acids (see Sections 12.4.2 and 12.4.3)], the flight muscle is highly tracheolated and contains mitochondria in quantity ranging from 30 to 45% of the total fiber volume (see Chapter 6). Glycogen, stored in large amounts in the muscle, is utilized at such a high rate during flight that it is almost totally depleted within 10–15 min. After that time, carbohydrate must be supplied from circulating blood sugar which derives, in turn, from sugar stored in the crop or glycogen stored in the fat body (see Section 12.3.2). A number of sugars such as fructose, mannose, sucrose, maltose, trehalose, and glucose are capable of maintaining flight when fed, but the time lapse between feeding and the onset of flight in a starved, exhausted fly indicates that each of these compounds must be converted into glucose before it is used.

The primary pathway of glycogen degradation is the same in flight muscle as in skeletal muscle. The reactions leading to pyruvate have been thoroughly investigated and certain of the control points carefully mapped in both *Phormia* and *Schistocerca gregaria*. The enormous change in metabolic rate from rest to flight (50 times in *Phormia*) requires a system that can massively respond to small cues within a short time period. The regulatory processes are remarkable in the variety of ways in which control is achieved.

Glycogen phosphorylase is a highly regulated enzyme, as might be expected in a muscle dependent on glycogen as a primary energy source. Present in two forms, *a* and *b*, in *Phormia,* it constitutes almost 1.5% of the total muscle protein.

Although the specific activities of phosphorylases *a* and *b* are the same when tested under optimal conditions, their requirements for maximal activity are very different. This means that the levels of substrates and activators in the muscle cell differentiate the state of activity of each form of the enzyme. There is a close interaction between AMP (an allosteric activator) and Pi with respect to the affinity of the enzyme for each in the presence of varying amounts of the other, and it is believed that the variation in cytoplasmic concentration of these components determines the *in vivo* activities of the two forms of the enzyme. In the resting fly, the activity of phosphorylase *b* has been shown to be negligible, whereas that of phosphorylase *a* is 50% of maximum.

Phosphorylase *b* may be converted to the *a* form by a phosphorylating enzyme, phosphorylase *b* kinase (see Section 12.5.3). The latter is stimulated by Ca^{2+} at levels as low at 10^{-8} *M* and by Pi and is probably activated by both of these ions (which would presumably increase in concentration) at flight inception. Kinase activation is, no doubt, responsible for the rapid conversion of phosphorylase *b* to *a* at that time, causing the amount of *a* to increase from 18 to 70% of the total phosphorylase activity. The higher concentration of phosphorylase *a,* operating at 50% of its maximum activity, is capable of splitting glycogen at a rate that will support flight energy requirements.

Also implicated in the control of the kinase in fat body and nervous tissue, but not in muscle, is the hypertrehalosemic hormone (see Section 12.6).

Another enzyme, phosphorylase *a* phosphatase, also demonstrated in *Phormia* flight muscle, catalyzes the return of phosphorylase to its *b,* or inactive, form. Its regulation has not been studied in detail.

The control of hexokinase, the enzyme through which glucose enters the glycolytic pathway, has been examined in locust muscle. The concentration and activity of the enzyme are such that it must operate at maximum capacity to move glucose into the pathway at a rate that will produce enough energy to support flight. As in other animals, the reaction product, glucose 6-phosphate, is an inhibitor of hexokinase, and shortly after flight commences, its concentration rises to an inhibitory level. This and other temporary problems occur because of the non-steady-state conditions developed in a number of systems at the inception of flight, and it, in particular, is compensated for through the allosteric activation of hexokinase by products of other flight-stimulated reactions. The products, alanine and Pi, both increase in concentration at this time and together completely reverse the glucose phosphate inhibition. Alanine, an amino acid derived from pyruvate by transamination, increases as a result of a slow acceleration in the Krebs cycle oxidation rate (see Section 12.4.2) as flight begins, leading to an accumulation of pyruvate, the end product of glycolysis. Pi arises from the breakdown of high-energy phosphate compounds (ATP) utilized to provide the initial boost for increased muscle contraction rates.

Further metabolism of glucose is rate controlled in blow fly muscle by the enzyme phosphofructokinase. In locust muscle, as well, this enzyme is inhibited by fructose 1,6-diP, the product of its reaction, and by ATP, one of its substrates. The ATP inhibition is reversed by AMP or Pi, both of which increase when the muscle becomes active.

Looking at these controls in qualitative terms, it may be said that flight-produced changes in metabolic product concentrations are most important in regulating the activities of key enzymes, and through them, glycolytic rates.

The large amount of energy (as ATP) required over extended time periods by activity such as sustained flight or locomotion does not allow utilization of the relatively inefficient anaerobic pathway of carbohydrate degradation in which the reoxidation of NADH is accomplished by the reduction of pyruvate to lactate. This, coupled with the problems of the very limited quantity of NAD and its compartmentalization within the cell due to its inability to cross the mitochondrial membrane, necessitates an alternative aerobic method for the regeneration of oxidized cytoplasmic dinucleotide from that reduced during glycolysis. The method utilized is a redox reaction involving the enzymic oxidation of NADH in the cytoplasm by dihydroxy-acetone P and movement of the resulting L-glycerol 3-P, to which the outer mitochondrial membrane is freely permeable, into the mitochondrion. At a site on the inner membrane the glycerol P is reoxidized to dihydroxyacetone P by a flavoprotein(FP)-linked glycerol 3-P dehydrogenase and returned to the cytoplasm, where it can again participate as an oxidant. The reduced flavoprotein is itself reoxidized through the mitochondrial electron transport system (see Section 12.4.2). Thus, a catalytic amount of dihydroxyacetone P can maintain glycogenolysis or glycolysis in the presence of active mito-chondria and O_2. It is worth noting that the redox potentials of the two electron-transport coenzymes (NAD^+, FP) are such that electrons are moved *into* the mitochondrion, indicating that function lies in that direction (see Section 12.4.2).

The series of reactions in the "glycerol 3-P shuttle" are:

1. Anaerobic production of L-glycerol 3-P (in cytoplasm):

Dihydroxyacetone P + NADH + H^+ →

   L-glycerol 3-P + NAD^+ (cytoplasmic glycerol 3-P dehydrogenase)

2. Aerobic restoration of dihydroxyacetone P (in mitochondrion):

L-Glycerol 3-P + FP → dihydroxyacetone P

      + FPH_2 (mitochondrial glycerol 3-P dehydrogenase)

Total: NADH + H^+ + FP → NAD^+ + FPH_2.

12.4.2. Two-Carbon Oxidation: Tricarboxylic Acid Cycle and Electron Transport System

The enormous increase in oxygen uptake at the inception of sustained activity in most organisms results from the fact that energy is, for the most part, generated by oxidation reactions in which molecular oxygen is the ultimate oxidant. No matter how an insect stores its fuel (carbohydrate, lipid, amino acid), the source of most of its high-energy phosphate compounds is oxidative phosphorylation accompanying the transfer of electrons along the respiratory chain in the mitochondrion. The respiratory chain is, itself, fed by the tricarboxylic acid cycle, also located in the mitochondrion, but the cycle derives its substrates from a variety of sources. If the insect is metabolizing carbohydrate, it generally does so through the glycolytic pathway in the cytoplasm, yielding pyruvate which is oxidatively decarboxylated in the mitochondrial matrix and delivered to the tricarboxylic acid enzymes as acetyl CoA. Lipid is utilized as fatty acid which is partially oxidized through the β-oxidative sequence of enzymes in the mitochondrial matrix and transferred to the cycle as acetyl CoA (see Section 12.4.3); proline can be incorporated into the cycle as α-ketoglutarate after oxidative-ring opening in the inner membrane of the mitochondrion and subsequent transamination. The ultimate disposition of proline consists of the oxidation of a two-carbon unit and production of alanine (see Section 12.4.4).

12.4.2.1. Tricarboxylic Acid Cycle.   The tricarboxylic acid cycle, the enzymes of which are found in the mitochondrial inner membrane and matrix, can be entered *in vitro* only by compounds to which the inner membrane is permeable. Thus, for example, blow fly mitochondria will only oxidize exogenous pyruvate and proline. Some lepidopteran mitochondria can oxidize long-chain acyl CoA derivatives and acetyl CoA (see Section 12.4.3); locust muscle will oxidize long-chain acyl carnitine compounds as well as glutamate and pyruvate; and preparations from periodical cicada flight muscle will oxidize glutamate and α-ketoglutarate. Aside from these compounds, no intermediates appear to be utilized by intact insect flight muscle mitochondria. Thus high degree of impermeability contrasts to that of fat body mitochondria but can be rationalized by a consideration of the functional specificity of the flight muscle.

Most insect enzymes of the tricarboxylic acid cycle have been examined in crude mitochondrial extracts or solubilized preparations, although a few, such as citrate synthase, have actually been purified to crystallization. The properties of the enzymes taken from various Diptera, Hymenoptera, and Lepidoptera resemble those found generally in other eukaryotes, and in recent years, interest has moved from structural investigation to the control of cycle activity.

Pyruvic acid oxidation proceeds through the tricarboxylic acid (Krebs) cycle in the following manner:

Pyruvate + NAD^+ + CoA $\rightarrow$ acetyl CoA
              + CO_2 + NADH + H^+ (pyruvate dehydrogenase complex)

Acetyl CoA + oxaloacetate + H_2O $\rightarrow$ citrate + CoA   (citrate synthase)

Citrate $\rightarrow$ cisaconitate $\rightarrow$ isocitrate   (aconitase)

Isocitrate + NAD^+ + CoA $\rightarrow$ α-ketoglutarate
                + CO_2 + NADH + H^+ (isocitrate dehydrogenase)

α-Ketoglutarate +NAD^+ + CoA $\rightarrow$ succinyl CoA + CO_2 + NADH
                  + H^+ (α-ketoglutarate dehydrogenase complex)

Succinyl CoA + GDP + Pi →
            succinate + CoA + CO_2 + GTP   (succinyl CoA synthetase)
Succinate + FAD → fumarate + $FADH_2$ (succinate dehydrogenase)
Fumarate + H_2O → L-malate   (fumarase)
L-Malate + NAD^+ →
            oxaloacetate + NADH + H^+ (malate dehydrogenase)

Total: Pyruvate + $4NAD^+$ + FAD + GDP + Pi + $2H_2O$ →
                        $3CO_2$ + 4NADH + $4H^+$ + $FADH_2$ + GTP.

Within the mitochondrion the rates of oxidation of the various compounds noted above are generally controlled by the relative concentrations of ADP, Pi, and ATP. Not only do these compounds act as substrates for respiratory chain activity, but they are allosteric effectors of some of the enzymes interacting with or in the tricarboxylic acid cycle. Thus, proline dehydrogenase (see Section 12.4.4) in the mitochondria of a number of Diptera is activated by ADP, whose effect is to decrease the "apparent K_m" for proline; NAD-linked isocitrate dehydrogenase, a major limiting enzyme in the cycle, is activated by ADP, Pi, and its substrate, isocitrate, and is inhibited by ATP and Ca^{2+}; and pyruvate dehydrogenase, a highly regulated enzyme, is inhibited by ATP. This last-named enzyme resembles glycogen phosphorylase in that it is activated by enzyme-catalyzed phosphorylation and seems to be hormonally controlled (see Section 12.4.1 and Section 12.6).

The respiratory chain (see below) is responsible for the collection of electrons from Krebs cycle intermediates, the subsequent flow of electrons to reactions with oxygen, and the conservation of free energy produced during electron transport in the form of phosphate bond energy.

It is generally accepted that three "energy-rich" phosphate bonds are generated as ATP during the reactions involved in the transport of electrons from a molecule of NADH to O_2, and two are generated from the similar reoxidation of reduced flavin. However, the method whereby the energy liberated in the oxidation reactions is conserved and converted to phosphate bond energy is still not entirely clear and will not be discussed.

Using the above values as a basis for the determination of energy production within the tricarboxylic acid cycle, we can calculate the total energy yield from the oxidation of one glucose unit derived from glycogen:

1. Glycogenolysis (Section 12.4.1); Glycerol 3-P shuttle (Section 12.4.1):
$(Glucose)_n$ + $2NAD^+$ + 3ADP + 3Pi →
            $(glucose)_{n-1}$ + 2 pyruvate + 2NADH + $2H^+$ + 3ATP + $3H_2O$
2NADH + $2H^+$ + 2FAD → $2NAD^+$ + $2FADH_2$
$2FADH_2$ + 4ADP + 4Pi + O_2 → 2FAD + 4ATP + $6H_2O$
2. TCA cycle:
2 Pyruvate + $8NAD^+$ + 2FAD + 2GDP + 2Pi + $4H_2O$ →
            $6CO_2$ + 8NADH + $8H^+$ + $2FADH_2$ + 2GTP

$$8NADH + 8H^+ + 24ADP + 24Pi + 4O_2 \rightarrow 8NAD^+ + 24ATP + 32H_2O$$
$$2FADH_2 + 4ADP + 4Pi + O_2 \rightarrow 2FAD + 4ATP + 6H_2O$$

Total: $(Glucose)_n + 35ADP + 2GDP + 37Pi + 6O_2 \rightarrow$
$$(glucose)_{n-1} + 6CO_2 + 35ATP + 2GTP + 43H_2O.$$

12.4.2.2. Electron Transport System.   The carriers in the respiratory (electron transport) chain of insect mitochondria and their locations may be represented as in (Fig. 12-1).

At the lowest oxidation level are the pyridine nucleotide-linked dehydrogenases, which transfer two reducing equivalents from the oxidized substrate, one to the nicotinamide ring of the pyridine nucleotide coenzyme and the second as a free hydrogen ion. These dehydrogenases may have NAD or NADP coenzymes. In the mitochondrial matrix are found NAD- and NADP-linked isocitrate dehydrogenase and malate dehydrogenase. The dihydrolipoyl dehydrogenase segment of the α-ketoglutarate dehydrogenase is located on the inner mitochondrial membrane.

At the next higher oxidation level are the flavin-linked dehydrogenases, the flavins being very tightly bound to the protein enzyme. The nature of the flavin prosthetic groups in insects is not known, but it is presumed that they are flavin mononucleotide (riboflavin phosphate) and flavin adenine dinucleotide as they are in mammals. In any case, it is the isoalloxazine ring of the riboflavin moiety to which the hydrogen atoms are transferred. Among these enzymes, succinate dehydrogenase and NADH dehydrogenase are found positioned on the inner membrane of the mitochondrion.

Coenzyme Q (ubiquinone), a benzoquinone with an isoprenoid side chain of carbon chain length 30–50, functions as an electron transport compound in insects and is located in the respiratory pathway between the flavin enzymes and the cytochromes. The length of the side chain varies among different species, and since insects do not synthesize isoprenoid units of chain length as great as C_{30}, it is presumed that these are provided by the hosts on which they feed.

The cytochromes of insects—b, c_{551}, c, a, and a_3—are responsible for electron transport from coenzyme Q to molecular oxygen. These heme-protein members of the respiratory chain are the compounds seen by Keilin in 1925 in the thorax of the wax moth, *Galleria mellonella*, and shown to increase and decrease in color intensity as the muscles were exercised. He named them cytochromes a, b, and c, and speculated that they carried electrons to O_2 when food was burned to provide energy. Since then a great deal of work has been done on these components of the chain, and it has been determined that transport is carried on similarly in insects and other eukaryotes, the differences, for the most part, being in the primary structures of the carriers. Thus, the compound represented in mammals as having its major absorption band in its reduced form at 554 nm, and given the appelation cytochrome c_1, is replaced in insect thoracic muscle by a compound which absorbs maximally at 551 nm. And cytochrome c, the amino

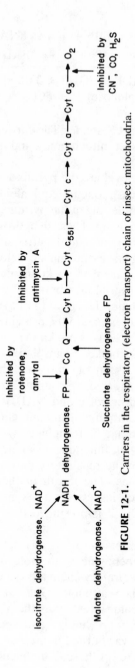

FIGURE 12-1. Carriers in the respiratory (electron transport) chain of insect mitochondria.

acids of which have been sequenced in a large number of animals to establish evolutionary kinship, shows varying degrees of similarity of amino acid sequences among species of insects and between insects and other organisms. Cytochromes a and a_3 are together called cytochrome oxidase and appear to function as components of a single unit responsible for the terminal reaction with O_2. Except for cytochrome c, which has been isolated from a number of insects, being the only easily solubilized heme-protein in the chain, none of the insect cytochromes have been characterized beyond their absorption spectra.

In insects, as in other animals, the locations of the components with reference to one another have been determined both by the use of inhibitors (Fig. 12-1) and by observations of changes in absorption of the colored compounds as respiratory demands change.

12.4.3. Utilization of Lipid

Insect species differ considerably with respect to the RQs (respiratory quotient = CO_2 liberated/O_2 consumed) measured during flight. If we assume that RQ differences represent variations in the types of fuel oxidized, it becomes important to examine the enzymes in a number of insect orders to see whether ordinal specificities exist. In the Hymenoptera, there seems to be some accord, since most of the species studied show RQs of 1.0, indicative of carbohydrate utilization:

$$\text{Carbohydrate (as glucose):} \quad C_6H_{12}O_6 + 6O_2 \rightarrow 6CO_2 + 6H_2O$$

and flight muscle enzyme patterns similar to that of *Apis mellifera* (Table 12-1) in which glycolytic enzymes are high and those concerned with lipolysis are low. However, in the past, Lepidoptera had been thought to burn fat exclusively, with an RQ of 0.7:

$$\text{Lipid (as palmitate):} \quad C_{16}H_{32}O_2 + 23O_2 \rightarrow 16CO_2 + 16H_2O$$

and although this is borne out in Table 12-1 by the enzymes in *Vanessa urticae*, it is not by *Plusia gamma*, which shows high activity of both carbohydrate-metabolizing and fat-splitting enzymes. With the investigation of larger number of Lepidoptera, RQs inconsistent with the old ideas have been demonstrated and, in fact, it has been proven that *Spodoptera frugiperda* (fall armyworm) uses carbohydrate as well as lipid during flight. It may be that adult Lepidoptera, which feed and thus have access to sources of carbohydrate, are able to burn carbohydrate for energy, whereas those that build energy storage depots during larval development and feed little or not at all as adults depend on lipid as a primary source of energy-rich compounds. This would be a logical consequence of the fact that lipid is a much more efficient storage compound than carbohydrate, liberating a much larger number of calories per gram (due to its highly reduced state), packaging better (less space taken up by H_2O of hydration in storage), and releasing more H_2O/mol oxidized. It may also be expected that every ordinal generalization will break down under the weight of examination of larger numbers of species and a recognition that selection for diet utilization should be operative across ordinal lines.

Most interesting is the fact that a number of insects (e.g., locusts and cockroaches), storing both carbohydrate and lipid will utilize carbohydrate first, and then, when called on for sustained flight, will switch to their lipid reserve. Nothing is known of the control of this switch in biochemical terms, although it has been shown to occur in the locust concomitant with the release of the adipokinetic hormone (see Section 12.6).

The metabolism of lipid by flight muscle follows a pattern generally similar to that found in other animals. Depending on the species, triacylglycerol, stored in the fat body, may be converted therein to one or another form and released into the hemolymph as diacylglycerol (*Hyalophora cecropia, Locusta migratoria*), free fatty acid (*Manduca sexta, Galleria mellonella, Leucophaea maderae*), or triacylglycerol (*Periplaneta americana*). In the hemolymph the lipid combines with one of several unique proteins and is transported to the muscle as lipoprotein. (At this time, the few insects in which transport has been studied are those that release diacylglycerol from the fat body. The lipoprotein carrier in these insects has been shown to be specific for diacylglycerols.)

Arriving at the muscle, the lipid is hydrolyzed by a lipase, liberating the fatty acid for use as an energy source (see Chapter 2, Section 2.5.4). (There is presently some confusion as to whether the lipases which have been studied in different insects hydrolyze mono-, di-, and triacylglycerols differentially and whether, in fact, the various enzymes have high enough turnover rates to support flight based on the use of lipid for energy. However, it is clear that in some of the insects in which diacylglycerol is the lipid transported from fat body to muscle, there is enough muscle diacylglycerol lipase to make free fatty acid available at a rate compatible with requirements.)

Once the lipid is degraded, the glycerol is almost entirely returned to the fat body for recycling, and the free fatty acid is oxidized within the mitochondrion. However, to get across the mitochondrial membrane the fatty acid must, in most insects, be transformed into an acyl CoA derivative and, from this, into an acyl-carnitine derivative. Carnitine (3-OH-4-trimethylamino butyric acid; vitamin B_t) is a compound found in high concentration in most animal muscle and in yeast. (The name Vitamin B_t derives from the fact that is required for growth, by a small number of beetle species in the family Tenebrionidae.) It is able to pass across mitochondrial membranes, whereas CoA, except as noted below, cannot. Therefore, acyl CoA-carnitine transferases are present on both sides of the mitochondrial membrane to catalyze the following reaction:

$$\text{Acyl CoA} + \text{carnitine} \rightleftharpoons \text{acyl carnitine} + \text{CoA}$$

providing for the translocation of the acyl moiety into the mitochondrion where it can be oxidized by the enzymes present therein.

The β oxidation pathway of fatty acid oxidation occurs in mitochondria as follows:

(C_{16}) hexadecanoyl CoA + FAD → Δ^2-*trans* hexadecenoyl CoA
+ $FADH_2$ (acyl CoA dehydrogenase)

Δ^2-*Trans* hexadecenoyl CoA + H_2O →
L-3-hydroxyhexadecanoyl CoA   (enoyl CoA hydratase)

L-3-Hydroxyhexadecanoyl CoA + NAD^+ → 3-ketohexadecanoyl CoA
+ NADH + H^+ (3-hydroxyacyl CoA dehydrogenase)

3-Ketohexadecanoyl CoA + CoA → (C_{14}) tetradecanoyl CoA
+ (C_2) acetyl CoA   (acetyl CoA acetyltransferase)

Total: (C_{16}) CoA + CoA + NAD^+ + FAD + H_2O →
(C_{14}) CoA + (C_2) CoA + NADH + H^+ + $FADH_2$

After which (C_{14}) CoA + FAD → Δ^2-*trans*-tetradecenoyl CoA + $FADH_2$ (acyl CoA dehydrogenase), and so on. Then, through the tricarboxylic acid cycle: (C_2) acetyl CoA + oxaloacetate + H_2O → citrate + CoA (citrate synthase), and so on. Continued cycling of the fatty acid, shortened by two carbons in each cycle, yields, in the case of a C_{16} fatty acid, eight acetyl CoA's which enter the tricarboxylic acid cycle, also located in the mitochondrion, and are oxidized to CO_2 and H_2O.

The complete oxidation and energy budget of one C_{16} fatty acid unit (palmitate) is outlined in the following sequence:

1. Activation of fatty acid (thioester link formed between CoA and fatty acid inside or outside of mitochondrion (see above), and the CoA derivative moved to its oxidation site within the mitochondrial matrix):

(C_{16}) Palmitic acid + CoA + ATP → (C_{16}) palmitoyl CoA + AMP
+ PPi (long-chain acyl CoA synthase)

ATP + AMP → 2ADP   (adenylate kinase)

PPi + H_2O → 2Pi   (inorganic pyrophosphatase)

Subtotal 1: C_{16} + CoA + 2ATP + H_2O → C_{16} CoA + 2ADP + 2Pi

2. β Oxidation sequence (see above):

C_{16} CoA + CoA + NAD^+ + FAD + H_2O →
C_{14} CoA + C_2 CoA + NADH + H^+ + $FADH_2$
(×7 as C_{16} is cleaved to eight C_2 units)

Subtotal 2: C_{16} CoA + 7CoA + $7NAD^+$ + 7FAD + $7H_2O$ →
$8C_2$ CoA + 7NADH + $7H^+$ + $7FADH_2$

3. Tricarboxylic acid cycle (see Section 12.4.2)

Subtotal 3: $8C_2 CoA + 24NAD^+ + 8FAD + 8GDP + 8Pi + 16H_2O \rightarrow$
$16CO_2 + 8CoA + 24NADH + 24H^+ + 8FADH_2 + 8GTP$

Subtotal 1–3: C_{16} (palmitic acid) $+ 2ATP + 24H_2O + 31NAD^+$
$+ 15FAD + 8GDP + 6Pi \rightarrow 16CO_2 + 2ADP + 31 NADH + 31 H^+$
$+ 15 FADH_2 + 8GTP$

If the assumption is made that the oxidation of each NADH through the electron transport system yields 3ATP and each $FADH_2$, 2ATP (see Section 12.4.2), then:

$31NADH + 31H^+ + 93ADP + 93Pi + 15.5O_2 \rightarrow$
$$31NAD^+ + 93 ATP + 124H_2O$$

$15FADH_2 + 30ADP + 30Pi + 7.5O_2 \rightarrow 15FAD + 30ATP + 45H_2O$

Overall total: $C_{16} + 121ADP + 8GDP + 129Pi + 23O_2 \rightarrow$
$$16CO_2 + 121ATP + 8GTP + 145H_2O.$$

Most of the enzymes in the β oxidation pathway have been shown to be present in the mitochondria of one or another of the species that use fatty acids as a source of energy. However, an apparent exception to the requirement of moving the fatty acid across the mitochondrial membrane as a carnityl ester has been demonstrated in two species of Lepidoptera. In the presence of CoA, and without added carnitine, thoracic mitochondria isolated from these moths can oxidize palmitic acid at rates approaching those *in vivo*. The carnitine-transferring enzyme is also absent, indicating that, at least in *Prodenia eridania* and *Trichoplusia ni,* CoA esters may be capable of moving across mitochondrial membranes.

12.4.4. Utilization of Amino Acids

Although the total amount of amino acid in the hemolymph of many insect species is comparatively high, no one compound can be considered as being present in "substrate" concentration. In fact, with a single exception, amino acids have never been demonstrated to be of any significance as metabolic fuels. That exception, proline, which arises from endogenous synthesis as well as nutrient absorption, has been shown to be important at two levels:

1. As a cosubstrate for pyruvate oxidation in some species, and
2. As the major source of flight energy in others.

In insects of a number of orders, the enzyme that catalyzes the first reaction in proline oxidation (proline dehydrogenase) is present in flight muscle mitochondria in an amount large enough to turn the amino acid over at a high rate, but not nearly sufficient to support the rate of energy production demanded by flight. What, then, is its function?

A look at the concentration of proline in blowfly flight muscle shows that it declines precipitously at the beginning of flight, leveling off after a short time. The dehydrogenase product (see below), is glutamate, which is then deaminated to α-ketoglutarate, a Krebs cycle intermediate. From this is produced an increased quantity of all of the other intermediates, leading to a higher Krebs cycle "concentration." An examination of glycolytic activity during the same few moments after flight commences shows that pyruvate piles up quickly and then disappears. Taken together, these results suggest that, in those insects with low, but significant, amounts of proline dehydrogenase, proline is utilized at the beginning of flight to bring up the Krebs cycle level to that sufficient to support pyruvate oxidation at a rate concomitant with its production from glycolysis.

In the tsetse fly *(Glossina morsitans)* and four beetle species *(Leptinotarsa decemlineata, Melolontha melolontha, Helicopris dilloni,* and *Popillia japonica)*, the level of proline dehydrogenase is much higher than is necessary to perform the previously described function, and, in fact, appears to operate as the primary enzyme in the utilization of proline as an energy source for flight. In these insects, other enzymes involved in proline oxidation (i. e., alanine aminotransferase and NAD-linked malic enzyme) are also found in high concentrations in flight muscle mitochondria, whereas Krebs cycle enzymes that do not participate in the proline pathway (i.e., citrate synthase and isocitric dehydrogenase) and some glycolytic enzymes may be as much as 10 times lower in concentration than they are in flies that oxidize carbohydrate to obtain flight energy.

The metabolism of proline in flight muscle mitochondria occurs in the following manner:

$$\text{Proline} + \text{FP} \rightarrow \Delta^1\text{-pyrroline 5-carboxylate} + \text{FPH}_2 \qquad \text{(proline oxidase)}$$

$$\Delta^1\text{-Pyrroline 5-carboxylate} + \text{H}_2\text{O} \rightarrow$$
$$\text{glutamate } \gamma\text{-semialdehyde} \quad \text{(spontaneous)}$$

$$\text{Glutamate } \gamma\text{-semialdehyde} + \text{NAD}^+ + \text{H}_2\text{O} \rightarrow$$
$$\text{glutamate} + \text{NADH} + \text{H}^+ \quad (\Delta^1\text{-pyrroline dehydrogenase})$$

$$\text{Glutamate} + \text{pyruvate} \rightarrow \alpha\text{-ketoglutarate}$$
$$+ \text{ alanine} \quad \text{(alanine aminotransferase)}$$

$$\alpha\text{-Ketoglutarate} + \text{CoA} + \text{NAD}^+ \rightarrow \text{succinyl CoA} + \text{CO}_2$$
$$+ \text{NADH} + \text{H}^+ \quad (\alpha\text{-ketoglutarate dehydrogenase complex})$$

$$\text{Succinyl CoA} + \text{GDP} + \text{Pi} \rightarrow$$
$$\text{succinate} + \text{CoA} + \text{GTP} \quad \text{(succinyl CoA synthetase)}$$

$$\text{Fumarate} + \text{H}_2\text{O} \rightarrow \text{L-malate} \qquad \text{(fumarase)}$$

$$\text{L-Malate} + \text{NAD}^+ \rightarrow \text{pyruvate} + \text{NADH} + \text{H}^+ + \text{CO}_2 \quad \text{(malic enzyme)}$$

Subtotal: proline + 3NAD$^+$ + FAD + FP + GDP + Pi + 3H$_2$O $\rightarrow$
             alanine + 2CO$_2$ + 3NADH + 3H$^+$ + FADH$_2$ + FPH$_2$ + GTP

In the respiratory chain

$$3NADH + 3H^+ + 9ADP + 9Pi + 1.5O_2 \rightarrow 3NAD^+ + 9ATP + 12H_2O$$
$$FADH_2 + 2ADP + 2Pi + 0.5O_2 \rightarrow FAD + 2ATP + 3H_2O$$
$$FPH_2 + 2ADP + 2Pi + 0.5O_2 \rightarrow FP + 2ATP + 3H_2O$$

Total energy yield: proline + 13ADP + GDP + 14Pi + 2.5O_2 $\rightarrow$
$$\text{alanine} + 2CO_2 + 13ATP + GTP + 15H_2O$$

There are many questions concerning the evolution of this pathway, but at present, our only clue to any answers is that protein is the major nutritional source for certain of these insects. High levels of protein in the food may have led to the utilization of some of the derived amino acids for energy production. However, it is important to note that even in these animals, the ultimate source of energy is C_2 units derived from stored fat (see Section 12.3.5).

12.5   OTHER METABOLIC TRANSFORMATIONS

12.5.1.   Pentose Phosphate Pathway of Carbohydrate Utilization

The enzymes concerned with the metabolism of carbohydrates have been examined in the fat bodies of a number of insects, but until recently there have been no attempts to systematically delineate the specific activities and locations of all of the glycolytic enzymes within that tissue. Based on assays of a relatively few enyzmes, it has generally been agreed that glycolysis proceeds in fat body as it does in muscle. However, a second pathway of glucose utilization, concerned more with the production of reduced NADP for lipid synthesis and pentoses for nucleic acid synthesis, is also present in fat body. The so-called "pentose phosphate pathway" or "pentose shunt" has been known in insects since the 1960s, when it was partially characterized in crude extracts of whole insects and fat body. The pathway has not been found in flight muscle of those insects (flies and locusts) which have been examined, an expected consequence of what is believed to be its metabolic function. The pathway in insects follows that established in other animal tissue:

6 Glucose 6-P + 6NADP$^+$ $\rightarrow$ 6 6-phosphogluconolactone
$\qquad\qquad$ + 6NADPH + 6H$^+$ (glucose 6-P dehydrogenase)
6 6-Phosphogluconolactone + 6H$_2$O $\rightarrow$
$\qquad\qquad\qquad\qquad\qquad$ 6 6-phosphogluconate   (lactonase)
6 6-Phosphogluconate + 6NADP$^+$ $\rightarrow$ 6 D-ribulose 5-P
$\qquad$ + 6CO$_2$ + 6NADPH + 6H$^+$ (6-phosphogluconate dehydrogenase)

2 D-Ribulose 5-P $\rightarrow$ 2 D-ribose 5-P   (ribose-P isomerase)

4 D-Ribulose 5-P $\rightarrow$ 4 D-xylulose 5-P   (ribulose-P 3-epimerase)

2 D-Ribose 5-P + 2 D-xylulose 5-P $\rightarrow$
$\qquad$ 2 D-sedoheptulose 7-P + 2 D-glyceraldehyde 3-P   (transketolase)

2 D-Sedoheptulose 7-P + 2 D-glyceraldehyde 3-P $\rightarrow$
$\qquad$ 2 D-fructose 6-P + 2 D-erythrose 4-P   (transaldolase)

2 D-Xylulose 5-P + 2 D-erythrose 4-P $\rightarrow$ 2 D-fructose 6-P
$\qquad$ + 2 D-glyceraldehyde 3-P   (transketolase)

D-Glyceraldehyde 3-P $\rightarrow$ diOH acetone P   (triose phosphate isomerase)

D-Glyceraldehyde 3-P + diOH acetone P $\rightarrow$ fructose 1,6-diP   (aldolase)

Fructose 1,6-diP $\rightarrow$ fructose 6-P + Pi   (hexose diphosphatase)

5 Fructose 6-P $\rightarrow$ 5 glucose 6-P   (glucose phosphate isomerase)

Total: 6 glucose 6-P + 12 NADP$^+$ + 6H$_2$O $\rightarrow$
$\qquad$ 5 glucose 6-P + 6CO$_2$ + 12NADPH + 12H$^+$ + Pi

The utility of the pathway as a source of reduced pyridine nucleotide and/or pentose phosphate is obvious.

A recent examination of the soluble carbohydrate-metabolizing enzymes of the American cockroach fat body has brought to light some interesting facts. It appears that in this insect, at least, the enzymes concerned with the glycolytic conversion of glucose 6-P to triose P (i.e., phosphoglucomutase, phosphofructokinase, and aldolase) (see Section 12.4.1) have very low activity compared to the other enzymes in the chain. Furthermore, glucose 6-P dehydrogenase, the primary and rate-limiting enzyme in the pentose phosphate pathway, and glucose phosphate isomerase, another of the same group of enzymes, are present in high activity. Since the major function of the pentose phosphate pathway is thought to be the provision of reduced NADP for lipogenesis (Section 12.3.3), and glycolytic activity is shown to be limited in this tissue, there is now direct enzymic evidence that lipid synthesis is probably the major metabolic concern of the fat body in the cockroach.

Measurements of the contribution of the pentose phosphate pathway to the total breakdown of glucose have been made in a number of insects, using the ratio of counts in CO$_2$ from metabolized C$_6$ and C$_1$–^{14}C-labeled glucose to indicate participation of the shunt. [As may be noted from the description of the glycolytic pathway (Section 12.4.1), CO$_2$ is first liberated from glucose at the glucose C$_{3-4}$ position when pyruvate is converted to acetyl CoA. After this, C$_2$–C$_5$ and C$_1$–C$_6$ are equivalently oxidized in the TCA cycle. The production of CO$_2$ in the pentose shunt derives entirely from the C$_1$ decarboxylation of 6-phosphogluconate. Thus, a C$_6$: C$_1$ ratio of less than 1 means that the latter pathway is being used—the lower the number, the greater the utilization. Although there are problems of interpretation in this method of analysis, it has been the source of most of the information regarding the ratios of use of the two pathways in insects.] The results of these experiments indicate that there are large variances in pentose phosphate utilization among species, ranging from 3% in female *Periplaneta americana* (21% in males) to 38% in *Melanoplus bivittatus,* which uses lipid for extended flight. It is also interesting to note that certain Diptera *(Phormia terranovae),* tolerant to low temperatures as adults, increase their pentose shunt activity at those temperatures. The biological significance of this finding is yet to be established.

12.5.2. Gluconeogenesis

Restoration of a store of glucose (as glycogen or trehalose) in an animal depleted of carbohydrate comes about through the process of gluconeogenesis. Thus, amino acids, which can be transaminated to pyruvate or oxaloacetate, and lipids, which, through the action of lipases, release free glycerol, provide the insect with carbohydrate by a general reversal of glycolysis. (In all insects studied up to the present time, fatty acids and amino acids which are converted to acetyl CoA by oxidative reactions, cannot be recovered as storage carbohydrate because of the irreversibility of the enzyme pyruvate dehydrogenase and the absence of a glyoxylate cycle.) The energy-releasing reactions of glycolysis, (1) catalyzed by pyruvate kinase:

$$\text{Phosphoenolpyruvate} + \text{ADP} \rightarrow \text{pyruvate} + \text{ATP}$$

and (2) catalyzed by 6-phosphofructokinase:

$$\text{D-Fructose 6-phosphate} + \text{ATP} \rightarrow \text{D-fructose 1,6-diP} + \text{ADP}$$

are not reversible (Section 12.4.1), so the substrates of these reactions must be recycled by indirect means.

Phosphoenolpyruvate, the substrate of reaction 1, can be recovered from a cytoplasmic reaction involving oxaloacetate, catalyzed by the enzyme phosphoenolpyruvate carboxykinase:

$$\text{Oxaloacetate} + \text{GTP} \rightarrow \text{phosphoenolpyruvate} + \text{CO}_2 + \text{GDP}$$

This enzyme has been found in fat body and muscle of a number of insects.

Oxaloacetate, a member of the Krebs cycle generally found in the mitochondrion, can be produced directly from pyruvate as well as through oxidative reactions in the cycle. Pyruvate carboxylase, the enzyme responsible for the reaction from pyruvate, has been described in the mitochondria of a number of insects:

$$\text{Pyruvate} + \text{CO}_2 + \text{ATP} + \text{H}_2\text{O} \rightarrow \text{oxaloacetate} + \text{ADP} + \text{Pi}$$

The mitochondrial membrane is not permeable to oxaloacetate but can be traversed by malate, so a pair of enzymes, analogous to those moving acyl

CoA across the mitochondrion, catalyze the reduction of oxaloacetate within the mitochondrion and its reoxidation in the cytoplasm:

$$\text{Oxaloacetate} + \text{NADH} + \text{H}^+ \rightarrow$$
$$\text{L-malate} + \text{NAD}^+ \quad (\text{malate dehydrogenase}_{\text{mito}})$$

$$\text{L-Malate} + \text{NAD}^+ \rightarrow$$
$$\text{oxaloacetate} + \text{NADH} + \text{H}^+ \quad (\text{malate dehydrogenase}_{\text{cyto}})$$

Total: $\text{Pyruvate}_{\text{mitro}} + \text{ATP} + \text{GTP} + \text{H}_2\text{O} \rightarrow$
$$\text{phosphoenolpyruvate}_{\text{cyto}} + \text{ADP} + \text{GDP} + \text{Pi}$$

Fructose 6-P, the substrate of reaction 2, is produced by the hydrolysis of fructose 6-diphosphate:

$$\text{D-Fructose 1,6-diP} \rightarrow \text{D-fructose 6-P} + \text{Pi}$$

The enzyme catalyzing this reaction, fructose 1,6-diphosphatase, is ubiquitous in insect fat body and muscle. Aside from its action in gluconeogenesis, it has been implicated in a metabolic cycle which generates heat in the thorax of certain bumblebees. In these insects, the heat required on cold days to raise the muscle temperature to a point at which the mitochondrial enzymes function at a rate sufficient to provide the energy for efficient contraction and, therefore, flight, is thought to be derived from a series of reactions involving 6-phosphofructokinase and fructose diphosphatase:

$$\text{Fructose 6-P} + \text{ATP} \rightarrow \text{fructose 1,6-diP} + \text{ADP}$$
$$\text{Fructose 1,6-diP} \rightarrow \text{fructose 6-P} + \text{Pi}$$

In this cycle, ATP is hydrolyzed and energy is released in the form of heat.

When the animal is in flight, thoracic temperatures are regulated without these reactions, but in intervals of rest the cycle is activated and muscle temperatures remain above ambient. Honeybees do not have this capability.

12.5.3. Production of Glycerol from Glycogen in Fat Body

Many insects respond to low temperatures by increasing their blood levels of cryoprotectants such as sugars (e.g., trehalose in *Eurosta solidagensis* (Diptera), sugar alcohols (e.g., sorbitol and threitol in *Upis ceramboides* (Coleoptera), or glycerol (e.g., adults of many Coleoptera, diapause pupae of many Lepidoptera). The methods whereby protection is obtained are not well known, but depressing the blood-freezing point and decreasing ice crystal size and rate of formation in the blood and tissue are two mechanisms presently under consideration. It may also be that the process varies from one to another of the compounds.

The factors controlling the metabolic changes leading to the production of the above-named compounds are similarly somewhat obscure, but glycerol synthesis has been studied, and it is known that it is formed in the fat body at the expense of glycogen.

The production of glycerol proceeds from the phosphorolytic cleavage of fat body glycogen (see Section 12.4.1) through the glycolytic pathway to diOHacetone-P and thence to L-glycerol 3-P, the latter reaction catalyzed by cytoplasmic glycerol 3-P dehydrogenase. L-Glycerol 3-P then undergoes hydrolysis to free glycerol and Pi in the presence of soluble glycerol 3-P phosphatase:

$$\text{L-Glycerol 3-P} \rightarrow \text{glycerol} + \text{Pi}$$

Part of the control of glycerol production must rest in the initial steps in the generation of glycolytic intermediates from the storage compound, glycogen. However, fat body glycogen phosphorylase has been studied in relatively few insects, and the enzymes controlling its activation and deactivation (phosphorylase kinase and phosphorylase phosphatase) have been examined in an even lesser number.

The phophorylase kinase in a crude preparation of cecropia larval fat body, in contrast to the kinase in dipteran flight muscle (see Section 12.4.1), becomes activated when incubated with ATP. Thus, it appears that, as in the mammalian liver cell, the enzyme must be phosphorylated before it can activate glycogen phosphorylase. The phosphorylase, itself, in the American cockroach fat body is activated when cAMP is added to whole fat body preparations, suggesting that cockroach fat body, as well as that of cecropia, contains the highly controlled glycogenolytic system originally described in mammals (Fig. 12-2).

An interesting recent finding concerned with insect response to cold stress is that the phosphorylase in whole cecropia diapause pupal fat body (and in fat body preparations from some orthopterans) is activated by low

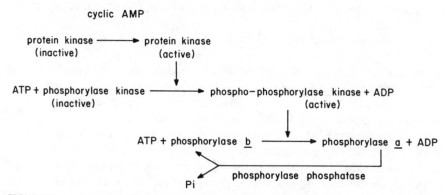

FIGURE 12-2. Suggested glycogenolytic system in the fat body of the American cockroach (*P. americana*).

temperature. The mechanism of activation is presently unknown, but it probably will be found within the control system shown in Fig. 12-2.

Having pointed out all of the possibilities surrounding the first step in glycerol synthesis, it must be said that where and how the glucose liberated in the phosphorylase reaction is shifted away from trehalose (see Section 12.3.2), and into glycerol, is yet to be completely established. In this case, as in so many others, it will probably be shown to depend on a complex of direct and indirect effects involving much of the metabolic machinery.

12.6. HORMONAL CONTROL OF METABOLISM

It has been known for some time that the major developmental hormones, ecdysteroids and juvenile hormones, are responsible for changes in blood and tissue levels of various storage compounds through their actions on general physiological states. However, there is now evidence for a number of hormones which control blood levels of specific compounds from which the insect may derive energy. Their names, hypertrehalosemic, hypotrehalosemic, and adipokinetic, describe general functions rather than specific actions. Two of them have been studied in detail, the hypertrehalosemic and adipokinetic hormones, and of these, the action of the former seems to be most well-understood.

12.6.1. Hormones and Carbohydrates

The hypertrehalosemic hormone has been known since 1961, when injections of *corpus cardiacum* were shown to increase the blood level of trehalose in cockroaches. Similar effects have been demonstrated in other insects in different orders, and it has been found that the initial response in all insects is the activation of fat body glycogen phosphorylase, resulting in an increased glucose phosphate level in that tissue.

This finding accounts for the fact that blood levels of trehalose increase in many insects when they are injected with glandular extracts. The close interaction between trehalose and glycogen metabolism (see Section 12.3.2), will, under certain circumstances of glycogen degradation, result in the production of trehalose, but hypertrehalosemia is not a necessary concomitant of the process. That the function of the hormone may simply be to release glucose from storage for any of a number of purposes is borne out in insects such as locusts, in which an injection of *corpora cardiaca* extract produces an activation of glycogen phosphorylase and a decrease in glycogen, but no increase in blood trehalose.

The action of the hormone (a low-molecular-weight peptide of unknown structure) has been duplicated in a number of insects by the addition of 3',5'-cAMP to whole fat body preparations, suggesting the attractive hypothesis that the hypertrehalosemic hormone behaves in fat body in the same way as does glucagon in mammalian liver; that is, it interacts with receptors responsible for activating adenyl cyclase. The cyclase then catalyzes the synthesis of cAMP, which in turn activates the protein kinase involved in the production of active phosphorylase kinase (see Section 12.5.3), yielding active glycogen phosphorylase. In this respect, it is of interest to note that peptides that immunologically cross-react with mammalian glucagon have been identified in some insect species.

A hypotrehalosemic hormone has also recently been discovered in members of the orders Diptera and Lepidoptera. The action of the hormone (a peptide that when isolated from *Calliphora vomitoria* (Diptera) heads and *Manduca sexta* (Lepidoptera) *cardiaca–allata* complexes exhibits immunological cross-reactivity with bovine insulin) is completely unknown, its bioassay based only on a depression of experimentally elevated blood levels of trehalose.

12.6.2. Hormones and Lipids

The adipokinetic hormone is chemically better known than the hormones involved in carbohydrate metabolism, having been isolated and purified from the glandular lobe of the *corpus cardiacum* of the locust, *Schistocerca gregaria*. In this animal it has been identified as a decapeptide consisting of the following sequence of amino acids:

Pyroglutamate-Leu-Asn-Phe-Thr-Pro-Asn-Trp-Gly-Threoninamide

The compound is structurally very specific, substitution in any amino acid leading to a notable decrease in activity.

Although its structure is known, there is little understanding of its action except that in the locust (and, with less evidence, in *Tenebrio molitor* and *Danaus plexippus*) it promotes the release of diacylglycerol from fat body into the hemolymph and, at the same time, stimulates the production of a specific lipoprotein system for transporting the lipid to the muscle. For example, the question of whether the release of diacylglycerol results from direct lipase activation or an increase in the rate of transport from the fat body has not yet been answered. To compound the confusion, recent experiments indicate that the same entity which, when injected, produces a hyperlipemic effect in locusts results in a hypolipemic effect in cockroaches. Cockroach extracts introduced into locusts increase diacylglycerol release from locust fat body, whereas locust or cockroach glandular extracts stimulate the uptake of di- and triacylglycerol from hemolymph into fat body in the cockroach.

Although at first glance, these activities would seem to be contradictory, a concept of the hormone as trigger rather than direct modifier permits evolutionary "necessity" to mold the system affected by the stimulus. Thus, the same hormone could, in different organisms, have very different effects on the same system. Whether this will prove to be the case in this situation awaits further investigation, but the idea is noteworthy.

Additionally, in the locust a relationship exists between the level of trehalose in the blood and the release of adipokinetic hormone, such that until the level of trehalose drops, the hormone is not released. This interaction probably permits the locust in migratory flight to exhaust its carbohydrate before utilizing diacylglycerol as a source of flight energy.

12.7. HOMEOSTASIS

In bringing this chapter to a close, it must be reiterated that the various pathways of metabolism do not operate independently but are totally interactive. The controls exerted among them are driven in the name of *homeostasis*. This principle, defined as "the maintenance of stability through coordinated response to disruption or stimulus," governs all living systems, and it is the loss of this capability which, in an organism leads to malfunction and death, and in a community, to decline and dissolution.

The capacity for adaptation to a new environment depends on the flexibility and sophistication of the physiological and chemical systems making up an organism and the ability, developed over evolutionary time, to integrate them into a functional unit. It is with these interactions that insect biochemistry is beginning to concern itself, and only through a systematic investigation of each of the contributing processes shall we finally achieve some semblance of understanding of the whole organism. Therefore, it is important in reading this chapter that we not lose sight of the functions that the described synthetic and catabolic reactions are meant to serve.

12.8. CONCLUSIONS

Students of biology are generally aware that the process of natural selection has resulted in a multitude of highly adapted organisms in which fundamental cellular reactions have an essential homogeneity. However, a knowledge of comparative biochemistry reveals the fact that the manifold adaptive solutions to physiological problems which are reflected in phenotypic differences have effected substantial and, as well, subtle differences in metabolic pathways, enzyme properties, and membrane characteristics. Thus, chemical dissimilarity of exoskeleton and endoskeleton is marked by unique enzymes and pathways of synthesis; divergence between homeothermy and poikilothermy gives rise to exquisite alterations in membrane composition, temperature optima of enzymes, and so on.

Even the applied entomologist, whose attention is not normally focused on the elegance and power of natural selection, must be drawn by self-interest to these variations, for it is only through investigation of the biochemical differences between insects and other organisms—plant and animal—that intelligent approaches to pest management practices can be generated.

From this congruence of basic and applied interests, we may look expectantly to a future replete with new perceptions and, hopefully, understanding of the metabolic processes of the insect.

REFERENCES

Reviews

E. Bailey, "Biochemistry of insect flight. Part 2: Fuel supply," in D. J. Candy and B. A. Kilby, Eds., *Insect Biochemistry and Function*, Chapman and Hall, London, 1975, pp. 91–176.

A. M. T. Beenakkers, D. J. Van der Horst, and W. J. A. Van Marrewijk, "Role of lipids in energy metabolism," in R. G. H. Downer, Ed., *Energy Metabolism in Insects*, Plenum Press, New York, 1981, pp. 53–100.

E. Bursell, "The role of proline in energy metabolism," in R. G. H. Downer, Ed., *Energy Metabolism in Insects*, Plenum Press, New York, 1981, pp. 135–154.

G. M. Chippendale, "The functions of carbohydrates in insect life processes," in M. Rockstein, Ed., *Biochemistry of Insects*, Academic Press, New York, 1978, pp. 1–55.

B. Crabtree and E. A. Newsholme, "Comparative aspects of fuel utilization and metabolism by muscle," in P. N. R. Usherwood, Ed., *Insect Muscle*, Academic Press, New York, 1975, pp. 405–500.

R. G. H. Downer, "The functional role of lipids in insects," in M. Rockstein, Ed., *Biochemistry of Insects*, Academic Press, New York, 1978, pp. 58–92.

S. Friedman, "Trehalose regulation, one aspect of metabolic homeostasis," *Ann. Rev. Ent.* **23**, 389–407 (1978).

J. Ilan and J. Ilan, "Protein synthesis in insects," in M. Rockstein, Ed., *The Physiology of Insecta*, 2nd ed. Academic Press, New York, 1974, pp. 356–422.

A. E. Kammer and B. Heinrich, "Insect flight metabolism," *Adv. Insect Physiol.* **13**, 133–228 (1978).

L. L. Keeley, "Endocrine regulation of fat body development and function," *Ann. Rev. Ent.* **23**, 329–358 (1978).

B. Sacktor, "Biological oxidations and energetics in insect mitochondria," in M. Rockstein, Ed., *The Physiology of Insecta*, 2nd ed., Academic Press, New York, 1974, pp. 272–353.

B. Sacktor, "Biochemistry of insect flight. Part 1: Utilization of fuels by muscle," in D. J. Candy and B. A. Kilby, Eds., *Insect Biochemistry and Function*, Chapman and Hall, London, 1975, pp. 1–88.

J. E. Steele, "Hormonal control of metabolism in insects," *Adv. Insect Physiol.* **12**, 239–323 (1976).

J. E. Steele, "Hormonal modulation of carbohydrate and lipid metabolism in fat body," in M. Locke and D. S. Smith, Eds., *Insect Biology in the Future, VBW 80*, Academic Press, New York, 1980, pp. 253–271.

J. V. Stone and W. Mordue, "Isolation of insect neuropeptides," *Insect Biochem.* **10**, 229–239 (1980).

Other Selected References

1. Carbohydrate Metabolism

M. Ashida and G. R. Wyatt, *Insect Biochem.* **9**, 403 (1979).

Y. Hayakawa and H. Chino, *Insect Biochem.* **11**, 43 (1981).

R. T. Mayer, A. C. Chen, and J. R. De Loach, *Insect Biochem.* **10**, 549 (1980).

R. Moreau, L. Gourdoux, and J. Dutrieu, *Comp. Biochem. Physiol.* **56B**, 175 (1977).

A. N. Rowan and E. A. Newsholme, *Biochem. J.* **178**, 209 (1979).

K. B. Storey, *Insect Biochem.* **10**, 637–645, 647 (1980).

K. B. Storey and E. Bailey, *Insect Biochem.* **8**, 73, 125 (1978).

R. Ziegler, A. Masaaki, A. M. Fallon, L. T. Wimer, S. S. Wyatt, and G. R. Wyatt, *J. Comp. Physiol.* **131B**, 321 (1979).

2. Lipid Metabolism

A. G. D. Hoffman and R. G. H. Downer, *Insect Biochem.* **9**, 129 (1979).

D. J. Van der Horst, J. M. Van Doorn, and A. M. TH. Beenakkers, *Insect Biochem.* **9**, 627 (1979).

W. J. A. Van Marrewijk, A. TH. M. Van den Broek, and A. M. TH. Beenakkers, *Insect Biochem.* **10**, 675 (1980).

E. Weeda, A. B. Koopmanschap, C. A. D. De Kort, and A. M. TH. Beenakkers, *Insect Biochem.* **10**, 631 (1980).

3. Protein and Amino Acid Metabolism

E. Bursell, *Insect Biochem.* **7**, 427 (1977).

M. A. Grula and R. F. Weaver, *Insect Biochem.* **11**, 149 (1981).

D. A. Norden and C. Matanganyidze, *Insect Biochem.* **9**, 85 (1979).

J. Nowock, S. Sridhara, and L. I. Gilbert, *Biochem. Biophys. Acta* **520**, 393 (1978).

D. J. Pearson, M. O. Imbuga, and J. B. Hoek, *Insect Biochem.* **9**, 461 (1979).

V. E. F. Sklar, J. A. Jaehning, L. P. Gage, and R. G. Roeder, *J. Biol. Chem.* **251**, 3794 (1976).

S. Sridhara and L. I. Gilbert, *Insect Biochem.* **9**, 467 (1979).

T. V. West, M. A. Grula, W. M. Wormington, and R. F. Weaver, *Insect Biochem.* **10**, 509 (1980).

4. Hormones and Metabolism

K. J. Kramer, H. S. Tager, and C. N. Childs, *Insect Biochem.* **10**, 179 (1980).

W. Mordue and J. V. Stone, *Insect Biochem.* **11**, 353 (1981).

T. E. Roche, K. J. Kramer, and D. W. Dyer, *Insect Biochem.* **10**, 577 (1980).

H. S. Tager, J. Markese, K. J. Kramer, R. D. Speirs, and C. N. Childs, *Biochem. J.* **156**, 515 (1976).

13

HORMONAL ACTION DURING
INSECT DEVELOPMENT

HERBERT OBERLANDER
Insect Attractants, Behavior, and Basic Biology Research Laboratory
Agricultural Research Service, USDA
Gainesville, Florida

CONTENTS

SUMMARY

The origins of insect endocrinology can be traced from nineteenth century studies of pioneers in vertebrate endocrinology to the work of Kopeč, who demonstrated the existence of hormones in insects in a series of classical surgical experiments. The role of hormones in insect development was elucidated by the key experiments conducted by Fraenkel, Wigglesworth, Fukuda, Williams, and others. The chemistry, biosynthesis, and inactivation of the major insect hormones—ecdysteroids, juvenile hormones (JHs), and prothoracicotropic hormone—are discussed in detail and the regulation of hormonal titer in the hemolymph emerges as a central aspect of hormonal action.

The brain, prothoracic glands, and *corpora allata* are the major endocrine organs in insects and their hormonal products have been characterized or identified in the last three decades. The biosynthesis and inactivation of ecdysteroids and JHs have been studied and the titers of these hormones are known to reflect the interplay of anabolism and catabolism. The interactions between each of the major hormones and the target tissues have been analyzed and the development (and loss) of competence to respond to the hormones is now recognized as a major feature of the relationship between ecdysteroid and JH action on tissues.

The mode of action of insect hormones can be analyzed in terms of the "second messenger" (cyclic nucleotide) and intracellular protein "receptor" models of hormonal action. For the most part, the evidence is consistent with the hypothesis that insect steroidal hormones (ecdysteroids) work via an intracellular protein receptor, and that peptide hormones (prothoracicotropic hormone, eclosion hormone, bursicon, etc.,) work via cyclic nucleotides. Juvenile hormone, though not a steroid, may act according to the steroid receptor model.

13.1. THE NEUROENDOCRINE SYSTEM IN INSECT DEVELOPMENT

13.1.1. Role of the Brain

The science of endocrinology can be traced to the experimental work of Berthold who in 1849 reported on the effects of removal and replacement of testes of cockerels. After castration, the sex combs of the fowl atrophied, although they grew normally in castrated males with a testicular graft. Berthold correctly concluded that the testes released a factor into the blood that maintained secondary sex characteristics in males. During the next half century a number of landmark experiments were conducted with vertebrates that clearly established the endocrine function of various tissues in vertebrates. This period of pioneering research in endocrinology culminated with the work of Bayliss and Starling (1902–1905) who coined the term *hormone* (from the Greek *hormon,* exciting) to describe "secretin," a secretion of the duodenal mucosa that is released when acidified food enters the stomach.

These early experiments on endocrine function in vertebrates were not lost on those interested in the biology of insects. However, a series of experiments on castration of both sexes in the commercial silkworm, *Bombyx mori,* the gypsy moth, *Lymantria dispar,* and the cricket, *Gryllus,* resulted in the erroneous conclusion that because the insect's sexual characteristics were unaffected by the surgery, insects did not have an endocrine system.

In 1912 it was reported that amphibian metamorphosis was influenced by a substance from the thyroid and it was observed that the brain (pituitary) was necessary for amphibian metamorphosis. These findings interested Kopeč, who sought to determine the basis of regulation of metamorphosis in insects.

Insect endocrinology begins with the classic experiments of Kopeč from 1917 to 1922. He used three basic (and still relevant) procedures to probe the control of metamorphosis in the gypsy moth: *ligation, extirpation,* and *transplantation.* Kopeč found that head ligation of caterpillars 7 days after the last larval molt prevented pupation posterior to the ligature. However, 10 days after the last molt, ligation of the caterpillars did not prevent the onset of metamorphosis. Kopeč suspected that the brain was involved and demonstrated this by extirpating the brain of larvae 7 days after the previous molt. Only 20% of the debrained larvae pupated compared with 92% of the operated controls. Conclusive evidence for the role of the brain in pupation was obtained in an experiment in which Kopeč restored the ability of debrained larvae to molt by transplanting a larval brain into the operated host.

In 1922 Kopeč wrote that the larval brain had "the function of an organ of internal secretion."

The next advances in insect endocrinology came in the 1930s when Fraenkel and Wigglesworth confirmed and extended Kopeč's findings on the endocrine role of the brain in larval molting and metamorphosis. Significantly, Wigglesworth utilized another basic endocrinological technique, *parabiosis*, to show that a "blood-borne factor" (i.e., a hormone) controlled molting in larval *Rhodnius prolixus*. *Rhodnius* larvae that were decapitated one day after feeding (*Rhodnius* is a blood-sucking bug) did not molt, while larvae decapitated seven days after the blood meal molted normally. When decapitated larvae of both ages were joined together in parabiosis, both the one-day and seven-day decapitated larvae molted. Wigglesworth's experiment clearly showed that a substance was passed from one insect to the other as would be expected with hormonal control of molting.

In all of these early studies it was clear that the source of hormone was required for a specific portion of an instar. This gland-dependent period has been called "the critical period." Thus, ligation or extirpation after the critical period is without effect because enough hormone to stimulate molting has already been released into the hemolymph.

13.1.2. Role of the Prothoracic Glands

First described in 1762 by Lyonet as "granulated vesicles," the prothoracic glands were given their present name by Ke in 1930 as a description of their location because their function had not yet been uncovered. Careful ligation experiments in the 1930s led scientists to conclude that, at least for Lepidoptera, the prothoracic region played a role, distinct from the brain's, in molting. Fukuda determined that it was the prothoracic glands that caused molting and metamorphosis in the silkworm, and Hadorn showed that the ring gland of *Drosophila melanogaster* (analogous in part to the prothoracic glands) was necessary for pupation.

Thus, by the early 1940s it appeared that two hormones, one from the brain and one from the prothoracic glands, were needed for molting and metamorphosis. This situation was resolved by Williams (1952) in a series of decisive experiments with diapausing pupae of the cecropia moth. Williams cut the pupae in half and sealed a plastic "window" over the wound with paraffin. In this manner "isolated abdomens" were produced that had neither brain nor prothoracic glands and, therefore, were ideal experimental subjects for testing endocrine activity of implants. The isolated pupal abdomens did not molt when either endocrinologically active brains or inactive prothoracic glands were implanted independently. However, if both an active brain and inactive prothoracic glands were implanted, the pupal abdomen metamorphosed. Also, active prothoracic glands stimulated development without a brain present. Thus, Williams concluded that insect metamorphosis, like amphibian metamorphosis, was controlled by a two-part system in which a brain hormone stimulated the prothoracic glands to release the molting hormone.

13.1.3. Role of the *Corpora Allata*

During the years when the hormonal regulation of molting and metamorphosis was established, the mechanism of directing a larval versus a pupal or adult molt was being investigated. Wigglesworth became intrigued with this problem when he noted that decapitated fourth-stage *Rhodnius* larvae developed adult characteristics precociously. Moreover, fifth-stage larvae that were decapitated and parabiosed to intact fourth-stage insects molted into larvae instead of adults. Wigglesworth found that the *corpora allata,* a paired structure located near the brain and connected to it with nerves, was responsible for preventing metamorphosis in larval *Rhodnius*. Transplantation of fourth-stage *corpora allata* into fifth-stage insects permitted molting but prevented metamorphosis. The hormone produced by the *corpora allata* was termed *juvenile hormone* (Wigglesworth, 1970).

The pioneering research of several investigators during the first half of this century set the stage for one of the most active areas of investigation in insect physiology (Fraenkel, 1935; Williams, 1952; Wiggleworth, 1970; Bodenstein, 1977). The endocrine scheme they unfolded is the core of our present understanding of the hormonal control of insect development (Fig. 13-1). In their scheme, the epidermis is viewed as responsible for the form of the insect through the production of larval, pupal, or adult cuticle and as the target tissue of the hormones.

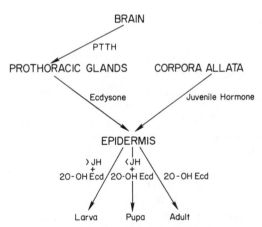

FIGURE 13-1.   Scheme for the hormonal control of molting and metamorphosis in endopterygote insects. JH, juvenile hormone; 20-OHEcd, 20-hydroxyecdysone.

13.2. ISOLATION AND IDENTIFICATION OF HORMONES

13.2.1. Prothoracicotropic Hormone (PTTH, Brain Hormone, Ecdysiotropin, Activation Hormone)

While prothoracicotropic hormone has been isolated and characterized, it has not yet been structurally identified. Two decades ago there was considerable disagreement on whether PTTH was a polypeptide or the steroid cholesterol. Both views resulted from extraction of hundreds of thousands of brains individually dissected from *B. mori* pupae. In one laboratory the methanol-soluble fraction of the brain extract was purified, and cholesterol was identified as the constituent that induced molting in debrained pupae. However, in another laboratory the *methanol-insoluble* fraction was purified and it was shown that a polypeptide caused molting in the debrained pupae used for bioassay. The proteinaceous nature of PTTH has been confirmed by recent work with *B. mori* and with the cockroach, *Periplaneta americana*.

Evidence for its proteinaceous nature includes its solubility in water and susceptibility to proteases; it is reported to be in the molecular weight range of 9000–40,000. This apparent molecular weight range is particularly perplexing because the absorption spectra of purified PTTH is not typical of proteins. This may be due to either impurities in the extract or a chemically bound sugar. The final elucidation of the structure of PTTH will depend in part on the development of highly sensitive bioassays. Recent research has demonstrated that exceedingly small amounts of PTTH may be assayed by determining the quantitative effect of the hormone on the production of ecdysone by isolated prothoracic glands *in vitro*. The use of the *in vitro* assay has led to the suggestion that there is one PTTH with a molecular weight of 22,000 and possibly a second with a molecular weight of 7,000 (King, 1983).

13.2.2. Ecdysteroids (Molting Hormones, Ecdysones)

Like PTTH, ecdysteroids (from ecdysis, the act of molting) were first isolated from *B. mori* pupae. However, instead of surgically removing hundreds of thousands of the small, transparent prothoracic glands, whole pupae were dried and extracted. The molting hormone activity was assayed for its ability to cause pupariation of ligated fly larvae. In 1954 Butenandt and Karlson reported the successful isolation of the molting hormone and named it *ecdysone*. Subsequently they obtained 250 mg of ecdysone, a highly polar steroid, from 1000 kg of silkworm pupae. A second isolated component, 20-hydroxyecdysone, is also known as β-ecdysone, ecdysterone, or crustecdysone. Recently it was agreed to refer to ecdysone analogues as "ecdysteroids." The original ecdysone (or α-ecdysone) has the structure $2\beta,3\beta,14\alpha,22R,25$-pentahydroxy-$5\beta$-cholest-7-en-6-one (Fig. 13-2).

Additional ecdysteroids have been identified in insects (26-hydroxyecdysone and 20,26-dihydroxyecdysone) and from plants. The original discovery

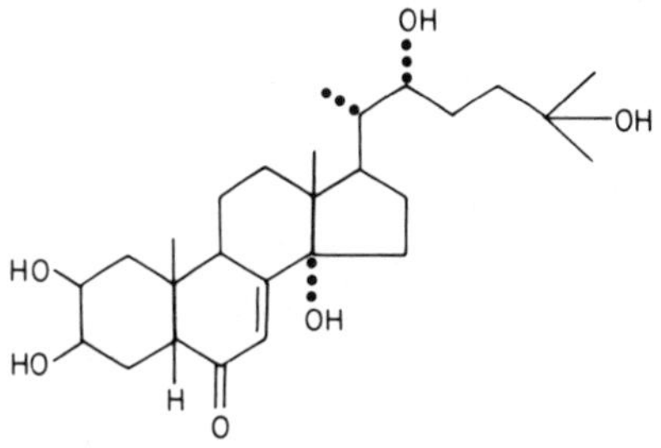

Ecdysone

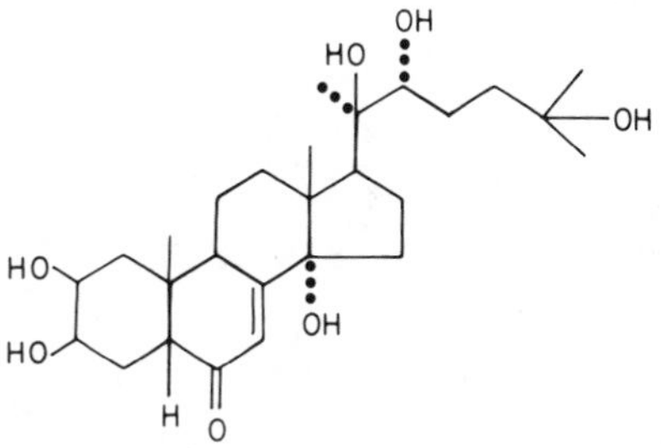

20-hydroxyecdysone

FIGURE 13-2. Structure of ecdysone and 20-hydroxyecdysone.

of "phytoecdysones" has led to the identification of more than 40 compounds from plants that are ecdysone analogues. Both ecdysone and 20-hydroxyecdysone are present in as many as 80 families of plants at concentrations up to 1% of the dry weight of the plant. In contrast, the dry weight of insects contains 10^{-5}–10^{-9} % ecdysteroids.

It was not clear from the early analytical research which (if any) of the ecdysteroids were synthesized and secreted by the prothoracic glands, because whole insects were used by the chemists. In recent years, it has been demonstrated that the prothoracic glands synthesize and release ecdysone, but not 20-hydroxyecdysone.

13.2.3. Juvenile Hormones (JHs)

The breakthrough in isolating and identifying JHs came many years after the *corpora allata* transplantation experiments with *Rhodnius* in the 1930s. In parabiosis experiments with the cecropia moth, Williams observed that a diapausing pupa joined to a female adult metamorphosed into an adult, whereas a pupa joined to a male adult formed a second pupa. These findings led Williams to conclude that male cecropia adults must contain large amounts of JH similar in activity to larval JH. Ether extracts of abdomens of

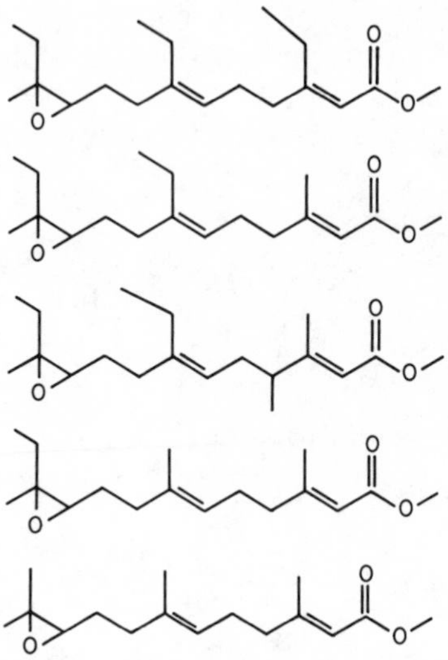

FIGURE 13-3.   Structure of juvenile hormones O, I, 4-methyl I, II, and III

adult males resulted in a "golden oil" that had JH activity in cecropia pupae and larval *Rhodnius*. Recent investigations have demonstrated that the JHs of the male adult cecropia are localized in the accessory sex glands.

Purification of lepidopterous extracts led to the identification of juvenile hormone (JH I) as methyl 10,11-epoxy-7-ethyl-3,11-dimethyl-*trans*-2,6-tri-decadienoate. JH II (a 17-carbon homologue) was identified shortly thereafter (Fig. 13-3). JH III, a 16-carbon homologue of JH III, was identified directly from material released by *corpora allata* cultured *in vitro*. Numerous investigators have now demonstrated that the isolated *corpora allata* of a variety of insects will produce JH I, JH II, and/or JH III *in vitro*. The discovery of two additional JHs in *Manduca* embryos raises the possibility of yet additional stage-specific JHs in other species (King, 1983).

13.3.   REGULATION OF HORMONE TITERS

13.3.1.   Prothoracicotropic Hormone

The titer of PTTH in insects during various stages of development has not been studied in detail, largely due to the absence of an ultrasensitive bioassay and the fact that the structure of the brain hormone has not yet been determined. The hormone is released from neurosecretory cells, specialized nerve cells that are adapted for secretion. The release of hormone from the brain is undoubtedly tied to the ability of neurosecretory cells to act on sensory information provided to the brain. In *Rhodnius* a blood meal that distends the abdomen results in the activation of stretch receptors that convey information to the brain to initiate the molting sequence. In *Manduca sexta* PTTH is released after a critical weight is reached and in relation to a circadian rhythm (daily clock). In *Manduca* the site of production of PTTH may be confined to a single lateral neurosecretory cell in each hemisphere of the brain as demonstrated by the *in vitro* culture of isolated neurosecretory cells in conjunction with prothoracic glands.

The specific proprioceptive or environmental stimuli that stimulate PTTH release will vary with the insect. However, it is important to recognize the role of the neurosecretory cells as intermediary centers of information processing. The *corpora cardiaca* had been believed to be the neurohemal organ that stores and releases PTTH. However, recent research has demonstrated that in *M. sexta* PTTH is stored in the *corpora allata* from which it is released to the hemolymph.

13.3.2. Ecdysteroids

13.3.2.1. Detection. Titers of ecdysteroids in insects have been measured by a variety of techniques: bioassay, radioimmune assay, and gas–liquid chromatography with electron capture detection. The advantage of the bioassay technique based on injection of ligated fly larvae is that little equipment is needed. However, distinct ecdysteroids are not separately measured, and the limit of detection is 1×10^{-1} μg. Radioimmune assay is a technique that has revolutionized endocrinology. Extracts need not be highly purified, and the detection of ecdysteroids approaches 1×10^{-4} μg. The technique is based on the immune response of an animal (e.g., a rabbit) to an ecdysteroid–protein complex. These antibodies react with radioactive ecdysteroid, and the fraction of label to the antibody decreases with the addition of increasing amounts of unlabeled ecdysteroid due to competition for the antibody. Comparison of suitable measurements of the radioactivity still bound to the antibody with a standard curve provides information on the amount of unknown ecdysteroid present in the extract. Despite its sensitivity, radioimmune assay does not readily distinguish between different ecdysteroid analogues. Therefore, sophisticated chemical techniques such as gas–liquid chromatography with electron capture detection or mass fragmentography are used in order to identify the ecdysteroids present (Morgan and Poole, 1977).

Representative results of ecdysteroid titer measurements are shown in Fig. 13-4. The concentration of hormone reaches a peak prior to each ecdysis. However, a double peak prior to pupation in lepidopterous species has been noted. In endopterygotes the larval titer of ecdysteroids is substantially lower than that found in pupae. In adults, hormone titers of some species are often quite high. In general, both ecdysone and 20-hydroxyecdysone have been detected, but 20-hydroxyecdysone appears to be the dominant ecdysteroid (Baehr et al., 1978; Bollenbacher et al., 1978).

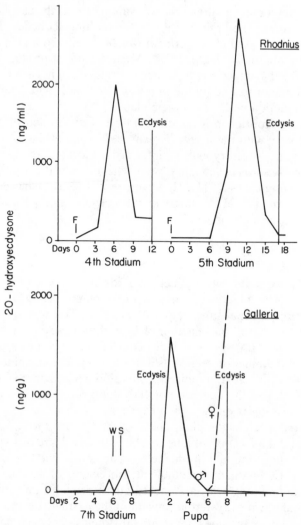

FIGURE 13-4. Titer of 20-hydroxyecdysone during development of an exopterygote insect, *Rhodnius prolixus,* and an endopterygote insect, *Galleria mellonella* (Baehr et al., 1978; Bollenbacher et al., 1978).

13.3.2.2. Biosynthesis. The synthesis of ecdysteroids in insects depends on the dietary intake of steroids since insects cannot synthesize these triterpenes. It is now well established that cholesterol can be converted by insects to ecdysteroids. The prothoracic glands are the site of synthesis of ecdysone, which is secreted into the hemolymph and converted by other tissues (fat body, midgut, epidermis) to the more active 20-hydroxyecdysone. For phytophagous insects the first step is to convert 29-carbon plant sterols to the 27-carbon precursors of ecdysone. The 27-carbon sterols are converted to cholesterol and then to 7-dehydrocholesterol, which is probably a primary precursor of $3\beta,14\alpha$-dihydroxy-5β-cholest-7-en-6-one. It has been demonstrated that about 40% of the cholestenone is converted to ecdysone by isolated prothoracic glands cultured *in vitro*. A proposal for ecdysone biosynthesis is shown in Fig. 13-5. The biosynthesis of 20-hydroxyecdysone from ecdysone is accomplished by a cytochrome-dependent hydroxylase (ecdysone 20-monooxygenase) associated with the mitochondria. It is generally accepted that 20-hydroxyecdysone need not be bound to protein or lipoprotein in the hemolymph, and that it may circulate as the free hormone until it interacts with target tissues.

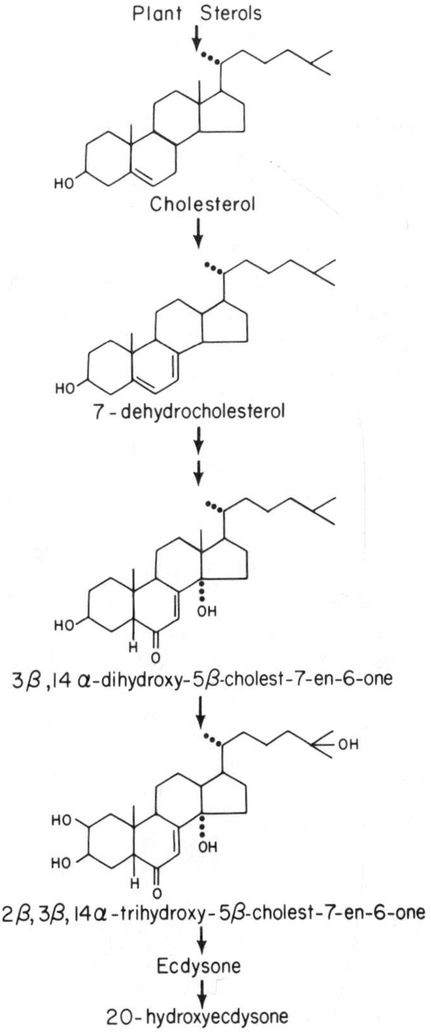

Plant Sterols

Cholesterol

7 - dehydrocholesterol

$3\beta,14\alpha$-dihydroxy-5β-cholest-7-en-6-one

$2\beta,3\beta,14\alpha$-trihydroxy-5β-cholest-7-en-6-one

Ecdysone

20-hydroxyecdysone

FIGURE 13-5. Biosynthetic scheme for ecdysone and 20-hydroxyecdysone (Morgan and Poole, 1977).

13.3.2.3. Inactivation. The primary mechanism responsible for the greater concentration of ecdysteroids prior to molting may be increased biosynthesis of the hormone. However, it is also theoretically possible to have a constant rate of biosynthesis and secretion and to regulate ecdysteroid titer by varying the rate of inactivation. In any event the insect must have a mechanism for eliminating ecdysteroids once the hormonal action has been accomplished. In some insects the half-life of injected ecdysteroids has been found to be as little as 3 hr. It now appears that ecdysone and 20-hydroxyecdysone are not excreted unchanged but are first conjugated with glucose, glucuronic acid, or sulfate. In addition, these ecdysteroids may be inactivated by being oxidized to different metabolites. A scheme that summarizes the principal modes of inactivation and excretion of ecdysteroids is shown in Fig. 13-6.

13.3.3. Juvenile Hormones

13.3.3.1. Detection. The most favored techniques for measuring JH are highly sensitive bioassays and gas–liquid chromatography with electron capture detection. Difficulties with the radioimmune assay of JH have prevented this method from being used. As with the ecdysteroids, the bioassay of JH is an advantageous approach because it is very sensitive, requires little equipment, and measures a functional property of the hormone. The *Galleria* wax test, for example, is based on the response of wounded pupal integument to a JH-wax seal. The extent and quality of the pupal wound patch in the resulting adults provides a measure of the JH activity. However, to obtain quantitative data on the relative amounts of JH I, JH II, or JH III, chemical procedures are required.

JH titers in exopterygotes and endopterygotes are elevated during larval life as predicted by Wigglesworth's early studies with the *corpora allata*. In *Rhodnius* the predominant hormone is JH I, which peaks on days 1 and 7 after feeding in the fourth instar. However, in the final larval instar, JH I only peaks one day after feeding, although small amounts of JH II and JH III are noted late in the instar. An analysis of the last larval instar of *Galleria* and *Pieris* also shows a pronounced reduction of JH titer, followed by a

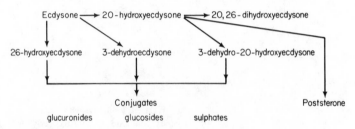

FIGURE 13-6. Inactivation of ecdysteroids (Morgan and Poole, 1977).

small peak of activity prior to pupation. JH titer falls to a low level after pupation but increases prior to completion of metamorphosis (Fig. 13-7).

13.3.3.2. Biosynthesis. Investigations of the biosyntheses of JHs have been greatly aided by the development of methods for successfully incubating *corpora allata in vitro* under conditions that permit continued syntheses of the hormones. The carbon skeleton of JH III is synthesized from acetate precursors that are converted to mevalonate and the various intermediates.

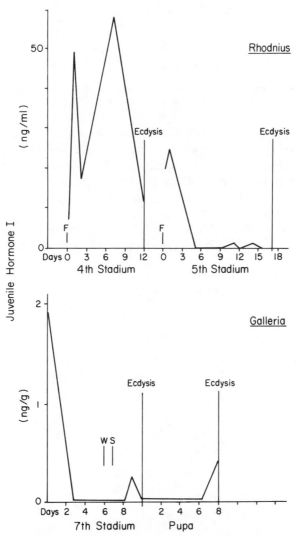

FIGURE 13-7. Titer of juvenile hormone I in an exopterygote insect, *Rhodnius prolixus,* and in an endopterygote insect, *Galleria mellonella* (Baehr et al., 1978; Hsiao and Hsiao, 1977).

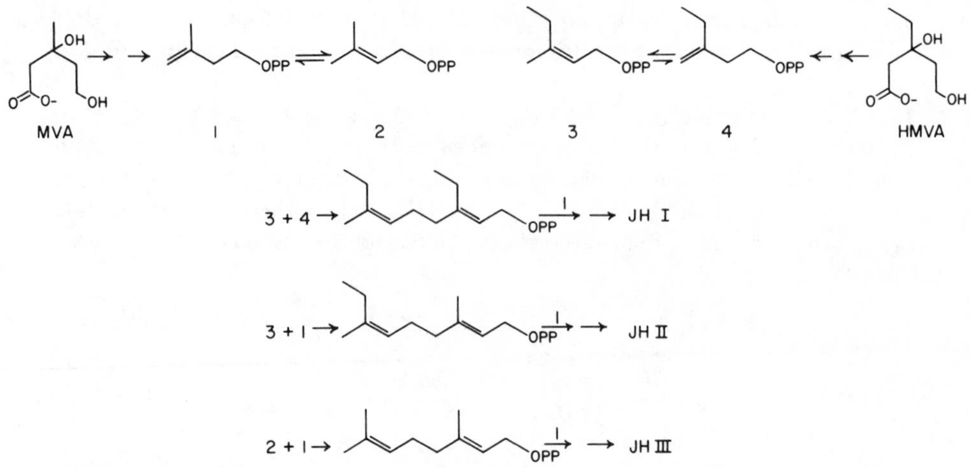

FIGURE 13-8. Biosynthetic scheme for juvenile hormones I, II, and III (From D. A. Schooley et al., in L. I. Gilbert, Ed., *The Juvenile Hormones,* Plenum Press, 1976.)

JH I and JH II utilize propionate as the source for the "extra" carbons that they possess (Fig. 13-3). Not all of the intermediates in JH biosyntheses have been determined, but an overall synthetic scheme is shown in Fig. 13-8. For each JH the methyl group of the ester is added to the complete carbon skeleton by transfer from 5-adenosyl methionine.

13.3.3.3. Inactivation. Manifestly, there must be an efficient mechanism for eliminating JH once it has served its purpose in maintaining larval life so

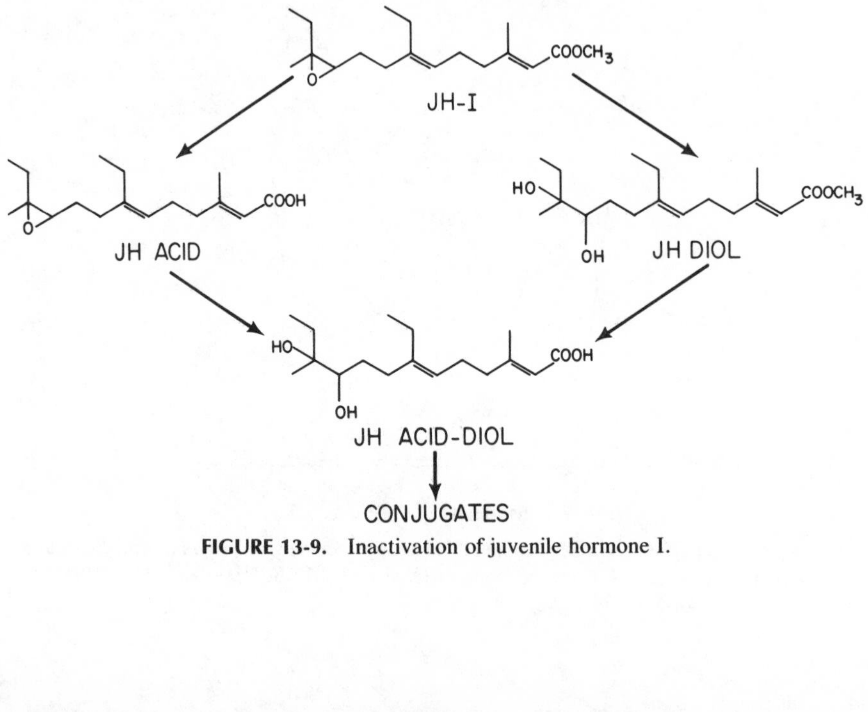

FIGURE 13-9. Inactivation of juvenile hormone I.

that metamorphosis can proceed. Biochemical inactivation of JH by esterases or by attacking the epoxide group occurs readily. During inactivation, JH is first enzymatically converted to either the JH diol or the JH acid and then to the JH acid–diol sulfate conjugate, which is excreted (Fig. 13-9).

To prevent inactivation of JH during the period when it is needed, many insects have specific low-molecular-weight protein in the hemolymph to which JH binds. Other insects rely on a less specific (and less protective) high-molecular-weight lipoprotein for this purpose. In both cases, in preparation for metamorphosis, a specific esterase is produced that can eliminate residual JH, even if it is bound to its carrier protein. Thus, decrease in JH titer can be accomplished by a cessation of biosynthesis and specific degradative processes that operate despite the presence of the protective protein.

13.4. HORMONE–TISSUE INTERACTIONS

13.4.1. Prothoracic Glands and Prothoracicotropic Hormone

The investigations of Fukuda, Williams, and others made it clear that the prothoracic glands were not only the source of a molting hormone but they were also the target of PTTH. Moreover, numerous experiments on the critical period for the brain in the initiation of molting suggested that the prothoracic glands required hormonal stimulation for several days to maintain the secretion of ecdysone. For example, histological changes in the epidermis of *Rhodnius* occur a few hours after a blood meal, which initiates molting in this insect, although removal of the brain as late as three days after feeding prevents molting. These experiments indicate that the prothoracic glands must secrete hormone soon after *Rhodnius* is fed, but that a sustained stimulus from PTTH is needed to stimulate sufficient secretion of ecdysone.

These early studies were based on *in vivo* experiments that could not directly reveal the relationship between the brain and prothoracic glands. Direct verification of the hypothesis that PTTH activated the prothoracic glands became possible with the advent of insect tissue culture and the development of ultrasensitive assays for ecdysteroids. First, bioassays of ecdysteroid activity were used to demonstrate that the brain stimulated the prothoracic glands *in vitro*. Definitive results were obtained by utilizing radioimmune assay to measure the secretion of ecdysone by *Manduca* prothoracic glands cocultured with surgically isolated neurosecretory cells from the brain. In these experiments the isolated prothoracic glands responded to whole-brain homogenates as well as portions of the brain that contained the appropriate neurosecretory cells. However, the glands did not respond to various ganglia, thus demonstrating the specificity of the stimulus from the brain. These experiments provided unequivocal evidence that the prothoracic glands are targets cells of PTTH, to which they react by secreting ecdysone (Bollenbacher, 1979; Bollenbacher and Bowen, 1983).

13.4.2. Dynamics of Ecdysteroid Action

It has been clearly demonstrated that active secreting prothoracic glands are required for extended periods if molting and metamorphosis are to proceed. For example, more than 30 years ago it was shown that active prothoracic glands of the cecropia moth were needed for two weeks to fully stimulate adult development. The hypothesis that the prothoracic glands provide a "sustained stimulus" rather than a "trigger" to development received further support from investigations of the metabolism of injected radioactive ecdysone. In these experiments with the flesh fly, *Sarcophaga peregrina,* 50% of the injected ecdysone was lost within 1 hr and 98% was not recoverable after 8 hr. From these studies it was proposed that continued secretion of ecdysone is required because it is rapidly inactivated. Moreover, at least in some insects, "covert effects" of the ecdysteroids on target tissues accumulate, although the hormone titer in the hemolymph may not reach a high level.

Recently various durations and dosage levels of ecdysteroids have been tested on target tissues cultured *in vitro.* Morphogenesis and differentiation have been examined in particular detail in cultured imaginal discs (the primordia of adult wings, legs, etc.) from *Sarcophaga, Drosophila,* and various Lepidoptera. In each of these systems ecdysteroid exposure *in vitro* was required for a minimum number of hours at a given concentration of hormone (typically 12–24 hr). Within certain limits there was an inverse relationship between the minimum concentration of hormone and the duration of exposure required for a response. There was, nevertheless, a minimum duration and concentration of ecdysteroid required and also a maximum level of hormone that could be used because excessive amounts of this hormone are inhibitory. Ecdysteroid concentrations of 10^{-8}–10^{-6} M were generally effective.

The response of insect tissues to hormones depends not only on the concentration and length of exposure of hormone but also on the receptiveness of the cells. Bodenstein (1977), in his early research on *Drosophila* imaginal discs, called this property of cells *competence.* He showed that the growth of the imaginal discs was related to their ability to respond to ring glands when both tissues were transplanted into adults. Recent work with *Drosophila in vivo,* and *Plodia interpunctella in vitro* have confirmed Bodenstein's hypothesis. It is clear now that at least for imaginal discs the acquisition of competence to respond to ecdysteroids requires a minimum number of cell divisions that permit the necessary cellular changes to occur. In *Drosophila,* once this tissue competence is achieved, the imaginal discs may be cultured indefinitely in adult flies and even after many years retain their ability to respond to ecdysteroids by completing metamorphosis.

13.4.3. Dynamics of Juvenile Hormone Action

Research on the interactions of JH with tissues of immature stages has been complicated by the fact that the effects of JH are seen only after an ecdysteroid-induced molt. Moreover, although Wigglesworth and others demonstrated that active *corpora allata* prevented metamorphosis, larvae of the cecropia moth failed to respond to transplanted *corpora allata* and did not produce supernumary larval instars when injected with JH extract. On the other hand, injections of JH readily induced cecropia pupae to molt into "second pupae" instead of adults.

A breakthrough in the study of JH action in lepidopterous larvae occurred when it was shown that injection of JH extract in *Galleria* larvae resulted in supernumary larval instars if the JH remained "in the body from the time of injection until the end of the period in which any cells are still sensitive to hormone." As a consequence of perceiving the importance of this sensitive period, JH effects were subsequently demonstrated in larvae of both cecropia and *Manduca*.

Investigations of the role of JH during the larval–pupal transformation show that the reduction in titer of JH in the last larval instar is a prerequisite for the secretion of brain hormone. This finding plus the research on specific degradation of JH in mature larvae demonstrate the existence of a "fail-safe" system in which pupation requires a reduced titer of JH, not only to determine that the quality of the molt is pupal but also to permit ecdysteroid secretion to occur only after excess JH has been eliminated. However, following the stimulation of the prothoracic glands by the brain hormone, there is a small increase in JH titer that may be necessary for the pupal molt.

The lack of detailed information on JH titers in the early larval instars makes it difficult to make any firm statements about the dynamics of JH action in young larvae. There is, for example, the conflicting evidence from later stages that the reduction in JH titer permits the brain to be turned on and secrete brain hormone, but increased JH (by injection) activates the prothoracic glands directly, bypassing the brain hormone. The hormonal and tissue interactions in the early instars remain among the least understood areas of endocrine research of insects.

The relationship of JH to tissue competence in the last larval instar has been studied successfully, particularly as a consequence of pure JH and *in vitro* bioassays becoming available. Two systems examined intensively in this regard are the epidermis of *Manduca* and the imaginal discs of *Plodia* (Riddiford, 1976; Oberlander, 1976). In *Manduca*, as with other larvae, the epidermis produces a larval cuticle and then is redirected at metamorphosis to pupal and then adult synthesis. The imaginal discs, on the other hand, produce cuticle for the first time at pupation. Until that point, imaginal discs are determined with respect to their developmental fate (wings, legs, etc.), but have not yet differentiated.

The epidermis that underlies the larval crochets of *Manduca* loses its ability to form crochets during the last (fifth) larval instar as the titer of JH declines. Surprisingly, larval crochet epidermis from young fourth-instar larvae formed crochets when cultured *in vitro* in the presence of 20-hydroxyecdysone but not JH. This finding emphasizes that, although JH is quickly metabolized in the hemolymph, its effects persist for several days. These persistent JH effects may be analogous to the "covert" effects described for ecdysteroids. Alternatively, some JH may be maintained in active form within the target cells even though it is inactivated in the hemolymph. However, after three days of culture *Manduca* epidermis would no longer respond to 20-hydroxyecdysone by making crochets unless JH was present in the medium. Thus, for larval crochet epidermis, responsiveness to ecdysteroids was sustained by exposure to JH.

In a related study the developmental capacity of dorsal abdominal epidermis from last-instar *Manduca* larvae exposed to 20-hydroxyecdysone in the absence of JH programmed the cells to engage in pupal synthesis with the next exposure to ecdysteroid. Coincubation of the epidermis with JH and 20-hydroxyecdysone resulted in additional larval synthesis after the next treatment with ecdysteroid. Thus, in these *in vitro* experiments the nature of the response of epidermis to 20-hydroxyecdysone is determined by the prior (recent) history of exposure to that hormone plus JH.

Wing imaginal discs from lepidopterous larvae may be arrested in their development by application of JH in the larval diet or in the culture medium for isolated tissues. Application of JH in the diet, however, provides a

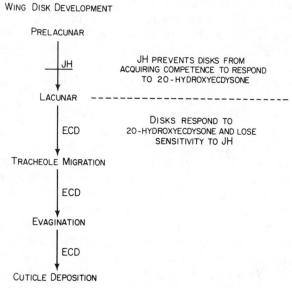

FIGURE 13-10. Competence to respond to either juvenile hormone (JH) or 20-hydroxyecdysone (Ecd) as a function of the stage of development for last-instar wing disks in Lepidoptera.

reliable, quantitative method for continuous JH exposure that results routinely in supernumary larvae. Wing imaginal discs from larvae treated in this manner from the beginning of the last larval instar are incapable of responding to ecdysteroids *in vitro* by producing a pupal cuticle. However, wing discs from mature larvae late in the last instar, though treated with JH and arrested in development, respond to 20-hydroxyecdysone *in vitro*. Thus, as the last larval instar progresses, the wing discs become more responsive to 20-hydroxyecdysone, but JH prevents the wing discs from developing competence to respond to ecdysteroids (Fig. 13–10). For imaginal discs, at least, JH may be viewed as a "status quo" hormone.

13.5. MECHANISMS OF HORMONAL ACTION

13.5.1. The Second Messenger and Protein Receptor Models

In 1957 it was proposed that some hormones, which act as a "chemical messenger" on target cells, require a "second messenger" within the cell to evoke the appropriate response. In an investigation of the action of glucagon on the liver, it was shown that the hormone stimulated the synthesis of intracellular cAMP (cyclic adenosine 3′,5′-monophosphate) by activating membrane-bound adenyl cyclase. Cyclic AMP (the second messenger) in turn activated the target cell to initiate the biochemical response. Since this pioneering work, cAMP and other cyclic nucleotides have been shown to be hormone mediators for many hormones, especially peptide hormones. Elucidation of a role for cAMP in insects requires a demonstration that cAMP levels rise in target cells upon exposure to hormone. In addition, inhibition of the enzyme that degrades intracellular cAMP should enhance the hormonal effects. Finally, it should be possible to substitute extrinsically applied cAMP (or an analogue) for the hormone itself in appropriate experimental systems (Pastan, 1972).

Numerous studies on the response of chick oviduct to vertebrate hormones have contributed greatly to our understanding of the mode of action of steroid hormones. Unlike peptide hormones, steroids act at the genetic level following binding of the hormone to intracellular protein receptors (O'Malley and Schrader, 1976).

Action of hormones at the genome involves regulation of "transcription" or the synthesis of RNA from a DNA template. There may also be regulation of the processing and transport of the RNA, as well as the "translation" of RNA into proteins. According to O'Malley, the selective transcription of portions of the DNA to messenger RNA (mRNA) that codes for key proteins requires direct interaction with the steroid–receptor complex. To establish whether insect steroid hormones act in this manner, it is necessary to show that the hormone stimulates the synthesis of new protein from the translation of mRNA, also newly synthesized. Moreover, the insect hormone should bind to an intracellular protein receptor, which then acts at the chromosomal level.

In Sections 13.2.1–13.2.3 it was pointed out that 20-hydroxyecdysone is a steroid, prothoracicotropic hormone is a peptide, and JH is a terpenoid. We can now consider the mode of action of these hormones in relation to the major theories of hormonal action.

13.5.2. Mode of Action of Prothoracicotropic Hormone

Investigations of the mode of action of PTTH have been hampered because it has not been identified and is not even generally available as a purified extract. However, the recent development of a radioimmune assay for ecdysone has made it possible to assess PTTH effects on isolated prothoracic glands. Early investigations established that prothoracic glands synthesized RNA during gland activation. Oberlander (1965) suggested that PTTH may act as a direct stimulus for the production of mRNA needed for the synthesis of enzymes in the ecdysone biosynthetic pathway. Confirmation that PTTH stimulated transcription of DNA to RNA in the prothoracic glands was obtained when it was demonstrated that brain "activation Factor I" stimulated RNA synthesis in prothoracic glands cultured *in vitro*. However, it is not entirely clear from studies of the effects of inhibitors of different classes of RNA what proportion of the newly synthesized RNA was mRNA. Actinomycin D inhibits DNA-dependent RNA synthesis in many systems, and in the *in vitro* experiments cited above, it completely blocked PTTH-induced RNA synthesis. Alpha-amanitin, which selectively inhibits RNA polymerase II, the enzyme responsible for mRNA synthesis, had only a small effect on the cultured prothoracic glands. Thus, while it is clear that the synthesis of RNA is a feature of PTTH activation of the prothoracic glands, the nature of the RNA has not been established.

Whatever actions PTTH may have on RNA synthesis in prothoracic glands, its peptide composition has led scientists to determine whether it may, like vertebrate peptide hormones, act by stimulating cAMP production. Vedeckis et al. (1974) showed that the prothoracic glands of *M. sexta* synthesize cAMP. Furthermore, inhibition of phosphodiesterase, which degrades cAMP, results in an increase in cAMP in active, but not inactive, prothoracic glands. Additional experiments showed that extrinsically applied cAMP stimulated secretion of ecdysone in cultured prothoracic glands. Stimulation of cAMP levels by application of PTTH to isolated prothoracic gland must be shown to confirm the hypothesis that PTTH acts via the second messenger model (Fig. 13-11).

13.5.3. Mode of Action of Ecdysteroids

The scientific literature is replete with studies on the actions of ecdysteroids in insects. The early experiments focused on the sequence of biochemical

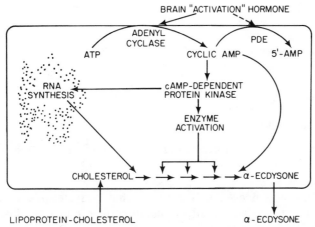

FIGURE 13-11. Model for the role of cAMP in prothoracic gland activation. [From W. V. Vedeckis et al., *Zool. Jb. Physiol.* **78,** 440 (1974). With permission from Cold Spring Harbor Laboratory.]

events during a molt cycle and on the response of an inactive insect to exposure to ecdysteroids provided by parabiosis, transplantation of active glands, or injection of the hormone itself. It is to be expected that a hormone that initiates molting, promotes metamorphosis, and terminates pupal diapause would stimulate many processes, including cell division, energy production, and the synthesis of various cellular constituents. The challenge is to distinguish between aspects of ecdysteroid-induced development and the mode of action of the hormone per se. Thus, early actions of ecdysteroids should be analyzed and related to subsequent developmental responses. In this section, we will discuss the actions of ecdysteroids on epidermis, imaginal discs, and salivary gland giant chromosomes. The results will be evaluated in terms of the major theories of hormonal action.

The hypothesis that an early action of ecdysteroids stimulates RNA synthesis originated with Wigglesworth (1970), who reported on the cytological changes that occurred in *Rhodnius* epidermis following injection of ecdysone. Within a few hours after injection the nucleoli in the epidermal cells enlarge, indicating increased RNA synthesis. Staining of the cells also shows an accumulation of RNA in the cytoplasm. These pioneering studies were confirmed and extended by other scientists who used radioautography or direct biochemical techniques. Subsequently, radioautography was used to localize and quantify isotopic precursors of RNA in saturniid moths during a larval molt cycle and during metamorphosis. It was concluded that the primary effect of ecdysteroids on the epidermis is to stimulate the synthesis or translation of mRNA necessary for molting. By this time Karlson and Sekeris (1966) were well into their landmark research on the effects of ecdysteroids on the synthesis of dopa decarboxylase, an enzyme utilized in cuticular

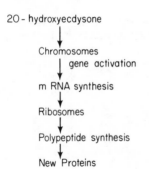

20 - hydroxyecdysone

↓

Chromosomes

↓ gene activation

m RNA synthesis

↓

Ribosomes

↓

Polypeptide synthesis

↓

New Proteins

FIGURE 13-12. Gene activation model of ecdysteroid action.

tanning. They advanced the hypothesis that ecdysteroids acted on the genome of target cells by directly stimulating the production of new mRNA (Fig. 13-12).

During the 1960s when the major studies on ecdysteroid action were undertaken, the effects of hormones on the "puffing" phenomenon in the giant chromosomes of dipteran salivary glands were also investigated. These chromosomes, which may be seen clearly with a light microscope, are 100 times thicker and 10 times longer than normal interphase chromosomes. They result from continued replication of the DNA, but without cell division. The chromosome filaments adhere to each other as replication proceeds and give rise to giant interphase chromosomes. During development, the appearance of the chromosomes changes as various portions of the banded structures thicken and then decrease in size. In the early 1950s it was concluded that the sequential appearance and disappearance of these chromosomal puffs represents differential gene activity. Subsequent research showed that the chromosomal puffs were sites of intense RNA synthesis and that this synthesis was blocked by actinomycin D.

Our understanding of the relationship of chromosomal puffs to ecdysteroid action was made possible by the intensive research undertaken by investigators in the last two decades. Their experiments on salivary glands *in vivo* and *in vitro* demonstrate that a specific sequence of chromosomal puffs is initiated in response to ecdysteroids. Induction of the later puffs is blocked by cycloheximide, an inhibitor of protein synthesis, while the early puffs are unaffected. Moreover, continued exposure to ecdysteroids inhibits the later puffs. These observations led to the hypothesis that an ecdysteroid–receptor complex acts positively at early puff sites and negatively at late sites. The activation of the early puffs leads to the synthesis of a regulatory protein that stimulates late sites and inhibits early sites (Fig. 13-13) (Ashburner et al., 1974). How ecdysteroids activate the genome is still not known, but it has been proposed, not without controversy, that hormonal effects are mediated through regulation of the ratio of intracellular ions such as K^+ and Na^+.

The mode of action of ecdysteroids has also been investigated in the imaginal discs of Diptera and Lepidoptera. The effects of ecdysteroids on

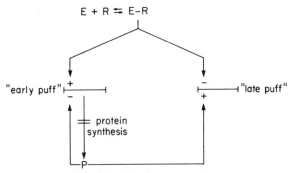

FIGURE 13-13. Gene activation in giant chromosomes of dipteran salivary glands. Ecdysone (E) binds to receptor (R) to form complex (E–R). This complex stimulates early chromosomal puff sites and inhibits late sites. The product (P) of the early sites stimulates late sites and inhibits early sites.

macromolecular synthesis have been examined in wing discs of the Indian meal moth, *P. interpunctella*. 20-Hydroxyecdysone stimulated chitin synthesis in cultured imaginal discs, and both RNA and protein synthesis were required during a limited hormone-dependent period. New protein synthesis started 2–3 hr after application of hormone (Oberlander, 1976). RNA synthesis is a very early response of *Drosophila* imaginal discs to ecdysteroids; the bulk of the RNA produced is ribosomal RNA although new mRNA may also be a factor (Fristrom, 1972).

Research on hormonal action on the epidermis, salivary glands, and imaginal discs points toward a primary effect of ecdysteroids at the genetic level. Whether this hormonal regulation of transcription is mediated through recognition and binding of ecdysteroids to a receptor protein or through stimulation of cAMP has been investigated separately. Several laboratories have considered the relationship of ecdysteroids to cAMP levels in insect tissues. In the mealworm, *Tenebrio molitor*, epidermal cAMP levels were lower in the presence of high titers of ecdysteroid. Other studies have also shown a variation in cAMP or cGMP (guanosine 3',5'-monophosphate) during insect development. In one direct experiment the injection of 20-hydroxyecdysone into cecropia pupae resulted in increased adenyl cyclase activity in the wing epidermis. On the other hand, a detailed study on ecdysteroid action on cultured *Drosophila* imaginal discs showed that cAMP could not substitute for hormone, nor did it enhance or repress hormonal action. Also, drugs that increase intracellular cAMP levels were shown to have no effect on the tissue's response to ecdysteroids. Although the evidence is incomplete, there is little support for a role for cAMP in ecdysteroid action.

There is good evidence that ecdysteroids may act by first binding to an intracellular protein receptor. A radiolabeled ecdysteroid bound to nuclear fractions of mass-isolated *Drosophila* imaginal discs, whereas in *Drosophila* cell line the radiolabeled ecdysteroid bound primarily to cytoplasmic fraction (Fristrom, 1972). In both systems the binding was high affinity and specific, and the binding sites could be saturated with a sufficient concentration of hormone. While isolation and characterization of the ecdysteroid receptors remains to be accomplished, clearly the main features of vertebrate steroid action apply to the ecdysteroids.

13.5.4. Mode of Action of Juvenile Hormone

The investigation of the mode of action of JH in immature insects has been one of the most challenging and intractable problems in insect physiology (Gilbert, 1976). For the most part the morphogenetic effects of JH are manifested only in the presence of ecdysteroids. The direct effects of JH on vitellogenesis and related processes is another matter and is considered separately in Chapter 11. In some experimental studies, stimulatory or inhibitory effects of JH on protein synthesis, RNA synthesis, and respiration have been observed. However, it has been difficult to relate such quantitative results to the actions of JH that determine the "quality" of a molt. Also, the terpenoid structure of JH makes neither the peptide nor steroid model of hormonal action a likely guide. In addition, there are technical difficulties that must be overcome in studying JH action. This hormone is unstable under many conditions, it has low solubility in aqueous media, and it adheres readily to many surfaces. Nevertheless, there is sufficient information to make at least a tentative assessment of whether JH acts in immature insects on transcription or translation and of whether the intracellular receptor–protein model is applicable.

In one series of experiments, investigators utilized the amino acid composition of stage-specific cuticle in the mealworm, *T. molitor,* as a criterion of the stability of adult versus pupal mRNA in insects treated with actinomycin D. It was concluded that both pupal and adult mRNA are present in the pupa and that regulation must take place at the translational level. Evidence was presented that a key role for JH was to control the availability of the specific transfer RNAs necessary in inserting the necessary amino acids into the stage-specific cuticular proteins under synthesis. However, this translational control mechanism has been challenged on experimental grounds. Thus, whether JH can act at the translational level is still an open question.

About a decade ago a theoretical scheme in which JH operated at the transcriptional level was presented. The model predicts selective effects of JH on transcription of larval, pupal, or adult gene sets; the presence of stage-specific RNA polymerases; and the binding of JH to intracellular receptors that can then attach to DNA. It is still not possible to decide to what extent the proposed model may be operational in insect cells. Research on multiple RNA polymerases is still inconclusive. However, JH-stimulated synthesis of qualitatively different RNAs in isolated nuclei has been reported. In addition, JH has been shown to have both inhibitory and stimulatory effects on puffing in giant chromosomes. Thus, there is accumulating evidence that JH may act by selectively affecting transcription.

Binding studies of JH to intracellular receptors are still in a preliminary stage. In one study, radiolabeled JH bound to high-affinity sites in the nuclear, mitochondrial, and microsomal fractions. Additional research is required to extend these findings and isolate and characterize the proposed JH receptors. Manifestly, the mode of action of JH in immature insects remains elusive, but the evidence to date lends tentative support to the relevance of the receptor–protein model of hormonal action.

13.5.5. Mode of Action of Eclosion and Tanning Hormones

The completion of adult development involves the onset of new insect behavioral patterns, such as eclosion from the cocoon. Truman (1980) demonstrated that eclosion of adult *M. sexta* is triggered by a peptide hormone released by the brain. The stereotyped behavioral program that ensues is mediated in the central nervous system by cAMP. Thus, the action of eclosion hormone appears to be in accord with that of other peptide hormones.

The newly eclosed adult has a soft cuticle that darkens and hardens in a process known as tanning or sclerotization. In the early 1960s it was reported that a proteinaceous material, probably a neurosecretory product, was necessary for cuticular tanning following the imaginal molt in *Calliphora*. It was subsequently demonstrated that the adult fly produces a tanning hormone, bursicon (Fraenkel and Hsiao, 1965). This hormone is believed to regulate tanning by initiating the transfer of tyrosine from the serum to the hemocytes where it is converted to dopamine. Subsequently, dopamine is taken up by the epidermis for conversion to the active cross-linking agents (see Chapter 4). Bursicon activity has been found in Orthoptera, Hemiptera, Coleoptera, and Lepidoptera. The production site of bursicon depends on the species, but it is usually located in neurosecretory cells of the midbrain or the abdominal ganglia.

The proteinaceous nature of bursicon and a related hormone, puparium tanning factor (PTF), suggests that the second messenger model of hormone action may operate here. In fact it has been demonstrated that neck-ligatured newly emerged flies will tan after injection of bursicon or cAMP. Also, in *Calliphora* PTF can be mimicked by application of cAMP. Additional evidence comes from experiments with *Tenebrio* pupae, where injection with bursicon produced a twofold increase in epidermal cAMP within 5 min. Thus, there is good evidence for the hypothesis that bursicon action is mediated through activation of cAMP synthesis in target cells (Reynolds, 1983).

13.5.6. Mode of Action of Diapause Hormones

In the life cycle of many insects the orderly process of development is arrested, which results in protecting a vulnerable stage from adverse environmental conditions. This period of developmental arrest is called *diapause*. Although both temperature and photoperiod are key environmental triggers in the regulation of diapause, the process is usually mediated by the neuroendocrine system. In pupal and adult reproductive diapause, the controlling system involves withholding the secretion of a stimulatory hormone. In adult diapause the failure to release JH will prevent both development of eggs in females and maturation of the accessory glands in males. Pupal diapause has been studied in detail in various Lepidoptera, where it is usually initiated by a short-day photoperiod. Either prolonged exposure to low temperatures or to a long-day photoperiod can terminate pupal diapause. It was shown long ago that pupal diapause resulted from both the failure to secrete brain hormone and the consequent inactivity of the prothoracic glands.

Unlike pupal and adult diapause, larval and egg diapause is caused by the presence of an inhibitory hormone. In Lepidoptera, including the southwestern corn borer, *Diatraea grandiosella,* the rice stem borer, *Chilo supressalis,* and the European corn borer, *Ostrinia nubilalis,* larval diapause has been linked to the maintenance of high JH titers. Regulation of larval diapause has been studied in greatest detail in *Diatraea* (Chippendale, 1977). It was demonstrated that allatectomized larvae terminate diapause prematurely, and that the termination of diapause is prevented by extrinsic application of a JH mimic. Measurements of JH titers in the hemolymph show that diapause is associated with the maintenance of a hormone titer intermediate between that of the high levels in fifth-instar larvae and the low levels normally found in nondiapause sixth- (last-) instar larvae. Apparently, the JH in diapausing larvae interacts with the system that controls secretion of brain hormone. However, even in diapausing larvae stationary molts (larval, not pupal) are observed. The biochemical effects of JH have not been probed in diapausing larvae except to note that a major fat body protein, called *diapauselin,* is present in such insects. Whether the mode of action of JH at the cellular level is the same or similar in larval diapause to its usual morphogenetic action in larvae is not known.

Diapause in eggs is common but has been examined primarily in the silkworm, *B. mori.* Nearly three decades ago it was reported that an injection of an extract of the subesophageal ganglion of female adults, reared on a short-day photoperiod as larvae, induced diapause in eggs produced by treated adults (nondiapause). Subsequent research on this maternally produced diapause hormone showed that it was a peptide hormone. The mode of action of the hormone is not known, but some biochemical and morphogenetic effects have been described. Pupal ovaries accumulate 3-hydroxykynurenine in response to diapause hormone, and the dark serosa of diapause eggs results from compounds synthesized from 3-hydroxykynurenine. Also, diapause termination may depend on low-temperature activation of an esterase, which lyses yolk cells and promotes embryogenesis. Whether the peptide diapause hormone's action involves cAMP has not yet been reported.

13.6. CONCLUSIONS

Over the next decade insect endocrinology should witness major advances in three areas: application of tissue culture systems for analyzing hormonal action; utilization of molecular genetics to understand the role of hormones in directing cellular activities (*D. melanogaster* cell lines that respond to ecdysteroids are being analyzed from this perspective of molecular genetics); and isolation and identification of neuropeptides that direct a wide variety of physiological and developmental events.

Finally, remember that a primary motivation for studying insect hormones is possible exploitation of the insect's neuroendocrine system to control pest species. Already hormonomimetic compounds that appear to mimic JHs have proved successful and anti-JHs have been tested in the laboratory. Biorational control of insect pests will require continuing effort both to understand the intricacies of hormonal action during insect development and to identify vulnerable sites of action for potential pest control strategies utilizing hormones or hormonal surrogates.

ACKNOWLEDGMENTS

I wish to thank the following authors and publishers for permission to reprint (or redraw) their published material: Fig. 13-8, Dr. David Schooley and Plenum Press; Fig. 13-11, Dr. Lawrence Gilbert and Cold Spring Harbor Laboratory. I thank Drs. Dwight E. Lynn, Kenneth W. Vick, and James L. Nation for their careful peer evaluations of this chapter.

REFERENCES

M. Ashburner, C. Chihara, P. Meltzer, and G. Richards, *Cold Spring Harbor Symp. Quant. Biol.* **38,** 655 (1974).

W. M. Bayliss and E. H. Starling, *J. Physiol.* **28,** 328 (1902).

J. C. Baehr, P. Porcheron, and F. Dray. *C. R. Acad. Sci. Ser. D.* **287D,** 523 (1978).

D. Bodenstein, *Milestones in Development Physiology of Insects,* Appleton-Century-Crofts, New York, 1977.

W. E. Bollenbacher, N. Agui, N. A. Granger, and L. I. Gilbert, *Proc. Natl. Acad. Sci. USA* **76,** 5148 (1979).

W. E. Bollenbacher and M. F. Bowen, "The prothoracicotropic hormone," in R. G. H. Downer and H. Laufer, Eds., *Endocrinology of Insects,* Alan R. Liss, New York, 1983, p. 89.

W. E. Bollenbacher, H. Zvenko, A. K. Kumaran, and L. I. Gilbert, *Gen. Comp. Endocrinol.* **34,** 169 (1978).

A. Butenandt and P. Karlson, *Z. Naturf.* **96,** 389 (1954).

G. M. Chippendale, *Ann. Rev. Ent.* **22,** 121 (1977).

G. Fraenkel, *Proc. Roy. Soc. Ser. B.* **118,** 1 (1935).

G. Fraenkel and C. Hsiao, *J. Insect Physiol.* **11,** 513 (1965).

J. W. Fristrom, in H. Ursprung and R. Nöthiger, Eds., Springer-Verlag, New York and Berlin, 1972, p. 109.

L. I. Gilbert, *The Juvenile Hormones,* Plenum Press, New York and London, 1976.

E. Hadorn, *Proc. Natl. Acad. Sci. USA,* **23,** 478 (1937).

P. Karlson and C. E. Sekeris, *Acta Endocrinol.* **53,** 505 (1966).

O. Ke, *Bull Sci. Fak. Terk. Ksisu Imp. Univ.* **4,** 12 (1930).

D. S. King, "Chemistry and metabolism of the juvenile hormones," in R. G. H. Downer and H. Laufer, Eds., *Endocrinology of Insects,* Alan R. Liss, New York, 1983, p. 57.

S. Kopeč, *Biol. Bull. (Woods Hole),* **42,** 323 (1922).

P. Lyonet, *Traité anatomique de la chenille qui ronge le bois de saule La Hage* (1762).

E. D. Morgan and C. F. Poole, *Comp. Biochem. Physiol.* **57B,** 99 (1977).

H. Oberlander, *In Vitro* **12,** 225 (1976).

H. Oberlander, S. J. Berry, A. Krishnakumaran, and H. A. Schneiderman, *J. Exp. Zool.* **159,** 1, 15 (1965).

B. W. O'Malley and W. T. Schrader, *Scientific Amer.* **234,** 32 (1976).

I. Pastan, *Scientific Amer.* **227,** 97 (1972).

S. E. Reynolds, "Bursicon," in R. G. H. Downer and H. Laufer, Eds., *Endocrinology of Insects,* Alan R. Liss, New York, 1983, p. 235.

L. M. Riddiford, *Nature* **259,** 115 (1976).

D. A. Schooley, K. J. Judy, B. J. Bergot, M. S. Hall, and R. C. Jennings, "Determination of the physiological levels of juvenile hormone in several insects and biosynthesis of the carbon skeletons of the juvenile hormone," in L. I. Gilbert, Ed., *The Juvenile Hormones,* Plenum, New York, London, 1976, p. 101.

J. W. Truman, "Eclosion hormone: Its role in coordinating ecdysial events in insects," in M. Locke and D. S. Smith, Eds., *Insect Biology of the Future,* Academic Press, New York, 1980, p. 385.

W. V. Vedeckis, W. E. Bollenbacher, and L. I. Gilbert, *Zool. Jb. Physiol.* **78,** 440 (1974).

V. B. Wigglesworth, *Insect Hormones,* W. H. Freeman, San Francisco, 1970.

C. M. Williams, *Biol. Bull. (Woods Hole)* **103,** 120 (1952).

14

EXOCRINE SYSTEMS

MURRAY S. BLUM
Department of Entomology
University of Georgia
Athens, Georgia

CONTENTS

SUMMARY

Exocrine glands, which may be distributed throughout the body of an insect, generally consist of epidermal cells with great biosynthetic capacities. These glands produce a variety of pheromones and allomones that subserve a multitude of intra- and interspecific functions. Pheromones consist of either primers, which cause an alteration of reproductive and endocrinological systems, or releasers mediating stimulus–response effects. Both classes of compounds have been evolved by presocial and eusocial species, and these exocrine products are particularly widespread in the latter insects.

In general, both pheromones and allomones are secreted in blends that frequently consist of compounds belonging to different chemical classes. For sex pheromones, the specificity of the message is the blend, specialized chemoreceptor cells on the antennae having been evolved to perceive the pheromonal molecules. Great sensitivity to the pheromones is predicated on the quantitative and qualitative characteristics of the blend.

Pheromones, which are generally synthesized *de novo* in exocrine glands, are frequently utilized parsimoniously in different intraspecific contexts. Allomones, on the other hand, function primarily as deterrents in interspecific contexts. These defensive compounds, the most complex exocrine products biosynthesized by insects, include a dazzling variety of hydrocarbons, 1,4-quinones, and steroids, many of which are novel natural products. The same compound may function both as an allomone and a pheromone, a fact that further emphasizes the adaptiveness of pheromonal parsimony. Preliminary studies on the modes of action of antagonistic allomones indicate that many of these compounds may interfere with the olfactory processes of predators.

14.1. INTRODUCTION

Among animals, insects have emerged as the natural products chemists par excellence. Their biosynthetic virtuosity reflects a remarkable variety of exocrine glands, structures that secrete their glandular components to the external environment. These glands may be found in all the major body regions, and probably they occur throughout the taxa in the class Insecta. Although such glands are all of ectodermal origin, most are derived from the epidermis, in contrast to various preoral organs or some of the abdominal glands associated with the genital apparatus. Some glands identified as non-exocrine structures have subsequently been demonstrated to possess exocrine roles as well. It has been suggested that the exocrine component of this duality in function was secondarily derived (Blum, 1974).

It has chiefly been in the last three decades that the roles and structures of insect exocrine compounds have been elucidated. A large number of these glandular products are utilized in defensive contexts, whereas others function as chemical signaling agents. Indeed, it has now been established that a multitude of intraspecific interactions among insects are chemisocial in nature, being triggered by volatile information-bearing compounds. Many aspects of insect behavior are now known to be regulated by external chemical signals, and it is very likely that a host of additional behaviors will be subsequently demonstrated to possess pheromonal bases as well.

Chemical investigations of insect exocrine compounds have led to the identification of a large variety of novel natural products. Although the natural products chemistry of relatively few species of insects has been examined, it is evident that these arthropods have the biosynthetic capacity to produce a great diversity of compounds belonging to many chemical classes. More than 50 compounds have been detected in some glandular secretions, and it will not prove surprising if insects continue to be a rich source of interesting exocrine products. Since little or nothing is known about the natural products chemistry of insect species in many taxa known to produce exudates, chemical analyses of these glandular products should prove fruitful for a long time to come.

14.2. TERMINOLOGY

Perhaps because insect exocrinology is a relatively young field, there have been problems with inexact terminology. Many terms have been used to describe exocrine secretions, particularly in regard to their origins and functions in chemical ecology. Indeed, more than 40 terms are presently used in this discipline, mostly with a considerable degree of inexactitude (Duffey, 1977). However, until the broad biological significance of these behavioral regulators is understood, it seems wiser to minimize the number of terms used to describe exocrine products. Consequently, in this chapter the terminology identified with exocrinology has been conservatively treated. This is to ensure that the roles of these glandular products in chemical ecology are not obscured by the use of inexact or misleading terms that imply that we understand the full biological significance of the systems being examined.

Insect exocrine compounds have been divided into categories based on the presumed adaptive value that these natural products may possess for individuals involved in either intra- or interspecific interactions (Brown et al., 1970). Intraspecific chemical stimuli are classified as *pheromones,* compounds that are considered to be adaptive for emitter individuals as a result of the favorable or developmental behavioral responses produced in perceiver individuals of the same species.

Pheromones are further fractionated into releaser and primer compounds. *Releasers* are defined as compounds that produce an immediate behavioral response in a receiver individual. The attraction of a male insect to a female, mediated by a sex pheromone, is a releaser response. *Primer* pheromones, on the other hand, produce a delayed behavioral reaction in the receiver that is mediated by physiological and endocrinological responses to the pheromonal stimulus. Inhibition of ovarian development and consequent suppression of reproductive activity by worker honeybees in response to ''queen substance'' constitutes an example of a response to a primer pheromone.

Most interspecific chemical signals are defined as *allomones.* These are also considered adaptive to the emitter because receiver individuals respond in a manner favorable to the signaler. Allomones, which may involve both animals and plants, may be characterized by either hostile (agonistic) or mutualistic relationships. Agonistic allomones are exemplified by the defensive secretions of many insects, whereas the floral volatiles of plants that attract pollinators are examples of mutualistic allomones.

The terms just described will be used throughout this chapter. However, the reader should be aware of a few other terms frequently encountered in the literature. The term *"ecomone"* (Florkin, 1965), which was created to encompass all exocrine compounds, includes compounds now classified as both pheromones and allomones. In addition, exocrine products that are considered to be signaling agents used in either intra- or interspecific contexts have been called *"semiochemicals"* (Law and Regnier, 1971). Although this term is convenient for lumping both pheromones and allomones, it suffers terminologically because intraspecific signals (pheromones), which constitute communication between conspecifics, are not distinguished from chemical stimuli (allomones), which initiate interaction between individuals of different species.

Another ''class'' of transpecific chemical messengers has been labeled *"kairomones"* (Brown et al., 1970). These compounds are regarded as nonadaptive or maladaptive to their emitters since it is the receiver, rather than the producer, that ultimately benefits as a consequence of the exocrine discharge. Kairomones include compounds that attract predators and parasites to the emitting individuals. However, this presumptive class of compounds often constitutes pheromones of great selective value to their producers and, as such, clearly does not constitute a separate class of compounds. Furthermore, these exocrine products are highly adaptive for their emitters, and to consider them maladaptive because they betray their producers to selected predators or parasites ignores their critical roles as sex pheromones, species recognition agents, aggregative stimuli, and so on. As Pasteels (1982) has emphasized, it is biologically questionable to consider kairomones as a valid class of exocrine compounds in the same sense as pheromones and allomones.

As stated in the present chapter, all insect-derived exocrine products will be referred to as either pheromones or allomones. Utilization of this dual classificatory system will hopefully permit these natural products to be discussed with some degree of biological precision, avoiding the terminological chaos that currently characterizes chemical ecology.

14.3. EXOCRINE GLANDS

Insect exocrine glands may be found in all body regions, and in many species they are present in the head, thorax, and abdomen (Fig. 14-1). However, whereas the distribution of these glands has been established for a few species, their roles, chemistry, and glandular structures are for the most part unknown. If the total exocrine chemistry of each species is idiosyncratic, which appears to be the case, the task of chemically characterizing these secretions is truly formidable. The same can be said for comprehending the functions of these exudates as well as appreciating their cytological subtleties.

The excellent classification of Noirot and Quennedey (1974) is utilized here to characterize the morphology of insect exocrine glands. This classificatory system is especially valuable since it accommodates both function and chemistry with cytology and ultrastructure. As these authors stress, insect epidermal cells can be remarkably versatile, first secreting the cuticle and subsequently other glandular products.

14.3.1. Class I Glands

Like common epidermal cells, class I gland cells are simply covered by the cuticle and the exudate must egress through this cuticular barrier. The gland cells secrete the cuticle that covers them. Class I gland cells are often in association with other classes of cells.

The sex pheromone glands of female lepidopterans are made up of class I gland cells, which really constitute a thickening of the epidermis. The glandular epidermis, which everts prior to pheromone release, is more-or-less folded (Fig. 14-2). Wax glands, typical of many homopterous species, possess a similar organization.

Many defensive glands contain class I gland cells. These saclike glands, whose cavities are filled with the secretions of the sac cells, are characteristic of the frontal glands of termite soldiers. The osmeteria of *Papilio* species and the prothoracic (Gilson's) glands of larval trichopterans are also en-

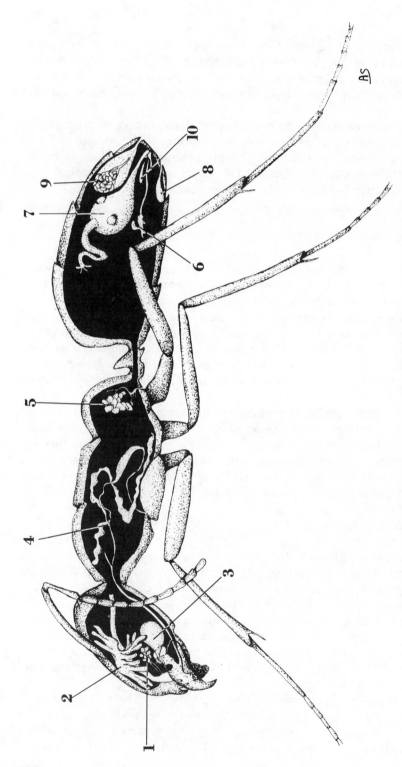

FIGURE 14-1. Exocrine glands of the Argentine ant, *Iridomyrmex humilis.* (1) Hypopharyngeal or maxillary gland; (2) postpharyngeal gland; (3) mandibular glands; (4) thoracic labial gland; (5) meta-pleural gland; (6) poison (venom) gland; (7) hindgut; (8) Pavan's (sternal) gland; (9) pygidial (anal) glands; (10) Dufour's gland. (Modified from Pavan and Ronchetti, 1955.)

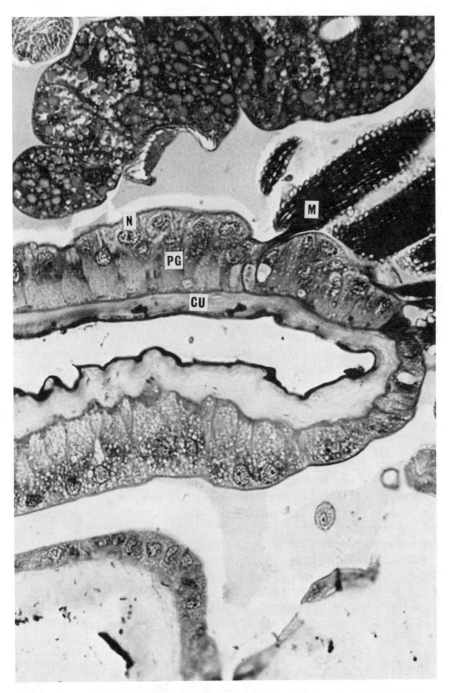

FIGURE 14-2. Type I gland cells from the female sex attractant gland of the noctuid *Euxoa messoria*. (N) Nucleus; (M) retractor muscle for the sex attractant gland; (PG) pheromone gland cells; (CU) unmodified cuticle. The vacuoles in the gland cells contain the sex pheromones or their precursors, ×560. (Courtesy of P. A. Teale.)

dowed with these gland cells. On the other hand, class I gland cells may be present in admixture with those of class III, as is the case for the metathoracic scent glands of some adult hemipterans and the tergal glands of some termites. The sternal glands of some termite species are also composed of class I and class III gland cells.

In the limited number of cases that have been examined by utilizing ultrathin sections, the egress of the secretion through the cuticle of class I gland cells occurs via cuticular pores. This has been established for the female sex pheromone gland of a lepidopteran, the prothoracic (Gilson's) gland of larval trichopterans, and the sternal and frontal glands of termites.

14.3.2. Class II Glands

Differentiated epidermal cells surround the class II gland cell (Fig. 14-3). The secretion from these cells is transferred to modified epidermal (columnar) cells before crossing the cuticle. Microvilli, often characteristic of class I and III gland cells, have never been observed associated with class II gland cells.

Class II gland cells are found in the glands associated with the tubular setae in the dorsal valves of many female lepidopterans (Fig. 14-3). They make up the sternal glands of some termites, whereas in species in other taxa they are in admixture with other classes of gland cells. Class II gland cells have also been found in the frontal glands of termite soldiers and in the tergal glands of some cockroaches.

14.3.3. Class III Glands

The gland cell of a class III unit is penetrated by a cuticular ductule or canal which is contiguous with the ductule or canal cell that has secreted it. In the dermal glands of *Tenebrio,* a simple case, one gland cell is connected with one canal cell. On the other hand, in the isolated dermal glands of *Rhodnius* and *Tenebrio,* three cells constitute the structural units. A canal cell, associated with a second cell containing a large central cavity, secretes the duct that connects the two cells. The cavity of the second cell may represent a saccule (reservoir) lined by a cuticular intima that is continuous with both the cavity and that of the third cell. In some cases (e.g., *Tenebrio*), the second cell is secretory and its cavity is crossed by the duct that abuts on the third cell. The third cell is glandular, and its secretion is collected in an extracellular vacuole (cavity) that communicates with the canal of the second cell.

Class III gland cells are characteristic of the abdominal two-unit reactor glands often found in adult tenebrionids (Happ, 1968) (Fig. 14-4). The gland cells in the anterior unit (Fig. 14-4B) are connected directly with a canal cell, whereas the posterior unit (Fig. 14-4A) contains three cells, a canal cell and two secretory cells, as in dermal glands. The unidirectional flow of the reac-

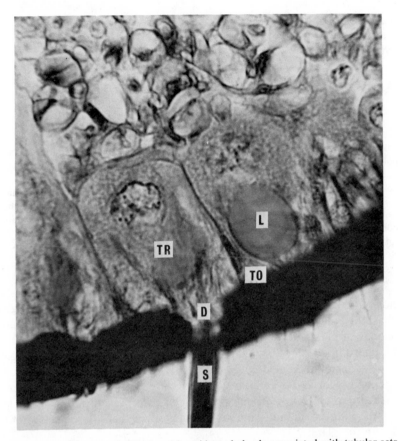

FIGURE 14-3. Type II gland cells found in epidermal glands associated with tubular setae in the dorsal valves of the noctuid *Euxoa messoria*. (TP) Trichogen; (TO) tormogen cell; (L) lipid droplet; (D) duct (infolded apical cell membrane); (N) nucleus; (S) tubular setae. The lipid droplets pass into the setae and are exuded through the tips of the setae, ×1000. (Courtesy of P. A. Teale.)

tion products from unit B to A (Fig. 14-4) through an efferent duct results in end products (quinones) being generated away from vulnerable secretory cells and their associated vesicles. The cuticular organelles in these reactor glands are isolated from the secretory cytoplasm, and the latter are thus insulated from possible biochemical lesions that could be induced by the highly reactive quinonoidal end products. This two-chamber reaction unit is ideally suited for the on-demand production of highly reactive and volatile compounds.

Class III gland cells are widely distributed throughout the Insecta. In particular, the defensive glands of a large variety of insects contain this class of gland cells. Class III cells have been described in the pygidial glands of dytiscids and carabids, the dorsal abdominal glands and metathoracic scent

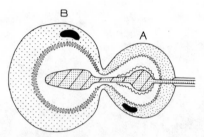

FIGURE 14-4. Type III gland cells found in the defensive (reactor) glands of the tenebrionid *Eleodes longicollis.* The two saclike reservoirs collect secretions of the gland cells which produce *p*-benzoquinones. The gland cells in A connect directly to a canal cell, producing quinol precursors, a portion of which is oxidized to *p*-quinones. The quinol–quinonoidal mixture is transferred along the efferent duct to secretory unit B, in which the glandular units contain a canal cell and two secretory cells. In the distal gland additional quinol is oxidized to quinones. The vesicular organelles in A and B are isolated from the sensitive secretory cytoplasm in these reactor glands. (Modified from Happ, 1968.) [See Noirot and Quennedey (1974) for details.]

glands of larval and adult hemipterans, respectively, the prothoracic glands of phasmids, and the anal glands of mole crickets. In addition, the metapleural glands of ants and the tergal glands of cockroaches contain this class of cells. Pheromone secretory glands, as exemplified by the abdominal glands of male scorpionflies, also contain class III gland cells.

Frequently class III cells are in admixture with those of the other two classes. In some termite species these cells are present along with class I cells in the sternal gland, whereas in other species all three classes of cells make up the gland.

14.3.4. Secondarily Adapted Glands

Not infrequently, glands not regarded as exocrine organs appear to have been adapted to subserve either pheromonal or allomonal functions. Such glandular parsimony may be widespread in the Insecta (Blum, 1981).

The labial (=salivary) glands of male bumblebees have been evolved to produce territorial pheromones that function as mating stations which attract individuals of both sexes. The labial glands of certain dermapterans, which penetrate into the abdomen, contain copious defensive secretions readily discharged from disturbed earwigs.

Hemipterans have evolved novel salivary defensive secretions. A reduviid spits its saliva for up to 30 cm, and this enzyme-rich secretion can cause intense pain and edema if it strikes the eye or nose membranes of vertebrates. On the other hand, aquatic hemipterans (*Velia* spp.) utilize their saliva to reduce the surface tension of the water behind them. This results in the veliids being rapidly propelled across the water, away from would-be predators.

On the other hand, salivary secretions of syrphid larvae and some termite soldiers form defensive glues that effectively entangle small predators such as ants.

Another part of the digestive tract, the hindgut, is the source of trail pheromones of formicine ants in several genera and defensive allomones in Thysanoptera. In thrips these compounds appear to be produced in the hindgut and are evacuated through the anus when these insects are tactually disturbed. These are probably not isolated examples of glandular parsimony, and it is likely that virtually any insect-derived gland whose secretion can be externalized can be regarded as a potential exocrine candidate.

In other cases, well-known exocrine glands have been demonstrated to possess unsuspected communicative functions that are probably derived. For example, the poison gland secretions of ants, well-established defensive exudates, contain releasers of trail following for species in a variety of genera. Significantly, the alarm pheromones of many larval hemipterans, eusocial hymenopterans, and isopterans are identical to the defensive allomones of a variety of solitary species. In these bugs, bees, ants, and termites, the alarm pheromones are obviously still functioning as defensive allomones, and it has been suggested that the pheromonal role was secondarily derived (Blum, 1974; Nault and Phelan, 1984).

Even reproductive glands have been demonstrated to be sources of exocrine compounds. The accessory glands of male meloids contain high concentrations of cantharidin, and this vesicant is transferred to the female in the seminal ejaculate. Since the adult female does not synthesize cantharidin, its availability to her constitutes a copulatory bonus that ensures that her tissues will be fortified with this vesicatory compound.

14.3.5. Characteristics of Secretions

The exocrine products of insects are externalized in a variety of forms that often can be correlated with their pheromonal and/or defensive functions. While many pheromones (e.g., sex attractants) are secreted in trace (nanogram) quantities, other types of pheromones (e.g., alarm pheromones) can be detected as discrete droplets. Some defensive allomones are delivered hypodermically (e.g., the sting-derived venoms of hymenopterans) or as contact deterrents that emerge as sprays, oozes, or froths. Characteristically, these defensive secretions are delivered from the glands closest to the point of tactile stimulation, thus conserving the insect's allomonal reserves.

14.3.5.1. Types of Discharge. Many defensive exudates are externalized by seeping from the glandular orifices. This delivery method is characteristic of the mandibular gland secretions of ants and bees, eversible pygidial glands of staphylinids, osmeteria of papilionid larvae, dorsolateral glands of chrysomelid larvae, and dorsolateral glands of adult cantharids (Eisner, 1970). Often the molested insect will rub its legs in the secretion and then wipe its allomone-fortified appendages on its assailant (e.g., adult Hemiptera). In other cases the insect's abdomen may be tilted so as to smear the opponent with the secretion, a defensive maneuver favored by many adult staphylinids, for example.

Many secretions are accurately sprayed at adversaries, but often only from the glands nearest to the point of attack. Spray delivery is characteristic of the abdominal glands of tenebrionids, pygidial glands of carabids, metathoracic scent glands of hemipterans, abdominal glands of cockroaches, and the poison glands of formicine ants. These sprays usually contaminate the body of the secretor, often providing it with a temporary defensive shield against the potential predator.

A wide variety of unrelated insects discharge their defensive exudates as froths which are usually fortified with air. Tracheal air mixed with the mesothoracic spiracular secretion of the grasshopper *Romalea microptera* is audibly discharged, probably adding to its deterrent effectiveness against vertebrates. The venoms of some ants (*Crematogaster* spp.) also emerge as froths which are painted with the spatulate sting on the bodies of arthropod adversaries. Adult arctiids forcefully expel the secretions of their paired cervical glands as froths which are often fortified with blood cells. The secretion of these froths by arctiids is accompanied by a hissing sound, which probably enhances the aposematism of these moths.

14.3.5.2. Reflex Bleeding.   The blood of many species of insects is utilized as a defensive exudate by being reflexively discharged after molestation. This is particularly true of many species of beetles and grasshoppers (Cuénot, 1896). In some species the blood is fortified with a variety of deterrent allomones which constitute some of the most novel natural products synthesized by arthropods. In other species, the reflexively discharged blood appears to be allomone-free, deterring small insect assailants by rapidly clotting and thus entangling them.

Adults of meloid and coccinellid beetles discharge blood from femorotibial joints, and those autohemorrhaging exudates, fortified with distasteful or emetic compounds, are highly repellent to a variety of predators. The same is true of autohemorrhage from the antennal sockets or lateral elytral margins of adults lampyrids, but in these insects the blood also rapidly coagulates and can entangle small predators. Blood, also fortified with distasteful compounds, is discharged reflexively from the mouths of adult chrysomelids in several genera.

Allomone-fortified blood is also discharged by grasshoppers in several genera, either from leg joints or from weakened spots at the base of the elytra. In the case of the former the blood can be accurately ejected for distances up to 50 cm (Cuénot, 1896).

Blood, accompanied by an audible sound, can also be ejected by some adult stoneflies after autohemorrhage from the coxal and the tibiofemoral joints. The rapidly clotting blood, which effectively entangles small insects, does not appear to contain distasteful compounds, and the stoneflies are palatable to vertebrates. The same is true for the blood delivered reflexively by some chrysomelid (*Diabrotica* spp.) larvae.

14.3.5.3. Conservation. Selection has favored the conservation of defensive allomones by insects, and various mechanisms have been evolved to achieve this end. Molested insects appear to secrete their defensive products frugally, often only discharging from glands proximate to the point of stimulation. In other cases, as in certain adult staphylinids, deterrent allomones are only secreted as a "last resort" after other defensive mechanisms have failed to discourage persistent assailants.

Chrysomelid larvae in a variety of genera can reclaim the droplets of their defensive secretions by inverted invagination of the glandular storage organs. The exudates, which are discharged on the ends of thoracic and abdominal tubercles, can be sucked back into the tubercle by muscle contraction. Similarly, blood discharged from the mouth of disturbed chrysomelid larvae can be withdrawn and possibly resorbed.

14.3.5.4. Age, Instar, and Sex. Various factors are correlated with the compositions of exocrine secretions. For both pheromones and allomones, quantitative analyses indicate that individual variation is the rule, and quantitation of pooled samples can be regarded as only a populational range. Furthermore, major compositional differences may characterize the secretions of individuals in different populations of the same species (see Section 14.4.3).

Both the qualitative and quantitative composition of a secretion may be critically correlated with age. For example, biosynthesis of glandular constituents may not occur until at least several days after adult ecdysis. Furthermore, different compounds may be produced sequentially, or in some instances the qualitative composition may be drastically altered in later instars. Even the physiological state of the individual versus its chronological age may be correlated with the glandular composition of its natural products.

Exocrine compounds, which are generally biosynthesized at specific times after eclosion, appear to reach a maximum at some age and then decline thereafter. Conceivably, observed individual variations may reflect age-related differences. For example, in male bumblebees synthesis of labial gland products is maximal at four days of age, the concentration of all constituents subsequentially diminishing. Similarly, a poison gland product of honeybee workers, histamine, is actively synthesized for about two weeks; in very old bees this biogenic amine cannot even be detected. The same is true for melittin, the major peptide present in the venom of mature workers, the guards and foragers. The synthesis of melittin is age dependent, being restricted to the poison glands of older workers that use the sting defensively. On the other hand, this peptide is not detectable as a venom product of young bees, which only produce promelittin, its precursor. Since young bees neither sting nor leave the hive, analysis of their promelittin-fortified venoms, *along with that of melittin-producing older bees,* will yield qualitative results that *do not reflect the true venom composition of functional stinging bees.* The same may be true for the exocrine secretions of other species of insects.

The age-dependent synthesis of exocrine compounds may be reduced, or suppressed, in at least some eusocial insects if they are subjected to stressful conditions. For example, the biogenesis of 2-heptanone, a mandibular gland product of honeybees that is normally produced when they begin to forage at about two weeks of age, is completely suppressed if the bees are caged (Boch and Shearer, 1967). Under similar conditions, synthesis of histamine in the poison gland of bee workers is greatly reduced. On the other hand, as described below, compounds may be produced either prematurely or after synthesis has normally ceased if the age composition of the workers in the colony is suddenly and drastically altered.

Eusocial insects exhibit temporal polyethism, a division of labor that reflects changes in the behavioral functions of workers as they age. For example, young worker bees are at first nurses, subsequently becoming foragers. Correlated with this polyethism are age-related changes in the synthesis of exocrine compounds. Young worker bees produce an acid in their mandibular glands, which they feed to larvae while they function as larval nurses. Subsequently, when they become foragers, they synthesize an alarm pheromone, 2-heptanone, in these glands, synthesis of the acid being reduced or suppressed. On the other hand, if young worker bees are suddenly removed from a colony, acid synthesis in foragers can be turned on again and they can revert to the status of nursing bees. Conversely, removal of older bees can result in the premature synthesis of 2-heptanone by young workers, which can become precocious foragers. Obviously, the age-related synthesis of natural products is susceptible to environmental perturbations.

Although larvae of many species of insects produce defensive allomones not produced by adults and vice versa, in some cases surprising changes in glandular chemistry occur between the larval instars themselves. For example, in some species of papilionids the osmeteria of last-instar larvae typically synthesize two short-chain fatty acids, whereas earlier instars produce a variety of terpenes in these organs, the acids being absent. These results, which presumably have major ecological correlates as yet unknown, emphasize the necessity of separately analyzing the secretions of different instars.

Sex-related differences in secretory chemistry, unrelated to sex pheromones, occur in the exocrine secretions of some species. Male meloids synthesize cantharidin, whereas females do not. In honeybees, workers—but not males—produce 2-heptanone in the mandibular glands. On the other hand, male bumblebees generate blends of territorial pheromones in their labial glands, but females do not produce these compounds. These sex-related differences in insect exocrine products may not be especially atypical.

In eusocial insects sex-linked natural products are often associated with organs possessed by one sex and not the other. For example, the alarm pheromones of honeybees are produced on the sting shaft; trail pheromones are produced in the sting-associated glands in many genera of ants. Males, which lack a sting apparatus, obviously do not produce these compounds.

Generalizations about the chemistry of eusocial insects are further complicated by exocrine differences between castes and subcastes. Worker honeybees produce 2-heptanone in their mandibular glands, whereas queens, members of the other female caste, do not synthesize this ketone. Similarly, fire ant workers produce toxic alkaloids in their venoms that are lacking in the venoms of their nonstinging queens. Beyond these caste differences, variations in the cephalic defensive compounds synthesized by minor and major soldiers of some termite and ant species demonstrate that there may be qualitative differences in the exocrine chemistry of subcaste individuals.

14.3.5.5. Series of Glands. Glands present in segmental series may exhibit both chemical and functional differences. For example, the secretion of the

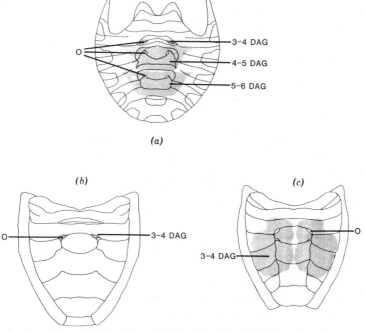

FIGURE 14-5. Exocrine glands of the pentatomid *Podisus maculiventris*. A, Dorsal abdomen of fifth-instar larva; B, Dorsal abdomen of female adult; C, Dorsal abdomen of male adult. (DAG) Dorsal abdominal gland; (O) Ostiole. Finely stipled areas outline underlying glands, and numbers indicate the position of intersegmental openings. (From Aldrich et al., 1984.)

third dorsal abdominal gland of pyrrhocorid larvae, discharged when the bug is disturbed, contains an aldehyde–alkane mixture. On the other hand, the two anterior glands produce only alkanes. The three glands are not discharged simultaneously. It appears that the exudate from the third gland releases alarm behavior in the aggregated larvae, whereas the secretions from glands 1 and 2 may be purely defensive in function.

Qualitative differences also characterize the secretions of the dorsal abdominal glands of lygaeid and pentatomid larvae. In the latter, additional differences are found in the exudates of homologous glands in the adult.

Both larvae and adults of the predatory pentatomid, *Podisus maculiventris,* possess dorsal abdominal glands. The larvae possess three pairs of glands (and reservoirs). The first pair, which is very small in comparison to the other two pairs (Fig. 14-5A), is homologous to the dorsal abdominal glands of both female and male adults (Fig. 14-5B,C). On the other hand, the secretion of the second pair of larval dorsal abdominal glands (3–4) differs from that of the glands in adults, and the male and female exudates differ from each other. In addition, the exocrine chemistry of larval glands 3–4 differs from those of glands 4–5 and 5–6 (Aldrich et al., 1984).

Clearly, then, all segmental glands of insects are not equal. Beyond that, it should be borne in mind that the chemistry of the exudates of glandular subdivisions may differ radically from each other. This has been demonstrated for the two reservoirs of the metathoracic scent glands of adult hemipterans and the paired sacs that make up the cervical gland of notodontid larvae. It will probably be found to be true for the subdivisions of other insect exocrine glands.

14.4. PHEROMONES

Pheromonal research has been an investigative explosion that essentially dates from the identification of the sex pheromone (bombykol) of the female silkworm, *Bombyx mori* (Butenandt et al., 1959). This landmark research constituted a model that subsequent investigators utilized too uncritically, with results that were often both misleading and frustrating. Unfortunately for future researchers, the sex attractant of the domesticated silkworm appeared to consist of a single compound, whose activity as a sexual releaser for males could easily be demonstrated by employing a bioassay for a short-range signaling compound.

Subsequent studies, searching for the pheromonal "magic bullet" functioning as a short-range attractant and releaser in other species, resulted in the identification of many one-compound sex pheromones. But these compounds did not trigger activity comparable to a live female. Eventually it became obvious that many pheromones consisted of blends of compounds, some of which functioned as *long-range* attractants. Indeed, even the original-model single-component sex pheromone, bombykol, was demonstrated to be accompanied by a second pheromonal compound.

The types of pheromones produced by species in many taxa have become evident and these compounds are now being catalogued in ever-increasing numbers. What has emerged is the realization that insects are remarkable synthesizers of natural products.

14.4.1. Functions

The pheromonal "language" of insects has been particularly difficult to decipher probably because its chemical syntax is so foreign to researchers, acultured in a world in which communication is mediated by acoustical and visual signals. Although the deciphering has been slow, investigators have now developed a pheromonal "search image" and many behaviors, previously described in anecdotal terms, have been demonstrated to possess pheromonal bases. Dazzling varieties of behaviors are being analyzed in terms of their release by chemical signals. While the particulars of each species' chemically mediated behavioral repertoire must be analyzed in detail, obviously insects have been incredibly versatile in exploiting behavior with their pheromones.

14.4.1.1. Primers. Several species of presocial insects have been found to produce primer pheromones, and these exocrine products are all associated with species that have aggregative propensities. The observed primer effects may result in developmental synchrony and may be reflected in either the acceleration or inhibition of maturation. Significantly, synchronous development of aggregated individuals promotes the possibility that the maximum number of conspecifics will be able to reproduce simultaneously, and is thus highly adaptive.

Adults of both male and female grasshoppers accelerate the growth of immatures (Table 14-1); the male effect is considerably stronger than that of the female. On the other hand, either young adults or last-instar larvae produce compounds that retard the maturation of adult males. Methyl esters have been reported to either increase the body weight of individuals or inhibit development of group-reared crickets. Another primer pheromone, locustol, is reported to be responsible for the transformation of individuals of *Locusta migratoria* from the solitary to the gregarious phase.

Primer pheromones secreted by male tenebrionids accelerate both oocyte development and female sex pheromone release; female primers also affect egg maturation. In the honeybee a pheromone produced by the queen suppresses the ovarian development of workers and their construction of queen cells (Table 14-1). Caste determination, induced in larvae of both termites and eusocial hymenopterans, appears to be mediated by primer pheromones. This effect also results from primers altering the feeding behavior of adults vis-à-vis the larvae.

14.4.1.2. Releasers. A multitude of behaviors are released in both presocial and eusocial species by pheromones. In particular, termites and eusocial

TABLE 14-1. Major Pheromonal Classes Identified in Insects

Perceiver Reaction	Chemical Stimulus Type[a]	Insect Taxa in which Identified
Primers		
Developmental synchrony (acceleration or inhibition)	Maturational pheromones	Orthoptera, Coleoptera
Ovarian inhibition	Queen substances	Dictyoptera (Isoptera) and eusocial Hymenoptera
Caste determination (in larvae) or change in worker behavior vis-à-vis larvae	Pheromonal caste determinators or nurse worker-feeding modifiers	Dictyoptera (Isoptera) and eusocial Hymenoptera
Releasers		
Attraction of individuals of opposite sex and possible release of sexual behaviors	Sex pheromones	Widespread
Aggregative attraction (both sexes)	Aggregation pheromones	Dictyoptera, Orthoptera, Hemiptera, Coleoptera, and Hymenoptera
Alarm (alerting, attracting, repelling)	Alarm pheromones	Dictyoptera (Isoptera), Hemiptera, and eusocial Hymenoptera
Recruitment or emigration	Trail pheromones	Dictyoptera (Isoptera), Lepidoptera, and eusocial Hymenoptera
Dispersion	Epideictic pheromones	Orthoptera, Homoptera, Diptera, Coleoptera, Lepidoptera, and Hymenoptera

[a] Blends of compounds usually present.

hymenopterans utilize these compounds to regulate a diversity of social interactions.

Aggregative pheromones have a widespread distribution in insects, having been detected in Orthoptera, Homoptera, Hemiptera, Coleoptera, and Hymenoptera (Table 14-1). Epideictic (dispersive) pheromones regulate the distribution of individuals in local populations of immature and adult moths and beetles. Oviposition by tephretid flies is also inhibited by these releasers. Alarm pheromones, which have also been identified in many aggregative species of treehoppers, aphids, and true bugs, are commonly produced by termites, wasps, bees, and ants.

Sex pheromones appear to be virtually ubiquitous in the Insecta. These compounds may act as short- or long-range attractants and may release precopulatory and copulatory behavior as well (Table 14-1). However, mating behavior may be regulated by pheromones that are different from the compounds functioning as sex attractants, and indeed in some cases the releasers of copulatory behavior appear to be contact pheromones. Both males and females may produce sex pheromones, and in many cases, mating behaviors are released by pheromones secreted by both sexes. The male-derived pheromones are often utilized when the male is very near to the female.

In many eusocial species, behaviors as diverse as emigration (Table 14-1) and female flight behavior are released by pheromones. In addition to the widespread alarm, aggregative, and sex pheromones, social interactions which include digging, grooming, and food exchange are pheromonally mediated. Clearly the development of a viable eusociality has reflected the evolution of a major pheromonal input.

14.4.2. Chemistry

The chemical virtuosity of insects as pheromone producers has become evident as a host of compounds, regulating various behaviors, have been identified. Most of the identified exocrine products have been sex pheromones, products of economically important species, particularly in the orders Coleoptera and Lepidoptera.

The sex pheromones of species in different orders are characterized by great structural eclecticism, and it appears that the members of some orders are biosynthetically more conservative than others. This conservatism appears to be particularly characteristic of species in many dipterous taxa, the females of which produce long-chain alkenes, such as (Z)-9-tricosene (Table 14-2), that function as very short-range excitants. There is also an indication that the pheromones secreted by female scale insects are primarily esters of branched and unsaturated alkenols (Table 14-2).

Female lepidopterans also appear to exhibit biogenetic conservatism vis-à-vis sex pheromones. Most of the identified compounds—derived from species in the families Tortricidae, Noctuidae, Pyralidae, and Arctiidae—consist of blends of aliphatic unsaturated alcohols, aldehydes, esters, or hydrocarbons (Table 14-2); some lymantriids utilize an epoxide (disparlure) as a sex pheromone.

Beetles appear to be the most unpredictable of the sex pheromone producers, species in each family tending to synthesize distinctive pheromones (Table 14-2). An allenic ester, alkenoic acids and aldehydes, branched alkanones, a furanone, and novel monoterpene alcohols and aldehydes have been identified as coleopterous releasers of sexual activity (Table 14-2). Analyses of these exocrine products in additional families should yield further pheromonal surprises.

The nonsexual pheromones of insects are too structurally varied to be susceptible to generalization except to note that species in certain taxa emphasize the production of particular classes of compounds. Most of the identified compounds are releasers rather than primers (Table 14-3), which may reflect the ease of bioassaying the former versus the latter.

Insect alarm pheromones are typical of the structural diversity that char-

TABLE 14-2. Characteristic Sex Pheromones Identified in Insects

Compound	Order	Genus	Family	Sex Producing
3,11-Dimethyl-2-nonacosanone[a]	Dictyoptera (Blattaria)	*Blattella*	Blattidae	♀
Periplanone-B[a]		*Periplaneta*	Blattidae	♀
3-Methyl-6-isopropenyl-9-decen-1-ol acetate[a]	Homoptera	*Aonidiella*	Diaspididae	♀
Syringaldehyde	Hemiptera	*Leptoglossus*	Coreidae	♂

Compound		Genus	Family	Order	Sex
(*Z*)-9-Tricosene (Muscalure) 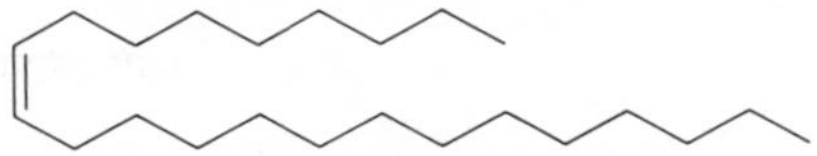		*Musca*	Muscidae	Diptera	♀
(*E,Z*)-3,5-Tetradecadienoic acid (Megatomoic acid) 		*Attagenus*	Dermestidae	Coleoptera	♀
Methyl (*E,E,E*)-2,4,5-tetra-decatrienoate 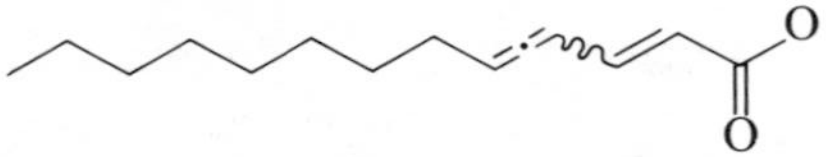		*Acanthoscelides*	Bruchidae		♂
2-(*Z*-isopropenyl-1-methylcyclobutyl)-ethanol[a] (Grandlure component) 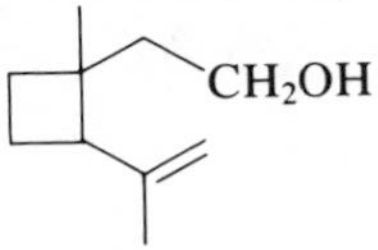		*Anthonomus*	Curculionidae		♂
10-Methyl-2-tridecanone 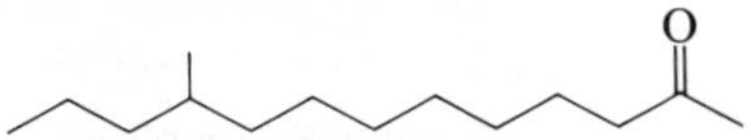		*Diabrotica*	Chrysomelidae		♀
(*Z*)-5-(1-decenyl)dihydro-2(3*H*)-furanone[a] (Japonilure) 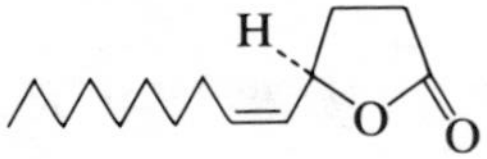		*Popillia*	Scarabaeidae		♀

CO$_2$H

CH$_2$OH

H

TABLE 14-2. (*Continued*)

Compound	Order	Genus	Family	Sex Producing
(Z,Z,Z)-3,6,9-Eicosotriene	Lepidoptera	*Caenurgina*	Noctuidae	♀
(Z)-9-Tetradecenal		*Heliothis*	Noctuidae	♀
1-Methylbutyl decanoate[a]		*Thyridopteryx*	Psychidae	♀
(Z)-11-Tetradecenyl acetate		*Argyrotaenia*	Tortricidae	♀
Methyl jasmonate[a]		*Grapholitha*	Tortricidae	♂
Undecanal		*Galleria*	Pyralidae	♂

Compound		Genus	Family	Sex
(E,Z)-10,12-Hexadecadien-1-ol (Bombykol)		Bombyx	Bombycidae	♀
2-Methylheptadecane		Holomelina	Arctiidae	♀
(Z)-7,8-Epoxy-2-methylheptadecane (Disparlure)[a]		Porthetria	Lymantriidae	♀
2,3-Dihydro-7-methylpyrrolizin-1-one		Danaus	Danaidae	♂
3,7-Dimethylpentadecan-1-ol-acetate[a]		Hymenoptera Neodiprion	Tenthredinidae	♀
9-Oxo-(E)-2-decenoic acid (Queen substance)		Apis	Apidae	♀

[a] Possesses a chiral center(s) and the pheromonal effluent contains at least one enantiomer.

TABLE 14-3. Nonsexual Pheromones of Insects and Their Glandular Sources

Compound	Pheromonal Function	Family	Order	Glandular Source
Primers				
2-Methoxy-5-ethylphenol (Locustol)	Phase transformer	Acrididae	Orthoptera	Crop
9-Oxo-(E)-2-decenoic acid (Queen substance)[a]	Ovarian inhibitor	Apidae	Hymenoptera	Mandibular glands (queen)
Releasers				
Germacrene-A[b]	Alarm releaser	Aphididae	Homoptera	Fat cells extending into cornicles
(E)-2-Hexenal	Alarm releaser	Pyrrhocoridae	Hemiptera	Larval dorsal abdominal glands

Compound		Function	Family	Order	Source
4-Methyl-3-heptanone[b]		Alarm releaser	Formicidae	Hymenoptera	Mandibular glands
2,5-Dimethyl-3-isopentylpyrazine		Alarm releaser	Formicidae	Hymenoptera	Mandibular glands
Isopentyl acetate		Alarm releaser	Apidae	Hymenoptera	Setaceous tissue on sting shaft (workers)
2-Methyl-6-methylene-7-octen-4-ol (Ipsenol)[b]		Aggregator	Scolytidae	Coleoptera	Gut-derived frass
2,4-Dimethyl-5-ethyl-6,8-dioxabicyclo[3.2.1]octane (α-Multistriatin)[b]		Aggregator	Scolytidae	Coleoptera	Gut-derived frass

559

TABLE 14-3. (*Continued*)

Compound	Pheromonal Function	Family	Order	Glandular Source
Geranylgeraniol	Territorial marker attractor	Apidae	Hymenoptera	Labial (= salivary) glands (males)
Methyl anthranilate	Female flight initiator	Formicidae	Hymenoptera	Mandibular glands (males)
Neocembrene-A (Nasutene)[b]	Trail-following releaser	Termitidae	Dictyoptera (Isoptera)	Sternal gland
Methyl 4-methylpyrrole-2-carboxylate	Trail-following releaser	Formicidae	Hymenoptera	Poison gland (worker)
Dimethyl disulfide CH_3SSCH_3	Releaser of digging behavior	Formicidae	Hymenoptera	Mandibular gland (workers)

[a] Also functions as a sex pheromone.
[b] Possesses a chiral center(s) and the pheromonal effluent contains at least one enantiomer.

acterizes these nonsexual releasers. Aphids utilize sesquiterpenes such as germacrene-A as alarm pheromones, whereas true bugs release alarm behaviors in conspecifics with alkenals [e.g., (E)-2-hexenal] (Table 14-3). Ants, on the other hand, often secrete alarm pheromones consisting of methyl or ethyl ketones or pyrazines. Honeybees generate alarm signals with simple esters such as isopentyl acetate and may utilize carbonyl compounds (e.g., 2-heptanone) as well. Termite soldiers secrete alarm pheromones that consist of monoterpene hydrocarbons such as α-pinene.

Coleopterous species in the family Scolytidae produce a great variety of monoterpenoid aggregation pheromones as exemplified by ipsenol and multistriatin (Table 14-3). These compounds are produced in mixtures often in combination with host-derived compounds. Terpenoid biosynthesis is also emphasized by bees and ants, some of which produce diterpenes (territorial attractants) (e.g., geranylgeraniol) as well as a variety of mono- and sesquiterpenes (alarm pheromones, territorial attractants). Methyl 4-methyl-pyrrole-2-carboxylate, a trail pheromone produced in the poison gland of ants (Table 14-3), may illustrate the ability of these insects to secondarily adapt a gland dedicated to the production of venoms containing nitrogenous allomones to also function as a social organ.

14.4.3. Blends, Specificity, Synergism, and Populational Differences

The specificity is the blend (Blum, 1974). Almost all insect pheromones are secreted as blends of compounds. Although a few sex pheromones appear to consist of single compounds (e.g., megatomoic acid, Table 14-2), these chemical messengers generally consist of mixtures of exocrine products. In most cases analyzed in detail, both specificity and sensitivity are correlated with the quantitative and qualitative composition of the secreted pheromone. Percentages of different geometric isomers and/or enantiomers are often key determinants of the degree of response exhibited by an insect to its pheromonal blend.

For example, many tortricid females secrete sex pheromones containing (Z)-11-tetradecenyl acetate (Table 14-2) in combination with varying percentages of the (E)-isomer (Roelofs, 1981). For 10 species, the proportions of the (E):(Z)-isomers vary from 88:12 to 3:97. The ratio utilized by each species appears to be optimal for that species, and a ratio different from that utilized by a species may be nonattractive or inhibitory to the males. In addition, for these male tortricids, additional compounds in the secreted blends increase specificity and/or sensitivity. In some cases, compounds such as dodecyl acetate, which are nonattractive in themselves, synergize the alkenyl acetates. Other geometric isomers [e.g., (Z)-9-tetradecenyl acetate] also may act as synergists, and in some cases the free alcohols [e.g., (Z)-11-tetradecen-1-ol] are present in the pheromonal effluent.

The concentration of a constituent may be a key determinant of its efficacy as a behavioral releaser. For example, comparison of the effects of sex pheromone concentration on male attraction for three species of moths demonstrated that each species exhibited a characteristic concentrational response. One species was attracted to low concentrations, one to high concentrations, and one to both low and high concentrations. Obviously concentrational effects are magnified considerably in terms of blends of pheromones.

Synergistic interactions of compounds in multicomponent pheromones are often required for these secretions to exhibit activity as either attractants or sexual releasers. This was first established by Silverstein et al. (1966), who demonstrated that three monoterpene alcohols (see ipsenol, Table 14-3) present in the frass of male bark beetles, *Ips paraconfusus*, were required for attraction of individuals of both sexes. This landmark investigation also established that an additional *Ips* species produces two of these alcohols but is inhibited by the third monoterpene produced by *I. paraconfusus*, thus demonstrating that the species specificity of pheromones could be qualitatively determined. Furthermore, it was shown that the three-component pheromone attracted bark beetle predators to the invaded host trees. Since that time, it has been demonstrated that species of parasitoids are frequently attracted by the sex pheromones of their hosts. In other words, in the animal world elegant mechanisms that evolved to promote reproductive isolation have been exploited by coevolved specialists. Even sex is not sacred!

Species specificity of pheromones may also be correlated with the biosynthesis of compounds with great enantiomeric exactitude. Enantiomers (=optical isomers) are chiral molecules not superposable on their mirror images; a chiral molecule and its mirror image are also referred to as *antipodes*. Many insect pheromones contain chiral centers, and laboratory syntheses of pure enantiomers have provided compounds that have been utilized to establish the relationship of enantiomeric composition and pheromonal activity.

It appears that most insect-derived chiral pheromones are produced as single enantiomers. For example, only one antipode of disparlure, the sex pheromone of the gypsy moth, *Lymantria dispar* (Table 14-2), occurs naturally and males only respond to this enantiomer; the unnatural isomer decreases the response. Similarly, in the case of the Japanese beetle, the sex pheromone japonilure (Table 14-2) consists of a single enantiomer, and the response of males is inhibited by very low concentrations of the unnatural antipode. On the other hand, an alarm pheromone of ants, 4-methyl-3-heptanone (Table 14-3), is produced as a single enantiomer, but the antipode, which is less active, does not interfere with the response to the natural isomer.

In the case of many other pheromones that are produced as chiral molecules, the response is to only one of the enantiomers, the other isomer being inert and not interfering with the response to the natural enantiomer (Silverstein, 1981).

Surprising populational differences characterize the compositions of the sex pheromones of several species of insects. Two populations of the European corn borer, *Ostrinia nubilalis,* have been identified, one using a 96:4 mixture of the (*E*):(*Z*)-isomers of 11-tetradecenyl acetate and the other a 3:97 mixture. Reproductively, these two populations are isolated from each other. In the case of two populations of the scolytid *Ips pini,* reproductive isolation is based on enantiomeric differences. A population in the western United States produces the (−) enantiomer of ipsdienol, 2-methyl-6-methylene-2, 7-octadien-4-ol, and is strongly inhibited by the (+) enantiomer. On the other hand, an eastern population of this species produces and uses a 65:35 mixture of (+) : (−) enantiomers, responding more strongly to the (−) compound. Since the response of the western population of *I. pini* is strongly inhibited by as little as 5% of the (+) enantiomer, the two populations are in effect reproductively isolated (Silverstein, 1982).

14.4.4. Perception

Pheromones are generally perceived by multiporous sensilla located on the antennae (see Chapter 8, Section 8.5). For sex pheromones these sensilla constitute numerous specialist chemoreceptors (50×10^3 in *Bombyx* males) that produce an intrinsic input (background) that can be overcome, at a behavioral threshold level, by an incredibly small number of impulses generated by sex pheromone molecules. For *Bombyx* males only 200 additional impulses, resulting from 300 cells hit by bombykol molecules, generate a behavioral threshold (Schneider, 1970). However, although the *Bombyx* system has been a very useful paradigm for analyzing sensory encoding of pheromonal signals, the subsequent realization that most sex pheromones were composed of blends (even bombykol is now known to be accompanied by a second compound) has necessitated additional analyses of this model. It now seems certain that specialist receptors are present on the antennae to accommodate the primary (long-range) and most of the secondary (close-range behavioral stimulants) pheromones.

The sex pheromone specialist cells on the male antennae of moths generally consist of two or five receptor cell types, each exhibiting a maximal response to a specific compound (Priesner, 1980). Sex pheromone molecules bind to acceptor molecules on the distal dendrites of the sensory neurons after diffusing via pore tubules. Sensory transduction has been depicted as reflecting activation and increase of conductance after binding has occurred, following adsorption and diffusion of the molecule (Kaissling, 1975). The rapid inactivation that terminates the transductive process cannot be readily explained in terms of any known physiological processes. For example, pheromonal hydrolysis by cuticular enzymes is far too slow to account for the rapid inactivation observed.

Besides the cells that mediate behavioral responses of males, inhibitory cell types that respond to compounds (e.g., bombykal) structurally related to primary compounds (e.g., bombykol) are also present. The precise functions of these inhibitory components of the pheromonal signal are not understood, but it is worth noting that they may be perceived by the same receptor trichodeum acceptor and neuron as a synergist (Seabrook, 1978). Information in single neurons is coded within single neurons and between a common sensillum, with interpretation occurring in the CNS.

Neurons in the deutocerebrum of male moths have been demonstrated to have congruent specificities with specialist cells in the antennae for specific sex pheromone molecules. Convergence of antennal receptors at central neurons ensures that a very high sensitivity will be maintained for primary pheromonal constituents. On the other hand, the sensitivities of the antennal receptors appears to be programmed, reaching a maximum some time after adult ecdysis and senescing subsequently.

Insects detect their pheromones along diffusive concentration gradients descending from the emissive source. There is a zone proximate to the point of emission, the active space, within which the concentration of pheromonal molecules is high enough to elicit a behavioral response (Bossert and Wilson, 1963). The shape of the active space varies with the physical environment in which the signal is generated, being an ellipsoidal plume when released into the wind from an elevated site, its long axis aligned downward. When released at ground level, the active space is semiellipsoidal in wind and hemispherical in still air. As the signal fades, the active space diminishes in size, reducing the zone, which is at or above the threshold for a behavioral response. However, the specifics relative to the nature of the odor plume actually encountered by an insect are difficult to predict and await further refinement (Elkinton and Cardé, 1984). The characteristics of such an olfactory communicative system are presumably consonant with the function of the signal, evolution having optimized the efficiency of the signal in an airborne channel.

Field data have demonstrated that our comprehension of chemical communication is still inadequate. For example, male moths of some species are not attracted into traps that utilize binary mixtures of sex pheromones at the same ratios secreted by their females. By constructing a diagram in which the ratio of isomers in the mixtures is plotted against the concentration (release rate), Roelofs (1978) has developed a threshold hypothesis for pheromone perception which attempts to explain trapping inconsistencies. This diagram envisages an attraction area bounded by the threshold for flight activation over the full range of binary mixtures and by the threshold for disorientation (alteration of in-flight behavior) of each pheromonal component. This hypothesis appears to be consonant with the observed attraction of male moths with both low and high release rates of a binary sex pheromone.

14.4.5. Behaviors Released

Pheromones release a diversity of behaviors that must be analyzed in terms of each insect and its specific chemical signaling system. Whereas in some lepidopterous species a single sex pheromone may function both as an attractant and a sexual stimulant (e.g., bombykol), other species utilize blends of pheromones. The individual components of these blends release a hierarchy of behavioral reactions, attractive and sexual responses often being elicited by different compounds. In one ant species a single pheromone may release the complete spectrum of alarm behaviors, whereas in other species a mixture of exocrine compounds is required to express the complete hierarchy of reactions eventuating in biting and/or stinging. These hierarchical reactions to multicomponent pheromones are particularly evident in the reactions of male moths to sex pheromones.

Male moths respond to a sex pheromone attractant by exhibiting optomotor-guided upwind anemotaxis involving internally generated zigzag movement (Cardé, 1984). When near the female the male may exhibit stereotyped sexual responses to other pheromonal components and may secrete sex pheromones himself. These appear to arrest female movement or elicit female sexual responses. The hierarchy of reactions exhibited by males may reflect the signal characteristic of each active pheromonal component, the diffusion rate and fade-out time of the active space of each being of great behavioral significance.

Insect responses to nonsexual pheromones often reflect the behavioral peculiarities of a particular species. For example, alarm pheromones may result in dispersion (aphids) or attack (bees), whereas the reactions to trail pheromones in ants may be contextually determined, resulting in either recruitment or emigration. In all cases, the magnitude of the response may be concentrationally dependent; ant alarm pheromones may function as simple attractants (low concentration) or alarm releasers (high concentration).

Detailed analyses of the behavioral responses of various species have demonstrated that pheromones control a variety of the observed idiosyncratic behaviors. For example, male carpenter ants induce flight behavior of females with mandibular gland pheromones (Table 14-3), synchronizing the airborne occurrence of the sexes as a mating prelude. Many other examples of idiosyncratic pheromones will undoubtedly be exposed by additional research.

14.4.6. Pheromonal Parsimony

The versatility of insects as chemical communicants is particularly evident as a consequence of our awareness of how different species have been able to regulate a diversity of behaviors with relatively few pheromones. This phenomenon, pheromonal parsimony (Blum and Brand, 1972), reflects the utilization of as few of these exocrine products as possible in order to subserve a multitude of functions. This development, possible because insects can exhibit a variety of responses to the same pheromone in different contexts or when exposed to different concentrations, is identified with the behavioral plasticity evolved by these invertebrates.

In the Douglas-fir beetle, *Dendroctonus pseudotsugae,* one pheromone, 3-methyl-2-cyclohexen-1-one (MCH), regulates a variety of behaviors related to aggregation and mating (Ryker, 1984). MCH in low concentration is part of the attractant pheromonal bouquet secreted by newly colonized females. However, when joined by a male, the female and her new partner emit high concentrations of MCH, which effectively inhibit aggregation by additional males. In addition, males give off high concentrations of MCH when they encounter other males at female galleries, much as they do when they encounter their prospective mates. In effect, this pheromone regulates population density because of the contextual responses to it of both males and females.

Pheromonal parsimony is particularly well established in eusocial insects. 9-Oxo-(E)-2-decenoic acid, the "queen substance" of the honeybee queen, is utilized as a drone sex attractant when she undertakes her nuptial flights, but in the milieu of the hive this compound functions as a primer pheromone that inhibits both ovarian development of workers and queen cell construction. Similar parsimonious utilization of alarm pheromones by ants and bees has been demonstrated. For example, a single alarm releaser may serve as a low-level attractant, a releaser of either attack or digging behavior, or a defensive allomone. Pheromonal parsimony may have been a key development in the evolution of a viable eusociality.

14.4.7. Metabolic Derivations

There is little evidence to indicate that insect pheromones are not biosynthesized *de novo*. For example, investigations with suitable precursors have demonstrated that for the biogenesis of typical moth sex pheromones—oxygenated aliphatic compounds (Table 14-2)—readily available lipid precursors are utilized. In ants, monoterpene alarm pheromones (e.g., citronellal) are derived via acetate condensation utilizing mevalonic acid as an intermediate, an example of the well-developed ability of insects to biosynthesize monoterpenoid constituents. Although insects cannot produce triterpenes such as cholesterol, they are eminently capable of synthesizing mono(C_{10})-, sesqui(C_{15})-, di(C_{20})-, and even sesterterpenes (C_{25}).

In some cases, insects convert simple monoterpenes, exogenously derived, into complex monoterpenoid pheromones. This is particularly true of bark beetles, which have been shown to produce terpenoid aggregation pheromones from the copious monoterpene hydrocarbons characteristic of the resins of their coniferous hosts. Similarly, male danaid butterflies derive their pyrrolizidine alkaloid sex pheromones, which are externalized on their hair pencils, from ingested pyrrolizidine alkaloid precursors.

14.5. ALLOMONES

The natural products potential of insects is best illustrated by their defensive allomones, many of which constitute novel animal products. More than 600 of these compounds, which are generally secreted as blends, have been identified (Blum, 1981), of which more than 100 are hydrocarbons. This well-developed capacity to synthesize hydrocarbons is probably correlated with the ability of the epidermal cells, from which exocrine glands are derived, to synthesize a great variety of cuticular hydrocarbons. However, these alkanes and alkenes usually accompany a potpourri of compounds belonging to other chemical classes ranging from novel steroids to the unique neuropeptides found in hymenopterous venoms.

14.5.1. Chemistry

Idiosyncratic compounds are produced by insects in a variety of taxa. 4-Oxo-(E)-2-octenal, a characteristic defensive compound in the metathoracic gland secretions of bugs in several families, has not been detected in any nonhemipterous species. The same is true of gyrinidone, one of several sesquiterpenes identified as pygidial gland products of Gyrinidae. Another sesquiterpene, dendrolasin, has only been identified in the mandibular gland secretion of one species of formicine ant, closely related forms producing a large number of terpenes such as citronellol (Table 14-4). Cyclopentanoid monoterpenes (e.g., iridodial and nepetelactone), on the other hand, have been identified as defensive allomones of phasmids, ants, and beetles. Novel diterpenes, the trinervitenes and kempenes, are only known to be produced by termite soldiers in the family Termitidae (Table 14-4).

Ants are distinctive in producing a large number of alkaloids in their poison glands. These compounds, which have only been identified in the venoms of myrmicine ants, include a variety of piperidines, pyrrolidines (Table 14-4), pyrrolizidines, and indolizidines.

Beetles produce a host of distinctive natural products, many of which are aromatic compounds. Benzaldehyde, salicylaldehyde, methyl p-hydroxybenzoate (Table 14-4), and methyl 2-hydroxy-6-methylbenzoate are illustrative of the diverse aromatic compounds synthesized in the exocrine glands of coleopterans. Simple acids such as tiglic and angelic have only been identified as pygidial gland products of carabids. Beetles also produce a large number of substituted 1,4-benzoquinones (Table 14-4), and in addition, some species synthesize naphthoquinones and anthroquinones as well.

The allomonal preeminence of beetles is also illustrated by the nonexocrine defensive compounds that they synthesize. These natural products, which are present in the hemolymph, include the novel terpenoid anhydride cantharidin, only known from meloid beetles, and the tricyclic alkaloids, the coccinellines, products of coccinelids (Table 14-4). The most complex nonprotein natural product identified in insects, pederin, is limited in its known insectan distribution to species in one staphylinid genus.

Beetles also produce an amazing variety of steroids derived from ingested sterols such as cholesterol. At least 20 steroids, many of which are identical to well-known vertebrate sex hormones such as testosterone, are synthesized in the prothoracic glands of adult dytiscid beetles (Table 14-4). In

TABLE 14-4. Defensive Allomones of Insects and Their Glandular Sources[a]

Compound	Family	Order	Glandular Source
n-Tridecane	Formicidae	Hymenoptera	Dufour's gland (worker)
α-Pinene	Termitidae	Dictyoptera (Isoptera)	Frontal gland (soldier)
Citronellol (CH_2OH)	Formicidae	Hymenoptera	Dufour's gland (worker)
4-Oxo-(E)-2-hexenal (CHO)	Pentatomidae	Hemiptera	Metasternal scent gland (adult)
Salicylaldehyde (CHO, OH)	Carabidae	Coleoptera	Pygidial glands
Iridodial (CHO, CHO)	Formicidae	Hymenoptera	Pygidial (= anal) glands (worker)

2-Tridecanone	Rhinotermitidae	Dictyoptera (Isoptera)	Frontal gland (soldier)
Gyrinidone	Gyrinidae	Coleoptera	Pygidial glands
Ethyl p-benzoquinone	Forficulidae	Dermaptera	Paired glands on the third and fourth tergites
6-Methyl-1,4-naphthoquinone	Tenebrionidae	Coleoptera	Paired abdominal sternal glands
Formic acid HCOOH	Notodontidae	Lepidoptera	Cervical gland (larva)
Tiglic acid	Carabidae	Coleoptera	Pygidial glands

569

TABLE 14-4. (*Continued*)

Compound	Family	Order	Glandular Source
p-Cresol	Limnephilidae	Trichoptera	Paired glands on the sternites of the seventh abdominal segment (adult)
(*E*)-2-Octenyl acetate	Coreidae	Hemiptera	Metasternal scent gland (adult)
Methyl *p*-hydroxybenzoate	Dytiscidae	Coleoptera	Pygidial glands
Nepetelactone	Phasmidae	Orthoptera	Paired dorsal prothoracic glands

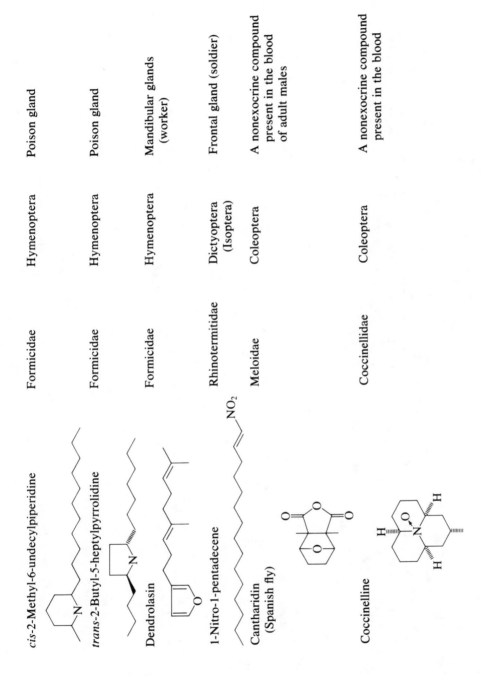

cis-2-Methyl-6-undecylpiperidine	Formicidae	Hymenoptera	Poison gland
trans-2-Butyl-5-heptylpyrrolidine	Formicidae	Hymenoptera	Poison gland
Dendrolasin	Formicidae	Hymenoptera	Mandibular glands (worker)
1-Nitro-1-pentadecene	Rhinotermitidae	Dictyoptera (Isoptera)	Frontal gland (soldier)
Cantharidin (Spanish fly)	Meloidae	Coleoptera	A nonexocrine compound present in the blood of adult males
Coccinelline	Coccinellidae	Coleoptera	A nonexocrine compound present in the blood

TABLE 14-4. (*Continued*)

Compound	Family	Order	Glandular Source
Trinervi-9β-ol	Termitidae	Dictyoptera (Isoptera)	Frontal gland (soldier)
Testosterone	Dytiscidae	Coleoptera	Prothoracic glands
Periplogenin	Chrysomelidae	Coleoptera	Pronotal and elytral glands

11-Oxo-5β,12β-dihydroxybufalin

Lampyridae Coleoptera A nonexocrine compound present in the blood of adults

Pederin

Staphylinidae Coleoptera A nonexocrine compound present in the blood of adults

[a] Many of these compounds are present in the defensive exudates of species in more than one family.

addition, cardenolides, steroids related to the compounds produced by milk-weeds, are produced by a variety of aposematic chrysomelids. Lampyrids, on the other hand, synthesize novel nonexocrine steroidal pyrones, the lucibufagins (Meinwald et al., 1979), that are related to the toxic bufodieno-lides synthesized by toads. Probably other families of beetles will be found to synthesize novel steroids that are derived from dietary sterols.

Although insects in many taxa appear to biosynthesize distinctive defen-sive allomones and pheromones, too few species have been analyzed to have confidence in the apparent restriction of any compound to species in particu-lar taxa. Undoubtedly, some of these exocrine compounds are true idiosyn-cratic natural products of selected species, but the frequent identification of the same distinctive allomone in unrelated insects demonstrates that, at this juncture, distributional generalizations are premature. Biosynthetic specia-lizations do not guarantee an insect a monopoly on the production of a particular natural product.

However, the distributional congruencies of defensive allomones in unre-lated species may not necessarily constitute random biosynthetic conver-gences. It has been suggested that, in some cases, identicalness of exocrine compounds in solitary insects and predatory ants may actually reflect the evolution of "cryptic" pheromones by the former (Blum, 1981). Cryptic pheromones are compounds synthesized by solitary arthropods that are identical to or structurally related to the alarm pheromones utilized by ants. Ketones such as 4-methyl-3-heptanone, 6-methyl-5-hepten-2-one, and 2-heptanone are typical releasers of alarm for formicids, and their production and secretion by multillids and cockroaches that encounter aggressive ants can temporarily disrupt the social cohesiveness of the formicids. With the ant workers exhibiting disorganized and frenetic alarm behavior, these fast-moving wasps or blattids can easily flee the scene of a confrontation, thus avoiding a fatal en masse attack by the formicids.

14.5.2. Modes of Action

Although a large number of insect exocrine compounds demonstrably deter at least some predatory species, their modes of action as defensive allo-mones are largely *terra incognita*. It is instructive to focus on some of the main groups of defensive allomones—hydrocarbons, unsaturated alde-hydes, and 1,4-quinones—to seek some insights into why these classes of allomones have been emphasized as predator deterrents by a wide diversity of insect species. Limited investigations point to these compounds as dis-arming agents for predators, their modes of action related to site-specific actions on the olfactory chemoreceptors of predatory species.

Hydrocarbons, while they could certainly facilitate the penetration of more polar allomones through an arthropod's lipophilic epicuticle, may pos-sess a more subtle role as agents of deterrence. Alkanes have been observed to alter the generator characteristics of antennal chemoreceptors. For the predator, partial and temporary anosmia may result, especially since normal chemosensory input has been interrupted. Since arthropod predators heavily utilize odor in prey detection, a hydrocarbon-blunted chemosensory system would be very maladaptive in terms of prey location. Indeed, under these circumstances, the prey could "hide" from the alkane-treated assailant whose predatory behavior was temporarily neutralized.

Defensive allomones of insects may also produce cryptic odors when they are unleashed against their omnipresent adversaries. Cryptic odors can be identified with at least two distinct neurophysiological phenomena, both of which reflect the presence of highly disruptive chemosensory inputs into the CNS. The chemical equivalent of a "white noise" (a large uncoordinated electrical input), a cryptic odor may generate an "uncoded" array of spikes in the chemosensory neurons or a "negative odor" altering the characteristics of the dendritic membrane after blocking its generator potential (Kittredge et al., 1974). Certain insect defensive exudates would seem to be eminently qualified to function as cryptic odors.

The multicomponent defensive secretions of many insects contain diverse compounds that belong to several chemical classes. These structurally eclectic exudates would appear to be ideally suited for overstimulating the chemosensory neurons of predators and thus assaulting the CNS with bursts of uncoded information. The presence in the exudate of defensive compounds belonging to different chemical classes would result in a variety of generalist receptors being stimulated, and the resultant sensory input would be, in effect, a highly disruptive chemical overload generating a maladaptive scrambled message. The Dufour's gland secretion of formicine ants would appear to be ideal for generating such a white noise, since it may contain more than 50 compounds belonging to five chemical classes. These heterogeneous blends, characteristic of many defensive secretions, may have been evolved to endow their possessors with cryptic odors when they are confronted with aggressive predators.

Both 1,4-quinones and α,β-unsaturated aldehydes (Table 14-4) may also function as cryptic odors, and it is likely that their efficacies as predator deterrents are correlated with their great reactivities. Both types of these carbonyl compounds are conjugated and can rapidly react with nucleophilic compounds, forming new compounds that are physiologically nonfunctional. The exposed chemoreceptor proteins in the olfactory pores may be prime targets for these quinones and aldehydes. If the olfactory proteins react with these carbonyl compounds, the resultant modified proteins may be inactivated, and olfactory activity may be drastically blunted. Under these circumstances, the predator may be rendered anosmic, enabling its carbonyl-producing prey to "hide" from it.

14.5.3. Avoidance of Autointoxication

Insects, producing a large variety of highly reactive and/or potentially toxic defensive allomones, have obviously evolved mechanisms for avoiding the intoxicative effects of these compounds. Once secreted, these natural products often drench the bodies of their producers and, while this may provide them with a protective chemical shield, it is also true that they must avoid the toxicity that many of these compounds are known to possess. Although the detoxicative abilities of insects vis-à-vis their own defensive compounds are largely unknown, it is now well established that morphological specializations have been evolved to isolate these toxins, away from sensitive tissues, in the bodies of their producers.

Reactive defensive allomones are stored in impermeable chitin-lined glands after their biosynthesis, removed from the sensitive glandular cells that produced the allomonal precursors. This is particularly true for the products of reactor glands, such as those found in tenebrionids (see Section 14.3.3), their highly reactive 1,4-quinones being generated in the glandular lumen at a safe distance from the secretory cytoplasm (Fig. 14-4). The same is true for the reactive aldehydes produced in the metathoracic scent gland reservoir of adult hemipterans. These carbonyl compounds are biosynthesized in the storage reservoir of the scent gland, removed from the sensitive glandular cytoplasm. The precursors of the aldehydes, the aliphatic esters, are also secreted into the storage reservoir, along with hydrolytic and oxidative enzymes. Thus, the toxic aldehydic end-products are generated from their ester precursors after hydrolysis to the alcohols (and acids) followed by oxidation. These final reactions are isolated away from tissues susceptible to the cytotoxic effects of the aldehydes.

In general, these reactive allomones never have access to the vulnerable tissues of their producers. On the other hand, if these glands are ruptured, their contents flood the hemocoele and death usually follows. By packaging their chemical arsenals in chitin-lined storage tanks, these allomonal chemists ensure that their natural product effronteries will be reserved for their persistent and omnipresent adversaries.

14.5.4. Metabolic Origins

Like pheromones, defensive allomones are generally synthesized *de novo* by their insect producers. Indeed, many defensive (agonistic) allomones [e.g., (*E*)-2-hexenal, 4-methyl-3-heptanone] constitute pheromones in eusocial species, and obviously, for these compounds there is really no valid distinction between pheromones and defensive allomones. The fact that several of the sex pheromones produced by female moths are identical to the defensive compounds of other insects further emphasizes the structural congruencies of pheromones and allomones.

There is little evidence that insects appropriate their exocrine compounds from their food, although in a few cases plant-derived compounds (cardenolides) have been demonstrated to be sequestered in defensive secretions. However, these plant natural products are unrelated to the exocrine compounds produced by these species. On the other hand, while many insects sequester compounds as diverse as pyrrolizidine alkaloids and cardenolides in their tissues after feeding on plants containing these compounds, these sequestered plant natural products clearly do not constitute exocrine compounds.

An examination of the diverse types of defensive allomones produced by insects (Table 14-4) testifies to the metabolic prowess of these arthropods. Additional studies will undoubtedly identify a host of additional novel exocrine compounds as products of insects.

14.6. CONCLUSIONS

Exocrinology, a research area that has really emerged only in the last 30 years, has rapidly expanded into a major discipline of insect physiology. The functions of a diversity of insect natural products have been probed in the last few decades, while at the same time a large number of these exocrine compounds, pheromones and allomones, have been chemically characterized. These glandular products have become key elements in a nascent chemical ecology, their recognition having resulted in significant developments in fields as diverse as neurophysiology, endocrinology, biochemistry, toxinology, and behavioral physiology.

However, the dizzying pace of exocrinological research has also served to raise a plethora of questions that beg to be answered by tomorrow's investigators. For example, the perception of pheromonal blends is barely understood, especially in terms of how the encoded information is processed in the CNS. Neurophysiological investigations have yet to establish how minor amounts of geometric isomers increase the male's sensitivity to the signal emitted by the female. The question of sensory transduction vis-à-vis pheromonal molecules remains to be determined, as well as how the transductive process is terminated so rapidly. Little information is available on the factors that regulate the biogenesis of pheromones or, for that matter, what metabolic pathways are utilized to biosynthesize these compounds.

Questions pertaining to allomones are equally daunting. The modes of action of these blends of defensive compounds are essentially *terra incognita,* and their biosyntheses have been barely illuminated. For the natural products chemist, these exocrine products constitute a virtual treasure trove of novel compounds. Since the secretions of so few species have been analyzed, the analytical challenge is considerable. It is fair to state that for behaviorists, physiologists, and biochemists, exocrinology promises to render insects more wondrous than ever.

ACKNOWLEDGMENTS

I thank A. A. Sorensen and L. F. Oertel for preparing Figs. 14-1 and 14-5, respectively. I am especially grateful to P. A. Teale and J. R. Aldrich for providing figures from their unpublished research. My thanks also go to R. M. Silverstein for providing valuable insights into the arcane world of chirality.

REFERENCES

J. R. Aldrich, J. P. Kochansky, W. R. Lusby, and J. D. Sexton, *J. Wash. Acad. Sci.* **74,** 39 (1984).

M. S. Blum, *Bull. Ent. Soc. Amer.* **20,** 30 (1974).

M. S. Blum, *Chemical Defenses of Arthropods,* Academic Press, New York and London, 1981, pp. 1–562.

M. S. Blum and J. M. Brand, *Amer. Zool.* **12,** 553 (1972).

R. Boch and D. A. Shearer, *Z. Vergl. Physiol.* **54,** 1 (1967).

W. H. Bossert and E. O. Wilson, *J. Theor. Biol.* **5,** 443 (1963).

W. L. Brown, Jr., T. Eisner, and R. H. Whittaker, *Bioscience* **20,** 21 (1970).

A. Butenandt, R. Beckmann, D. Stamm, and E. Hecker, *Z. Naturforsch. B.* **14,** 283 (1959).

R. T. Cardé, "Chemo-orientation in flying insects," in W. J. Bell and R. T. Cardé, Eds., *Chemical Ecology of Insects,* Chapman and Hall, London and New York, 1984, pp. 111–124.

L. Cuénot, *Arch. Zool. Exp. Gen.* **4,** 655 (1896).

S. S. Duffey, *Proc. 15th Int. Congr. Ent.* (*1976*), Washington, D.C., 1977, pp. 376–378.

T. Eisner, "Chemical defense against predation in insects," in E. Sondheimer and J. B. Simeone, Eds., *Chemical Ecology,* Academic Press, New York and London, 1970, pp. 159–175.

J. S. Elkinton and R. T. Cardé, "Odor dispersion," in W. J. Bell and R. T. Cardé, Eds., *Chemical Ecology of Insects,* Chapman and Hall, London and New York, 1984, pp. 73–91.

M. Florkin, *Bull. Cl. Sci. Acad. Roy. Belg.* **51,** 239 (1965).

G. M. Happ, *J. Insect Physiol.* **14,** 1821 (1968).

K.-E. Kaissling, *Verh. Dtsch. Zool. Ges.* **67,** 1 (1975).

J. S. Kittredge, M. Terry, and F. T. Takahashi, *U.S. Fish. Bull.* **69,** 337 (1974).

J. H. Law and F. E. Regnier, *Ann. Rev. Biochem.* **40,** 533 (1971).

J. Meinwald, D. F. Wiemer, and T. Eisner, *J. Amer. Chem. Soc.* **101,** 3055 (1979).

L. R. Nault and P. L. Phelan, "Alarm pheromones and sociality in pre-social insects," in W. J. Bell and R. T. Cardé, Eds., *Chemical Ecology of Insects,* Chapman and Hall, London and New York, 1984, pp. 237–256.

C. Noirot and A. Quennedey, *Ann. Rev. Ent.* **19,** 61 (1974).

J. M. Pasteels, *J. Chem. Ecol.* **7,** 1079 (1982).

M. Pavan and G. Ronchetti, *Atti. Soc. Ital. Sci. Nat.* **93,** 377 (1955).

E. Priesner, "Sensory encoding of pheromone signals and related stimuli in male moths," in *Insect Neurobiology and Pesticide Action* (*Neurotox 79*), Soc. Chem. Ind., London, 1980, pp. 359–366.

W. L. Roelofs, *J. Chem. Ecol.* **4,** 685 (1978).

W. L. Roelofs, "Attractive and aggregating pheromones," in D. A. Nordlund, R. L. Jones, and W. J. Lewis, Eds., *Semiochemicals,* Wiley, New York, 1981, pp. 215–235.

L. C. Ryker, *Scientific Amer.* **250,** 112 (1984).

D. Schneider, "Olfactory receptors for the sexual attractant (bombykol) of the silkmoth," in F. O. Schmitt, Ed., *The Neurosciences: Second Study Program,* Rockefeller University Press, New York, 1970, pp. 511–518.

W. D. Seabrook, *Ann. Rev. Ent.* **23,** 471 (1978).

R. M. Silverstein, *Science* **213,** 1326 (1981).

R. M. Silverstein, *Pure Appl. Chem.* **54,** 2479 (1982).

R. M. Silverstein, J. O. Rodin, and D. L. Wood, *Science* **154,** 509 (1966).

INDEX

581